Electrons in Fluids

The Nature of Metal—Ammonia Solutions

Edited by
J. Jortner and N. R. Kestner

With 271 Figures

Springer-Verlag
New York · Heidelberg · Berlin 1973

Professor Dr. JOSHUA JORTNER
Institute of Chemistry, Tel-Aviv University,
Ramat Aviv, Tel-Aviv, Israel

Professor Dr. NEIL R. KESTNER
Department of Chemistry, Louisiana State University,
Baton Rouge, LA 70803, USA

ISBN 13: 978-3-642-61964-9   e-ISBN 13: 978-3-642-61962-5
DOI: 10.1007/978-3-642-61962-5

Preface

Colloque Weyl I was convened in June 1963 at the Catholic University of Lille to commemorate one hundred years of the study of metal–ammonia solutions. This memorable event, which involved a "single-particle excitation", inspired Gerard Lepoutre to assemble an international group of physicists and chemists to discuss the nature of metal–ammonia solutions. Colloque Weyl II, which took place at Cornell Universtiy, Ithaca, N.Y. in June 1969, was initiated as a "cooperative interaction" between M. J. Sienko, J. L. Dye, J. J. Lagowski, G. Lepoutre and J. C. Thompson. That meeting made it clear that Colloque Weyl should be continued in order to promote the fruitful exchange of ideas set in motion at Lille and at Cornell.

Colloque Weyl III came into being as the result of a resolution passed at the Cornell meeting, Tel-Aviv University being the suggested site. The Organizing Committee consisted of E. D. Bergmann, J. Jortner, J. J. Lagowski, G. Lepoutre, U. Schindewolf and M. J. Sienko, reflecting the international and interdisciplinary aspects of the field.

The original aim of the Colloque Weyl was to contribute to the understanding of metal–ammonia solutions. Colloque Weyl III extended the scope of the meetings to include related work on electrons in polar and non-polar liquids and in dense fluids, the electronic states of disordered systems, and the metal–insulator transition, thus the chemical and physical properties of electrons in fluids from a general point of view. The interdisciplinary nature of the field and of related areas was emphasized by an intensive exchange of ideas between chemists and physicists, experimentalists and theoreticians.

This volume provides a record of the lectures, communications and discussions that took place during the week of Colloque Weyl III. The contributions are not necessarily presented in chronological order. Some changes have been made to afford a more coherent presentation of certain subjects.

Colloque Weyl III was held at Kibbutz Hanita, Upper Galilee, Israel on June 19–23, 1972. The conference was organized by Tel-Aviv University and sponsored by the Israeli Academy of Sciences and Humanities, the Israeli Foundation for Research Grants, the Israeli National Council for Research and Development, Tel-Aviv University as well as the U. S. Army European Research Office. We are very grateful to these agencies for their support and cooperation in making this a truely international conference.

JOSHUA JORTNER
NEIL R. KESTNER

Tel-Aviv, June 1973

Contents*

* Plenary discussions appear in **heavy type**.

List of Contributors

ACRIVOS, J. V.   Department of Chemistry, San Jose State College, San Jose, CA 95114/USA

VAN ANTWERP, C. L.   Juniata College, Chemistry Department, Huntingdon, PA/USA

BAR-ELI, K.   Department of Chemistry, Tel-Aviv University, Tel-Aviv/Israel

BELLONI, J.   Laboratoire de Physico-Chimie des Rayonnements, associé au C.N.R.S., Faculté des Sciences, 91400 Orsay/France

BOLL, H.   Department of Chemistry, Tufts University, Medford, MA 02155/USA

BOWEN, D. E.   Physics Department, The University of Texas at El Paso, El Paso, TX 79968/USA

BREITSCHWERDT, K. G.   Institut für Angewandte Physik der Universität Heidelberg, 6900 Heidelberg/Germany

BROWN, R. G.   Department of Chemical Engineering, The University of Minnesota, Minneapolis, MN 55455/USA

COHEN, M. H.   University of Cambridge, Department of Physics, Cavendish Laboratory Free School Lane, Cambridge, GB 2 3RQ/Great Britain

COTTS, R. M.   Department of Physics, Cornell University, Ithaca, NY 14850/USA

CZAPSKI, G.   Department of Physical Chemistry. The Hebrew University, Jerusalem/Israel

DAMAY, P.   Faculté Catholiques de Lille, 59046 Lille Cedex/France

DAVID, TH.   Baker Laboratory of Chemistry, Cornell University, Ithaca, NY 14850/USA

DAVIS, H. T.   Departments of Chemical Engineering and Materials, Science, University of Minnesota, Minneapolis, MN 55455/USA

DEBACKER, M. G.   Laboratoire de Chimie Physique Faculté Catholiques de Lille, 59046 Lille Cedex/France

DEBETTIGNIES, B.   Laboratoire de Spectroscopie Raman, Université des Sciences et Techniques, 5900 Villeneuve d'Ascq/France

DELAHAY, P.   Department of Chemistry, New York University, 4
                           Washington Place, New York, NY 10003/USA

DEWALD, R. R.   Department of Chemistry, Tufts University, Medford,
                           MA 02155/USA

DORFMAN, L. M.   Department of Chemistry, The Ohio State University,
                           Columbus, OH 43210/USA

DYE, J. L.   Chemistry Department, Michigan State University, East
                           Lansing, MI 48823/USA

EVEN, U.   Department of Chemistry, Tel-Aviv University, Tel-Aviv/
                           Israel

GAATHON, A.   Department of Physical Chemistry, The Hebrew Uni-
                           versity, Jerusalem/Israel

GARROWAY, A. N.   Department of Physics, University of Nottingham,
                           Nottingham, N 672 RD/Great Britain

GEBALLE, T. H.   Department of Applied Physics, Stanford University,
                           Stanford, CA 94305/USA

GEDANKEN, A.   Department of Chemistry, Tel-Aviv University, Tel-Aviv/
                           Israel

GILEACHI, E.   Department of Chemistry, Tel-Aviv University, Tel-Aviv/
                           Israel

GLAUNSINGER, W. S.   Assistant Professor of Chemistry, Department of Che-
                           mistry, Arizona State University, Tempe, AZ 85281/USA

HENSEL, F.   Institut für Physikalische Chemie und Elektrochemie,
                           Universität Karlsruhe, 7500 Karlsruhe/Germany

HAMILTON, J. A.   Juniata College, Chemistry Department, Huntingdon,
                           PA/USA

ICHIKAWA, K.   Physics Department, The University of Texas at Austin,
                           Austin, TX 78712/USA

JONES, R. L.   Department of Chemistry, Tufts University, Medford,
                           MA 02155/USA

JORTNER, J.   Department of Chemistry, Tel-Aviv University, Tel-Aviv/
                           Israel

JOU, F. Y.   Department of Chemistry, The Ohio State University,
                           Columbus, OH 43210/USA

KESTNER, N. R.   Department of Chemistry, Louisiana State University,
                           Baton Rouge, LA 70803/USA

KOEHLER, W. H.   Department of Chemistry, Texas Christian University,
                           Fort Worth, TX 76129/USA

KOMMANDEUR, J.   Department of Chemistry, Groningen University,
                           Groningen/The Netherlands

LAGOWSKI, J. J.   Department of Chemistry, The University of Texas at
                           Austin, Austin, TX 78712/USA

LAMBERT, C.   Service de Chimie-Physique, Centre d'Etudes Nucléaires
                      de Saclay B.P.N° 2, 91 Gif-sur-Yvette/France

LELIEUR, J. P.   Physics Department, University of Texas at Austin,
                      Austin, TX 78712/USA

LEPOUTRE, G.   Laboratoire de Chimie-Physique, Faculté Catholiques
                      de Lille, 59046 Lille Cedex/France

LOGAN, J.   Department of Chemistry, Louisiana State University,
                      Baton Rouge, LA 70803/USA

MATHESON, M.   Argonne National Laboratory, Argonne, JL/USA

MEYERS, S. F.   Department of Applied Physics, Stanford University,
                      Stanford, CA 94305/USA

NEHARI, S.   Department of Chemistry, Tel-Aviv University, Tel-Aviv/
                      Israel

RADSCHEIT, H.   Institut für Angewandte Physik, Universität Heidelberg,
                      6900 Heidelberg/Germany

RAZ, B.   Department of Chemistry, Tel-Aviv University, Tel-Aviv/
                      Israel

RENTZEPIS, P. M.   Bell Laboratories, Murray Hill, NJ 07974/USA

ROBERTS, J. H.   Department of Chemistry, University of Texas at Austin,
                      Austin, TX 78712/USA

RUSCH, P. F.   Laboratoire de Chimie Physique EHEI, 5900 Lille/France

SAITO, E.   D.R.A./S.R.I.R. Ma, C.E.N./Saclay, B.P.N° 2,
                      91190 Gif-sur-Yvette/France

SCHETTLER, P. D.   Chemistry Department, Juniata College, Huntingdon,
                      PA/USA

SCHINDEWOLF, U.   Institut für physikalische Chemie und Elektrochemie,
                      Universität Karlsruhe, 7500 Karlsruhe/Germany

SCHMIDT, L. D.   Department of Chemical Engineering, University of
                      Minnesota, Minneapolis, MN 55455/USA

SIENKO, M. J.   Baker Laboratory of Chemistry, Cornell University,
                      Ithaca, NY 14850/USA

SMITH, B. L.   Department of Chemistry, Texas Christian University,
                      Fort Worth, TX 76129/USA

SPEAR, J. D.   Juniata College, Chemistry Department, Huntingdon,
                      PA/USA

THOMPSON, J. C.   Physics Department, The University of Texas at Austin,
                      Austin, TX 78712/USA

THILLY, J. E.   Juniata College, Chemistry Department, Huntingdon,
                      PA/USA

WEBMAN, I.   Department of Chemistry, Tel-Aviv University, Tel-Aviv/
                      Israel

ZOLOTOV, S.   Baker Laboratory of Chemistry, Cornell University,
                      Ithaca, NY 14850/USA

Theory of Electrons in Polar Fluids

NEIL R. KESTNER

Abstract

Recent model calculations of an electron in a polar fluid are reviewed with primary emphasis on electrons in ammonia and water. These static calculations are compared with each other and with experimental data. These models are capable of explaining qualitatively, and in most cases, quantitatively, the effects of variables such as temperature, pressure, and the molecular structure and density of the solvent. These theories are then used to make brief qualitative statements about electrons in polar vapors, in solutions of higher metal (and electron) concentrations, and in a wide variety of fluids. Predictions are also made of the location of higher excited states and of the stability of two-electron cavity species.

Due to the success of these static calculations, more attention must be paid to the limitations of these approaches. Discussion of alternative ways to improve the treatment of electrons in liquids is considered briefly. In particular such approaches may be necessary to explain the observed line shapes.

I. Introduction

In this work we shall review the recent theoretical attempt to calculate the properties and to examine the physical state of the electron in polar fluids. For the most part this discussion will be limited to ammonia and water solutions at normal liquid densities and temperatures. A few statements will be made towards the end, of what might be expected in other fluid ranges. Much of the discussion will center around the work — both published and unpublished — of Jortner, this author, and our associates. Work of other authors will be referred to when it complements or extends our results. Thus this article does not exhaustively cover the literature but hopefully covers all major developments.

Our dicussion shall be further limited predominantly to very dilute solutions, solutions in which spin pairing is unimportant. Not only does this simplify the discussion, but it is the only concentration range where theory has been able to make substantial progress in the last few years. For a general discussion of the dilemma at higher concentrations the reader is referred to the article by Dye from Colloque Weyl II [1].

To understand the recent work it is advisable to review from an historical perspective the steps leading up to our present theory. By choice some alternative approaches will not be mentioned since they have not, at present, yielded such extensive agreement with experimental data.

II. Historical Background

From 1940 to 1946 Ogg [2] proposed models for an electron in liquid ammonia. He assumed an infinite repulsion between the electron and the ammonia molecule and thus had the electron dig a hole in the liquid, the size of the hole being determined by the balance among the energy of a particle in a box, the Born-Landau polarization energy, and the surface tension energy needed to form the cavity. While we now know that there are serious flaws in many of his assumptions and that his model is better suited to some nonpolar fluids, nonetheless since that time most (but not all) theories have assumed that the electron resides in some sort of cavity, with its spectrum given by the lowest allowed excitation energy.

Minor improvements of the Ogg model were made by Lipscomb [3] and Stairs [4], but the major developments involved the application of Landau's polaron theory [5] originally used for F-centers in ionic crystals, to the liquid state. The original work was done by Davidov [6], Dergen [7], and Platzman and Franck [8]. However the paper which developed the theory most completely and included a detailed comparison with the experimental data was the paper by Jortner [9], which has been the reference point for even the most recent work. In this work the adiabatic approach [10] is used in which only the inertial (or low frequency solvent modes) polarization is responsible for localizing the electron. Jortner also assumed a definite finite cavity and thus the binding potential is

$$
\begin{aligned}
V(r) &= -\frac{\beta e^2}{r .} \quad \text{for } r > R_0 \\
&= -\frac{\beta e^2}{R_0} \quad \text{for } r < R_0
\end{aligned}
\tag{1}
$$

where R_0 is a cavity size picked to agree with experimental data and

$$
\beta = (1/D_{\text{op}} - 1/D_{\text{s}})
\tag{2}
$$

where D_{op} is the optical dielectric constant and D_{s} is the solvent static dielectric constant. When the other polarization terms are added after the variational solution of the ground and first excited states are calculated, agreement with the experimental transition energies is reasonable for reasonable choices of the cavity size, R_0, i.e. about 3.2 Å. In the above work the solvent is assumed to be a continuum with properties the same as those of bulk solvent.

Land and O'Reilly [11] made an important modification of these theories. They assumed that there is a first coordination layer of the cavity in which one must treat the solvent as discrete. Beyond this first layer one again can use the continuum ideas of Jortner. They also assumed that the solvent molecules in this first coordination layer had a dipole and a quadripole moment whose orientation with respect to the radius vector was determined by the temperature dependent Langevin function. They assumed that the solvent density in the first layer was only slightly different from that in the bulk. From this work one arrives at a much more elaborate expression for the energy. This expression contains the size of the cavity as a parameter. This can be picked to agree with volume expansion or other data. The importance of this work is the recognition of the discrete nature of the first coordination layer and the essentially attractive nature of the interaction. They also used the adiabatic theory, *but* the inertial polarization which represents most of the binding is due to charge-multipole interactions in the first coordination layer as well as the continuum contribution from the solvent beyond this first layer.

Another development which we will consider later concerns the validity of the adiabatic approach. For a discussion of this point the papers of Jortner and Jortner, Rice, and Wilson [10] should be consulted. We will discuss this again when the recent calculations are presented in the next section.

In all of the above work, no attempt has been made to include all of the factors responsible for the cavity stability. The Land and O'Reilly [11] work did include an estimate of the loss in solvent-solvent interactions assuming that these interactions are the same as those in bulk solvent. Apparently the color of the solutions and the agreement of the observed spectrum with simple models made the assumption of a cavity very reasonable. In a related research area, however, this assumption did not seem so reasonable. Naïve theories of electrons in liquid helium suggested that the electron resides in a cavity of about 20Å radius. This was hard to believe and the low concentration of electrons which could be introduced made any measurement of its absorption spectrum impossible. Therefore in a series of papers the equations which must be obeyed by an electron in the medium were derived. The first paper by Kestner, Jortner, Cohen, and Rice [12] used a simple pseudopotential [13] form of the electron-helium atom interaction. The second paper used a theory of Bose fluids to eliminate the arbitrary use of surface tension [14]. However, several flaws remained and the comprehensive work is that of Springett, Cohen, and Jortner [15] in which they apply the Wigner Seitz theory to electrons in nonpolar fluids and arrive at simple criteria for the stability of a cavity. These results include the screening of the long range polarization interaction by the liquid as developed by Lekner [16]. In the Springett et al. paper [15] it is shown that under most conditions one can assume that the interaction of the electron with the solvent in the continuum region, i.e. beyond the first layer, can be represented by the energy of a quasi-free electron, V_0, which is not a function of the distance from the cavity. This term includes multiple scattering and electronic polarization effects. Its dependence on solvent and solvent density can be calculated [15]. Values of V_0 have been estimated by many authors [15, 17].

The most important result of the Springett et al. [15] paper is its generation of stability criteria for the cavity (or bubble) state as compared with the quasi-free state of the electron. The total energy of a cavity state, E_T, is composed of two contributions, one electronic, E_{el}, and one the energy to reorganize the medium and create the cavity, E_m,

$$E_T = E_{el} + E_m. \tag{3}$$

If we assume a spherical cavity then E_T is a function of the cavity size, R, and the most stable cavity is that at which E_T is a minimum, i.e.

$$\left(\frac{\partial E_T}{\partial R}\right)_{R=R_0} = 0 \tag{4}$$

corresponding to a cavity radius R_0. However, the electron will be localized only if

$$E_T(R_0) < V_0. \tag{5}$$

The important point is that it is the total energy and not the electronic energy which determines the stability. To be sure, we have ignored entropy but this should be minor at liquid helium temperatures and probably under most situations. To be precise one should in principle use a proper free energy function.

Elaborations on the theory of electrons in nonpolar media by this approach can be found elsewhere [18—20]. As to its implications in the calculation of the properties of electrons in polar liquids, these were first reviewed at Colloque Weyl II [19]. The most naïve approach one can take is to add to Eq. (1) in the continuum region a contribution V_0. Secondly, one must calculate the medium reorganization energy, E_m, before one can discuss cavity radius or the existence of a cavity. This was done in the Colloque Weyl II paper [19]. In that work we also attempted to put in a discrete first coordination layer of the solvent [19].

Because our recent work includes all of the above factors, we will end this brief historical review and use the ideas presented above to justify our comprehensive model of the theory of electrons in polar liquids.

More comprehensive reviews of these and other models are available. They should be consulted in regard to alternative proposals and for a review of the problems in other concentration ranges [21].

III. Calculations of the State of an Electron in Liquid Ammonia

A. Temperature Dependent Potential Calculations

Based on the factors shown to be important in the last section we propose the following model [22] for the electron in a polar fluid:

1. We assume that in the first coordination layer there is a small fixed number of solvent molecules. In this work we will assume values of $N = 4, 6, 8$, and 12, although the first two seem to be most important.
2. The solvent molecules in the first layer interact with the electron via their dipole moment and with other solvent molecules in the first coordination layer via their repulsive forces, i.e. primarily hydrogen-hydrogen repulsions.
3. The electron interacts with the continuum as in polaron theory except for the presence of the V_0 term.
4. The energy to form the cavity will involve surface tension work, E_{ST}, hydrogen-hydrogen repulsions, E_{HH}, dipole-dipole repulsions in the first layer, E_{dd}, pressure volume work, E_{PV}, and, since we will use the adiabatic theory, an energy to polarize the medium, Π.

The electron density will be assumed determined by the solution of the adiabatic calculation

$$(-\tfrac{1}{2}\nabla^2 + V(r))\,\psi_i(r) = W_i\psi_i(r) \tag{6}$$

where

$$
\begin{aligned}
V(r) &= -N\mu e/r_d^2 - \beta e^2/r_0 & 0 < r < R \\
&= -N\mu e/r_d^2 - \beta e^2/r_c + V_0 & R < r < r_d \\
&= -\beta e^2/r + V_0 & r_d < r
\end{aligned}
\right\} \tag{7}
$$

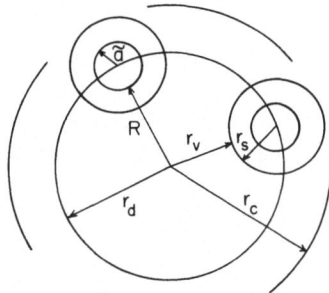

Fig. 1. Definitions of the distances involved in the molecular models. r_v is the void radius of the cavity, r_s the effective solvent radius, and $\tilde{a}$ the effective hard core of the molecules located at a distance r_d from the center of the cavity. The continuum begins at r_c (see Ref. [22])

where the coordinates are defined as in Fig. 1. In this model (referred to as Model 3 in our papers) r_d is the distance to the dipole from the center of the cavity, and r_c is the distance to the start of the continuum which lies beyond the first coordination layer. In this model it is assumed that the parameter, V_0, the energy of the quasi-free electron in the medium, also represents the interaction of the electron with the molecules in the first layer beyond that due to inertial polarization effects. V_0 is not known experimentally. It has been estimated to be about $-0.5\,\text{eV}$ in liquid ammonia. Since its value is not known exactly we will do calculations for $V_0 = 0.5$ to $-0.5\,\text{eV}$.

The value of the effective dipole moment, μ, is equal to

$$\mu = \mu_0 \cos \Theta \tag{8}$$

where Θ is the angle between the radius vector and the dipole moment vector. In the simplest form of the theory we assume that the cosine can be replaced by its average value which can be calculated by the Langevin function, i.e.

$$\langle \cos \Theta \rangle_{1s} = \coth X - X^{-1} \tag{9}$$

where

$$X = \mu_0 e C_{1s}/k T r_d^2 \tag{10}$$

and

$$C_{1s} = \int_0^R |\psi_{1s}|^2 \, d\tau . \tag{11}$$

This form of the theory will have a temperature dependent potential and for each temperature a different equation is solved. This is slightly inconsistent since one should average over temperature at the end of the calculation. Nonetheless this is a very minor error as regards the energy calculation. The ground state (1s) and all vertically excited states (2s, 2p, etc.) will involve the same value of μ since absorption of light is assumed to obey a Franck-Condon restriction.

For all energy states the electronic energy will be given by

$$E_{\text{el}(i)} = W_i + S_i \tag{12}$$

where the polarization is

$$S_i = -N\alpha C_i^2/r_d^4 - e\gamma_0 C_i^2/(2r_c) \tag{13}$$

where C_i is the charge enclosed within radius R for the state i in question [23].
The medium reorganization energy for the electronic state i can be constructed from the following process:
a) We create a cavity of radius R which requires the expenditure of the surface tension energy

$$E_{ST} = 4\pi\gamma R^2 \tag{14}$$

where $\gamma \cong 40$ dyn/cm and of the pressure volume work

$$E_{PV} = \tfrac{4}{3}\pi R^3 P. \tag{15}$$

In most cases at normal pressures this term is negligible.
b) Under the influence of the charge distribution of the system we allow the dipoles in the first coordination layer to rotate under the field of the trapped electron. All rotation is dominated by the charge enclosed in the ground state, but electronic polarization effects can respond to the charge enclosed by the state in question.
When the molecules in the first coordination layer adjust to the influence of the enclosed charge, two energy contributions become important. First of all, the dipole moments of all molecules in the first coordination layer point at roughly the same angle with respect to the radius vector; this means they will tend to repel each other. This is a repulsion not present in the polar liquid where the molecules are more randomly oriented. We will refer to it as E_{ddi}. Secondly, when these molecules are properly aligned, in the ammonia molecule and to a lesser extent in molecules of water and alcohols, the hydrogen atoms which now must point inwards also find themselves closer to hydrogens on a nearby molecule which is also located in the first coordination layer. This leads to an additional repulsive contribution to the energy not found in the bulk liquid. This energy will be called E_{HH}. In addition, there is a loss of some energy due to the breaking of hydrogen bonds. For the moment, we will neglect this contribution as being small in ammonia and very difficult to calculate for water and alcohols.
c) Finally we let the continuum outside the first solvation layer readjust to the field of the charge distribution. This is the Π energy which has the same origin as in the polaron theory.
In order to calculate the contributions which come about because of the rotation of the molecules in the first layer we need some very specific models. We consider only coordination numbers of $N = 4, 6, 8,$ and 12 with the molecules being symmetrically distributed at a radius r_d from the center of the cavity. The form of the repulsions are

$$E_{ddi} = \frac{D_N \mu_T^2}{r_d^3} \tag{16}$$

where

$$\mu_T = \mu_0 \langle \cos\Theta \rangle_{1s} + e\alpha C_i/r_d^2 \tag{17}$$

Table I. Constants for medium reorganization energy

N	D_N [a]	$C_{HH}^{(N)}$ [eV]	A_N	B_N (Å)
4	2.2964	2602.4	1.633	0.471
6	7.1140	5204.7	1.414	0.600
8	12.820	6940.0	1.155	0.752
12	41.074	10416.0	1.000	0.843

[a] Numbers from Buckingham, A.D.: Discuss. Farady Soc. **24**, 151 (1967).

includes the induced dipole moment also. The constants D_N are listed in Table I.

Using the hydrogen-hydrogen interaction judged best by Eisenberg and Kauzman from their studies on water [24], we can arrive at a total hydrogen-hydrogen repulsion for the first coordination layer in ammonia of

$$E_{HH} = C_{HH}^{(N)} \cdot \exp(-4.60(A_N R - B_N)) \cdot \langle \cos \Theta \rangle \tag{18}$$

with the constants defined in Table I. For details as to how these constants are obtained, see Ref. [22]. This contribution is the hardest to obtain and subject to the most criticism. We will continue to use this one form in all of our work, realizing that a slight alteration of the constants in Eq. (18) could produce even better agreement with experimental values. When $\langle \cos \Theta \rangle = 0$, $E_{HH} = 0$, i.e. the same value as in the bulk liquid.

The last contribution to be evaluated is the energy to polarize the medium outside the first coordination layer by the charge density of the electron in the ground state. The formula was originally proposed by Jortner [9] and corrected by Land and O'Reilly [11], namely

$$\Pi_s = \frac{\beta}{2} \left[\int_{R_c}^{\infty} G_0(r) \, \psi_{1s}^2 r^2 \, dr + G_0(r_c) \, P(r_c) \right] \tag{19}$$

where

$$G_0(r) = \frac{1}{r} \int_0^r |\psi_{1s}^{(t)}|^2 \, t^2 \, dt + \int_r^{\infty} |\psi_{1s}^{(t)}| \, t^2 \, dt \tag{20}$$

and

$$P(r_c) = e \int_0^{R_c} |\psi_{1s}^{(t)}|^2 \, t^2 \, dt \,. \tag{21}$$

The only change from the original continuum model is that this contribution comes only from r_c (the start of the continuum) and beyond.

Finally we obtain the state i

$$E_{Mi} = E_{ST} + E_{PV} + E_{ddi} + E_{HH} + \Pi_s \,. \tag{22}$$

In our calculations we assume that the 1s and 2p functions can be represented by a single Slater type function whose exponents can be determined by the variational theorem. The electronic energy is minimized for a fixed temperature and then the polarization and medium reorganization energy terms are added. The optimum cavity size is determined by the minimum of the total energy, Eq. (3).

Table II. Results of model 3 calculations on the one-electron cavity in ammonia (203 °K)

		$V_0 = 0.5$ eV	$V_0 = 0.0$ eV	$V_0 = -0.5$ eV
$N = 4$	$E_t (= \Delta H_1)$	-0.537 eV	-0.909 eV	-1.30 eV
	E_{el}	-1.668 eV	-2.010 eV	-2.404 eV
	R_0	1.75 Å	1.75 Å	1.70 Å
	R_0^{eff}	3.1 Å	3.1 Å	3.0 Å
	$h\nu_{max}$	1.16 eV	1.03 eV	0.94 eV
$N = 6$	E_t	-0.678 eV	-0.972 eV	-1.294 eV
	E_{el}	-2.069 eV	-2.326 eV	-2.603 eV
	R_0	2.20 Å	2.15 Å	2.15 Å
	R_0^{eff}	3.1 Å	3.0 Å	3.0 Å
	$h\nu_{max}$	1.30 eV	1.15 eV	0.99 eV

R_0^{eff} is the effective cavity radius measured by volume expansion experiments and $h\nu_{max}$ is the lowest allowed optical transition.

The results of these calculations at 203 °K and $N = 4$ and 6 are listed in Table II. These results supercede those published in Ref. [22].
We have also used the same model to calculate the 2s state (using hydrogenic terminology) i.e. the next highest s state orthogonal to the ground state. To do this trial wave functions for all s states consisted of three Slater type functions of the 1s, 2s, and 3s type. All exponents and coefficients were optimized to yield the best energy. The results are summarized in Table III. The change in the $1s^2 2p$ transition energy simply corresponds to the change in the calculated 1s energy with the better wave function. It is also apparent that the one-term wave function is reasonably accurate and thus conclusions drawn from its use for the strongly bound electronic states should be reliable. For weakly bound states one should accept its predictions cautiously. Fueki [25] et al. found much larger effects when they studied the simple continuum model.

Table III. The energy of the 1s and 2s states of the electron in ammonia solutions using a three-term wave function [based on work of Kestner, N. R., Logan, J.: J. Phys. Chem. (in press)] $(T = 203 °K)$

V_0(eV)	R_0(Å)	Total energies (eV)		Transition energy (eV)	
		1s state	2s state	$1s \rightarrow 2p$	$1s \rightarrow 2s$
$N = 4$					
0.5	1.75	$-0.677 \, (-0.537)$[a]	0.869	1.328	1.546
0.0	1.70	$-1.053 \, (-0.909)$[a]	0.394	1.200	1.447[b]
-0.5	1.70	$-1.454 \, (-1.30)$[a]	-0.114	1.093	1.340[c]
$N = 6$					
0.0	2.15	$-1.156 \, (-0.972)$[a]	$+0.533$	1.369	1.689
-0.5	2.10	$-1.480 \, (-1.294)$[a]	$+0.0675$	1.211	1.548

[a] Calculated using a one-term wave function (see Table II).
[b] Estimated half-line width is 0.124 eV with temperature dependence of the band maximum of -8.1 cm^{-1}/°K.
[c] Estimated half-line width is 0.112 eV.

Table IV. Metal–ammonia solutions relaxed states at 203 °K

	$V_0 = 0.5$ eV	$V_0 = 0.0$ eV	$V_0 = -0.5$ eV
$N = 4$			
$E_t(\overline{2p})$	0.3160 eV	-0.1826 eV	-0.6811 eV
$h\nu$ (emission)	0.594 eV	0.548 eV	0.506 eV
$E_t(Ts)$	-0.2780 eV	-0.7301 eV	-1.1873 eV
R	1.55 Å	1.55 Å	1.55 Å
Cs	0.197	0.09	0.08
Cp	0.007	0.007	0.007
$N = 6$			
$E_t(\overline{2p})$	0.3839 eV	-0.1126 eV	-0.6093 eV
$h\nu$ (emission)	0.627 eV	0.535 eV	0.467 eV
$E_t(Ts)$	-0.2429 eV	-0.6477 eV	-1.0765 eV
R	2.05 Å	2.05 Å	2.00 Å
Cs	0.181	0.153	0.126
Cp	0.020	0.019	0.017

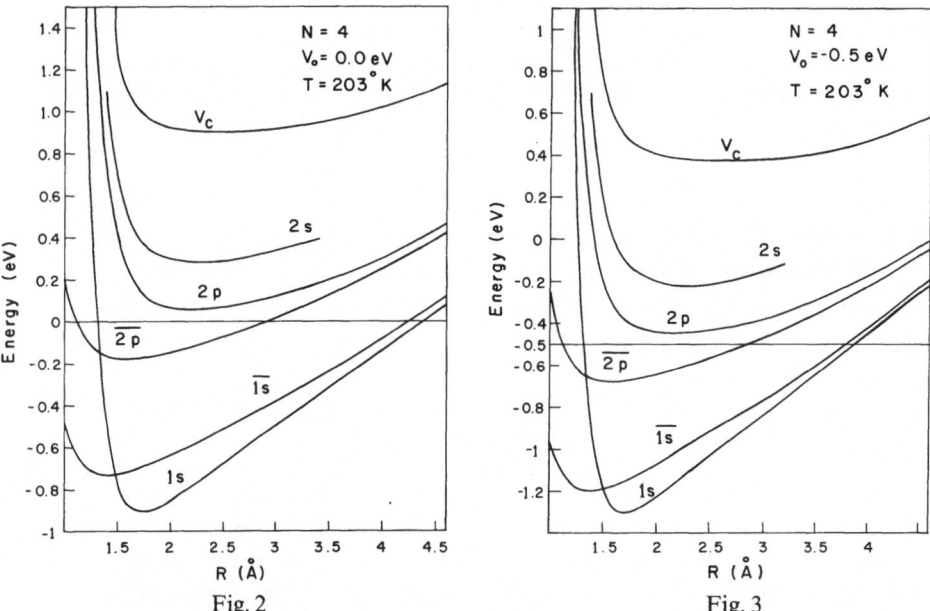

Fig. 2

Fig. 3

Fig. 2. Configurational diagrams for the total energy as a function of the radius R for various electronic states when $N = 4$, $V_0 = 0.0$ eV, and $T = 203$ °K. The ground state is denoted by 1s and the two bound excited states and the continuum level which exist when the polarization field is determined by the ground state are denoted by 2s, 2p, and V_c, respectively. The $\overline{2p}$ state is the lowest p energy state when the polarization is determined by the 2p excited state, and the $\overline{1s}$ state is the lowest energy level under the same polarization

Fig. 3. Configurational diagrams for the total energy as a function of the radius R for various electronic states when $N = 4$, $V_0 = 0.5$ eV, and $T = 203$ °K. The notation is the same as in Fig. 2

The effects of temperature and pressure can be studied by evaluating the energy by introducing the correct temperature and pressure dependent properties of the medium. Temperature occurs explicitly in the energy expressions. We have studied temperature dependences in metal–ammonia solutions and Feng, Fueki, and Kevin [26] using a closely related formalism have studied the pressure dependence of the spectrum of electrons in methanol and ethanol. In all cases agreement with experimental data is very good.

In the above calculations we have calculated the energy and properties of the excited states as they arise in a Franck-Condon transition from the ground state, i.e., the inertial polarization of the first coordination layer, and the continuum is fixed at the value appropriate for the charge density of the ground or 1s state. If the cavity is sufficiently long-lived the inertial polarization could relax to a value appropriate for the excited state in question. We have evaluated the relaxed 2p states and the relaxed 1s state where the inertial polarization is now determined by the charge density of the relaxed 2p state, i.e. we evaluate the average of $\cos\Theta$ using the 2p charge density, and Π using the relaxed 2p wave functions. The results are summarized in Table IV where the energy for the 2p→1s emission is also listed. It is significantly red-shifted from the absorption.

The curves for the total energy are plotted for the states discussed above in Fig. 2 and 3 for two typical cases. Also shown in these figures is the vertical continuum level, V_c, i.e. the energy of a quasi-free electron if the cavity and medium have their intertial polarizations fixed at the values dictated by the charge density of the ground state:

$$V_c = E_{ST} + E_{HH} + E_{PV} + E_{dds} + \Pi_s + V_o . \tag{23}$$

B. Comparison of Temperature Dependent Potential Calculations with Experimental Data

The principle results of our calculations are compared with experimental data in Tables II, IV, V, an VI. We will now review some of the major accomplishments of the theory.

a) A *stable cavity* does indeed exist in liquid ammonia of a radius R_0 of about 1.7—2.2 Å. It is stable because in addition to Eqs. (3) and (4) we have

$$(\partial^2 E_T/\partial R^2)_{R=R_0} > 0 . \tag{24}$$

b) The optimum *coordination number* in the first layer is about 4 but the model is not sufficiently accurate to rule out coordinations of 6, 8 or numbers intermediate. Nuclear magnetic relaxation data has been interpreted by Catterall [27] to be consistent with

$$3 \le N \le 13$$

completely in line with our predictions. The recent work of Pinkowitz and Swift [28] suggests a coordination number of around 30 but this must involve molecules beyond the first coordination layer.

A related measure of the total coordination number of the sodium ion and the solvated electron can be estimated from the determination of the high frequency permittivity of metal–ammonia solutions. In moderate concentration regions Breitschwerdt and Radsheit [29] estimate that about 15 ammonia molecules are

Table V. Results of model 3 calculations one-electron cavity (ammonia) 203 °K

	$V_0 = 0.5$ eV (in eV)	$V_0 = 0.0$ eV (in eV)	$V_0 = -0.5$ eV (in eV)
$N = 4$			
ΔH_1	-0.537	-0.909	-1.302
Vertical continuum level	1.558	1.038	0.544
Relaxed continuum level (vertical)	0.749	0.245	-0.242
Totally relaxed level	$+0.500$	0.000	-0.500
Vertical photoelectric threshold, P	2.095	1.946	2.348
Continuum relaxed photoelectric threshold	1.286	1.153	0.044
Vertical photoconductivity threshold, I	2.095	1.946	1.848
$N = 6$			
Vertical photoelectric threshold, P	2.863	2.168	1.967
Vertical photoconductivity threshold, I	2.863	2.168	1.467

Table VI. Comparison of experimental and theoretical properties of metal–ammonia solutions (240 °K)

	Experimental	Theoretical
$h\nu_{max}$	0.80 eV[a]	0.916 eV (1.32)[b]
ε_{max} extinction coefficient	49000 M^{-1} cm^{-1} [a]	—
f oscillator strength	0.77[a]	0.50
$W_{1/2}$ half-line width	0.46 eV[a]	0.105 eV (0.14 eV)[c]
$dh\nu_{max}/dT$	$-(1.5 \pm 0.2) \times 10^{-3}$ eV/deg	-5.3×10^{-4} eV/deg (-1.97×10^{-4} eV/deg)[c]
ΔH	1.7 ± 0.7 eV[d]	1.28 eV (1.43 eV)[b]
P, photoelectric threshold	1.6 eV[e]	1.85 eV
Symmetrically made vibration	—	62.4 cm^{-1}
forbidden 1s → 2s transition	—	1.34 eV[b]

[a] Quinn, R. K., Lagowski, J. J.: J. Phys. Chem. **73**, 2326 (1969).
[b] Using 2-term ground state basis set.
[c] Includes the statistically weighted $N = 6$ contributions.
[d] Jortner, J.: J. Chem. Phys. **30**, 839 (1959).
[e] Teal, G. V.: Phys. Rev. **71**, 138 (1948).

irrotationally bound to some charge center per dissolved sodium *atom*. This number is again in reasonable agreement with our coordination numbers.

c) *Volume expansion* data is explained. When the ammonia molecules are oriented under the field of the trapped electron the molecules in the first coordination layer are separated much more than in the bulk, i.e. the density of molecules in the first coordination layer is much lower than the bulk density. This phenomenon, commonly called "structure breaking" in electrolyte chemistry, means that in calculating the effective volume expansion of the liquid per electron one needs to include not only the actual void volume but the extra volume occupied by the oriented first coordination layer molecules. When this is done [22], one arrives at an effective cavity radius of about 3.0—3.1 Å which compares favorably with volume expansion data which have been interpreted in terms of a cavity of about 3.2 Å [9].

This large "structure breaking" effect was also found by Lepoutre and Demortier [30] in their recent study of the entropies of transfer and entropies of reaction of electrons in ammonia.

d) *Mobility* data are also consistent with a very loose first coordination layer. The loose structure allows the relatively high mobility of the electron (see Table V) and the extremely short lifetime of the trapped species [27]. The cavity does not move as a whole but must proceed by an "amoeba type" of motion.

e) *Viscosity* data are also consistent with extensive structure breaking of the solvent structure by the electron. Experimental values for the viscosity of metal–ammonia solutions are considerably lower than those of the pure solvent [31].

f) The *heat of solution* is also reasonably given by the present model since

$$\Delta H = -E_T(R_0) \tag{25}$$

using the total ground state energy. The present experimental value of 1.0 to 1.7 eV is uncertain because we do not know the absolute heat of solution of the proton in liquid ammonia. As shown in Table VI, even the use of more elaborate trial functions does not destroy this good agreement.

g) The *photoconductivity threshold* involves the photoconductivity onset, I, which is simply the energy needed to go from the ground state to the vertical continuum level or

$$I = V_{cs}(R_0) - E_T(R_0). \tag{26}$$

Since noninertial polarization effects are small we have (very roughly)

$$I \approx V_0 - E_{el}(R_0)$$

for the ground electronic state. The actual experimental value is likely to be much smaller since we have calculated this at the equilibrium cavity radius and these effects will be smeared by the thermal motion of the cavity.

h) The *photoelectric threshold*, P, for electron emission into the gas phase involves the further excitation of the electron into the gas phase. If the energy of a quasi-free electron is negative, additional work is required to remove the electron from the liquid into the gas phase. If the V_0 energy is positive, then it already has sufficient energy to escape. All of this is based on an idealized model with no electron-surface interaction. In actual experimental situations there are interactions between electrons in the gas and the surface layer which will modify the value of of the measured value P. Thus it is not surprising that our values of P do not correspond to experimental numbers. In addition the actual photoelectrically produced electrons do not all originate at the equilibrium configuration (specified by R_0) but from any thermally accessible configurations.

i) The *thermal ionization threshold* for the electron cavity would involve a transition from the ground state to the energy of the quasi-free energy, V_0,

$$\text{or} \quad \Delta E = V_0 - E_T(R_0).\tag{27}$$

This energy would range from 1.04 eV ($N = 4$, $V_0 = 0.5$) to 0.80 eV ($N = 4$, $V_0 = -0.5$ eV). Thermal ionization of the 2p electron to the continuum would involve almost no energy since the 2p level is very close to V_0.

j) The frequency of the *totally symmetric vibration* can be calculated from the energy curves which are potential curves for the ammonia molecules. The second derivative of the energy curve at the minimum yields the force constant for the totally symmetric mode. Predictions [22] are that this vibration will have an energy in the range of 25—100 cm^{-1}. If the species lasts long enough it is possible to monitor this by Raman Scattering from metal–ammonia solutions.

C. Optical Properties of the Electron as Calculated by the Temperature Dependent Potential Model

The configuration diagrams for the electronic states, of which Figs. 2 and 3 are examples, allow us to calculate the line shapes and other properties of the optical transitions. Involved in this assumption is the hope that the shape and temperature dependence of the optical transitions will be dominated by totally symmetric mode and by one configurational coordinate, namely R. If other modes are involved, we would then require energy curves for each normal coordinate.

In Table V we observe several results of the model calculation.

a) The values for the 1s → 2p excitation correspond reasonably well with the maximum in the experimentally observed absorption band. In contrast with earlier work, the correct expression is

$$hv = E_T^{2p}(R_0) - E_T^{1s}(R_0).\tag{28}$$

b) The oscillator strength for the above transition is much less than that obtained experimentally. This simply indicates that the experimental curve is more complicated than one simple transition.

c) There is a temperature dependence of roughly the correct magnitude and of the right sign for the maximum of the absorption band. This can be traced primarily to the role of temperature in not allowing the dipole moments to completely align in the field of the trapped electron. This shift is obtained without any ad hoc introduction of the dependence of any parameter on temperature, as in the original work of Jortner [9].

We can get the complete line shape by making two assumptions.

a) We will assume the high temperature approximation so as to apply classical statistics to the ground state population. This is good for 200—300 °K in our case.

b) We will assume the semi-classical Condon approximation in which the electronic transition moment is independent of the nuclear configuration, i.e. R.

With these assumptions the line shape is determined by the thermal population of the ground state and the dependence of the vertical transition energy on R. If $X = R - R_0$ and

$$A(X) = E_T^{1s}(R) - E_T^{1s}(R_0) \tag{29}$$

then the unnormalized intensity distribution function $F(E)$ for the optical excitation at energy E is

$$F(E) = \exp\left(-\frac{A(X)}{kT}\right)\left|\frac{dX}{dE}\right| \tag{30}$$

These functions are plotted for two values of T and N on the left in Figs. 4 and 5. These figures show that:

a) The values of $h\nu$ and $h\nu_{\max}$ are very close together [22], indicating that a harmonic approximation for the ground state is reasonably accurate at the temperatures involved.

b) Although the line widths of individual transitions are quite wide, 0.096 to 0.148 eV, nonetheless they are smaller than the experimental width by a factor of three or four. The line widths of individual transitions also obey the harmonic approximation quite closely since they are roughly proportional to the $\sqrt{T}$.

c) The line shapes are slightly asymmetric but to the low energy side, contrary to the experimental data.

The actual absorption band predicted by this model is more complicated since, at least, the $N = 6$ and $N = 4$ cavities contribute. In general

$$F(E) = \sum_N f_N F_N \exp[-E_T^N(R)/kT] / \sum_N \exp[-E_T^N(R)/kT] \tag{31}$$

where f_N is the (mean) oscillator strength of the cavity species with a coordination number N, F_N is $F(E)$ for the same species, and E_T^N is the ground state energy of the same species. If we have only $N = 4$ and 6 and further $f_4 = f_6$ (see Ref. 22) then the formula is relatively simple

$$F(E) = F_4(E) + F_6(E) \cdot \exp[-(E_T^6(R) - E_T^4(R))/kT] \tag{32}$$

neglecting a bunch of normalization factors. This is plotted on the right side of Figs. 4 and 5. The composite curve is slightly broader, in general, but has the same general features as the individual profiles. Thus the cause of the very asymmetric band is not known.

In Fig. 6 we show a typical comparison of the experimental and theoretical line shapes with intensity arbitrary. Also listed in this figure is the estimated transition energies to the 2s, 3p, 4p, and continuum levels as predicted by the model. For the 3p and 4p states, the *electronic energies* should correspond to

$$E_n \to -\frac{\beta^2}{2n^2} + V_0 \quad \text{(when } n > 3) \tag{33}$$

in atomic units since the electron is in a very large Rydberg-like state. Thus the model is not consistent with a large contribution of the oscillator strength being carried by the higher excited states. This represents one of the major

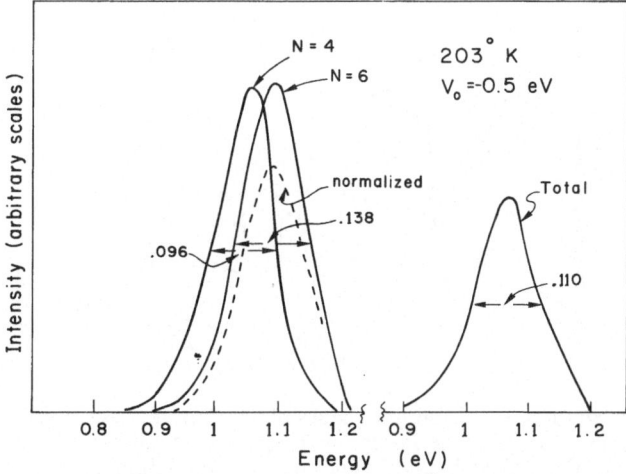

Fig. 4. Line shapes calculated at 203 °K and $V_0 = -0.5$ eV by the temperature dependent potential model. On the left are the line shapes for the $N = 4$ and $N = 6$ cavities including the $N = 6$ results normalized by its relative Boltzman factor. On the right is the composite line shape obtained from the $N = 4$ and $N = 6$ cavities following Eq. (32) in the text

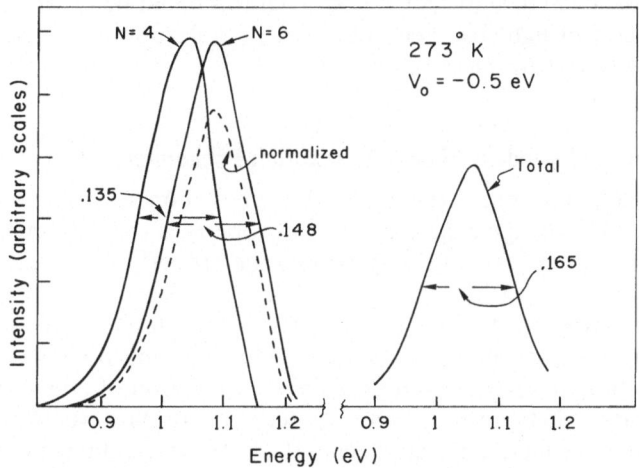

Fig. 5. Line shapes calculated at 213 °K and $V_0 = 0.5$ eV by the temperature dependent potential model. The notation is the same as in Fig. 4

failures of the model. Before modifying the model greatly it is important to look at the contributions from other normal modes of vibrations, not only totally symmetric ones. It is also possible, but unlikely, that with a huge vibronic mixing some of the intensity in the high energy side of the absorption band is due to the normally forbidden 1s → 2s transition.

Two further optical transitions are of interest. It should be possible to observe the 1s → 2s transition via two-photon absorption especially in deuteroammonia [32]. Also in Table IV the energy for the emission from the relaxed 2p state is listed.

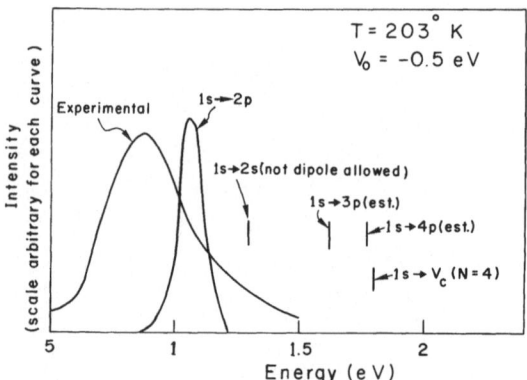

Fig. 6. Typical transition energies and line shapes obtained from these models in comparison with typical experimental data. The 1s→2p line shape is the total curve in Fig. 4. The other lines indicate where the various bound-bound and bound-continuum (1s→V_c) levels are located for this model. Although the location of the theoretical maximum does not agree with experimental data, it is the shape and width which are in most serious disagreement

If the cavity species were sufficiently long-lived this could be observed. However, since no emission of light has been detected the relaxation of the excited state must go via other radiationless routes.

D. Temperature Independent Theory-Relaxation Phenomena [33]

A better approach to the entire theory would be based on a temperature independent potential. The average expected for any property is then obtained by weighing its value for any one state of the system by the properly normalized Boltzman factor.

It is a simple matter to convert our previous work into this new formation. Instead of using a temperature averaged cosine function in Eq. (7), we now use the full cosine function and present our energies as a function of two coordinates, R and $\cos \Theta$. In principle there is still another set of coordinates representing quasi-polaron modes of the medium. The effect of the latter are contained in the Π term, for example, which vanishes for a quasi-free electron but exists for any localized state.

In Fig. 7, 8, and 9 we list a few of the infinite number of curves one obtains from these calculations. In particular we only show curves for $\cos \Theta = 0.0$ and 1.0. The last subscript on these curves refers to the value of the $\cos \Theta$. At small radii the one-term wave function is inadequate and thus the crossing of the $2P_0$ and V_{m_0} curves is probably coincidental. In all cases the Π term added is of a value appropriate to the state in question. This means for the 1s state we use Π_s [calculated using $\psi_{1s}(R, \cos \Theta)$], for the 2p state use Π_p [calculated using $\psi_{2p}(R, \cos \Theta)$], and for the continuum levels we use $\Pi = 0$. In addition we have the factor of $\cos \Theta$ in the hydrogen-hydrogen repulsion so that when the first coordination layer relaxes this contribution vanishes also.

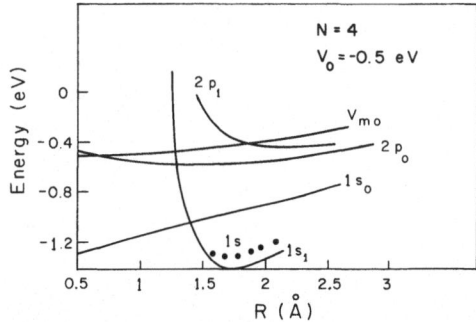

Fig. 7. Temperature independent potential model configuration diagrams for the total energy as a function of the radius R for $N = 4$ and $V_0 = -0.5$ eV. The last subscript now refers to the value of $\cos\Theta$. Only curves for two values of $\cos\Theta$ are plotted. Each state has the bulk medium polarized according to its electron density (see text). The dots labelled 1s denote the temperature averaged result where $\langle\cos\Theta\rangle_{1s} = 0.8$ to 0.9 V_M is the medium or continuum level

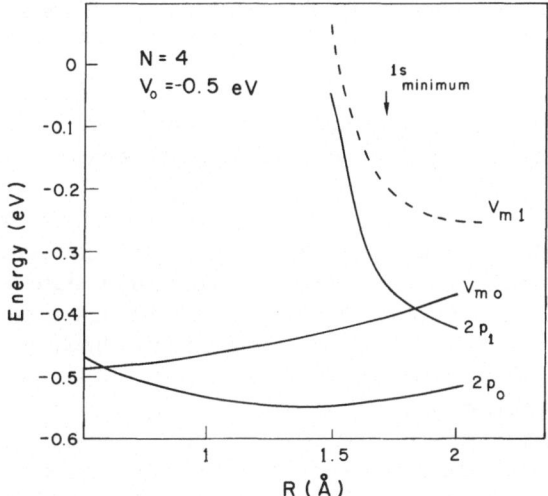

Fig. 8. An expanded view of the upper portions of Fig. 7 including the completely oriented dipole curve for the continuum or medium level. The arrow indicates the position of the minimum for the 1s state

To relate this to our previous work it should be remembered that previously we used $\langle\cos\Theta\rangle_{1s}$ for all Franck-Condon transitions and $\langle\cos\Theta\rangle_{1s}$ is about 0.9. On Fig. 6 we also indicate where the temperature-averaged result occurs relative to $1s_1$.•

The use of this temperature independent potential to calculate the thermally averaged energy of the bound states leads to almost exactly the same answers as in our previous calculations, since for values of $\cos\Theta$ above about 0.6 the energy is a linear function of $\cos\Theta$ and averaging yields the same Langevin result as before. However, for weakly bound states, i.e. small values of $\cos\Theta$, this approximation is poor and thus our relaxed 2p states would probably change if we were

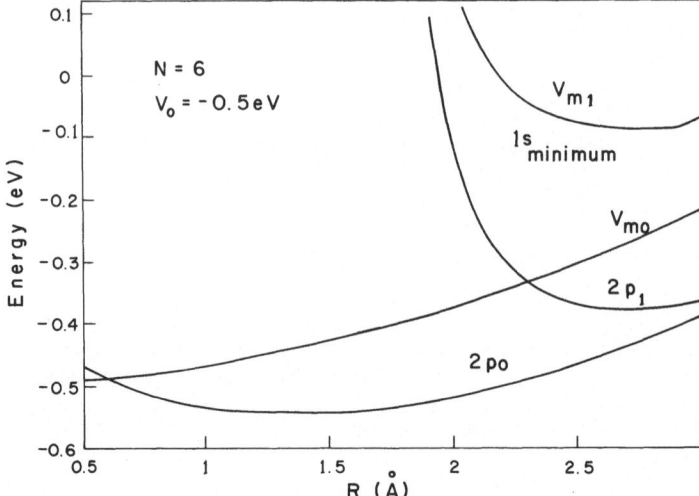

Fig. 9. Temperature independent potential model configuration diagrams for the total energy as a function of the radius R for $N=6$ and $V_0 = -0.5\,\text{eV}$. Notation is the same as in Fig. 7 and 8

to perform the proper averaging as opposed to using a temperature dependent potential. It is unlikely that the relative behavior of the previous relaxed 2p and 1s states would change greatly.

The importance of these results lies in what they can suggest concerning possible relaxation mechanisms i.e. radiationless transitions, of this one isolated cavity. It is possible for the excited 2p state to relax to a continuum state (or vice versa). There are even ways for the excited 2p state, within the few examples of levels presented here, to decay to the ground state. In all cases however these are not Frank-Condon transitions since in almost all cases the curves cross only when $\cos\Theta$ is different for the two states.

It is also possible for a thermal flucuation to change a large to a small cavity and vice versa with the net expenditure of very little energy (see $1s_0$ and $1s_1$ curves). It is possible that this provides one mechanism for the mobility of the electron. The activation energy from these curves corresponds to about 6 kcal/mole but a search of the entire energy surface may yield smaller numbers. For reference the activation energy involved in the viscosity of potassium–ammonia solutions is about 1.6 kcal/mole [31]. In addition it is not clear how accurate our model is. At the moment, our analyses of these results are very preliminary. They are presented to indicate the type of attack that will be necessary to explain the many radiationless processes which must occur. At the moment we do not even have estimates for the rates of possible processes suggested by the energy surfaces.

In addition alternative energy surfaces may be needed in order to consider some properties of the liquid. For example, for mobility studies one needs to consider some sort of electron transfer process in the manner of Levich and his co-workers [34] but complicated by the discrete solvation layer and the lack of a definite location to which an electron must transfer. Nonetheless the discussions of Levich on electron transfer processes provide a reasonable starting point.

IV. Calculations and Results in Other Systems

A. Hydrated Electron

In Table VII we present the results of this model for an excess electron in water. Since the water molecule can rotate to reduce the amount of hydrogen-hydrogen repulsion we have assumed it is negligible in these calculations. The small, almost zero, cavity size is consistent with pressure effect measurements and with the entropy calculations of Lepoutre and Demortier [30].

More elaborate calculations have been performed on this species by Fueki, Feng, Kevin, and Christoffersen [35]. They have deviated from our model in two ways. First, they use the SCF (Self-Consistent Field or independent particle) procedure and not the adiabatic procedure we have employed in defining our $V(r)$ [10]. Especially for water this procedure is to be preferred [10]. Secondly, they minimize the total enery E_T and not simply as we did. This latter difference is not important as the lower half of Table VII indicates. However, the use of the SCF procedure greatly improves their agreement with the experimental data. The experimental data and typical results found by our calculation are listed in Table - VIII. Fueki *et al.* find good agreement with most experimental data but again their optical spectrum is much too narrow. They used a V_0 in the range 0.5 to -0.5 eV but the latest work of Gaathon and Jortner [36] suggests that it could be as large as 1.6 eV.

Several calculations have been made attempting to study the electronic spectra of $(H_2O)_N^-$. The first calculations by Natori and Watanabe [37] were followed by the molecular orbital work of Weissmann and Cohan [38], extended Hückel calculations by Aldrich, Gary, and Cusachs [39], and extensive calculations of molecular models by Ray [40]. In general all calculations support our model in suggesting almost no void space and only a minor disturbance in the liquid structure. They usually are not able to introduce all medium and temperature effects.

Recently an extensive calculational effort has been made by Newton on this problem using a double-zeta LCAO level gaussian basis set calculation with four molecules in the first coordination layer with dipoles directed toward the center [41]. The medium is carefully introduced using the independent particle or SCF model [10]. He has found the latter to be extremely important (without this contribution the electron would not localize). His results are very impressive and indicative of what can be done. He finds a free energy of formation of about 1 eV and an optimal value of $r_d = 2.45$ Å (corresponding to a void radius of 1.0—1.4 Å). About 80—84% of the electronic charge is within the first solvation sphere. His extra electronic energy, including the continuum contributions and with fixed dipoles ($\cos \Theta = 1.0$), is -1.25 eV. Work is now underway to relax the orientation of the water molecules and calculate the energy of the excited state.

B. Other Solvents

The model presented in this paper allows one to make qualitative predictions of the effect of solvent structure on the maximum of the absorption band. If we consider the alcohols and remember that the shape of the molecule which makes up the cavity walls greatly influences the size of the cavity and thus, inversely the

Table VII. Results of model 3 calculations on the one-electron cavity in water (298 °K)

	$V_0 = 1.5$ eV	$V_0 = 1.0$ eV	$V_0 = 0.5$ eV	$V_0 = 0.0$ eV	$V_0 = -0.5$ eV
N = 4					
$E_T(= -\Delta H)$	-0.3219 eV	-0.7616 eV	-1.221 eV	-1.685 eV	-2.228 eV
E_{el}	-2.286 eV	-3.002 eV	-3.594 eV	-4.096 eV	-4.603 eV
R_0	0.90 Å	0.70 Å	0.60 Å	0.56 Å	~0.50 Å
$h\nu$	2.37 eV	2.57 eV	2.68 eV	2.70 eV	2.74 eV
Total minimization					
E_T		-0.756 eV	-1.214 eV	-1.678 eV	
E_{el}		-2.959 eV	-3.652 eV	-4.138 eV	
R_0		0.75 Å	0.60 Å	0.55 Å	

For reference: Fueki, Feng, Kevan, and Christoffersen [J. Phys. Chem. **75**, 2297 (1971)] obtained for $N = 4$ and $V_0 = 0.0$ eV by an SCF method and total minimization $E_T = 1.94$ eV $h\nu = 1.38$ eV.

Table VIII. Comparison of experimental and theoretical properties of hydrated electrons (300 °K)

	Experimental	Theoretical (this paper) $N = 4$, $V_0 = 1.0$ eV
$h\nu_{max}$	1.72 eV[a]	2.57 eV
ε_{max} extinction coefficient	15800 M^{-1} cm^{-1}[a]	—
f oscillator strength	0.65[a]	—
$W_{1/2}$ half-line width	0.92 eV[a]	—
$dh\nu_{max}/dT$	-2.9×10^{-3} eV/deg[a]	—
ΔH	1.7 eV[b]	0.76 eV
P, photoelectric threshold	unknown	4.00 eV
Electron mobility	2.5×10^{-3} cm^2/volt-sec[c]	—

[a] Hart, E.J., Gottschall, W.C.: J. Am. Chem. Soc. **71**, 2102 (1969).
[b] Baxendale, J.M.: Radiation Res. Suppl. **4**, 139 (1964). — Jortner, J., Noyes, R.M.: J. Phys. Chem. **70**, 770 (1966).
[c] Schmidt, M., Buck, W.L.: Science **151**, 70 (1966).

transition energy, we see at once that it is not surprising that tertiary butanol has a transition energy similar to 2-propanol, but very different from 1-butanol in accord with experimental data [42, 43]. Any attempt to relate them to the dielectric constant fails [43]. Because of the increased repulsion between solvent molecules, electrons in 2-propanol have a much smaller transition energy than electrons in 1-propanol [42]. Considering the differences in the liquid ranges and other variations, ethanol, methanol, and 1-butanol lead to remarkably similar transition energies for the solvated electron. Small amines should provide another test of these qualitative ideas as could studies in inorganic liquids related to ammonia and water.

Fueki et al. [44] have made quantitative studies on a number of alcohols in the liquid and glass state. Again very good qualitative agreement is found using small coordination numbers. In a recent study they have also been able to obtain good agreement with the shift of the absorption maximum with pressure in methanol and ethanol [45].

In mixed solvents the transition energy appears to shift continuously with charge in composition in most cases [46, 47]. This indicates that the first coordination layer is essentially a random mixture and there are few specific interactions. An exception might be in tetrahydrofuran and water mixtures.

C. Low Density Polar Gases

In early work [19] we speculated that it might be possible to observe a solvated electron even in reasonably low density polar gases. In recent work [48] the solvated electron was found in ammonia vapor with a density at least as low as 0.1 g/ml. Theoretical work indicated it could exist at even lower densities based on energy considerations. However at much lower densities the entropy contributions could be important, and this author [49] using very simple arguments to calculate these contributions has estimated that the trapped species might be unstable below 0.03 g/ml. It is interesting to speculate on the density dependence of the properties of excess electrons in media with various electron-molecule interactions from the strongly repulsive helium [29, 20] example to ammonia or water until finally we have a reasonably stable negative ion (such as SF_6).

D. One Possible Spin-Pairing Species $(e_2^=)_{am}$

Using the above models, Copeland and this author [50] have calculated the stability of the two-electron cavity species. The total energy of the optimal two-electron cavity in ammonia solutions is listed in Table IX. The electron-electron interaction was handled in the independent electron scheme following the work of Land and O'Reilly [11]. Correlation energy corrections were included via a configuration interaction scheme. In Table IX the one- and two-electron species are compared and it is shown that the two-electron species is unstable relative to two cavities by about 1 eV per electron. Thus this species is probably not responsible for the spin pairing observed in moderate concentration solutions. Much more work is needed on alternative schemes of spin pairing, considering related species by similar models as well as beginning new approaches to the entire theory of moderate concentration solutions.

Table IX. Results of model 3 calculations two-electron cavity in ammonia ($N = 12$, 203 °K)

	$V_0 = 0.5$ eV	$V_0 = 0.0$ eV	$V_0 = -0.5$ eV
E_t (eV)	0.8773	0.3687	-0.1985
E_{el} (eV)	-2.9855	-3.4399	-8.9336
$E_t + E_{corr} (= +\Delta H_2)$ (eV)	0.467	-0.031	-0.580
$E_{el} + E_{corr}$ (eV)	-3.396	-3.840	-4.215
R_0 (Å)	3.50	3.45	3.45
R_0^{eff} (Å)	4.12	4.02	4.02
Optimum orbital exponent	0.4063	0.3780	0.3364
Charge enclosed within radius R	1.082	0.967	0.819
Average value of the cosine of the dipole angle	0.925	0.918	0.903

Table X. Relative stability of the two-electron cavity (ammonia) (203 °K)

	$V_0 = 0.5$ eV	$V_0 = 0.0$ eV	$V_0 = -0.5$ eV
$\Delta H_1 (N = 6)$	-0.678 eV	-0.972 eV	
$E_{el} \quad (N = 6)$	-2.069 eV	-2.326 eV	
$\Delta H_1 (N = 4)$			-1.302 eV
$E_{el} \quad (N = 4)$			-2.404 eV
$\Delta H_2 (N = 12)$	0.467 eV	-0.031 eV	-0.580 eV
$E_{el} \quad (N = 12)$	-3.396 eV	-3.840 eV	-4.215 eV
$\Delta H_{21} = \frac{1}{2} \Delta H_2 - \Delta H_1$	0.91 eV	0.96 eV	1.01 eV

Very likely the correct theoretical approach in the moderate concentration region will involve techniques similar to those being used currently to study liquid metals and amorphous semi-conductors. As Dye [1] has indicated it is very hard to deal with this concentration range using ordinary chemical ideas which involve well-defined species and well-defined equilibrium constants.

V. Summary — Accomplishments and Limitations of the Models

The basic models presented in this paper are able to explain most of the major qualitative features of excess electrons in fluids — when their concentration is small. When pushed the models can explain most of the basic experimental data quantitatively. Nevertheless more work is needed to even further test the basic features of the model and explain its serious failings, as for example, in the case of line shapes.

One point we have not discussed in detail concerns our use of the weak coupling polaron or adiabatic approach as compared with the better independent particle or self-consistent field (SCF) picture [10]. With our introduction of the discrete first coordination layer, we only include those contributions from the bulk solvent in our continuum region which lies beyond the first layer. Thus electron-bulk solvent interactions are much weaker in our models than in the older continuum treatments. For this reason we can use (at least, for ammonia and with only a few difficulties, for water) the simpler adiabatic treatment. The strong interactions are introduced explicitly by our first coordination layer.

Theoretically we need further studies of these models on

a) electrons in mixed solvent systems,

b) the role of the nonsymmetrical modes of the cavity in determining the line shape and various radiationless processes,

c) relaxation paths for the excited state with estimates of the rates of these processes,

d) electrons in low density polar vapors including entropy contributions with the object to predict the dependence of properties on molecular and bulk parameters.

Further theoretical work is also needed

a) to determine V_0 accurately since it may not be obtained experimentally

b) to study the effect of the short lifetime of the cavity species on its properties. The mixed solvent results are of interest in this regard as are studies of two photon absorptions,

c) to explain the very assymetric absorption bands. Possibly the deficiency is in our models and could be explained if several coordination numbers were equally likely, but it may be more complicated and involve the dynamic character of the cavity. An analysis of line widths in low density gases may provide further clues.

d) to predict transport properties e.g. the mobility of the electron and the viscosity of the solution

e) to begin to study moderately concentrated solutions without the need to speak in terms of ordinary chemical species or ordinary equilibria.

Experimentally there are still a few major problems. We need

a) an experimental method which can yield V_0 for a polar fluid, especially if it can be done as a function of density

b) observations of two photon transitions in the solution from the 1s to 2s states, for example. This could confirm the basic model since the lower levels have an ordering closer to that of a particle in a box than in a hydrogen atom.

c) better studies of the optical spectrum, e.g. saturation techniques with lasers. (Despite the initial failure to observe any effect in water [51], further attempts should be made.) There must be ways to break the broad absorption into its individual components.

d) mobility studies of excess electrons in low density gases as a function of density using various polar and nonpolar molecules with widely different dipole moments and electron affinities. Mixed gases would be extremely interesting.

If the progress made in the experimental and theoretical studies of low concentrations of excess electrons in various fluids since our last Colloque Weyl continues, it may well be that when we meet again all of our present problems will be understood and we can tackle the more complex problems.

Acknowledgement

The initial stages of this work were supported by National Science Foundation Grants.

References

1. Dye, J. L.: Proc. Colloque Weyl II, Lagowski, J. J., Sienko, M. J., (Eds), pp. 1—17. London: Butterworths 1970.
2. Ogg, R. A.: J. Am. Chem. Soc. **68**, 155 (1940); — J. Chem. Phys. **14**, 114 (1946); **14**, 295 (1946); — Phys. Rev. **69**, 243 (1946); **69**, 668 (1946).
3. Libscomb, W. N.: J. Chem. Phys. **21**, 52 (1953).
4. Stairs, R. A.: J. Chem. Phys. **27**, 1431 (1957).
5. Landau, L.: Physik. Z. Sowiet Union **3**, 664 (1933).
6. Davydov, A. S.: J. Exp. Theoret. Phys. U.S.S.R. **18**, 913 (1948).
7. Deigen, M. F.: Zhur. Exp. Theoret. Phys. U.S.S.R. **26**, 300 (1954).
8. Platzman, R. L., Franck, J.: Z. Physik **138**, 411 (1954).
9. Jortner, J.: J. Chem. Phys. **30**, 839 (1959).
10. For a comparison of the adiabatic and self-consistent methods for treating electrons in various media, see Jortner, J.: Mol. Phys. **5**, 257 (1962) for the theory, and Jortner, J., Rice, S. A., Wilson, E. G.: In: (Eds.) Lepoutre, G., Sienko, M. J.: Solutions metal–ammoniac: Colloque Weyl, pp. 222—276. New York: Benjamin 1964. We will return to this point later in this paper.
11. O'Reilly, D. E.: J. Chem. Phys. **41**, 3736 (1964). — Land, R. H., O'Reilly, D. E.: J. Chem. Phys. **46**, 4496 (1967).
12. Kestner, N. R., Jortner, J., Cohen, M. H., Rice, S. A.: J. Chem. Phys. **43**, 2614 (1965).
13. Kestner, N. R., Jortner, J., Cohen, M. H., Rice, S. A.: Phys. Rev. **140**, A56(1956). — Jortner, J., Kestner, N. R., Cohen, M. H., Rice, S. A.: In: Sinanogln, O. (Ed.): Modern quantum chemistry, Vol. 2, p. 129. New York: Academic Press 1966.
14. Hiroike, K., Kestner, N. R., Rice, S. A., Jortner, J.: J. Chem. Phys. **43**, 2625 (1965).
15. Springett, B. E., Cohen, M. H., Jortner, J.: Phys. Rev. **159**, 183 (1967).
16. Lekner, J.: Phys. Rev. **158**, 130 (1967). — Cohen, M. H., Lekner, J.: Phys. Rev. 158, **305** (1967).
17. For hydrocarbon liquids, see: Holroyd, R. A., Allen, M.: J. Chem. Phys. **54**, 5014 (1971). — Davis, H. T., Schmidt, L. D., Minday, R. M.: Phys. Rev. **43**, 1027 (1971). — Fueki, K., Feng, D.-F., Kevin, L.: Chem. Phys. Letters **13**, 616 (1972). For inert gas solids, see: Jortner, J.: Berichte der Bunsen Gesellschaft, **75**, 696 (1971). — Raz, B., Jortner, J.: Chem. Phys. Letters **4**, 155 (1969). — Proc. Roy. Soc. A317, 113 (1970). For polar liquids V_0 is not known, but can only be estimated.
18. Jortner, J., Rice, S. A.: In: Solvated electrons, p. 7. Hart, E. J., Ed., American Chemical Society, Washington, D. C., 1965.
19. Jortner, J., Kestner, N. R.: In: Metal–ammonia solutions — Colloque Weyl II, Lagowski, J. J., Sienko, M. J., Eds., p. 49. London: Butterworths 1970.
20. Jortner, J.: Ber. Bunsen-Gesellsch. **75**, 696 (1971).
21. (a) Lepoutre, G., Sienko, M. J.: Metal ammonia solutions. New York: Benjamin 1964. — (b) Thompson, J. C.: In: Lagowski, J. J. (Ed.): Chemistry of non-aqueous solvents, p. 265. New York: Academic Press 1967. — (c) Solvated electron — Advances in Chemistry Series No. 50. Washington D. C.: 1904. — (d) Lagowski, J. J., Sienko, M. J.: Metal–ammonia solutions — Proceedings of Colloque Weyl II. London: Butterworths 1970. — (e) Hart, E. J., Anbar, M.: The hydrated electron. New York: Wiley 1970. — (f) Cohen, M. H., Thompson, J. C.: Advances in Physics **17**, 857 (1968). — (g) Catterall, R., Mott, N. F.: Advan. Phys. **18**, 665 (1969).
22. Copeland, D. A., Kestner, N. R., Jortner, J.: J. Chem. Phys. **53**, 1189 (1970).
23. The formula equivalent to Eq. (13) in Ref. [22] contains an error, since the polarization of the continuum was counted twice in S_i and in V_0. The present equation puts the continuum polarization completely in V_0, where it belongs to.
24. Eisenberg, D., Kauzmann, W.: The structure and properties of water, p. 46. New York: Oxford University Press 1968.
25. Fueki, K., Feng, D.-F., Kevan, L.: Chem. Phys. Letters **4**, 313 (1969).
26. Feng, D.-F., Fueki, K., Kevan, L.: J. Chem. Phys. (in press).
27. Catterall, R.: In: Lepoutre, G., Sienko, M. J. (Eds.): Metal–ammonia solutions, p. 41. New York: Benjamin 1964. — Catterall, R.: Nature **229**, 10 (1971). — Catterall, R.: In: Lagows-

ki, J. J., Sienko, M. J. (Eds.): Metal–ammonia solutions — Colloque Weyl II, p. 105. London: Butterworth 1970.

28. Pinkowitz, R. A., Swift, T. J.: J. Chem. Phys. **54**, 2858 (1971).
29. Breitschwerdt, K. G., Radscheit, H.: Ber. Bunsen-Gesellsch. **75**, 644 (1971).
30. Lepoutre, G., Demortier, A.: Ber. Bunsen-Gesellsch. **75**, 647 (1971).
31. Hutchison, C. A., Jr., O'Reilly, D. E.: J. Chem. Phys. **52**, 4400 (1970).
32. For more discussion of the calculations and importance of the 2s state, see Kestner, N. R., Logan, J.: J. Phys. Chem. **76**, 2738 (1972).
33. This section is based on initial results of a study of radiationless transitions in metal–ammonia solutions by Jortner, J., and the author. J. Phys. Chem. **77**, 1040 (1973).
34. See, for example, Levich, V. G.: In: Erying, H., Henderson, D., Jost, W. (Eds.): Physical chemistry — an advanced treatise, Vol. IX B. New York: Academic Press 1970. Levich, V. G.: Advan. Electrochem. Eng. **4**, 249 (1965).
35. Fueki, K., Feng, D.-F., Kevan, L., Christoffersen, R.: J. Phys. Chem. **75**, 2291 (1971).
36. Gaathon, A., Jortner, J. (to be published). It is referred to in Olinger, R., Schindewolf, U., Gaathon, A., Jortner, J.: Ber. Bunsen-Gesellsch. **75**, 690 (1971).
37. Natori, M., Watanabe, T.: J. Phys. Soc. Japan **4**, 1573 (1966).
38. Weissmann, M., Cohan, N. V.: Chem. Phys. Letters **7**, 445 (1970).
39. Aldrich, M. S., Gary, L. P., Cusachs, L. C.: (unpublished work done in 1970).
40. Ray, S.: Chem. Phys. Letters **11**, 573 (1971).
41. Newton, M.: (unpublished work, 1972). We are grateful for the extensive details of his work which were sent to us prior to publication. See J. Chem. Phys. **58**, 5833 (1973).
42. Matheson, M. S., Dorfman, L. M.: Pulse radiolysis. Cambridge: M.I.T. Press 1969.
43. Saver, M. C., Jr., Arai, S., Dorfman, L. M.: J. Chem. Phys. **42**, 708 (1965).
44. Methanol: Fueki, K., Feng, D.-F., Kevan, L.: Chem. Phys. Letters **10**, 504 (1971). Glassy ethanol (including possible slow dipole orientation effects): Fueki, K., Feng, D.-F., Kevan, L.: J. Chem. Phys. **56**, 575 (1972).
45. Feng, D.-F., Fueki, K., Kevan, L.: J. Chem. Phys. **57**, 1253 (1972).
46. Dorfman, L. M., Jou, F. Y., Wageman, R.: Ber. Bunsen-Gesellsch. **75**, 681 (1971) (tetrahydrofuran and other solvents).
47. Olinger, R., Schindewolf, U.: Ber. Bunsen-Gesellsch. **75**, 693 (1971) (ammonia and water mixtures at various temperatures and pressures); Dye, J. L., De Backer, M. G., orfman, L. M.: J. Chem. Phys. **52**, 6251 (1970) (ammonia-water and ethylene diamine-water solutions).
48. Olinger, R., Schindewolf, U., Gaathon, A., Jortner, J.: Ber. Bunsen-Gesellsch. **75**, 690 (1971).
49. Kestner, N. R.: (unpublished).
50. Copeland, D. A., Kestner, N. R.: J. Chem. Phys. **58**, 3500 (1973).
51. Kenney-Wallace, G., Walker, D. C.: Ber. Bunsen-Gesellsch. **75**, 634 (1971).

Discussion

K. BAR-ELI What is the effect of the value chosen for V_0 on the absorptions obtained in your calculation?

N. R. KESTNER For ammonia, the absorption maximum shifts to lower energies as V_0 decreases. However the change is only 0.22 eV when V_0 changes from $+0.5$ eV to -0.5 eV when we use $N = 4$. For $N = 6$ the changes are only slightly larger. Therefore agreement with experimental data for metal–ammonia solutions is not sensitive to the exact value of V_0 used for these conditions. In the case of electrons in low density polar gases the value of V_0 is more important and one really needs two values of V_0, one for the continuum and one for the first layer as will be reported in the work of Gaathon and Jortner later in this conference. As regards line width, the value of V_0 does not have much effect.

K. BAR-ELI   What is the effect of the size of the dipoles on the absorption?

N. R. KESTNER   In these calculations, we have assumed that the dipoles can be approximated by point dipoles. This is the simplest approximation and is but one of the many assumptions in the model. As with many simple models, correcting one point such as this is not fruitful since often the results become poorer. In principle to do the calculation properly one needs to carry out extensive quantum mechanical calculations such as those of Newton where none of these approximations regarding the first coordination layer were made. Nevertheless, due to the magnitude of the problem the molecules had fixed orientations and the outer medium was treated by a continuum approximation.

The dependence on the magnitude of the dipole moment is as would be expected: the larger the magnitude of the dipole moment, the deeper is the trap, and the higher the excitation energy. All of this assumes that there are no steric repulsions forcing the cavity to be larger, in which case the two factors would compete.

P. DELAHAY   At large distances the potential in your model varies as $1/r$. Then, no matter how the potential varies at short distances, one would expect an infinite number of stationary states. In making comparisons with experimental absorption spectra one should then include not only the $1s \rightarrow 2p$ transition but also higher transitions, including transition to the continuum. The latter may be significant in some solvents.

N. R. KESTNER   In principle there are an infinite number of bound states as you say but in our calculations for metal–ammonia solutions the higher states are predicted to lie at such high energies that they should not be considered as part of the commonly observed broad absorption. Accurate calculations of the oscillator strength for the $1s \rightarrow 2p$ transition made by Gaathon and Jortner suggest that most of the observed line intensity is accounted for by this transition. These arguments are not conclusive but in this case seem to be consistent with several pieces of data. In other solvents, especially those with shallow traps and small β values, I readily agree that one could observe higher transitions including those to the continuum.

P. DELAHAY   What is a quasi-free electron? You mentioned that "it essentially does not disturb the medium". I suppose you include electronic polarization.

N. R. KESTNER   A quasi-free electron is an electron moving in a medium which is not distorted due to the presence of the charge. Its energy includes the kinetic energy due to scattering from atoms, including multiple scattering, and the electronic polarization interactions with the atoms, all properly screened.

J. L. DYE   As I recall the work of Newton, most of the electron density resided in the cavity. Could you comment on this and give your comparison?

N. R. KESTNER   In the unpublished work which Newton kindly sent me he quotes that for water 80—84% of the electron density is within the radius defined by the start of the continuum beyond the first coordination layer, i.e., r_c. In our model for ammonia we have about 84% of the charge within the radius defined by the center of the dipoles and thus our comparable values may even be slightly larger than his. For water we obtain similar results. In general the charge is primarily localized within the first coordination layer. Nevertheless the electron wave function is very expanded compared with most ions and even most anions. It is worth pointing out again that this localization results only when both the first coordination layer *and* the continuum contributions are included.

J. V. ACRIVOS   Given that E_{Ti} depends strongly on E_{HH}, what is the protonproton intermolecular separation in a cavity for $N = 4$ (and the given value of R) and how does this affect the free rotation of the molecules in the first solvation sphere?

N. R. KESTNER   Using the formulas listed in our original paper, the protonproton internuclear separation in the most favorable configuration of the ammonia molecules is about 2.4 Å. This distance would decrease rapidly of the ammonia molecules were to rotate and thus the coordinated ammonias cannot rotate. They can, however, change their angle of orientation with respect to the radius vector if the molecules perform the sort of cooperative motion we have included in our temperature independent calculations. It is important to remember that one of the basic goals of our model is to include such repulsive interactions properly.

M. H. COHEN   You have found that the total energy is closely similar for two quite different configurations, 4 and 6 nearest-neighbors, whereas the electronic component changes somewhat more. This suggests as a possible explanation for the line width, the contributions of relatively large fluctuations in configuration of the individual near neighbors (involving changes in orientation and position of all of them instead of just the symmetric mode). The total energy changes involved may also be substantially less than the effects on the optical transition energy.

N. R. KESTNER   I agree, and this type of effect was implied when I said that we need to study asymmetric modes of vibrations. Possibly even rather large distortions may be important. This seems more likely at the moment to be the explanation of the line shape and width than the possibility of many symmetric cavities with various coordination numbers. We plan to look into your suggestion more thoroughly.

J. JORTNER   I would like to comment on the line width of the optical absorption band of the solvated electron. It is now obvious that calculations based on radial displacements of a single configuration yield to a value which is lower at least by a factor of 2 as compared with the experimental value. The suggestion that several configurations (characterized by different N values) contribute to the optical spectrum is very attractive. The experimental data of Olinger and Schindewolf on the absorption spectra of the localized electron in low density $(0.4—0.15 \, \text{gm cm}^{-3})$ supercritical ammonia provide tentative evidence that the line widths may be narrower than in the normal liquid. This may indicate that in a lower density a single configuration dominates. To confuse the issue, Gaathon, Czapski and myself have observed that in supercritical water the absorption line width does not exhibit appreciable narrowing relative to the spectrum of the hydrated electron in normal water. It is apparent that the model based on radial displacements only is oversimplified and that angular displacements have to be introduced to explain the optical line shapes and relaxation processes of electrons in liquids.

U. SCHINDEWOLF   In our experiments we restricted ourselves to the ammonia density dependence of the position of the absorption spectrum and on the minimum density necessary for electron solvation. We did not stress the half width of the absorption spectrum. If this is of relevance to theory, our experiments have to be repeated with an equipment of higher sensitivity, unless theory is revised to fit the experimental data, which are coauthored by Jortner.

J. JORTNER   The problem of the relaxation of the quasi-free electron to the localized excess electron state is very exciting. This problem should be handled as a nonradiative relaxation process induced by the breakdown of the Born-Oppenheimer approximation resulting in scrambling of different electronic configurations. One should not consider the process just in terms of crossing of potential curves as suggested by Seitz many years ago. Rather a "nuclear tunnelling" process between a given electronic manifold and a dissipative continuum of another electronic manifold should be considered. This problem has been treated in solid state physics, where Kubo and Toyozawa considered electron capture by a hole in semiconductors, and has been recently very popular in molecular physics regarding radiationless transitions in large molecules. Concerning the relaxation of the solvated electrons two questions are pertinent: (a) Is the relaxation process adiabatic (i.e. proceeding on a compound single potential surface) or is it nonadiabatic? (b) Does the relaxation of the quasi-free electron proceed directly to the ground electronic state or does it cascade (nonradiatively) via intermediate electronically excited bound states?

Metal-Ammonia Solutions: The Dilute Region

J. J. LAGOWSKI

Abstract

This paper is a discussion of the physical properties of dilute metal–ammonia solutions. The paper is subdivided as follows: 1. The nature of M^+. 2. The nature of e^-. 3. Optical properties. 4. The nature of M. 5. The nature of the solvent.

Introduction

The charge from the Organizing Committee for Colloque Weyl III for this paper was to discuss the physical properties of dilute metal–ammonia solutions. Two problems, which may not necessarily be separable, are immediately apparent: (1) what does "dilute" mean, and (2) how much information to include. For brevity we shall limit our discussion mainly to work that has appeared since Colloque Weyl II. Dye's [1] lucid exposition, which appears in the Proceedings of that conference, of the properties of metal–ammonia solutions at low to moderate concentrations (up to 0.1 M) as they relate to the models describing the species in solution serves as reference to previous work. We shall define solutions $<10^{-3}$ M as dilute for the purposes of this discussion. Thus, this paper is concerned primarily with the properties of dilute solutions as they illuminate the nature of the species shown in Eq. (1) [2].

$$M^+ + e^- \rightleftarrows M \quad M = Na, \quad K_1 \, (-34\,^\circ C) = 3.41 \times 10^{-3} \tag{1}$$

At higher analytical concentrations of metal, other equilibria and species have been postulated to account fot the magnetic and conductance behavior of the more concentrated solutions, e.g., Eqs. (2) and (3) [3].

$$M + e^- \rightleftarrows M^- \quad M = Na, \quad K_2(33^\circ\,C) = 2 \times 10^5, \tag{2}$$

$$M^- + M^+ \rightleftarrows M_2, \quad K_3 \simeq K_1 \tag{3}$$

Presumably these are considered in other papers at this meeting. In addition to the species described in Eq. (1), an attempt will be made to discuss the role of the equilibrium process shown in Eq. (1).

The Nature of M^+

Although there is little direct evidence, several independent and indirect lines of arguments lead to a reasonably consistent description of the nature of cation solvation in liquid ammonia. The magnitudes of the dipole moment and polarizability of liquid ammonia suggest that cation solvation in this solvent should be greater than in water. The dipole moment of ammonia (1.49 D) is only about 80%

that of water (1.84 D) suggesting that electrostatic interactions of the ion-dipole type are not as important in the solvation of cations in the former solvent. However, the polarizability of ammonia (2.21×10^{-25} cm^3) is considerably greater than that of water (1.48×10^{-25} cm^3) which should lead to the existence of greater dispersion forces between ammonia molecules and solute species. Indeed, London has shown that the dispersion effect is nearly twice as large for ammonia as it is for water [4]. The larger dispersion forces available when ammonia molecules interact with solute species are undoubtedly the reason that the alkali and alkaline earth metal halides form ammoniates (Table I) containing more solvent molecules than can be accounted for by considering the solute molecules to be associated with metal ions using the commonly found coordination numbers for these families of elements, e.g., 4 and 6. Although little direct structural information is available on these ammoniates, it is likely that the solvates containing more than 4 ammonia molecules for lithium and 6 ammonia molecules for the other alkali and alkaline earth halides contain ammonia molecules associated with the anion present.

An additional piece of tangential evidence concerning the solvation of ions in liquid ammonia comes from the theoretical treatment of the viscosity of salt solutions [5]. Moderate success has been achieved in fitting the observed viscosity characteristics of fully ionized, univalent salts in liquid ammonia with a model in which the cations are octahedrally coordinated and occupy vacancies in the liquid.

An estimate of the number of ammonia molecules associated with Mg^{2+} in solution has been made from line shape analysis of the proton nmr spectrum of solutions of magnesium salts in ammonia [6]. The results indicate a primary solvation number of 6, presumably octahedrally coordinated, in this solvent and are consistent with the coordination number expected for this species from other arguments.

The unity of these arguments concerning the number of ammonia molecules associated with the alkali and alkaline earth ions in liquid ammonia extends to the stoichiometry of the metal solvates obtained from concentrated metal–ammonia solutions. Thus, lithium forms a tetramine $Li(NH_3)_4$, and calcium a hexamine, $Ca(NH_3)_6$. In addition, europium and ytterbium, the general chemistry of which mimics that of the alkali earths, also yield hexamines, $M(NH_3)_6$. In agree-

Table I. Ammoniates formed by the alkali and alkaline earth halides; the number of molecules per mole of compound is indicated.

Ion	Cl$^-$	Br$^-$	I$^-$
Li$^+$	5, 4, 3, 2, 1	5, 4, 3, 2, 1	7, 4, 3, 2, 1
Na$^+$	5	5	6
K$^+$	—	—	6, 4
Rb$^+$	—	3	6
Ca^{2+}	8, 4, 2, 1	8, 6, 2, 1	8, 6, 2, 1
Ba^{2+}	8	8, 4, 2, 1	10, 9, 8, 6, 4, 2
Sr^{2+}	8, 1	8, 2, 1	8, 6, 2, 1

ment with the results of previous arguments, lithium is tetrahedrally coordinated in $Li(NH_3)_4$ [7, 8] and the hexamines contain an octahedral arrangement [9] of ammonia molecules.

In the light of the continuity of the evidence presented here it is highly likely that the cations in metal–ammonia solutions are strongly solvated at all analytical concentrations of metal.

The Nature of e^-

Considerable evidence is available from several different types of investigations that the solvated electron in dilute liquid ammonia solutions is associated with a large, but perhaps unspecifiable, number of solvent molecules. Thus, a high degree of electron solvation has been deduced [5] from a study of the viscosities of metal–ammonia solutions [10, 11]. The viscosity data for potassium solutions up to $\cong 1.5$ M was treated by assuming a model in which the cation is octahedrally coordinated and the electron occupies a vacancy associated into vacancy clusters. Specifically, the best model requires $\cong 3.5$ vacancies per cluster which then has between 22 and 26 nearest-neighbor solvent molecules.

Electron spin resonance studies on solutions of the alkali metals in the liquid in the intermediate concentration range have also been interpreted [12] in terms of a valency cluster model for the solvated electron with a coordination shell containing 24 solvent molecules. The electron correlation times in dilute solutions correspond to the self-diffusion of the valency cluster in the liquid. In addition to these measurements, the solvation number of the electron has been estimated from a study of the high resolution proton *nmr* spectra of dilute potassium-ammonia solutions [13]. Line shape analysis suggests that a large spin-lattice relaxation occurs in these solutions through magnetic-dipolar interactions with unpaired electrons. The results of these calculations reveal electron solvation numbers of 20—49 in the concentration range 0.06—0.15 M over the temperature range $-37.8°$ to $-28.0°$ C. An earlier analysis of magnetic resonance measurements suggested that the electron is associated with from three to 13 ammonia molecules [14].

Optical Properties

The optical properties of dilute metal–ammonia solutions alone provide little direct insight into the nature of the solvated electron. The fact that dilute solutions containing equivalent amounts of alkali and alkaline earth metals give virtually the same near-infrared absorption spectra indicates the presence of a common absorbing species which must be described without reference to the cation. Indeed, dilute solutions of solvated electrons electrochemically generated in the presence of tetraalkylammonium ions with widely-varying structural parameters are also optically indistinguishable from those formed by the dissolution of metal atoms [15]. These results do not support the Becker, Lindquist, and Alder description of the monomer, at least in dilute solutions [16].

Several features of the near-infrared optical spectrum of metal–ammonia solutions are of special interest. The band attributed to the solvated electron at $\sim 1.5\ \mu$ is unusually broad and asymmetric toward the high-energy side and the position

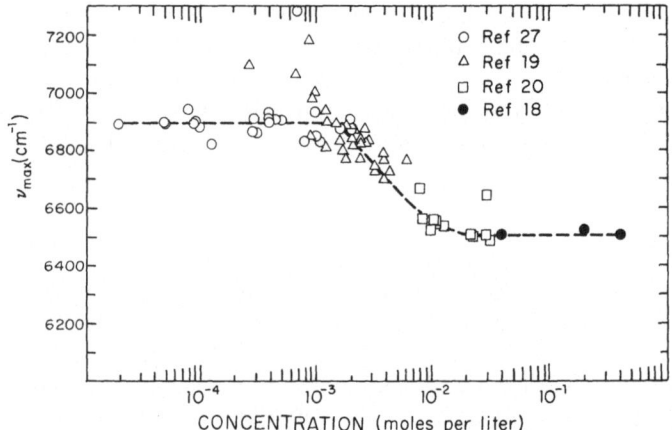

Fig. 1. Position of the near-infrared band maximum as a function of concentration

of the absorption maximum moves to lower energy with increasing metal concentration [17]. The available data show (Fig. 1) that the position of the band maximum is relatively unchanged below $\sim 10^{-3}$ M and above $\sim 10^{-2}$ M. The band position appears to reach a limiting value of ~ 6500 cm^{-1} [18] and ~ 6900 cm^{-1} [15, 19, 20] in concentrated and dilute solutions, respectively. Recent data [21] suggest that the limiting value in dilute solutions should be ~ 7000 cm^{-1}. In the intermediate concentration range, i.e., between 10^{-3} M and 10^{-2} M, the band energy decreases rapidly with increasing concentration. As an aside, this is the concentration range in which spin pairing increases markedly with increasing concentration [14], which suggests that the near-infrared transition might arise from a fundamentally different electron-containing species than is present in the dilute range. By the constraints placed upon this paper in the introduction, it would be improper to continue to discuss the optical properties of solutions more concentrated than 10^{-3} M, since such discussion is reserved for another paper in this symposium.

The most successful models for the solvated electron in dilute ammonia solutions involve a self-trapping mechanism in solvent voids or cavities [22]. Although the details vary for different approaches, it is generally conceded that the molecules in the first solvation layer are oriented by the electric field of the electron, and solvent molecules in subsequent layers are polarized. Qualitatively, the cavity can be imagined as arising from the balance of repulsion of the oriented solvent dipoles and hydrogen-hydrogen interactions on the one hand, and the interaction of the solvent dipoles with the electron on the other hand. Detailed analysis [22] of a molecular model for the solvated electron indicates that a range of radii for stable cavities is possible; electron solvation numbers of 4 and 6 gave acceptable results, with the lower value forming the basis for the better model. The effective radius of the cavity ranges from 3.05 Å for coordination number 4 to 4.2 Å for coordination number 6. The volume expansion of sodium-ammonia solutions and pressure dependence of equilibrium constants [24] indicates that the effective cavity radius is about 3.2 Å. The temperature dependence of the spectrum in the most current model arises from a change in the thermal motion of the solvent molecules in the first solvation sphere as well as from the temperature dependence of the dielectric constant.

The asymmetry of the solvated electron band has been attributed to the contribution of transitions involving higher excited states to the spectrum [22], or the existence or a distribution of electron species with varying numbers of solvent molecules [23]. Presently [22], theoretical considerations indicate that the major factor determining the line shape is the separation of various transition energies rather than the broadening of any one transition.

The Nature of M

For dilute solutions, the weight of evidence suggests that M corresponds to a solvent-separated ion-paired species [M^+ (am), e^- (am)] as would be expected for a solvent with a moderately low dielectric constant. Thus, conductivity data for metal–ammonia solutions yield an equilibrium constant for the association of the solvated cation and electron of the same order of magnitude observed for the association of conventional salts in this solvent. The change in apparent ion-pairing constants (K_1) in dilute solutions of sodium [2] and cesium [25] at several temperatures appears in Table II. If K_1 reflects the magnitude of association of charged species in a solvent with a moderate dielectric constant, the variation of K_1 with temperature is not that predicted on the basis of accepted theories of ion-pairing. It has been suggested [2, 25] that the variation in the distance of closest approach with the temperature shown in Table II is related to the thermal expansion of the solvated electron.

The optical properties of ammonia solutions also are compatible with the ion-pairing interpretation. The position of the near-infrared band is relatively insensitive to changes in the analytical concentration of metal in the dilute region (Fig. 1). It might be tempting to attribute the slight change in band position which occurs between the dilute ($< 10^{-3}$ M) and the beginning of the intermediate concentration range to the presence of two or more species in equilibrium. However, attempts to deconvolute the experimental band envelope into an internally consistent set of bands using computer techniques were unsuccessful [18]. If, for example, two absorbing species were present in equilibrium, it should be possible to resolve the envelope into two bands; the intensities of these bands should vary regularly with the analytical concentration of the metal, but the band positions

Table II. Temperature variation of the dielectric constant and viscosity for liquid ammonia, and values of K_1 and the distance of closest approach (a_0) for solutions of sodium and cesium [2, 25]

t, °C	D	$\eta \times 10^3$ poise	$K_1 \times 10^{-3}$		a_0, Å	
			Na	Cs	Na	Cs
-33.9	21.8	2.558	3.41	4.84	4.46	4.89
-45.0	22.7	2.992	2.48	3.30	4.20	4.48
-65.0	25.1	4.320	1.88	2.19	3.95	4.30

should be constant within experimental error. It is, of course, mathematically possible to decompose the broad near-infrared band into any number of bands; consistent results as a function of concentration could not be obtained with respect to position nor were trends in resolved band intensities recognizable when 2, 3, or 4 bands were resolved.

An alternative interpretation for the change in band position with concentration involves a non-specific change in the environment of the solvated electron; that is, as the analytical concentration of metal increases, the concentration of bulk ammonia decreases and the solution becomes more ordered. At higher concentrations, the solvated cations and anions could conceivably become more tightly associated, and in the same instances the solvating molecules would be expected to come under the influence of both the cation and electron. In effect, this process would be analogous to that suggested for the formation of solvent-shared ion-pairs. Such nonspecific environmental changes affecting the optical properties of solvated electrons have been reported [26] for electrons in ammonia-water mixtures (Fig. 2). These experiments reveal a continuous change in the position of the solvated electron band as the concentration of water in the system increases. The results are in agreement with a model for the solvated electron in which the energy of the electron is determined by the total environment of its surroundings rather than by specific near-neighbor interactions.

The spectral parameters for dilute solutions of solvated electrons (8—2000×10^{-6} M) are virtually unchanged in the presence of excess alkali-metal iodides (0—1.68 M) [27]. These results are consistent with the suggestion that the absorbing species is associated with the cation through a simple electrostatic interaction to form solvent-separated ion-pairs. The spectroscopic indistinguishability of a charged absorbing species, whether it is a free ion or is involved as a solvent-separated ion-pair has been established previously in liquid ammonia

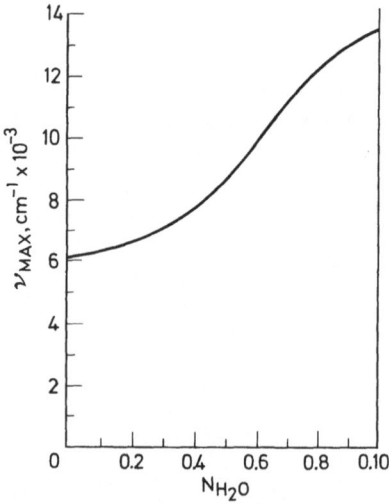

Fig. 2. Position of the infrared band maximum as a function of the mole fraction of water in water-ammonia mixtures. (After Ref. [26])

solutions for several systems involving charge-transfer-to-solvent transitions [28—30], as well as intramolecular transitions [31, 32]. The position of spectral bands attributed to charge-transfer-to-solvent processes in liquid ammonia can be altered by introducing a large excess ($\sim 100x$) of highly polarizing cations, such as Li^+ [28], which probably alter the potential field of the cavity by forming solvent-shared or contact ion-pairs. In effect, the solutions desolvate, i.e., the concentration of free solvent decreases by the addition of polarizing cations. The larger alkali metal cations at the same concentrations as Li^+ do not affect the optical parameters of charge-transfer-to-solvent spectra.

Nature of the Solvent

Compared to the information available on liquid water, relatively little data which might be useful in describing the structure of liquid ammonia are available. Specifically, information obtained from a study of the vibrational spectra of ammonia and its solutions containing different types of solvents is potentially useful for understanding the nature of the solute-solvent interaction in metal–ammonia solutions. The existence of a relatively large number of papers devoted to vibrational spectroscopy of liquid ammonia solutions attests to the intensity of interest in this area. We shall have to await the details of these reports to see whether the promise is fulfilled.

Unfortunately, meaningful results are not expected from conventional techniques involving vibrational spectroscopy on dilute solutions because of the inherently low intensities of the transitions which occur in this spectral region. Such experiments on intermediate and more concentrated solutions are expected to give more definitive results.

The vibrational spectrum of liquid [33] and solid [34] ammonia has been interpreted in terms of relatively strong hydrogen-bonded interactions. The presence of salts in relatively large concentrations affects the structure of the liquid in specific ways [35]. Large cations or anions such as tetraalkylammonium cations or NO_3^- apparently have no effect on the fundamental frequencies of the solvent. However, halide ions strongly perturb the stretching frequencies, presumably through hydrogen bonding of the type $H_2NH \cdots X^-$. Monatomic cations increase the symmetrical bending frequency of the solvent molecules because they interact with the free electron pair on the nitrogen atom $M^+ \cdots NH_3$. The magnitude of the shift of the symmetrical bending frequency increases with charge density of the cation. The infrared spectra of concentrated solutions ($\sim 10^{-1}$ M) of the alkali metal iodides and bromides have been interpreted in terms of a solvent-shared ion pair $M^+ \cdots NH_3 \cdots X^-$, which is consistent with the model described previously in this paper.

Acknowledgement

The generous and continuing assistance of the Robert A. Welch Foundation and the National Science Foundation is gratefully acknowledged.

References

1. Dye, J. L.: In: Lagowski, J. J., Sienko, M. J. (Eds.): Metal–ammonia solutions, p. 1. London: Butterworths 1970.
2. Dewald, R. R., Roberts, J. H.: J. Phys. Chem. **72**, 4242 (1968).
3. Golden, S., Guttman, C., Tuttle, T. R., Jr.: J. Chem. Phys. **44**, 3791 (1966).
4. London, F.: Trans. Faraday Soc. **33**, 8 (1937).
5. O'Reilly, D. E.: J. Chem. Phys. **50**, 5378 (1969).
6. Harrison, L. W., Swift, T. J.: J. Am. Chem. Soc. **92**, 1963 (1970).
7. Mammano, N., Sienko, M. J.: J. Am. Chem. Soc. **90**, 6322 (1968).
8. Kleinman, L., Hyder, S. B., Thompson, C. M., Thompson, J. C.: In: Lagowski, J. J., Sienko, M. J. (Eds.): Metal–ammonia solutions, p. 229. London: Butterworths.
9. Mammano, N.: In: Lagowski, J. J., Sienko, M. J. (Eds.): Metal–ammonia solutions, p. 369. London: Butterworths.
10. Hutchinson, C. A., O'Reilly, D. E.: J. Chem. Phys. **52**, 4400 (1970).
11. Nozaki, T., Shimoji, M.: Trans. Faraday Soc. **65**, 1489 (1969).
12. O'Reilly, D. E.: J. Chem. Phys. **50**, 4743 (1969).
13. Pinkowitz, R. A., Swift, T. J.: J. Chem. Phys. **54**, 2858 (1971).
14. Catterall, R.: In: Lagowski, J. J., Sienko, M. J. (Eds.): Metal–ammonia solutions, p. 105. London: Butterworths 1970.
15. Quinn, R. K., Lagowski, J. J.: J. Phys. Chem. **72**, 1374 (1968).
16. Becker, E., Lindquist, R. H., Alder, B. J.: J. Chem. Phys. **25**, 971 (1956).
17. Koehler, W. H.: Ph. D. Dissertation. The University of Texas, Austin, Texas, 1969.
18. Koehler, W. H., Lagowski, J. J.: J. Phys. Chem. **73**, 2329 (1969).
19. Douthit, R. C., Dye, J. L.: J. Am. Chem. Soc. **82**, 4472 (1969).
20. Gold, M., Jolly, W. L.: Inorg. Chem. **1**, 818 (1962).
21. Hurley, I., Tuttle, T. R., Golden, S.: In: Lagowski, J. J., Sienko, M. J. (Eds.): Metal–ammonia solutions, p. 503. London: Butterworths 1970.
22. Copland, D. A., Kestner, N. R., Jortner, J.: J. Chem. Phys. **53**, 1189 (1970).
23. Rusch, P. F., Koehler, W. H., Lagowski, J. J.: In: Lagowski, J. J., Sienko, M. J. (Eds.): Metal–ammonia solutions, p. 41. London: Butterworths 1970.
24. Schindewolf, U.: In: Lagowski, J. J., Sienko, M. J. (Eds.): Metal–ammonia solutions, p. 199. London: Butterworths 1970.
25. Dewald, R. R.: J. Phys. Chem. **73**, 2615 (1969).
26. Dye, J. L., DeBacker, M. G., Dorfman, L. M.: J. Chem. Phys. **52**, 6251 (1970).
27. Quinn, R. K., Lagowski, J. J.: J. Phys. Chem. **73**, 2326 (1969).
28. Nelson, J. T., Cuthrell, R. E., Lagowski, J. J.: J. Phys. Chem. **70**, 1492 (1966).
29. Nelson, J. T., Lagowski, J. J.: Inorg. Chem. **6**, 862 (1967).
30. Caruso, J. A., Takemoto, J. H., Lagowski, J. J.: Spectrosc. Letters **1**, 311 (1968).
31. Fohn, E. C., Cuthrell, R. E., Lagowski, J. J.: Inorg. Chem. **4**, 1002 (1965).
32. Cuthrell, R. E., Fohn, E. C., Lagowski, J. J.: Inorg. Chem. **5**, 111 (1966).
33. Corset, J., Lascombe, J.: J. Chim. Phys. **64**, 665 (1967).
34. Reding, F. P., Hornig, D. F.: J. Chem. Phys. **19**, 594 (1951).
35. Corset, J., Huong, P. V., Lascombe, J.: Spectrochim. Acta **24**A, 1385 (1968).

Discussion

P. DELAHAY: When you say that the solvation number is, for instance, 24, what do you exactly mean by solvation number?

J. JORTNER   The experiments which monitor the coordination number of the solvated electron do not probe just the first coordination layer but also the molecules outside it (in the vicinity of the electron localization center). To quote an example from a related filed, ENDOR experiments of the F center in ionic crystals probe the charge density up to the fifth coordination layer. Thus there is no contradiction between experiment and the theoretical models for the localized electron in polar liquids. It should be noted, however, that the theoretical model presented by Kestner introduces an element of "coarse graining" putting the onset of the continuum just beyond the first coordination layer.

W. H. KOEHLER   You have suggested that we might need to consider solvent molecules in "coordination layers" in addition to the primary coordination sphere. Do you envision solvent molecule exchange between the primary sphere and say the secondary layer?

J. J. LAGOWSKI   Yes.

K. BAR-ELI   (1) What is the exact definition of the term "desolvation"? (2) Do you mean that there is a change of symmetry from a species which is dipole shaped, to a species which is spherically symmetric. If so, why is there no sign of a hyperfine splitting in the ESR spectra?

J. J. LAGOWSKI   In my discussion, I considered metal–ammonia solutions to contain two kinds of ammonia molecules: (a) those associated with charged species and (b) those which form the bulk of the solvent, i.e., those that are associated with each other. In this context the term „desolvation" means a process which decreases the number of bulk, or solvent, ammonia molecules.

J. C. THOMPSON   The shift in the location of the absorption peak which occurs near 10^{-3} molar is better correlated with the onset of spin pairing rather than with ion pairing.

J. J. LAGOWSKI   I agree that it is more proper to correlate the ONSET of the spin pairing process with the shift in the absorption peak.

U. SCHINDEWOLF   If ion pairing is made responsible for the shift of the solvated electron spectrum with increasing metal concentration we should expect differences for different metals. Experimental data do not give any evidence for this. The shifts of the spectra of solvated electrons and of ions like iodide with electrolyte addition also is explained here by ion pairing. Shifts in one direction or the other can be observed by change in temperature or mixing solvents. We do not assume that these shifts are due to a change of equilibrium of ion pairs but rather to environmental or structural changes of the solvent. I would think, therefore, that electrolytes also yield structural changes which are responsible for the observed spectral shifts, ion pairing not being necessary for the explanation of the effect.

J. J. LAGOWSKI   I prefer the "environmental" explanation for the shift in the spectrum. We already have the example of a spectral shift which occurs because of a change in solvent in the recent work of Dye, DeBacker and Dorfman [J. Chem. Phys. **52**, 6251 (1970)].

J. L. DYE   The shift in the position of the $e_{NH_3}^-$ absorption band with concentration occurs in a region in which ion-pair formation becomes most important. It is possible that the shift is caused by formation of the ion-pair.

J.J.LAGOWSKI   There is some evidence that the shift does not arise from ion-pairing. Quinn and Lagowski [J. Chem. Phys. **73**, 2326 (1969)] have shown that the position of the 1.5 μ band is not affected by the presence of high concentrations of alkali metal cations.

J.C.THOMPSON and G.LEPOUTRE   One must be careful to separate regions of *onset* of a property such as ion-pairing and the region in which most of the change occurs. Between 10^{-3}M and 10^{-2}M the ion-pairing varies from only about 15% (dependent on temperature and activity coefficient effects) to about 60%.

The Effect of Electrolytes on the Hydrogen-Bonded Structure of Liquid Ammonia

JOHN H. ROBERTS and J. J. LAGOWSKI

Abstract

Extensive studies are reported on the effect of electrolytes on the near-infrared overtones and combination bands of liquid ammonia. The largest shift (to higher frequency) relative to pure ammonia occurred in a solution of sodium perchlorate. These results are interpreted in terms of a model for the effect of electrolytes on the ammonia solvent structure.

Introduction

Experiments designed to measure a wide variety of physical properties have been performed on solutions in liquid ammonia [1—3]. Although there is now agreement on many points regarding the basic experimental evidence and a growing consensus concerning the theoretical interpretation of this evidence, many points are still in contention.

The experimental investigations have been directed almost exclusively at gaining an understanding of the nature of the species in solution. As more experimental work has been brought into agreement and the theoretical interpretation has achieved a degree of self-consistency, it has become clear that an understanding of the species in solution is not the whole story.

Relatively little attention has been paid to the interaction of solutes with the solvent and almost none to the effect of solutes on the structure of the liquid itself. It is possible that an understanding of solution phenomena from the point of view of the solvent and the structure of the solution may permit a more complete description of these phenomena. It is with this view in mind that we have undertaken this research.

The advances that have been made recently in understanding hydrogen-bonded liquids have employed a wide variety of techniques [4]. An extensive body of literature has accumulated on the subject, which promises to become even larger as interest in this aspect of the liquid state increases, and except for the most relevant investigations no attempt will be made to describe this material here. It must be pointed out that there is almost a complete absence of studies on liquid ammonia in this literature [4—6], just as there are few studies directed toward these general goals in the liquid ammonia literature.

At present there is considerable controversy over models which describe water. The two main divisions of opinion are generally represented by advocates of the mixture model [7—9], and the continuum model [10—12]. At this time there appears to be more experimental evidence which is compatible with the mixture model but it remains to be seen if either model can fully describe aqueous phenomena. One of the aims of this research has been to see if either of these models is general enough to describe solution phenomena in liquid ammonia in addition to water, and if neither is, then to provide the background for a more general model of hydrogen-bonded liquids.

It has been demonstrated repeatedly that there are many similarities between liquid ammonia and water. These similarities in some instances have been over-emphasized and in drawing on the experiences of researchers who have worked with water careful attention must be paid to the differences between the two liquids. This research has been directed towards explaining the differences as well as the similarities. In particular one must consider the rather narrow liquid range of ammonia at temperatures much lower than those for water and the differences in solvation characteristics of the two liquids.

Vibrational spectroscopy has been used extensively to study hydrogen-bonded solvent systems [4—6]. The sensitivity of the infrared spectrum of a liquid to perturbations to the structure of the liquid has been demonstrated in many investigations [5]. In liquid ammonia pioneering work has been done by Corset [13], and others will present papers on their work at this symposium.

Although the theoretical basis for understanding vibrational spectra of liquids has not advanced to a level comparable to that with which we can explain similar phenomena in gases and solids [14], there are certain fundemantal characteristics that have emerged from the bulk of previous work. The formation of a hydrogen bond results in a shift to lower frequencies for stretching modes and to higher frequencies for bending modes [5]. Moreover combination bands shift by at least the sum of the shifts of the fundamentals. This enhanced sensitivity of combination bands has been used recently by a number of authors to study aqueous systems [10. 15—19]. In this work we have used the near-infrared bands of liquid ammonia to study the structure of the solvent and the behavior of charged solutes in liquid ammonia.

Experimental

It was our aim in these experiments to examine a number of possibilities which would simplify the experimental procedures used. Because of the relatively high concentration of salts required in studies of this nature, and the relatively low solubilities of many interesting solutes, it was decided to use cells capable of holding ammonia solutions at ambient temperatures. Cells of the conventional type [20] could not be used and cell designs similar to previous high pressure cells [13] were rejected as too cumbersome. Cells were constructed of quartz as shown in Fig. 1. The cell dimensions are 6 mm by 25 mm with a 1 mm path length and 1 mm wall thickness. Calculations made before construction indicated these cells could hold 10 atmospheres pressure with an adequate safety factor and indeed none has ever failed under experimental conditions.

The NH_3 used was Matheson (99.99%) and the ND_3 used had been prepared earlier [20] and stored under anhydrous conditions. ND_2H and NDH_2 mixtures were prepared by condensing the appropriate amounts of the pure starting materials over sodium. The solvent purification vessel contained sufficient catalysts to ensure complete exchange.

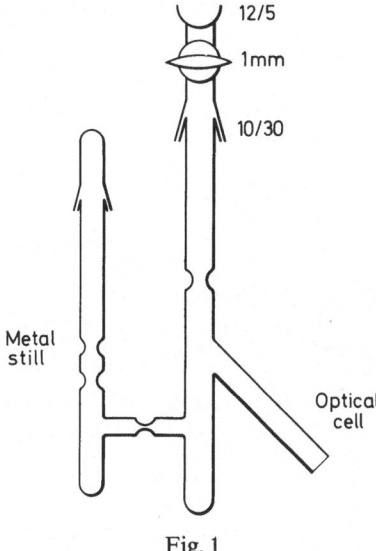

Fig. 1

Solvent purification and handling procedures which have been described elsewhere [20, 21] were used. The salts used were reagent grade with no further
purification. Extreme precautions were used to maintain the dryness of all materials. The transfer of solutes was done in an inert atmosphere [20] or by distillation
[21].

The spectra were recorded on a Cary 14 spectrophotometer which was left unmodified. It was decided that the usual modification of the Cary 14 which allows
monochromation of the light before it passes through the sample would result in
too great a compromise of the instrument's spectroscopic capabilities. The resulting heating problems were solved through the use of a modified Haake Model FK
low-temperature circulator and a commercial cell holder supplied by Cary. The
temperatures selected were maintained to better than 1° C, and since the cells
were very small there was practically no temperature gradient across the light
beam. A continuous stream of dry nitrogen was passed through the sample compartment to prevent condensation during a run. Temperature calibration was
done under operating conditions and the resetability of the temperatures was
checked repeatedly.

Results

Liquid Ammonia

Figure 2 shows the resolved near-infrared spectrum of liquid ammonia at −35° C.
The method of non-linear, damped least squares is used by the resolution program [22]. The program allows the use of linear combinations of Gaussian and
Lorentzian distribution functions to represent the resolved bands. Separate base
lines for the experimentally observed spectrum, top line of squares, and the resolved bands allow a clear display of the results. An error curve is also included

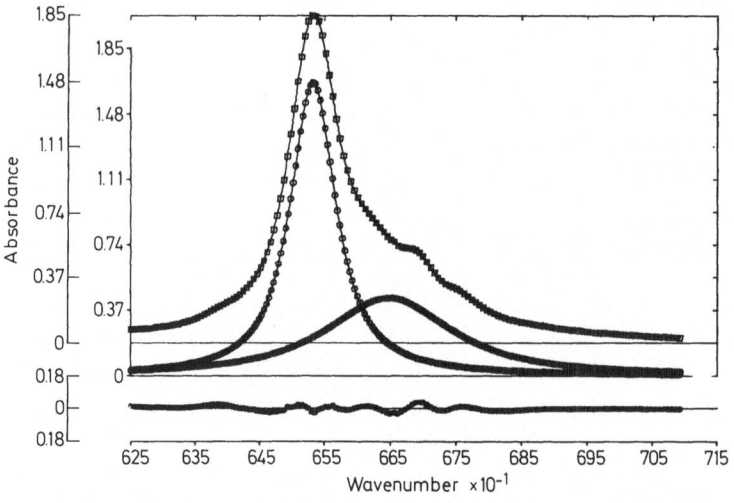

Resolved near infrared spectrum of liquid ammonia at -35°C
Combination bands

Fig. 2

which provides a visual criterion for the goodness of fit. The comment "combination bands" below the title of the plot indicates that the program was given the
choice of any combination of Gaussian and Lorentzian distributions to achieve
the best fit. In some specific cases which will be pointed out the best fit used pure
Lorentzians, although in nearly all cases the bands are of predominantly Lorentzian character.

Even though this area of the spectrum is well separated from other absorptions,
Fig. 2 shows that there are at least two bands in this region. These bands have
been tentatively assigned to $2v_1$, the overtone of the symmetric stretching mode
and to the combination $v_1 + v_3$, v_3 being the degenerate, asymmetric stretching
mode. In the fundamental region there is some question concerning the assignment of bands since v_1, v_3, and $2v_4$ occur very closely to one another and the
situation is complicated by Fermi resonance [23]. Since the interpretation of
changes in the vibrational spectrum is dependent upon the assignment of the
bands, the importance of cerifying these assignments cannot be overemphasized.
Table I gives the parameters of the two near-infrared absorption bands of liquid
ammonia at $-35°$ C, $-10°$ C, and $25°$ C. The changes in position, intensity and
width at half-height with changing temperature are all consistent. A greater extent
of hydrogen bonding is expected at lower temperatures and thus the shift to lower
frequency with decreasing temperature is in agreement with the assignment of
these bands to $2v_1$ and $v_1 + v_3$. Positions for fundamentals calculated from these
results are in agreement with other work [13]. The changes in intensity are
consistent with the change in density of liquid ammonia with temperature. The
areas listed are not integrated absorption coefficients and thus cannot really be
compared for solutions of different density. The most significant feature of these
results is the degree to which the spectrum of liquid ammonia remains constant

Table I. Parameters for the resolved bands of the near-infrared spectrum of liquid ammonia: t is the temperature in °C, X is the position in cm^{-1}, $Y1$ is the Gaussian intensity, $Y2$ is the Lorentzian intensity, W is the width at half-height in cm^{-1}, and A is the relative area

Band	t	X	$Y1$	$Y2$	W	A
$2v_1$	25	6541	0.016	1.421	95	214
	−10	6536	0.137	1.430	88	211
	−35	6532	0.144	1.515	84	213
$v_1 + v_3$	25	6651	0.080	0.318	236	138
	−10	6649	0.039	0.386	231	150
	−35	6648	0.046	0.392	228	151

with temperature. Certainly the ten atmosphere overpressure at 25° C has no effect, and none is to be expected as indicated by other work [17, 24]. But one might expect to have a larger change due to the temperature dependence of the hydrogen bond as is observed for water over a similar change in temperature [16, 19].

Difference spectra have been employed to make visual comparisons between two solutions [19, 25], and Fig. 3 shows a computed difference spectrum for liquid ammonia. Since a comparison between the same concentrations of ammonia molecules was desired, the spectrum of ammonia at the higher temperature was multiplied by a density factor before being subtracted from the spectrum at the lower temperature. Thus a negative difference in absorption at a given frequency reflects an increase in the molar absorptivity at that frequency with increasing temperature. A general increase in molar absorptivity with increasing temperature is observed. A line drawn through zero would show that most of the change in area is in the negative direction and therefore the integrated absorption coefficient increases with increasing temperature.

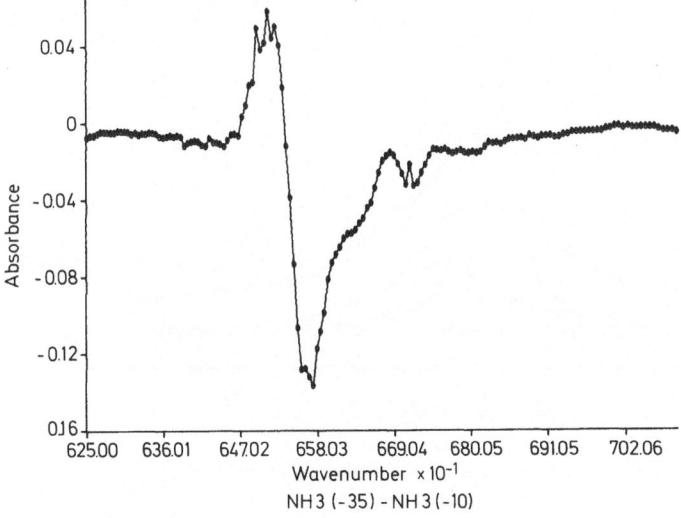

Fig. 3

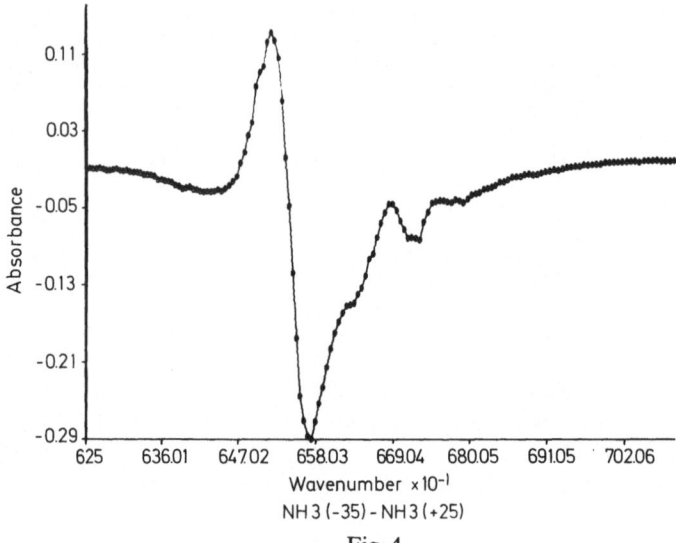

Fig. 4

The shift of the differences from a maximum of 6905 cm^{-1} to a minimum at 6568 cm^{-1} is consistent with the shift of $2v_1$. However, the overlapping minimum in the 6600—6620 cm^{-1} region and a third minimum at 6717 cm^{-1} suggest that the broad $v_1 + v_3$ band is complex. Fig. 4 shows the difference between liquid ammonia at $-35°$ C and $25°$ C. The differences are more negative yet indicate a further increase in molar absorptivity at $25°$ C. The maximum at 6514 cm^{-1} and minimum at 6573 cm^{-1} are again equidistant about the frequency maximum of $2v_1$. The minima at 6717 cm^{-1} and 6622 cm^{-1} are also evident. These features and

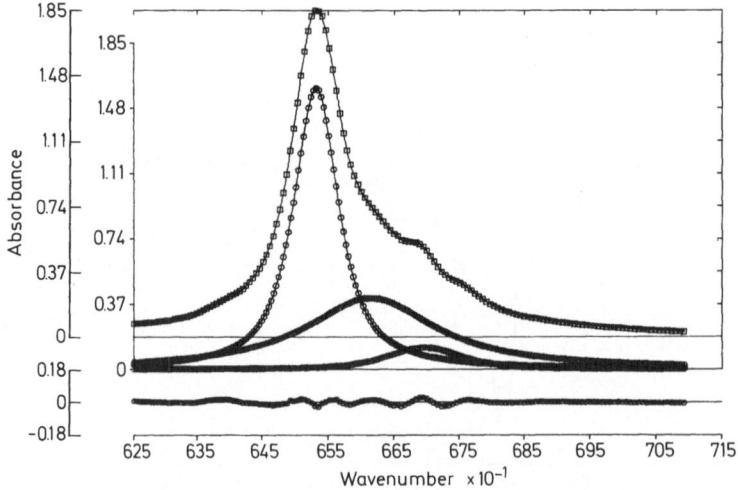

Resolved near infrared spectrum of liquid ammonia at -35°C
Combination bands

Fig. 5

Table II. Parameters for the resolved bands of the near-infrared spectrum of liquid ammonia: t is the temperature in $°C$, X is the position in cm^{-1}, $Y1$ is the Gaussian intensity, $Y2$ is the Lorentzian intensity, W is the width at half-height in cm^{-1}, and A is the relative area

Band	t	X	$Y1$	$Y2$	W	A
$2v_1$	25	6540	0.092	1.291	92	196
	−10	6536	0.221	1.281	85	192
	−35	6532	0.235	1.360	81	193
$(v_1 + v_3)_A$	25	6620	0	0.380	231	138
	−10	6618	0	0.403	241	153
	−35	6615	0	0.404	239	152
$(v_1 + v_3)_B$	25	6709	0.051	0.060	146	22
	−10	6700	0.046	0.061	134	19
	−35	6697	0.055	0.069	135	22

the much greater width at half-height of $v_1 + v_3$ suggested that perhaps the combination band at least could be resolved into components.

Figure 5 shows the spectrum of liquid ammonia resolved into three bands. The band for $2v_1$ remains at $6532\ cm^{-1}$. The positions of the components of $v_1 + v_3$ at $6615\ cm^{-1}$ and $6697\ cm^{-1}$ correspond closely to the positions of the minima in the difference spectra. Table II gives the parameters of the resolved bands at three temperatures. The low frequency component of $v_1 + v_3$ is pure Lorentzian in character while the high frequency component is almost half-Gaussian in character. The low frequency component is much broader than the other component. A comparison of the relative areas shows that there is a small increase in the high frequency component relative to the low frequency component with increasing temperature.

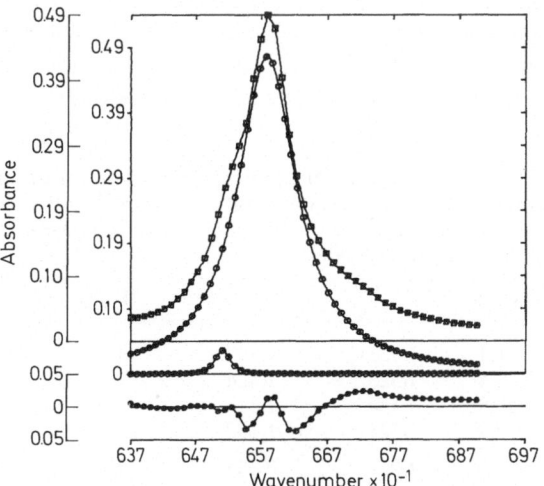

Resolved near infrared spectrum of a mixture of ND2H and NDH2 in ND3 at 25°C. Combination bands

Fig. 6

The small shifts of $2v_1$ and its extreme sharpness indicate that the components of which it is composed are close together and also very sharp. It is beyond the state of the art to resolve them if indeed they can ever be resolved.

Investigations in water have made use of the characteristics of HOD to study changes in the structure of water [16, 23, 26, 28]. The normal modes are described as being decoupled and bands in the *OH* overtone region can be unambiguously assigned to various hydrogen-bonded components. Fig. 6 shows the overtone region of ND_2H in excess ND_3. A single sharp band at 6578 cm^{-1} is observed; the small peak at 6510 cm^{-1} is probably due to NDH_2. Either most of the ammonia molecules are in the same state, or the N-H band of ND_2H is insensitive to the different states, or both of these possibilities are true.

Salt Solutions

Investigations of the effects of solutes upon the structure of the liquid have used several types of schems to classify the results obtained. Salts have been described in terms of structure-making or structure-breaking ability, or have been assigned a structural temperature [16], and in a few cases more sophisticated descriptions have been attempted [17].

As is shown in Tables III and IV, relatively high concentrations of salts are required. The increased temperature range made possible by the cells used permitted the study of solutes at 25° C which were not soluble enough at lower temperatures. Since there is relatively little change in the spectrum of the solvent with temperature, and the changes upon addition of a given salt relative to liquid ammonia were nearly the same for each temperature, only the data for $-35°$ C will be compared here.

All of the salts studied had a marked effect upon the near-infrared spectrum of liquid ammonia. These effects can be divided into two categories as noted above. Fig. 7 shows the resolved spectrum of a solution of sodium iodide. Although alkali metal halides are considered to be structure-breakers in aqueous solution, there is a marked shift of the bands in liquid ammonia to lower frequencies indicating an enhancement of the structure. For the iodides, the differences upon changing the cation are quite small. Aqueous studies have also reported very small changes

Table III. Parameters for the resolved bands of the near-infrared spectrum of solutions of alkali metal halides in liquid ammonia: X is the position in cm^{-1}, $Y2$ is he Lorentzian intensity, W is the width at half-height in cm^{-1}, and A is the relative area. NaBr was run at 25° C. Concentration in mole ratio NH_3/salt

Band	Salt	Conc.	X	$Y2$	W	A
$2v_1$	NaI	11.4	6523	1.342	97	204
	RbI	19.4	6526	1.424	92	206
	CsI	15.8	6524	1.314	95	197
	NaBr	6.9	6524	1.192	107	201
$v_1 + v_3$	NaI		6651	0.348	177	97
	RbI		6655	0.360	192	109
	CsI		6654	0.345	197	107
	NaBr		6646	0.273	206	88

Table IV. Parameters for the resolved bands of the near-infrared spectrum of salt solutions in liquid ammonia: X is the position in cm^{-1}, $Y1$ is the Gaussian intensity, $Y2$ is the Lorentzian intensity, W is the width at half-heigth in cm^{-1}, and A is the relative area. Concentration is in mole ratio NH_3/salt

Band	Salt	Conc.	X	$Y1$	$Y2$	W	A
$2v_1$	$NaClO_4$	7.3	6548	0.113	1.299	91	197
	$LiClO_4$	6.0	6542	0.063	1.384	82	184
	$NaNO_3$	7.8	6533	0.146	1.232	84	177
	$NaSCN$	11.3	6528	0.138	1.216	89	183
$(v_1 + v_3)_A$	$NaClO_4$		6632	0.003	0.118	43	8
	$LiClO_4$		6611	0.001	0.182	108	31
	$NaNO_3$		6595	0.007	0.182	196	58
	$NaSCN$		6600	0	0.184	207	60
$(v_1 + v_3)_B$	$NaClO_4$		6703	0	0.229	174	63
	$LiClO_4$		6688	0	0.210	207	68
	$NaNO_3$		6682	0	0.222	187	66
	$NaSCN$		6674	0	0.225	162	57

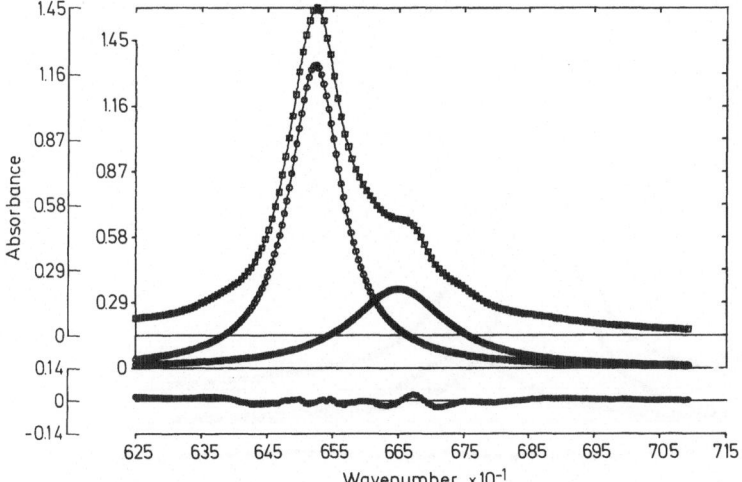

Resolved near infrared spectrum of a solution of sodium iodide
in liquid ammonia at -35°C. Combination bands

Fig. 7

upon changing the cation for a given anion and have ascribed this to the screening effect of the oxygen atom in the water molecule [16, 19]. We observed a similar screening effect due to the nitrogen atom of amonia. Cesium iodide, Fig. 8, which would be expected to produce the largest change relative to sodium iodide has basically the same effect upon the spectrum. Sodium bromide, Fig. 9, which was run at 25° C because of its low solubility at lower temperatures, shows the same structure-making effect as the iodides. The parameters of the resolved bands of the alkali metal halides, Table III, show that the bands are pure Lorentzian in character. $2v_1$ in the alkali metal halide solutions shows a small but definite shift

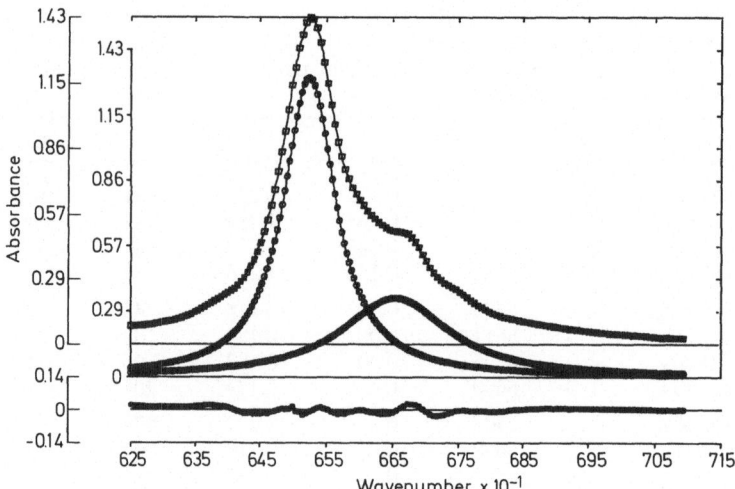

Resolved near infrared spectrum of a solution of cesium iodide
in liquid ammonia at -35°C. Combination bands.

Fig. 8

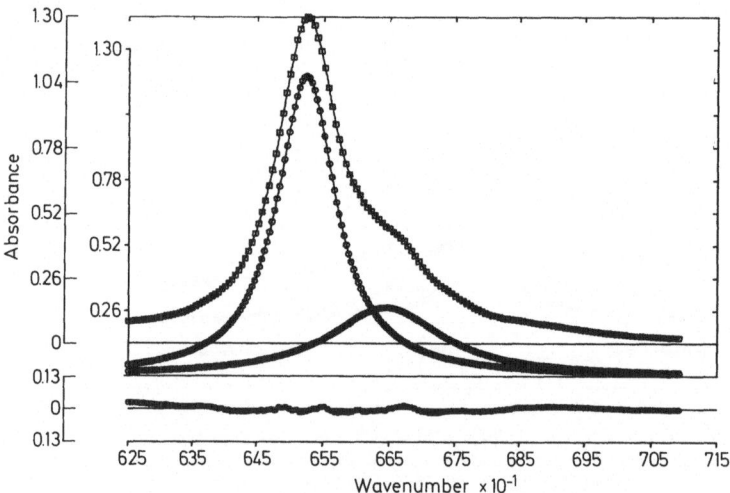

Resolved near infrared spectrum of a solution of sodium bromide in
liquid ammonia at 25°C. Combination bands.

Fig. 9

to lower frequency, while $v_1 + v_3$ is nearly the same frequency as the $v_1 + v_3$ band of liquid ammonia. Even though a nearly perfect fit using pure Lorentzians was achieved with only two bands, an attempt was made to resolve these spectra into three bands. In all cases the intensity of the third band went to zero leaving just two bands.

A solution of sodium perchlorate in liquid ammonia has a spectrum, Fig. 10, which is typical of a number of other salts. The bands of the solvent are shifted to

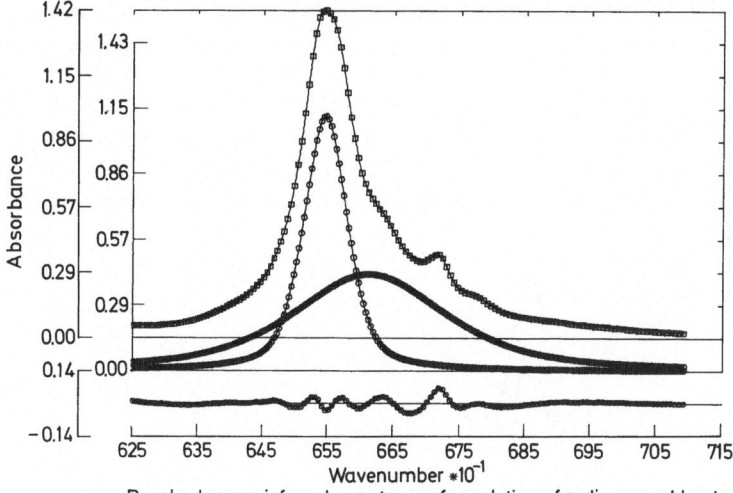

Resolved near infrared spectrum of a solution of sodium perchlorate
in liquid ammonia at 35°C, combination bands

Fig. 10

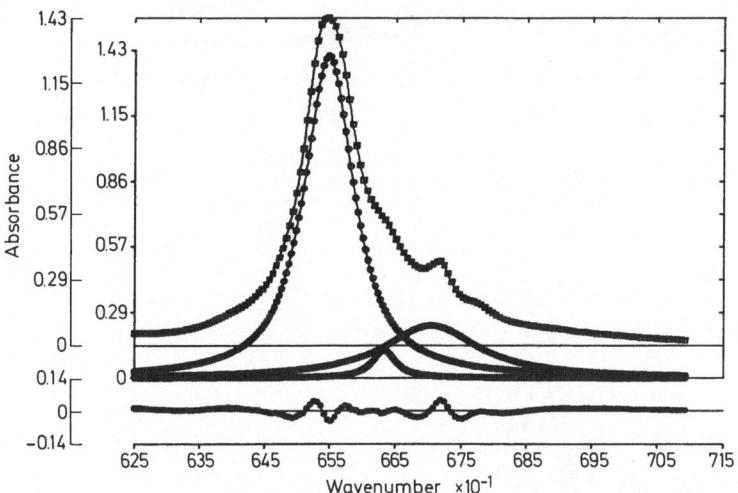

Resolved near infrared spectrum of a solution of sodium perchlorate
in liquid ammonia at -35°C. Combination bands.

Fig. 11

much higher frequencies. $2v_1$ is shifted to higher frequency relative to liquid ammonia. With $v_1 + v_3$ considerably displaced from the experimentally observed maximum and the large fluctuations in the error curve, the fit using two bands is unacceptable.

Fig. 11 shows the spectrum resolved into three bands. The positions of the components of $v_1 + v_3$ at $6632\,\mathrm{cm}^{-1}$ and $6703\,\mathrm{cm}^{-1}$ in the sodium perchlorate solution correspond closely to the positions of the components in the spectrum of liquid ammonia at $6615\,\mathrm{cm}^{-1}$ and $6697\,\mathrm{cm}^{-1}$. However, the relative intensities have reversed. The high frequency component in the sodium perchlorate solution is much greater in intensity and area than the low frequency component, whereas the reverse was true in pure ammonia.

Fig. 12 shows the effect of sodium nitrate and Fig. 13 shows that of sodium thiocyanate upon the solvent bands. Both salts show effects similar to sodium perchlorate but less pronounced. The effect of lithium perchlorate, Fig. 14, is also less than that of sodium perchlorate but greater than that of sodium nitrate or sodium

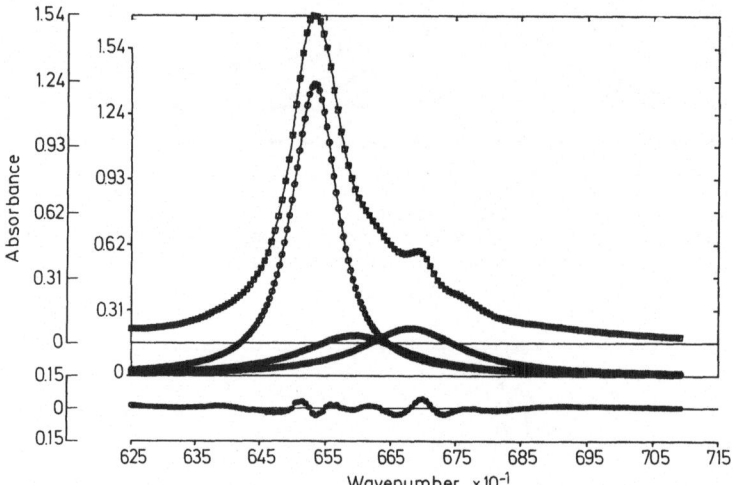

Resolved near infrared spectrum of a solution of sodium nitrate in liquid ammonia at -35°C. Combination bands.

Fig. 12

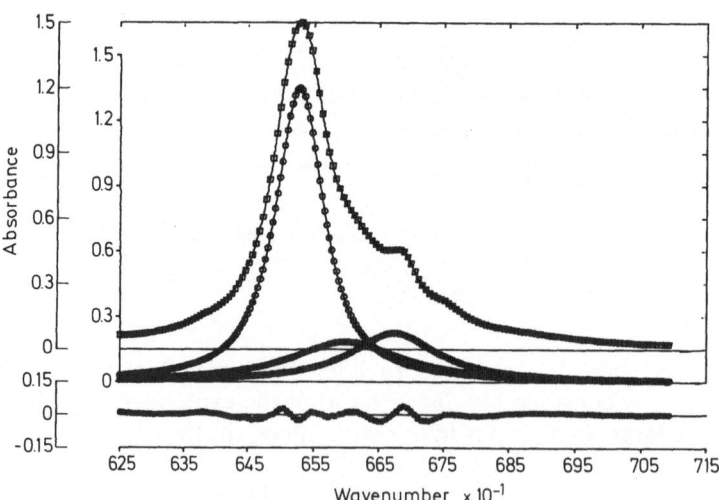

Resolved near infrared spectrum of a solution of sodium thiocyanate in liquid ammonia at -35°C. Combination bands.

Fig. 13

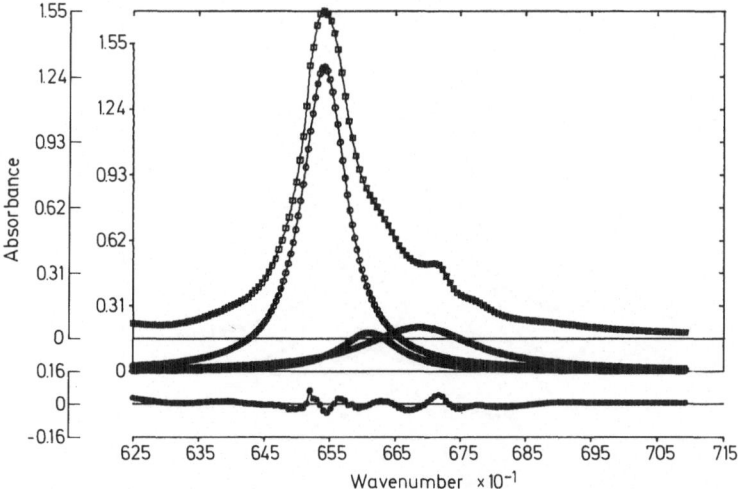

Resolved near infrared spectrum of a solution lithium perchlorate in liquid ammonia at -35°C. Combination bands.

Fig. 14

thiocyanate. This may be due to the effect of the lithium ion which is the most strongly solvated alkali metal ion in liquid ammonia [31].

The parameters of the resolved bands, Table IV, show that $2\nu_1$ is shifted to higher frequency relative to pure liquid ammonia, Table II, in the perchlorate solutions, but stays at about the same position in the nitrate and thiocyanate solutions. There is some variation in the positions of the components of $\nu_1 + \nu_3$, but they remain in the same areas and are well separated. The frequencies of the components are close to the frequencies of the resolved components of the $\nu_1 + \nu_3$ band of liquid ammonia (Table II). However, the relative areas of the peaks are inverted. In all cases the high frequency peak has grown at the expense of the low frequency peak, with the perchlorate solutions showing the greatest effect.

Metal–Ammonia Solutions

Encouraged by the results of our studies of salt solutions in liquid ammonia, we have attempted to apply this technique to metal–ammonia solutions. Fig. 15 shows the resolved near-infrared bands of liquid ND_3 in a 2.6×10^{-4} M solution of sodium in ND_3. The spectrum of the solvent was obtained by subtracting the absorption tail of the solvated electron band. It was assumed that changes in the solvent spectrum were minimal at the end points and in the areas between the peaks. Six points in these regions and the maximum at higher frequency due to the solvated electron were used to define the solvated electron absorption curve. The concentration was calculated from the absorbance maximum of the solvated electron and from data in the literature [20]. At this concentration the absorbance due to the electron was 0.68 at the high-frequency end. Although higher concentrations were examined, most of the peaks were off scale and the noise level from the increased absorption tail of the electron made any attempt at analysis

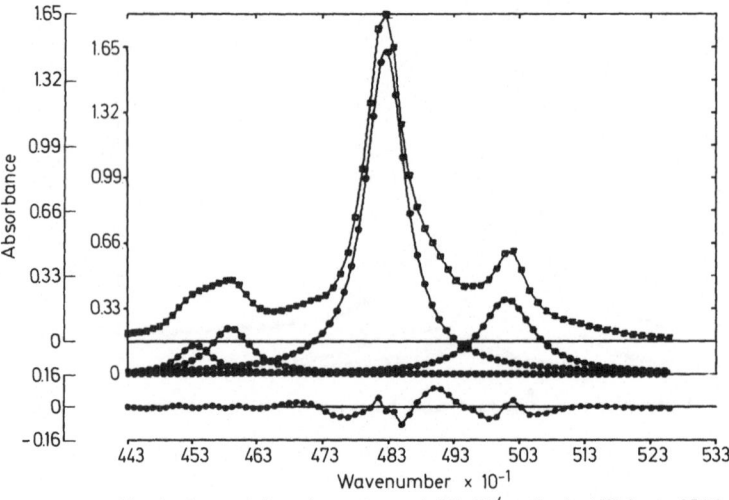

Resolved near infrared spectrum of 2.6×10^{-4} m Na in ND 3 at -35°C
Combination bands.

Fig. 15

unreasonable. If the path length were decreased, the metal concentration could be increased but the solvent absorptions would be much smaller.

Table V gives the band parameters of liquid ND_3 and of a sodium-ND_3 solution. The band at $4595\,cm^{-1}$ is an N-H band due to residual hydrogen. There are distinct differences in the spectra but the shifts are quite small. All peaks decrease in intensity, part of which can be attributed to the change in concentration of the solvent due to the volume expansion upon formation of a metal–ammonia solution. Although the most widely accepted model for the solvated electron [32] suggests considerable orientation of the solvent molecules around the solvated electron, there has recently been a suggestion that entropy effects can be explained if the orientation of solvent molecules around the electron is less ordered than in

Table V. Parameters for the resolved bands of the near-infrared spectrum of ND_3 and Na–ND_3 solutions: X is the position in cm^{-1}, $Y1$ is the Gaussian intensity, $Y2$ is the Lorentzian intensity, W is the width at half-height in cm^{-1}, and A is the relative area

Band	Solution	X	Y	W	A
$2v_1$	ND_3	4542	0.153	66	16
	Na–ND_3	4539	0.145	65	15
$v_3 + v_4$	ND_3	4595	0.240	74	28
(N–H)	Na–ND_3	4594	0.232	77	28
$v_1 + v_3$	ND_3	4826	1.670	71	185
	Na–ND_3	4830	1.635	73	187
$2v_3$	ND_3	5014	0.384	101	61
	Na–ND_3	5015	0.375	98	58

the pure liquid [33]. The small shifts observed here are in the direction to be expected for an enhancement of structure relative to the pure solvent. A decrease in intensity for overtones and combinations of stretching modes also suggests that the solute has a structure-promoting effect, even though some of the decrease in intensity must be due to volume expansion. Observations at higher concentrations might provide greater insight into this question. Workers in other areas of the vibrational spectrum have encountered similar problems [22, 34]. Possibly more information could be gained from the near-infrared spectrum of metal–ammonia solutions with an increase in the signal-to-noise ratio which would permit resolution of the bands with greater certainty.

Discussion

The near-infrared spectrum of liquid ammonia from 6250 cm^{-1} to 7092 cm^{-1} has two absorption bands which have been assigned to $2v_1$ and $v_1 + v_3$. $2v_1$ is very sharp in comparison to the bands of most liquids and shifts very little with changes in temperature or with the addition of salts. The greatest change was a shift of 16 cm^{-1} to higher frequency relative to pure ammonia in a solution of sodium perchlorate.

$v_1 + v_3$ is much broader than $2v_1$ and has been resolved into two components for the pure liquid and for solutions of structure-breaking salts. The positions of the components vary only a small amount depending upon the salt, but the relative intensities and the areas change dramatically. An excellent fit for the spectra of solutions of alkali metal halides is achieved using one pure Lorentzian for $v_1 + v_3$. Very small changes are observed upon changing alkali metal cation, and this can be attributed to the screening effect of the nitrogen atom.

There are several possible explanations for the observation of the two bands which have been labeled $(v_1 + v_3)_A$ and $(v_1 + v_3)_B$. The appearance of $2v_3$ can be eliminated since it would be expected to occur close to 6800 cm^{-1}.

Another possibility is that some strong interaction has lowered the symmetry of the ammonia molecule and the degeneracy of v_3 has been lifted. If this were the case, the combination bands $(v_1 + v_{3a})$ and $v_1 + v_{3b})$ might be observed. Some experimental data have been explained with this interpretation [13]. However, lifting the degeneracy of v_3 would not account for the intensity changes that are observed in solutions of structure-breaking salts relative to pure ammonia. Also, one would expect a greater possibility for disturbing the symmetry of the ammonia molecule in solutions where it is in a more restrictive situation, i.e., in solutions of structure-making salts. In these solutions only one pure Lorentzian intensity is observed.

An alternative explanation would attribute the two components to the $v_1 + v_3$ band of ammonia molecules in different environments, just as various bands in water are split when the environment is perturbed.

The structure-breaking properties of the perchlorate ion have been extensively studied in aqueous solution [8, 16, 28, 29] and in methanol [26]. The most prevalent view is that the splitting of the stretching bands is due to the formation of free OH groups at the expense of hydrogen-bonded OH groups [8]. The effect is similar to that observed in pure water simply by changing the temperature [17,

19]. These studies have provided the strongest experimental evidence for the mixture model of water. It has also been noted that this effect is limited to perchlorate in aqueous solutions and that the effect of other large ions is much less or even in the opposite direction [17, 27, 30].

The results of this study are not inconsistent with a very simple mixture model comparable to the basic considerations used to explain similar phenomena in aqueous solutions. The low frequency component of $v_1 + v_3$ can be assigned to hydrogen-bonded ammonia and the high-frequency component to free ammonia. Pure liquid ammonia appears to be extensively hydrogen-bonded even at 25° C. In solutions of alkali metal halides the hydrogen bonding is even more complete since the high-frequency component of $v_1 + v_3$ is absent. In solutions of other salts this component increases relative to the low-frequency component until it predominates in perchlorate solutions.

The narrow liquid range of ammonia at what are considered to be normal pressures can then be explained in terms of breaking the hydrogen-bonded structure. Although solid ammonia has an octahedral arrangement of hydrogen atoms around the nitrogen atom, solid water has a tetrahedral structure. If the lone pair of electrons on the nitrogen atom of liquid ammonia participates in three hydrogen bonds, they must be weaker than the hydrogen bonds of water where each molecule has two lone pairs and two OH groups to form hydrogen bonds. On the other hand, ammonia may only be capable of forming one strong hydrogen bond on the average per molecule. If this is the case, nearly every broken hydrogen bond would form free ammonia and the vapor pressure would increase rapidly as hydrogen bonds are broken in the pure liquid. Studies of hydrogen bonding in water have shown that in the lower temperature range a significant portion of the molecules participate in two hydrogen bonds, and that practically no monomeric water is present [18]. At higher temperatures the vapor pressure of water increases rapidly as the proportion of monomeric water increases. Ammonia could be viewed as having a smaller liquid range and one that occurs at lower temperatures, due to the fact that nearly every broken hydrogen bond produces monomeric ammonia.

A mechanism for solvation in liquid ammonia which is consistent with the above-mentioned observations must strongly deemphasize ionic forces. The apparent dependence of solubilities in liquid ammonia upon polarizability has been noted by many authors [35]. Indeed investigations of aqueous solutions have recently given more importance to non-ionic forces, especially in explaining the effects of large ions upon the structure of water. What emerges is a view of these solutions in which the ions of the solute are enclosed in a dynamic clathrate-type structure. Some ions permit the solvent cage to form with preservation or even enhancement of the hydrogen-bonded structure of the liquid, whereas other ions disrupt the hydrogen bonds of the liquid and the solvent molecules are arranged in such a manner that new hydrogen bonds do not form.

Acknowledgement

We gratefully acknowledge the financial assistance of the Robert A. Welch Foundation and the National Science Foundation.

References

1. Metal–ammonia solutions. Colloque Weyl II, Lagowski, J. J., Sienko, M. J. (Eds.). London: Butterworths 1970.
2. Solvated electron. Advances in Chemistry Series No. 50, Gould, R. F., (Ed.), ACS, 1965.
3. Metal–ammonia solutions. Colloque Weyl I, Lepoutre, G., Sienko, M. J., (Eds.). New York: Benjamin 1964.
4. Hydrogen-bonded solvent systems. Covvington, A. K., Jones, P., (Eds.), London: Taylor and Francis Ltd. 1968.
5. The hydrogen bond. Pimentel, G. C., McClellan, A. L., (Eds.), San Francisco: W. H. Freeman & Co. 1960.
6. Hydrogen bonding. Hadzi, D., (Ed.), Pergamon Press 1959.
7. Walrafen, G. E.: In: Covvington, A. K., Jones, P. (Eds.): Hydrogen bonded solvent systems, p. 9. London: Taylor & Frances 1968.
8. Walrafen, G. E.: J. Chem. Phys. **52**, 4176 (1970).
9. Senior, W. A., Verrall, R. E.: J. Phys. Chem. **73**, 4242 (1969).
10. Franck, E. U., Roth, K.: Disc. Faraday Soc. **43**, 108 (1967).
11. Wall, T. T., Hornig, D. F.: J. Chem. Phys. **43**, 2079 (1965).
12. Falk, M., Ford, T. A.: Can. J. Chem. **44**, 1699 (1966).
13. Corset, J., Houng, P. V., Lascombe, J.: Spectrochim. Acta **24**A, 1385 (1968).
14. Herzberg, G.: Molecular spectra and molecular structure, Vol. II. Infrared and raman spectra of polyatomic molecules. New York: Van Nostrand 1945.
15. Buijs, K., Choppin, G. R.: J. Chem. Phys. **39**, 2035 (1963).
16. Worley, J. D., Klotz, I. M.: J. Chem. Phys. **45**, 2868 (1966).
17. Luck, W.: Ber. Bunsenges. Phys. Chem. **69**, 626 (1965); — Z. Naturforsch. B. **24**, 482 (1969).
18. Luck, W. A. P., Ditter, W.: J. Phys. Chem. **74**, 3687 (1970).
19. McCabe, W. C., Surbramanian, S., Fisher, H. F.: J. Phys. Chem. **74**, 4360 (1970).
20. Burow, D. F., Lagowski, J. J.: In: Gould, R. F. (Ed.): Solvated electron. Advances in Chemistry Series No. 50, p. 125. ACS 1965.
21. Dewald, R. R., Roberts, J. H.: J. Phys. Chem. **72**, 4224 (1968).
22. Rusch, P. F.: Ph. D. Dissertation. The University of Texas at Austin, 1971.
23. DeBettignies, B., Wallart, F.: Compt. Rend. Ser. B, 640 (1970).
24. Walrafen, G. E.: J. Chem. Phys. **55**, 5137 (1971).
25. Bonner, O. D., Woolsey, G. B.: J. Phys. Chem. **72**, 899 (1968).
26. Hartman, K. A., Jr.: J. Phys. Chem. **70**, 270 (1966).
27. Brink, G., Falk, W.: Can. J. Chem. **48**, 3019 (1970).
28. Kecki, Z., Dryjanski, P., Kozlowska, E.: Roczniki Chem. **42**, 1749 (1968).
29. Schultz, J. W., Hornig, D. F.: J. Phys. Chem. **65**, 2131 (1961).
30. Bunzl, K. W.: J. Phys. Chem. **71**, 1358 (1967).
31. Nelson, J. T., Cuthrell, R. E., Lagowski, J. J.: J. Phys. Chem. **70**, 1492 (1966).
32. Copeland, D. A., Kestner, N. R., Jortner, J.: J. Chem. Phys. **53**, 1189 (1970).
33. Lepoutre, G., Jortner, J.: J. Phys. Chem. **76**, 683 (1972).
34. Koehler, W. H.: Private communication.
35. Lagowski, J. J.: Pure Appl. Chem. **25**, 429 (1971).

Magnetic Relaxation Properties of Dilute Solutions of Sodium in Liquid Ammonia

C. Lambert

Abstract

The nuclear spin relaxation phenomenon is briefly reviewed and the cases of the proton and nitrogen in dilute metal–ammonia solutions are discussed.

I. Introduction

An ordinary magnetic resonance experiment gives the average value of the magnetic field acting at the site of the nucleus. For metal–ammonia solutions, such measurements provide the values of the paramagnetic shift of the resonance lines (analogous to the knight shift in metals) and subsequently with the values of the spin densities at each nucleus site. This paramagnetic shift also indicates the existence of paramagnetic species containing metal atoms (monomers of Becker *et al.* [1]).

Nevertheless, these experiments give no information about fluctuating magnetic fields in the liquid.

Measurement and analyses of spin relaxation times yield information concerning:

1. the nature of the fluctuating magnetic interaction;
2. the strength of the modulated interaction;
3. the correlation time which is, roughly speaking, the mean frequency of the fluctuation.

These fluctuations are related to the motions of the spin carriers in relation to each other. By motion, we mean:

— actual motion of diffusion (translation, rotation)

— chemical exchange of an atom (or a group of atoms) between solute species (or between solute species and solvent)

— quantal motion of a spin from one of its eigenstates to another (flip-flop).

We shall discuss these phenomena for dilute solutions of sodium in ammonia, but first we shall review the nuclear relaxation process and how it works.

II. Nuclear Relaxation

The relaxation time of a nuclear spin system is the time constant of the motion of the nuclear magnetization towards its thermal equilibrium value I_0. When this equilibrium is reached, the magnetization is directed along the axis (Oz) of the magnetic field H_0 selected for the resonance experiment. From a theoretical point of view, it is useful to distinguish between the longitudinal relaxation time T_1, and the transverse relaxation time T_2:

— T_1 is the time constant for the component of the magnetization, $\langle I_z \rangle$, parallel to H_0.

— T_2 is the time constant for the other components, $\langle I_x \rangle$ and $\langle I_y \rangle$, normal to H_0.

In many cases, one can write with very good approximation:

$$\langle \dot{I}_z \rangle = -\frac{1}{T_1}(\langle I_z \rangle - I_0); \quad \langle \dot{I}_x \rangle = -\frac{\langle I_x \rangle}{T_2}; \quad \langle \dot{I}_y \rangle = -\frac{\langle I_y \rangle}{T_2}. \tag{1}$$

These equations define [2] T_1 and T_2.

The transverse relaxation time is related to the width of the resonance line $\Delta H_{1/2}$ by the formula:

$$T_2 = 2/\gamma \cdot \Delta H_{1/2}$$

where γ is the gyromagnetic ratio of the nucleus.

As a rule, $T_2 \leqq T_1$, but, in liquids the so-called "extreme narrowing" condition often exists, in which case:

$$T_2 = T_1 .$$

Relaxation Mechanism

In the case of spin $I = 1/2$ located in a magnetic field H_0, the energy is quantized, and the system possesses two eigenstates $|a\rangle$ and $|b\rangle$ separated by an energy gap of magnitude $\hbar\gamma H_0$ (as sketched in Fig. 1). The magnetization of the system is proportional to the excess, n, of the population n_a of the lower state, respective to the population n_b of the upper state:

$$\langle I_z \rangle \propto n = n_a - n_b .$$

Thus, Eq. (1) is equivalent to

$$\dot{n} = -\frac{n - n_0}{T_1} . \tag{2}$$

n_0 is the thermal equilibrium value of the difference $n_a - n_b$ and corresponds to Boltzmann's ratio:

$$n_b/n_a = \exp(-\hbar\gamma H_0/kT) .$$

Eq. (2) shows that T_1 is related to the probabilities of transition between the states $|a\rangle$ and $|b\rangle$, and it is possible to derive [2], by the detailed balance method, the following equation for T_1:

$$T_1^{-1} = W_{a \rightarrow b} + W_{b \rightarrow a}$$

where $W_{i \rightarrow j}$ is the probability for the transition from the state $|i\rangle$ to the state $|j\rangle$.

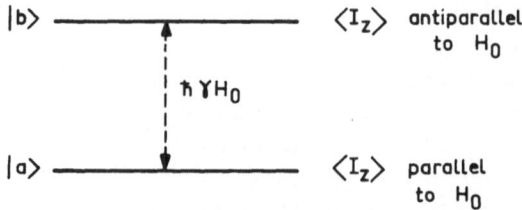

Fig. 1. Magnetic energy levels for a spin $I = \frac{1}{2}$ in a magnetic field H_0

The transition between the states $|a\rangle$ and $|b\rangle$ are induced by the fluctuating magnetic fields normal to H_0. The magnetic energy fluctuating with these fields is weak compared with the Zeeman energy of the spin levels. Thus it is possible to use the perturbation theory and to compute the probabilities of transition with the aid of the "golden rule". In this way, the relaxation rate appears always as a product:

$$T_1^{-1} \propto A^2 \cdot J(\omega_0, \tau_c). \tag{3}$$

In this formula, A is the average value of the fluctuating magnetic energy, $\omega_0 = \gamma H_0$, τ_c is the correlation time of the fluctuation, and J is a frequency function having the dimension of time. (Expressing A in frequency units, one gets T_1^{-1} in frequency units.)

More exactly, J is the Fourier transform of the autocorrelation function [3]:

$$G(\tau) = \langle f(t+\tau) \cdot f(t) \rangle$$

of the function $f(\tau)$ describing the fluctuation law:

$$J(\omega_0, \tau_c) = \int\limits_{-\infty}^{+\infty} G(\tau) \cdot e^{i\omega\tau} \cdot d\tau.$$

It often happens that the correlation function of the fluctuation is of the form

$$G(\tau) = \exp(-|\tau|/\tau_c),$$

thus:

$$J(\omega_0, \tau_c) = \frac{2\tau_c}{1 + \omega_0^2 \tau_c^2}$$

Detailed computations [3] show that Eq. (3) can be expressed as a function of the components h_x, h_y, h_z of the fluctuating magnetic field. The following equations are obtained for T_1 and T_2:

$$\left.\begin{aligned}
T_1^{-1} &= \gamma^2 (\langle h_x^2 \rangle + \langle h_y^2 \rangle) \frac{\tau_c}{1 + \omega_0^2 \tau_c^2} \\
T_2^{-1} &= \gamma^2 \left[\langle h_z^2 \rangle \tau_c + \frac{1}{2} \left(\langle h_x^2 \rangle + \langle h_y^2 \rangle \right) \frac{\tau_c}{1 + \omega_0^2 \tau_c^2} \right]
\end{aligned}\right\}. \tag{4}$$

In a liquid, the fluctuating magnetic field* is isotropic:

$$\langle h_x^2 \rangle = \langle h_y^2 \rangle = \langle h_z^2 \rangle = \langle h^2 \rangle$$

and in the case of extreme narrowing: $\omega_0^2 \tau_c^2 \ll 1$, thus:

$$T_1^{-1} = T_2^{-1} = 2\gamma^2 \langle h^2 \rangle \tau_c.$$

This equation takes the form of formula (3) with $A^2 = \gamma^2 \langle h^2 \rangle$ and $J(\omega_0, \tau_c) = \tau_c$. Practically, the fluctuating local field is produced by the neighboring spins, and their interaction with the spin of interest is, in most cases, either of the scalar type or of the dipolar type.

---

* The bracket indicates the average value.

Dipolar Relaxation Mechanism

In dipolar relaxation, the local field at the site of spin I is produced by the magnetic dipole of the spin S located at a distance r from I:

$$\langle h^2 \rangle \simeq \frac{\hbar^2 \gamma_S^2 \, S(S+1)}{r^6}.$$

The fluctuation is produced by the motion of S respective to I. The intermolecular motion (translation) produces a variation of r for its orientation as well as for its modulus.

The molecular rotation produces only a variation in the orientation of the vector r and thus modulates the angular part of the local field. (For the sake of simplicity, the dependence on angular variables is not included in the formula above.)

In the case of extreme narrowing, one gets:

$$T_1^{-1} \propto p \cdot \gamma_I^2 \gamma_S^2 \frac{\hbar^2 \, S(S+1)}{r^6} \cdot \tau_c$$

(for like spins, $\gamma_I = \gamma_S$ and $p = 3/2$; for unlike spins, $\gamma_I \neq \gamma_S$ and $p = 1$).

For the molecular rotation, τ_c is a third of the Debye correlation time:

$$\tau_c = \frac{1}{3} \cdot \frac{4\pi \eta a^3}{kT}.$$

a being the molecular radius and η the viscosity of the liquid. Full computation gives [4]:

$$T_{1R}^{-1} = p \cdot \frac{16}{9} \cdot \frac{\gamma_I^2 \gamma_S^2 \hbar^2 \, S(S+1)}{r^6} \cdot \frac{\pi \eta a^3}{kT}. \tag{5}$$

For intermolecular motion of translation, $\tau_c \simeq r/2D$, D being the diffusion coefficient: $D = kT/6\pi a\eta$. Full computation, taking into account the concentration of spins producing relaxation, yields [4]:

$$T_{1T}^{-1} = p \cdot \frac{16\pi}{15} \cdot \gamma_I^2 \gamma_S^2 \hbar^2 \, S(S+1) \frac{N}{6aD}, \tag{6}$$

where N is the number of spins S per cubic centimeter. For rotation as well as for translation, the relaxation rate obeys the law:

$$T_1^{-1} \propto \eta/T. \tag{7}$$

Scalar Relaxation Mechanism

This type of relaxation mechanism arises from the modulation of the scalar coupling energy between the two spins I and S. This energy is of the form:

$$\mathscr{H}_{HF} = \hbar A \cdot \boldsymbol{I} \cdot \boldsymbol{S}. \tag{8}$$

which describes the case for indirect interaction, via electrons, for two nuclear spins of the same molecule. It also describes the case for hyperfine interaction between a nuclear spin and the spin of an unpaired electron.

Neither translation nor rotation can modulate this interaction; but chemical exchange acting as a chopping mechanism for the coupling constant A, and the transitions of the spin S between its eigenstates are operative processes for the production of fluctuating magnetic fields at the site of spin I.

In the case of scalar interaction, the relaxation rate is expressed as [2]:

$$T_1^{-1} = \frac{2}{3} A^2 S(S+1) \frac{\tau_c}{1 + (\omega_I - \omega_S)^2 \tau_c^2}.$$ (9)

For chemical exchange, τ_c is the mean lifetime of the atom in the species; for the transitions of the spin S, τ_c is the relaxation time of S.

III. Proton Relaxation in Sodium–Ammonia Solutions

The protons in metal–ammonia solutions are subjected to several relaxation mechanisms:
— dipolar interaction with unpaired electrons
— hyperfine interaction with unpaired electrons
— interaction with other nuclear spins.

All of these processes are independent from each other, and the over-all relaxation rate is the sum of partial relaxation rates relevant to each of them:

$$T_1^{-1} = T_{1D}^{-1} + T_{1HF}^{-1} + T_{1n}^{-1}.$$ (10)

T_{1D}, T_{1HF}, T_{1n}, are the relaxation times for the three mechanisms above, respectively. Whatever the technique used for measurements, one obtains only T_1, but not the separate values of each term of Eq. (10).

Meanwhile, it is easy to compute T_{1n}, because purely nuclear relaxation processes in dilute solutions are practically the same as in pure solvent. Thus, T_{1n} can be deduced from the measurement of the relaxation time $T_{1,0}$ in pure solvent, by applying a correction for viscosity according to Eq. (7):

$$T_{1n}^{-1} = T_{1,0}^{-1} \cdot \frac{\eta}{\eta_0}.$$

η is the viscosity of the solution, η_0 the viscosity of the pure solvent.

Only the techniques of dynamic polarization can provide detailed information on the other terms, T_{1D} and T_{1HF}. Indeed, the magnetization $\langle I_z \rangle$ of nuclear spins is coupled with the magnetization $\langle S_z \rangle$ of electronic spins. Instead of Eq. (1), it is useful to write Solomon's equation:

$$\langle \dot{I}_z \rangle = - \frac{1}{T_1} (\langle I_z \rangle - I_0) + \left(\frac{1}{T_{1HF}} - \frac{1}{2} \cdot \frac{1}{T_{1D}} \right) (\langle S_z \rangle - S_0).$$ (11)

If the electronic spin system is in thermal equilibrium, $\langle S_z \rangle = S_0$ and the formula above reduces to Eq. (1). For the steady state, $\langle \dot{I}_z \rangle = 0$, and Eq. (11) is satisfied by: $\langle I_z \rangle = I_0$ and $\langle S_z \rangle = S_0$ (thermal equilibrium for both of the spin systems). Nevertheless, by applying a strong radio-frequency field oscillating at the Larmor frequency of electrons, it is possible to saturate the resonance of the system S,

$\langle S_z \rangle = 0$; with this condition, one obtains [2] for the steady state:

$$\frac{\langle I_z \rangle}{I_0} = 1 - f\,\frac{S_0}{I_0} = 1 - f\,\frac{\gamma_e}{\gamma_I} \simeq f\,\frac{|\gamma_e|}{\gamma_I} \qquad (12)$$

with:

$$f = \frac{T_{1\,\mathrm{HF}}^{-1} - \dfrac{1}{2}\,T_{1\,\mathrm{D}}^{-1}}{T_1^{-1}}. \qquad (13)$$

The quantity f is called the "leakage factor". Generally, $f \simeq 0.1$—1 and $|\gamma_e|/\gamma_I \simeq 10^2$—$10^4$*, so that Eq. (12) indicates that saturating the resonance of electrons results in an enhancement of the NMR signal: this phenomenon is called the Overhauser effect. (If hyperfine interaction plays the prominent part $\langle I_z \rangle/I_0 > 0$, the signal is enhanced, if dipolar interaction plays the prominent part $\langle I_z \rangle/I_0 < 0$ the signal is reversed and enhanced.)

Table I. Proton dipolar and hyperfine relaxation rates, computed from experimental values of T_1 and f (R = the mole ratio NH_3/Na)

C (mole/I)	R	T_1 (sec)	f	$T_{1\,\mathrm{HF}}^{-1}$ (sec^{-1})	$T_{1\,\mathrm{D}}^{-1}$ (sec^{-1})	$T_{1\,\mathrm{HF}}$ (sec)	$T_{1\,\mathrm{D}}$ (sec)
0	∞	22.5	0				
0.037	1.000	14.8	0.05	0.010	0.013	100	77
0.057	695	9.23	0.07	0.026	0.037	38.5	27
0.098	412	4.27	0.108	0.080	0.110	12.5	9
0.3	133	2.60	0.175	0.160	0.180	6.25	5.55

The measurement of f is carried out by extrapolating the enhancement value for the infinite value of the power used for saturation of the ESR line. After computation of $T_{1\,n}$ and measurement of T_1 and f, Eqs. (10) and (13) allow determination of $T_{1\,\mathrm{D}}$ and $T_{1\,\mathrm{HF}}$. This result for room temperature is shown in Table I; the computation is carried out from Sher's results [5] for T_1 and Itoh and Takeda's results [6] for f.
Theoretical computations enable us to find the results shown in Table I with some approximation.

Relaxation in Pure Solvent

The main relaxation mechanism is the intramolecular dipolar interaction. The relaxation rate is computed (7) with the aid of previous equations:

$$T_{1,0}^{-1} = \left(\frac{3\gamma_H^4}{d_{H-H}^6} + \frac{8}{3}\,\frac{\gamma_H^2\gamma_N^2}{d_{N-H}^6} \right) \hbar^2\,\tau_c. \qquad (14)$$

The first term stands for the interaction with the other two protons of an ammonia molecule; the second stands for the interaction with nitrogen (d_{H-H} and d_{N-H} are the corresponding internuclear distances: 1.4 Å and 1 Å). The correlation

* $|\gamma_e|/\gamma_I = 660$ for proton, and 9100 for nitrogen.

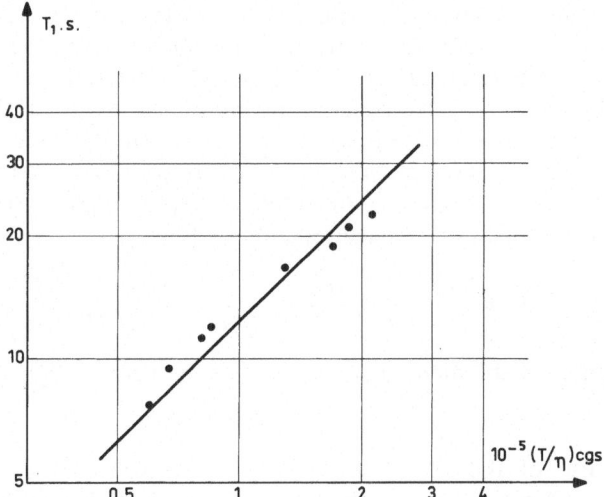

Fig. 2. Relaxation time of the proton in pure ammonia, as a function of temperature and viscosity

time τ_c is computed, as stated previously, with the aid of the formula:

$$\tau_c = \frac{4\pi}{3} \cdot \frac{\eta a^3}{kT},$$

where $a = d_{\text{H-H}}$ and η is given by Kikuchi's measurements [8]. At room temperature, the following values are obtained:

$$\tau_c = 0.6 \cdot 10^{-12}\,\text{sec}; \qquad T_{1,0} = 16.6\,\text{sec}.$$

Nitrogen contributes 8% to this relaxation rate. Intermolecular mechanisms are only weakly operative in this situation Sher's experimental value (22 sec) is thus explained within a factor of 1.5. There is good agreement with Eq. (7), as can be seen in Fig. 2. (This log plot is based on Sher's data, but Newmark [9] and co-workers reported the same results.)

The agreement between theory and experimental results is good if one keeps in mind the roughness of the model, in which molecules are considered to be little macroscopic spheres tumbling in a continuous viscous medium. Neither for the form of the molecule nor for the continuity of the medium is this quite the physical reality.

Hyperfine Relaxation in the Solution

Applying Eq. (9) to the nuclei included in the solvation shell of solvated electrons and to those included in the monomer, one obtains [7]:

$$T_{1\,\text{HF}}^{-1} = \frac{2}{3} S(S+1) \left(\frac{c_- n_-}{Rc} A_-^2 \tau_- + \frac{c_0 n_0}{Rc} A_0^2 \tau_0 \right). \tag{15}$$

This equation corresponds to the extreme narrowing case. R is the mole ratio $[NH_3]/Na$; Rc, c_-, c_0 are concentrations of solvent, solvated electrons and monomers, c is the over-all metal concentration in the solution. τ_- and τ_0 are the lifetimes of the solvated electron and the monomer. n_- and n_0 are the number of solvent molecules included in a solvated electron and in a monomer. A_- and A_0 are the hyperfine interaction constants for protons in the solvated electron and in the monomer, and both are related to the corresponding spin densities P_-^H and P_0^H, deduced from knight shift measurements by [7]:

$$A_- = \frac{8\pi}{3}\gamma_e\gamma_H\hbar\,\frac{P_-^H}{3n_-}; \qquad A_0 = \frac{8\pi}{3}\gamma_e\gamma_H\hbar\,\frac{P_0^H}{3n_0}. \tag{16}$$

Relaxation studies of solvated electrons give us the value [10]:

$$\tau_- = 10^{-12}\ \text{sec}.$$

Line-width study of the NMR signal of Na by O'Reilly [11] shows that $\tau_0 \simeq 4.10^{-12}$ sec, ($T_1 \simeq 10^{-3}$ sec). Taking $n_- = 23$, $n_0 = 6$ [7], with $|P_-^H| \simeq 10^{23}$ [7] and $|P_0^H| \simeq 10^{22}$ [12], the formula above becomes:

$$T_{1\,HF}^{-1} = \frac{100}{R}\left[4\,\frac{c_-}{c} + 2.7\,\frac{c_0}{c}\right]\text{sec}^{-1}.$$

The concentrations c_- and c_0 are computed on the basis of the model of Becker et al. [7] with the following values for equilibrium constants:

$$K_1 = 2.7 \cdot 10^{-3}\ \text{Mole/kg}; \qquad K_2 = 1.6\ (\text{Mole/kg})^{1/2}.$$

The results thus obtained for $T_{1\,HF}^{-1}$ are presented in Table II: they are not in good agreement with the experimental results in Table I. The reason for this divergence is probably the fact that a slight uncertainty in the values of spin densities results in a large error for the relaxation rate.

Table II. Theoretical values of the proton hyperfine relaxation rate

C	0.05	0.06	0.1	0.2
R	800	660	400	200
$T_{1\,HF}^{-1}$ (sec^{-1})	0.35	0.41	0.59	0.97

Electron Dipolar Relaxation in Solution

The correlation time pertinent for the modulation by the rotation of species can be computed on the basis of the third of the Debye relaxation time. For $R_- = 5$ Å and $R_0 = 4.5$ Å (R_- and R_0 being the radii of solvated electron and monomer, including the solvation shell), we obtain:

$$\tau_- \simeq \tau_0 \simeq 10^{-11}\ \text{sec}\quad (\text{for rotation}).$$

The lifetime of solvated electrons and monomers is shorter (10^{-12} sec) so that these species cannot perform a complete rotation during their life. Thus, the rotation is not an operative mechanism for relaxation, and only translation is able to modulate the dipolar interaction and to produce relaxation.

In the case of translation we have to apply Eq. (6) after correction for:

— the difference between the radii of the species [2], and between their diffusion coefficients [2];

— the extreme narrowing condition which is not fulfilled (because, for translation $\tau_{+\,\mathrm{Dif}} \simeq 3 \cdot 10^{-11}$ sec, $\tau_{-\,\mathrm{Dif}} \simeq 6 \cdot 10^{-12}$ sec as $\omega_e \simeq 6 \cdot 10^{10}\, Hz$).

Taking into account these corrections, complete computations [7] give us:

$$T_{1\mathrm{D}}^{-1} = \gamma_{\mathrm{H}}^2 \gamma_e^2 \hbar^2\, S(S+1)\, \frac{\eta N\, 10^{-3}}{kT} \tag{17}$$

$$\cdot \left\{ \left[\frac{8\pi^2}{25} + \frac{56\pi^2}{75}\, \frac{1}{1 + \omega_e^2 \tau_{-\,\mathrm{Dif}}^2} \right] c_- \varphi_- + \left[\frac{8\pi^2}{25} + \frac{56\pi^2}{75}\, \frac{1}{1 + \omega_e^2 \tau_{+\,\mathrm{Dif}}^2} \right] c_0 \varphi_0 \right\}.$$

S is the spin number of the electron, index e refers to the electron. The first term in the brackets is the contribution of the drift of the solvated electron to the relaxation, the second term is the contribution of the drift of the monomer. Quantities φ_- and φ_0 are the size effect correction factors [2]; for φ_- the diffusion coefficient is computed from the theory of tunnelling of the solvated electron [7]. With the aid of the magnetic specific susceptibility χ_S expressed as

$$\chi_S \cdot c \cdot 10^{-3} = (c_- + c_0) \cdot N \cdot 10^{-3} \cdot \gamma_e^2 \hbar^2\, \frac{S(S+1)}{3kT} \tag{18}$$

and putting down all of the numerical values, Eq. (17) is transposed as follows:

$$T_{1\mathrm{D}}^{-1} = \gamma_{\mathrm{H}}^2 \eta \chi_S \cdot c \cdot 10^{-3} \cdot \frac{4c_- + 12.5\, c_0}{c_- + c_0}\, \mathrm{sec}^{-1}.$$

Table III. Theoretical values of the proton dipolar relaxation rate

C	0.03	0.06	0.1	0.2
R	1.330	660	400	200
$T_{1\mathrm{D}}^{-1}$ (sec^{-1})	0.07	0.46	0.47	0.8

Computing the concentration, as was done previously for the case of the scalar interaction, one again obtains results (Table III) in disagreement with the experimental values in Table I. The roughness of the model used for the size correction is presumably responsible for the divergence.

IV. Nitrogen Relaxation in Sodium Ammonia Solutions

The same argument is applicable to the proton and leads to Eq. (10). The same reasoning can be performed step by step.

Relaxation in Pure Solvent

In pure solvent, the relaxation of nitrogen is governed by quadripolar interaction. This sort of relaxation process was not discussed previously in this paper; the nucleus possessing a quadrupole moment can be rocked by the gradient of the local electrical field, which tumbles with the molecule.

In this motion, the spin is reoriented, together with the quadrupole of the nucleus.

The relaxation rate for this mechanism can be expressed [2], in the case of extreme narrowing, as

$$T_{1,0}^{-1} = \frac{3}{8} \left(\frac{eq}{\hbar} \frac{\partial^2 V}{\partial z^2} \right)^2 \cdot \tau_c \ . \tag{19}$$

q is the quadrupole moment of nitrogen, and $\dfrac{\partial^2 V}{\partial z^2}$ is the gradient of the electrical field at the nitrogen site. τ_c is the correlation time for the rotation of the ammonia molecule and has been previously computed:

$$\tau_c = \frac{4\pi}{3} \frac{\eta a^3}{kT} = 0.6 \cdot 10^{-12} \sec \quad \text{at room temperature}\,.$$

The quadripolar coupling constant can be measured [13] from hyperfrequency spectroscopy of gases:

$$\frac{1}{2\pi} \left(\frac{eq}{\hbar} \frac{\partial^2 V}{\partial z^2} \right) = 4 \,\text{MHz}\,.$$

Thus, we obtain for room temperature

$$T_{1,0}^{-1} = 140 \sec^{-1}; \qquad T_{1,0} = 7 \,\text{msec}\,.$$

The measurement carried out by O'Reilly [14] at 240 °K yields $T_{1,0} = 4.4$ msec; this value converted to room temperature by the law $T_1 \propto T/\eta$ gives $T_{1,0} = 10$ msec, which is in agreement with the computation above. These values are in agreement within a factor of 2 with measurements performed by Atkins [15] and co-workers.

Hyperfine Relaxation in Solutions

Equation (15) applies, but instead of Eq. (16), we obtain [7] for nitrogen:

$$A_- = \frac{8\pi}{3} \gamma_e \gamma_N \hbar \frac{P_-^N}{n_-}; \qquad A_0 = \frac{8\pi}{3} \gamma_e \gamma_N \hbar \frac{P_0^N}{n_0} \tag{20}$$

then:

$$T_{1\,\text{HF}}^{-1} = \frac{2}{3} S(S+1) \left(\frac{8\pi}{3} \right)^2 (\gamma_e \gamma_N \hbar)^2 \frac{1}{R} \left[\frac{(P_-^N)^2}{n_-} \tau_- \frac{c_-}{c} + \frac{(P_0^N)^2}{n_0} \tau_0 \frac{c_0}{c} \right]. \tag{21}$$

For the electronic spin system, the relaxation rate as a result of hyperfine interaction [7] is

$$T_{1\,\text{HF},e}^{-1} = \frac{2}{3} I^N(I^N+1) \left(\frac{8\pi}{3} \right)^2 \frac{(\gamma_e \gamma_N \hbar)^2}{c_- + c_0} \left[\frac{(P_-^N)^2}{n_-} \tau_- c_- + \frac{(P_0^N)^2}{n_0} \tau_0 c_0 \right] \tag{22}$$

in such a way that

$$\frac{T_{1\,HF}}{T_{1\,e}} = \frac{I^N(I^N+1)}{S(S+1)} \frac{Rc}{c_- + c_0}.$$ (23)

Table IV. Ratio of relaxation rates as a result of hyperfine interaction, for nitrogen and the electron

C (mole/1)	R	$T_{1\,HF}/T_{1\,e}$
0.01	4000	11100
0.02	2000	6500
0.03	1330	4680
0.04	1000	3780
0.05	800	3180
0.06	660	2790
0.07	570	2510
0.08	500	2330
0.09	445	2160
0.10	400	2040
0.20	200	1475
0.30	133	1405
0.35	114	1440

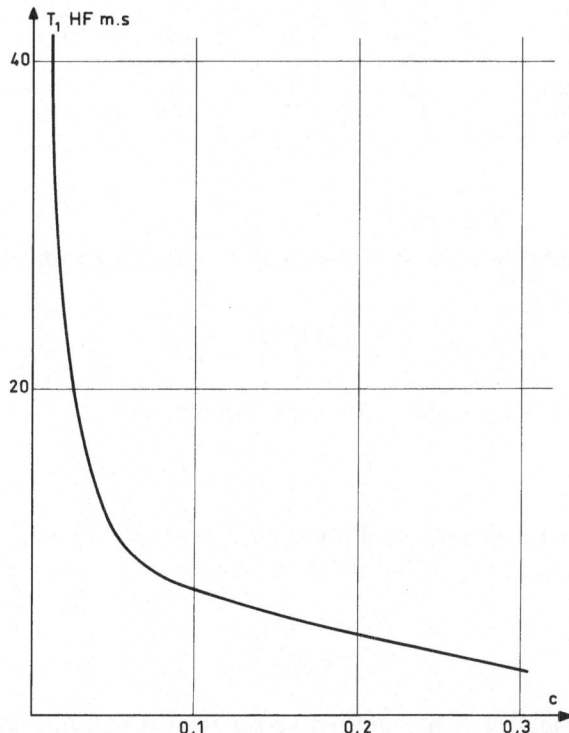

Fig. 3. Nitrogen relaxation time due to hyperfine interaction as a function of metal concentration, at room temperature [Eq. (23)]

Here T_{1e} is the electron relaxation time and I^N the spin of nitrogen, since, in dilute solutions, the whole relaxation of the electron is produced only by hyperfine interaction with nitrogen [10, 16].

Equation (23) enables us to compute T_{1HF}, for nitrogen, from the values of T_{1e}: Table IV gives the values of the ratio T_{1HF}/T_{1e} for room temperature; and Table V gives the values of T_{1HF} after a slight correction, of T_{1e}, for dipolar interaction between electronic spins. The relaxation time of nitrogen as a result of hyperfine interaction, versus the concentration of the solution, is plotted in Fig. 3.

Full computation of Eq. (21), yields values for T_{1HF} which are in agreement with those given in Table V, within a factor of 2. The theory gives better results for nitrogen than for the proton, presumably because the values of the spin densities are more accurate.

Table V. Relaxation time as a result of hyperfine interaction for nitrogen, computed from the relaxation time T_{1e} of electrons. $T_{1e'}$ is the value of T_{1e} corrected for the effect of the electron dipole-dipole relaxation process [7] $[T_{1e'}^{-1} = T_{1e}^{-1}$-(dipole-dipole rate)]. $T_{1e'}$ is used in Eq. (23) instead of T_{1e}

C (mole/1)	R	T_{1e} (μsec)	$T_{1e'}$ (μsec)	T_{1HF} (msec)
0.01	4000	3.12	3.36	37.3
0.02	2000	3.11	3.57	23.8
0.05	800	3.00	3.64	11.6
0.10	400	2.78	3.85	7.9
0.20	200	2.16	3.44	5.1
0.30	133	1.27	1.86	2.6

Dipolar Relaxation in Solutions

Equation (17) is applicable in this case, with γ_N substituted for γ_H. Then

$$(T_{1D}^{-1})_N = \left(\frac{\gamma_N}{\gamma_H}\right)^2 \cdot (T_{1D}^{-1})_H = \frac{1}{200}(T_{1D}^{-1})_H$$

this relaxation rate is negligible with respect to the others.

Overhauser Effect

With the aid of the data presented above, it is possible to compute the leakage factor f_N for nitrogen [17]. The dipolar rate being negligible, it reduces to

$$f_N = \frac{T_{1HF}^{-1}}{T_{1HF}^{-1} + T_{1,0}^{-1}}. \tag{24}$$

The numerical values of f_N are plotted versus the over-all metal concentration, for room temperature (Fig. 4). This result was verified [7, 17] at room temperature for a pumping frequency of 10 GHz.

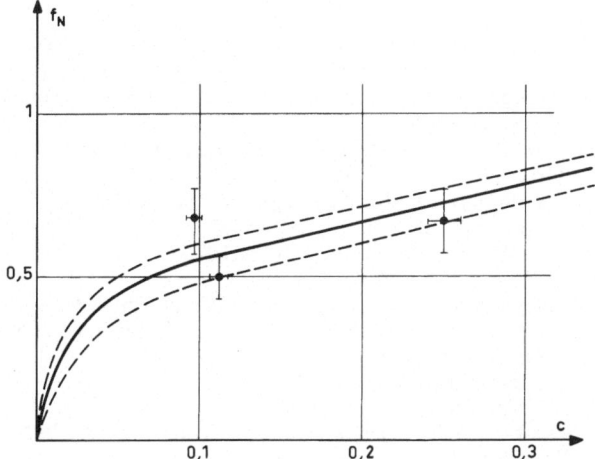

Fig. 4. Leakage factor of nitrogen as a function of metal concentration at room temperature. The computation is carried out with the aid of Eq. (24). Dotted lines indicate the uncertainty in f_N arising from the error in $T_{1,0}$ measurements in Ref. [34]. Experimental points are taken for a pumping frequency of 10 GHz [17]

The leakage factor of the proton [6, 18] is five times lower than the corresponding quantity for nitrogen. The reason is that the dipolar relaxation rate of the proton is great enough for cancellation of the hyperfine relaxation rate in the numerator of formula [13].

V. Conclusions

Because of their complexity, metal–ammonia solutions do not seem to be the proper system for studying magnetic relaxation laws. Lack of accuracy in the knowledge concerning the concentration of each solute species and also in the knowledge of spin densities at the proton site involves rather large errors in the results obtained from theoretical computations of partial relaxation rates (mainly in regard to the proton).

On the other hand several mechanisms contribute to the whole relaxation process. Thus, dynamic polarization measurements are necessary, with over-all relaxation rate measurements, for the computation of each of the partial rates from experimental data.

Nevertheless, the experiment involving the Overhauser effect clearly indicates, for each sort of nuclei, the respective importance of the different types of spin interactions.

Finally, it will be noted that even for sodium, for which few qualitative data [11] are available, relaxation time studies in dilute solutions lead to valuable information concerning the lifetime of the solute species.

References

1. Becker, E., Lindquist, R. H., Alder, B. J.: J. Chem. Phys. **25**, 971 (1956).
2. Abragam, A.: Les Principes du magnétisme nucléaire. P.U.F. Paris 1961.
3. Carrington, A., McLachlan, A. D.: Introduction to magnetic resonance. New York: Harper International Edition 1957.
4. Pople, J. A., Schneider, W. G., Bernstein, H. J.: High resolution nuclear magnetic resonance. New York: Mc Graw Hill 1959.
5. Sher, A.: Thesis, Washington University, Dept. of Physics (1959).
6. Itoh, J., Takeda, T.: J. Phys. Soc. Jap. **18**, 1560 (1963).
7. Lambert, C.: Thèse – Paris (1968) C.N.R.S. N° AO 1732.
8. Kikuchi, S.: J. Soc. Chem. Ind. Jap. **47**, 488 (1944).
9. Newmark, R. A., Stephenson, J. C., Waugh, J. S.: J. Chem. Phys. **46**, 3514 (1967).
10. Lambert, C.: In: Metal–ammonia solutions, Colloque Weyl II, J. J. Lagowski, M. J. Sienko, Eds., p. 301. London: Butterworths 1970.
11. O'Reilly, D. E.: J. Chem. Phys. **41**, 3729 (1964).
12. O'Reilly, D. E.: J. Chem. Phys. **41**, 3736 (1964).
13. Townes, C. H., Schawlow, A. L.: Microwave spectroscopy. New York: Mc Graw Hill 1955.
14. O'Reilly, D. E.: Phys. Rev. Letters **11**, 545 (1963).
15. Atkins, P. W., Loewenstein, A., Margalit, Y.: Mol. Phys. **17**, 329 (1969).
16a. Hutchison, C. A., O'Reilly, D. E.: J. Chem. Phys. **34**, 1279 (1961).
    b. Pollak, V. L.: J. Chem. Phys. **34**, 864 (1961).
17. Lambert, C.: J. Chem. Phys. **48**, 2389 (1968).
18. Carver, T. R., Slichter, C. P.: Phys. Rev. **102**, 975 (1956).

Chemical Equilibria between Anions and the Ammoniated Electron

J. BELLONI and E. SAITO

Abstract

It is well known that the decomposition of solutions of metals in liquid ammonia is limited by the reverse reaction between the amide ion and hydrogen gas leading to the formation of solvated electrons. We show that such an equilibrium may exist between other anions and the solvated electrons in the presence of hydrogen.

In a series of papers, reviewed in the Proceedings of Colloque Weyl II [1], Jolly and co-workers demonstrated that the ammoniated electrons could exist in stable equilibrium with amide ions in liquid ammonia under hydrogen pressure. These observations were discussed within the framework of thermodynamic equilibria among a strong base, hydrogen, and the solvated electron.

In this paper, we shall discuss equilibria in liquid ammonia between e_{am}^- and anions other than NH_2^-.

Fradin [2] in his thesis, presented observations on the γ-irradiation of sodium amide solutions for long periods of time. In a solution initially 10^{-2} mole $\cdot$ l^{-1} of $NaNH_2$, he found that ammoniated electrons exist at a stationary concentration* of 8.5×10^{-6} mole $\cdot$ l^{-1}. After radiolysis, the ampoule contained hydrogen and nitrogen at pressures of 4 and 1.2 atm, respectively, and hydrazine at a total concentration of 6×10^{-2} mole $\cdot$ l^{-1}. Because of its acidic properties in liquid ammonia [4, 5], N_2H_4 is neutralized by the base NH_2^- as soon as it is formed and is thus replaced by the hydrazide ion $N_2H_3^-$ which attains a concentration also of 10^{-2} mole $\cdot$ l^{-1}.

These ampoules did not contain any metal catalyst which, by accelerating equilibration, would have ensured that we are really dealing with a thermal phenomenon. Nevertheless, after a short period of irradiation the system returned to its initial equilibrium state. This seemed sufficiently conclusive for Fradin to propose that an equilibrium, analogous to that shown by Kirschke and Jolly [6] to exist among NH_2^-, H_2 and e_{am}^-, also exists with the other anion $N_2H_3^-$, present in this case:

$$N_2H_3^- + 1/2\,H_2 \rightleftharpoons e_{am}^- + N_2H_4 \qquad (1)$$

This equilibrium will be more favorable to the formation of e_{am}^- than that involving amide, for which at the same concentration as that of hydrazide, i.e., 10^{-2} mole $\cdot$ l^{-1}, $[e_{am}^-]$ would be only 1.8×19^{-7} mole $\cdot$ l^{-1} (compared with 8.5×10^{-6} mole $\cdot$ l^{-1}). This concentration of the solvated electron has been calculated using

---

* The value obtained by Fradin at 20° C for $\varepsilon_{max}(e_{am}^-)$ is not very different from other values measured at low temperatures and given in the literature. For this reason we have recalculated the concentration of e_{am}^- taking $\varepsilon_{max} = 4.8 \times 10^4$ l $\cdot$ mole$^{-1} \cdot$ cm^{-1} [3] (instead of Fradin's value 4.2×10^4 l $\cdot$ mole$^{-1} \cdot$ cm^{-1}).

the equilibrium constant given by Kirschke and Jolly [6] for low amide concentration and taking an extinction coefficient at 1365 nm of 2.4×10^4 mole$^{-1} \cdot$ cm^{-1} ($\varepsilon_{1365}(e_{am}^-) = 0.5 \times \varepsilon_{1750}(e_{am}^-)$) rather than 1.1×10^4 l mole$^{-1} \cdot$ cm^{-1} [7].

Moreover, values of K_c (the equilibrium constant calculated using concentration instead of activities) which we obtained in another series of experiments [8] agree with those of the preceding authors [6] after normalization of the extinction coefficient.

We also tried to determine whether the formation of e_{am}^- observed in the case of NH_2^- - or $N_2H_3^-$ solutions also occurs for other anions. Using a technique almost identical to that described in another paper [8], we prepared ampoules in which the potassium and ammonia are condensed at one end of the tube and a product, alcohol or water, in a side arm. After being sealed off, the apparatus is warmed to room temperature and the added compound is allowed to react with the metal. Platinum foil in another side arm can be used to accelerate the reaction.

In the presence of potassium metal (0.5 mole $\cdot$ l^{-1}) and a slightly greater concentration of ethanol, the deep blue color took more than 24 h to disappear. The hydrogen formed developed a pressure of 5 atm. Under these conditions, after the solution had been kept for a few days in the dark, in contact with the platinum foil, the absorption spectrum was recorded with pure ammonia in the reference cell and found to be stable. Between 600 and 2500 nm, we observed a series of absorption bands separated by more transparent regions, the whole being much more complex than the spectra obtained in the presence of KNH_2. Nevertheless an absorption envelope could be identified with a maximum at 1750 nm (OD $\geqslant 2$, path length $= 2.5$ cm). After a short exposure of the solution to daylight, all absorption bands, especially that at 1750 nm, were greatly enhanced. Leaving the solution in contact with the platinum restores the spectrum to its original form. Although the transparent "bands" present on both sides tend to reduce the optical density at 1750 nm, nevertheless, assuming that the extinction coefficient of e_{am}^- remains unchanged, the concentration of e_{am}^- will be $\geq 1.7 \times 10^{-5}$ mole $\cdot$ l^{-1}. Under the equivalent conditions with KNH_2 we would have found only 3.3×10^{-6} mole $\cdot$ l^{-1}. Thus we are led to suppose the existence of another equilibrium:

$$CH_3CH_2O^- + 1/2\ H_2 \rightleftharpoons e_{am}^- + CH_3CH_2OH \tag{2}$$

Other experiments were carried out in which the potassium metal attacks water, which in liquid ammonia is a weak acid. The initial content of the solutions was 1.8 mole $\cdot$ l^{-1} in HOH and about 0.8 mole $\cdot$ l^{-1} in potassium. After the reaction was complete the final pressure of H_2 was 1.5 atm. There was also a precipitate of KOH, because its solubility in liquid ammonia is only about 10^{-2} mole $\cdot$ l^{-1}. The spectrum measured with pure ammonia as reference did not show any particular band in the range 500 to 2200 nm. Nevertheless a short exposure to a high-pressure mercury lamp gave an absorption which decreased with a decay time T of ~ 1 min.

In another experiment in which the quantity of potassium was close to the stoichiometric amount (1.8 mole $\cdot$ l^{-1}) necessary to react with the H_2O present, the hydrogen pressure developed was 3 atm and the spectrum showed a series of bands whose envelope presented a maximum at 1750 nm. The OD was less

Table I. Equilibrium concentration of e_{am}^- in presence of various anions

	N_2H_4 (2)	CH_3CH_2OH	HOH
H_2	4 atm	1.6 atm	3 atm
AK	10^{-2}	0.5	saturated
mole 1^{-1}			$\sim 10^{-2}$
			total 1.8
AH	$5 \cdot 10^{-2}$	~ 0	~ 0
mole 1^{-1}			
$[e_{am}^-]_{A^-}$	$8 \cdot 10^{-6}$	$\ll 1.7 \cdot 10^{-5}$	10^{-5}
mole 1^{-1}			
$[e_{am}^-]_{NH_2^-}$ [a]	$1.8 \cdot 10^{-7}$	$1.6 \cdot 10^{-6}$	$\sim 2 \cdot 10^{-7}$ [b]
mole 1^{-1}			

[a] $[e_{am}^-]_{NH_2^-}$ is calculated employing the constant K_c of Kirschke and Jolly [6] normalized with the extinction coefficient which we have used:

$$K_c(\text{corr.}) = K_c \frac{2.4 \cdot 10^4}{1.1 \cdot 10^4}.$$

[b] $[NH_2^-] \sim OH_{sat}^- \sim 10^{-2} \, mol \cdot l^{-1}$.

than in the case of ethanol. Here also, the transparent regions on both sides of λ_{max} may reduce the value of ε_{max}, but the equilibrium concentration of e_{am}^- was at least 10^{-5} mole $\cdot l^{-1}$. An equivalent concentration of amide at the same hydrogen pressure would have led to an equilibrium concentration of e_{am}^- not greater than 2×10^{-7} mole $\cdot l^{-1}$. We thus also postulate the existence of the equilibrium:

$$OH^- + 1/2 \, H_2 \rightleftharpoons HOH + e_{am}^-. \qquad (3)$$

The above comparison shows that, in the absence of an excess of water, the concentration of e_{am}^- in equilibrium with a given concentration of OH^- may be much greater than with that in equilibrium with the same concentration of NH_2^-. This remark which applies to N_2H_4 and CH_3CH_2OH as well as to HOH, may seem paradoxical if one considers the acidic properties of these molecules relative to NH_3. Their dissociation, leading to the formation of NH_4^+ which will react with the solvated electron to give H_2 [1, 4, 9—11] should displace the equilibria (1), (2) and (3) to the left. It seems therefore, provided the undissociated acid is in slight excess, the formation of e_{am}^- is favored.

For a comparison of the equilibria (1), (2) and (3) with (4) involving NH_2^-:

$$NH_2^- + 1/2 \, H_2 \rightleftharpoons e_{am}^- + NH_3 \qquad (4)$$

$$K_{c(4)} = \frac{[NH_2^-] [P_{H_2}]^{1/2}}{[e_{am}^-] [NH_3]} = K_c/[NH_3]$$

one must take into account the concentration of the undissociated acids (N_2H_4, alcohol or water) as compared with that of NH_3. The high concentration of NH_3 relative to NH_2^- makes the reverse reaction in (4) more efficient than the equivalent process in the equilibria (1), (2) and (3).

A comparison of the equilibrium constants of (1), (2) and (3) with $K_{c(4)}$ requires a knowledge of the concentrations of the respective anions and of the degree of dissociation and complexation of the species present. The technique employed here was inadequate for these determinations. A precise study of the influence of each of the constituents has been undertaken.

Note Added in Proof: Experiments which we have continued in this field have shown that these solutions present a greater sensitivity to light than suspected. The "paradoxical" effect which we have described above may have come from a pseudo stationary concentration of e_{am}^- higher than that for thermodynamical equilibrium. The relative long life of e_{am}^- obtained in our transparent quartz cells may have led to such a situation.

References

1. Jolly, W. L.: In: Lagowski, J. J., Sienko, M. J. (Eds.): Colloquium Weyl II. London: Butterworths 1970.
2. Fradin de la Renaudiere, J.: Thesis, Paris 1971.
3. Douthit, R. C., Dye, J. L.: J. Am. Chem. Soc. **82**, 4472 (1960). — Jolly, W. L., Hallada, C. J., Gold, M.: In: Lepoutre, G., Sienko, M. J. (Eds.): Solutions metal–ammoniac. London: Benjamin 1964. — Quinn, R. K., Lagowski, J. J.: J. Phys. Chem. **73**, 2326 (1969).
4. Belloni, J.: Int. J. Rad. Phys. Chem. **1**, 441 (1969).
5. Thiebault, A., Herlem, M., Belloni, J.: J. Electroanal. Chem. **32**, 456 (1971).
6. Kirschke, J. J., Jolly, W. L.: Inorg. Chem. **6**, 855 (1967).
7. Corset, J., Lepoutre, G.: In: Lepoutre, G., and Sienko, M. J. (Eds.): Solutions Metal-ammoniac, p. 156. London: Benjamin 1964.
8. Belloni, J., Saito, E.: Coloquium Weyl III. This volume, p. 460.
9. Jolly, W. L.: Advan. Chem. **50**, 27 (1965).
10. Dewald, R. R., Tsina, R. V.: J. Phys. Chem. **72**, 4520 (1968).
11. Jolly, W. L., Prizant, L.: Chem. Commun. **1968**, 1345.

Discussion

U. SCHINDEWOLF   When studying the pressure dependence of the equilibrium of the "Jolly reaction", we also tried to investigate the corresponding aqueous system and the ammonia system with OH^- instead of NH_2^-. Even at H_2 pressures of 100 atm and up to 300° C (2000 at over-all pressure), no solvated electrons were observed. Thermodynamic estimations reveal that all solvated electron formation by $H_2 + OH^-$ in H_2O and NH_3 is not favored. The Jolly reaction can be split up as follows:

$$1/2\, H_{2\,gas} \rightarrow H_{gas}\ ;\tag{1}$$

$$H_{gas} \rightarrow H_{gas}^+ + e_{gas}^-\ ;\tag{2}$$

$$H_{gas}^+ \xrightarrow{NH_3} NH_{4\,solu}^+\ ;\tag{3}$$

$$NH_{4\,solu}^+ + NH_{2\,solu}^- \rightarrow 2\,NH_3\ ;\tag{4}$$

$$e_{gas}^- \xrightarrow{NH_3} e_{solu}^-\ .\tag{5}$$

In the case of the $H_2 + OH^-$ reaction, step 4 with equilibrium constant K around 10^{+30} has to be replaced by $NH_{4\,solu}^+ + OH^- \rightarrow NH_3 + H_2O$ with $K = 10^{12}$, everything else being the same. Then for K for $1/2\, H_2 + OH^- \rightarrow e_{solu}^- + H_2O$ has to be smaller by 18 orders of magnitude than K for $1/2\, H_2 + NH_2^- \rightarrow e_{solu}^- + NH_3$. Equilibrium concentration therefore should be far below detectibility.

Note Added in Proof: Mrs. Nagendrappa in my laboratory has reinvestigated the equilibrium $OH^- + 1/2\,H_2 \rightarrow e^- + H_2O$ in ammonia under the following conditions: ammonia in equilibrium with solid potassium hydroxide, equilibrated with 50 atm hydrogen, temperature range between 20° and 100° C. With an optical path length of 2 cm, no absorption could be detected which can be ascribed to solvated electrons. Even considering the low solubility of potassium hydroxide in amonia, we conclude from this that the equilibrium constant of said reaction is smaller than that of the corresponding reaction with potassium amide.

Metal Solutions in Amines and Ethers

J. L. DYE

Abstract

This paper reviews the most recent studies of metal solutions in amines and ethers. The metal solubility can be greatly enhanced by the addition of "crown" and "cryptate" to the solution. Studies are also presented which provide evidence for the existence of the monomer (M) and alkali anion (M$^-$) species. Finally a general equilibrium scheme for all species is presented and the influence of the solvent is discussed.

Introduction

In any description of metal solutions we invariably return to the ammonia system for guidance. Perhaps this is because we "understand" metal-ammonia solutions best. Certainly it is here that we have the most data about physical and chemical properties. Yet, behind the apparent simplicity of these solutions lies a set of complex interactions which influence the solution properties [1]. The absence of specific optical bands or ESR patterns for the various "species" makes the characterization of these solutions difficult and has led to the development of a number of models [2—8]. Most investigators agree that at infinite dilution the species are e_{solv}^- and M^+ and their interaction at higher concentrations can be described by the formation of a species of stoichiometry M, probably best characterized as an ion-pair, $M^+ \cdot e_{solv}^-$. The nature of the interaction which leads to spin-pairing is far less clear, however, and the species $e_2^=$, M^- and M_2 have been invoked to describe this phenomenon. Particularly puzzling is the fact that so many properties do not seem to be affected by the spin-pairing process [1]. In addition, the cation generally plays a secondary role. For example, the shape of the optical spectrum is unaffected by the counter-ion [9—11]. Although the optical spectrum has been decomposed into two bands [12] on the basis of the concentration dependence of the spectrum, the presence of the underlying bands is by no means obvious. Of course, the simplicity of the ESR spectrum could be attributed to rapid electron-exchange phenomena. Knight shifts [13—15] do indicate contact of the electron at the metal nucleus but once again, the electron exchange is very rapid.

We must conclude that, in spite of the intensity of effort, metal-ammonia solutions have not yielded enough specific information about the nature of the species which are present. In contrast, metal solutions in amines and ethers are rich in information about distinguishable species. Admittedly, some of these individual "species" may be as complex as metal-ammonia solutions, but we can detect by optical and ESR spectra at least three distinct reducing "species". By studying the behavior of these species as a function of temperature, solvent and metal, it has become possible to identify their stoichiometry and to measure some of their properties. Some of the most obvious differences between solutions in these solvents and in ammonia are the drastic reduction in solubility in the former, the appearance of metal-dependent optical absorption bands, and the identification by ESR of a monomer, M. These differences suggest that new species are present that are not important in ammonia. Alternatively, of course, the same species might be present in ammonia but obscured by band-overlap [12, 16] and exchange processes. Even if this is the case, the availability of specific spectroscopic information makes the study of metal-amine and metal-ether solutions perhaps even more profitable than the study of metal-ammonia solutions.

In order to be useful in the study of metal solutions a potential solvent should be:

1. kinetically inert to reduction,
2. reasonably effective as a cation-solvating agent,
3. readily purified.

To date the only non-amine or ether which qualifies is hexamethyl phosphoric triamide (HMPA) [17—19]. The increase in cation solvation which is provided by the use of certain cyclic polyethers has considerably extended the range of solvents which can be used [20—22]. Included for the first time are aliphatic mono-ethers and secondary amines. Solubilities in other solvents such as the primary amines [23—25], polyethers [26—28] and THF [29] are greatly enhanced by the use of "crown" (I) [20—22] or "cryptate" (II) [21, 22].

The optical spectra of metal solutions and the species responsible will be considered in detail later. It should be noted, however, that general confusion existed prior to the demonstration by Hurley, Tuttle, and Golden [30] that potassium in solutions can exchange with sodium from borosilicate glass, an observation which has been confirmed in our laboratory [31]. After this source of confusion had been removed, it seemed clear that for a given metal, at most two optical bands needed to be considered. Lately, this simple picture has been complicated by reports of additional bands [32—34]. However, their existence has not been confirmed nor their identity established.*

The discovery of metal hyperfine splitting in metal-amine solutions [35—40] gave the first unambiguous identification of stoichiometry. The absorptions could only

---

* *Note added in proof:* Since the time of this presentation, these bands have been observed as transients by several investigators and the originat assignment to species of stoichiometry M has been verified.

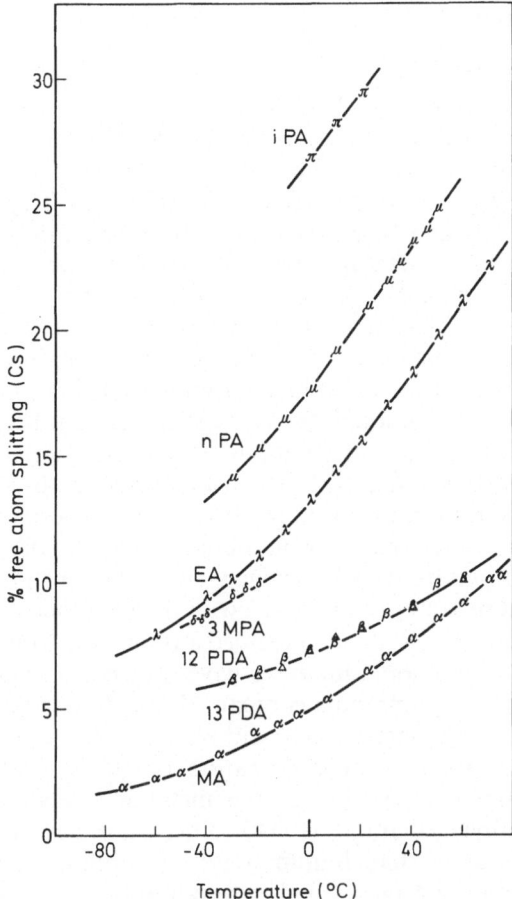

Fig. 1. Variation of the hyperfine splitting of the Cs monomer with solvent and temperature [41]. Solvent identification: iPA = isopropylamine; nPA = n-propylamine; EA = ethylamine; 3 MPA = 3-methoxy propylamine; 1,2 PDA and 1,3 PDA = 1,2- and 1,3-propanediamine; MA = methylamine

come from a monomeric species, M. However, elucidation of the nature of this species has not been so simple. The marked increase in hyperfine splitting with an increase in temperature and with a decrease in solvent polarity (see Fig. 1 for an example taken from work in our laboratory [41]) suggests that the monomer becomes "atom-like" under certain conditions [36]. However it is not possible at this time [40] to distinguish between a rapid equilibrium among distinct species [36] and a single species whose electron distribution changes drastically with solvent and temperature [37]. One fact is clear, however. Under all conditions studied so far, the concentration of the monomer is much less than that of the optically absorbing species — the monomer is a minor, but easily detectible constituent of these solutions.

The Solvated Electron

Preliminary Identification

The existence of a blue color is not diagnostic for e_{solv}^-. Nevertheless, it was natural to conclude that the broad, intense absorption bands in metal solutions in amines and in ethers were absorptions of the solvated electron. The presence of two bands in the same solution prompted Blades and Hodgins [24] to postulate the presence of "amine traps" and "aliphatic traps". Once the bands had been divided into metal-dependent and metal-independent bands it was easy to assign the metal-independent IR band to the solvated electron and the other bands to metal-containing species. The correlation between the presence of a narrow ESR singlet and the observation of an IR band further strengthened this assignment [42]. The intensity of the infrared absorption and that of the ESR singlet can be correlated [36, 41]. Although the correlation has not been quantitatively studied, it should be noted that for solutions of K in ethylamine-ammonia mixtures the spin concentrations are typically of the order of 10^{-6} M for solutions which have a total concentration of infrared absorbing species of the order of 10^{-4} M. Just as in ammonia, then, it would appear that the infrared absorption must include contributions from both paramagnetic and diamagnetic constituents. It was found [43] that mixtures of ammonia with diethyl-ether gave rise to a new optical band, but that the IR band remained prominent. More recently [36], studies with ethylamine-ammonia mixtures have shown a prominent increase in the IR band with increasing ammonia content. Although the relative intensity of this band depends upon the metal used, neither the position of this absorption, nor its shape depend significantly upon the cation. Finally, it is interesting to note the rarity of this band compared with the metal-dependent bands. Only in ammonia, HMPA, ethylenediamine (EDA), other diamines, methylamine and ethylamine do metals other than lithium yield an appreciable IR band in the absence of cation-complexing agents. This generalization is reinforced by the fact that solvents which show little or no IR band have only a weak ESR singlet.

Conductance

Although a number of studies of the conductivity of metal–ammonia solutions have been made [44, 45], conductance studies of other metal solutions have been limited to methylamine [46, 47] and ethylendiamine [25]. However, the results provide not only further verification that the species responsible for the IR band is e_{solv}^- (and loosely-bound aggregates?) but also that the conduction mechanism is viscosity-dependent. Dilute cesium solutions in ethylenediamine show only the IR band and yield a limiting conductance [25], which obeys Walden's rule when compared with metal-ammonia solutions. The values of $\Lambda_0 \eta$ are 3.1 and 2.9 in EDA and NH_3, respectively. In Cs–$MeNH_2$ solutions, which also show the metal-dependent band, the value of $\Lambda_0 \eta$ is 1.3 [47]. Similarly, solutions of Rb in EDA which show both bands give $\Lambda_0 \eta = 1.8$. In the region of concentration which has been studied, the apparent conductance is decreased by the presence of the lower-conducting metal-dependent species.

The variation of conductance with concentration can be used to determine approximate values for the association constants for the reaction

$$Cs^+ + e^- \rightleftharpoons Cs^+ \cdot e^- \tag{1}$$

ranging from 7×10^3 in $MeNH_2$ and EDA [47] to 2×10^2 in NH_3 [45]. These values are similar to the association constants of normal salts in these solvents, which lends credence to the postulate that ion pairs are formed. For example, the observed value of K_a in EDA would be predicted by the Fuoss expression [48]

$$K_a = \frac{4\pi a^3 N}{3000} \exp\left[\frac{e^2}{aDkT}\right] \tag{2}$$

for a closest-approach distance of 4.2 Å. The calculation of valid association constants from conductance might better be done by using the Fuoss-Onsager approach [49] as modified, for example, by Justice [50]. However, the results quoted here indicate that the association of a cation and a solvated electron can be viewed as a "normal" ion-pairing process with association constants not much different from those of salts in the same solvents.

Optical Spectrum

Of the arguments used to assign the IR absorption to the solvated electron, perhaps the strongest is that derived from a comparison of the spectra of metal solutions with those obtained by pulse radiolysis. The band shapes and peak positions for both ammonia [51, 52] and ethylenediamine [52, 53] are independent of the method used, metal solution or pulse radiolysis. This leaves little doubt that the same species is formed by either method. It should be noted, however, that the spectra obtained by pulse radiolysis were *not* completely free of alkali cations since it was necessary to make the solution basic ($\approx 10^{-4}$ M) in order to prevent rapid reaction of e_{solv}^- with the oxidizing species produced by the pulse.

Although the spectrum of e_{solv}^- in a number of solvents has been studied by pulse radiolysis [54], it has not been possible until recently to compare the spectra with those produced by dissolving metals. The reason for this is that most of the solvents which dissolve alkali metals without reaction show only the metal-dependent absorption. With the discovery [20—22] that the polycyclic ethers, "crown" [55] (I) and "cryptate" [56] (II), can be used to dissolve metals in a number of solvents and that an excess of II (and in some cases also I) gives rise to the metal-independent IR band, it became possible to extend the comparison of the spectrum of e_{solv}^- to a number of solvents [22]. The comparison of the position of the peak for a number of solvents is shown in Fig. 2. Clearly, the agreement indicates that the IR band can indeed be attributed to e_{solv}^-. However, as will be emphasized later, the shape of the band may be concentration-dependent.

Common to all solvents is the characteristic asymmetry of the band of e_{solv}^- with its high-energy "tail". This has been atttributed [57] to the onset of the bound-to-continuum transition. Also characteristic is the large temperature dependence of the peak position. The temperature coefficient varies from one solvent to another. For example, $-d\nu_{max}/dT$ for e_{solv}^- varies from 12 cm^{-1} deg^{-1} for diethyl

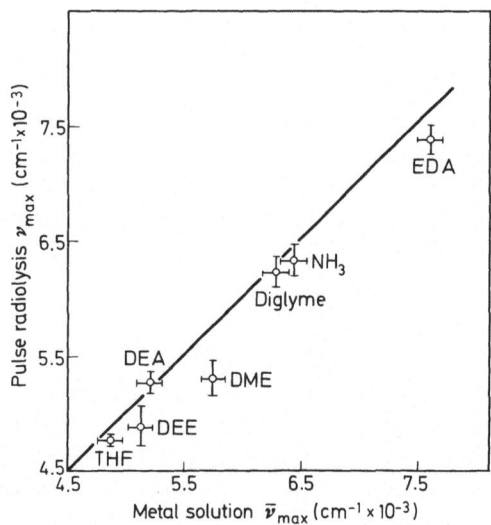

Fig. 2. The relation between the peak position of e_{solv}^- obtained by dissolving metals and by pulse radiolysis [54]. The straight line has unit slope. Comparison in the case of diglyme is with flash photolysis [32] rather than pulse radiolysis

ether to 36 cm^{-1} deg^{-1} for 1, 2 propanediamine [22]. The most surprising feature of these spectra is the sensitivity of the band-width to solvent and to temperature. The variation of the shape and position of the IR band with solvent is shown in Fig. 3 while Fig. 4 shows an example of the sensitivity of the shape to temperature. Based upon comparison with the results of pulse radiolysis shown in Fig. 5 we feel that the variation in shape and band width is probably dependent upon concentration, either because of the formation of spin-paired species or of ion-paired species. The new bands observed in the flash photolysis [32] of metal-

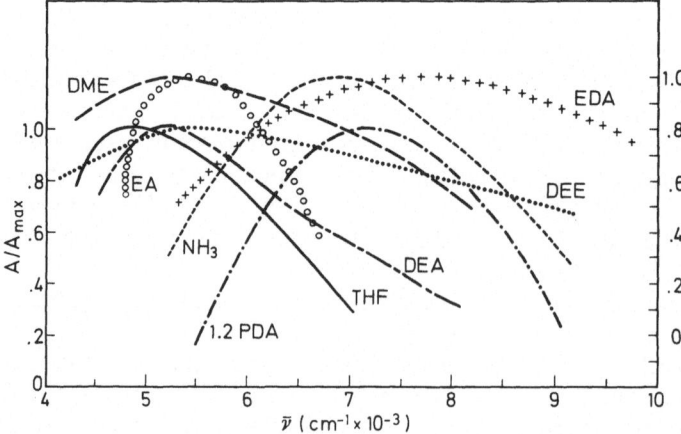

Fig. 3. The spectra of the solvated electron at 25° C in various solvents. Note the vertical displacement of some of the curves

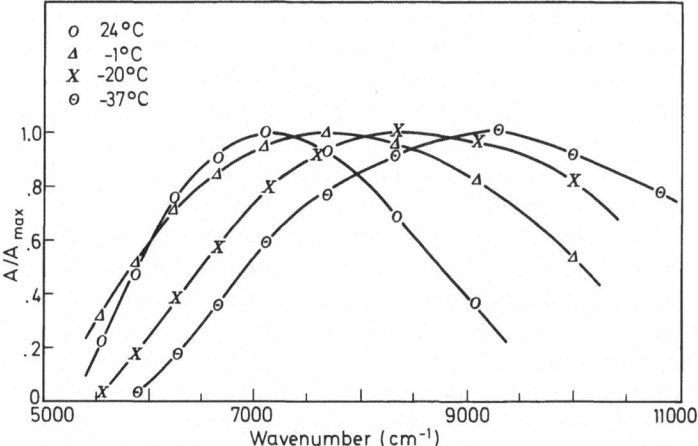

Fig. 4. The spectrum of the solvated electron in 1,2 propanediamine at various temperatures

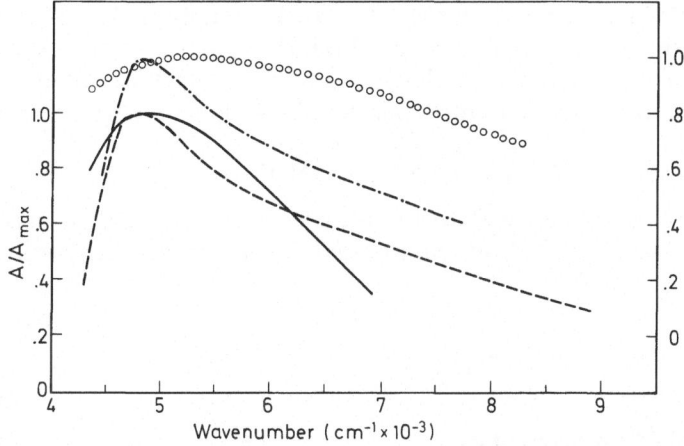

Fig. 5. Comparison of the band shape of e^-_{solv} at 25° C in DEE and THF obtained by dissolving metals and by pulse radiolysis. ······ K in DEE with cryptate; —·—·—·— pulse radiolysis of DEE; ——— K in THF with cryptate; ———— pulse radiolysis of THF

ether solutions and attributed to cation-electron pairs may be indicative of this behavior. Probably the best way to resolve this question would be to study the spectra obtained by pulse radiolysis of solutions which contain salts of the alkali metals as a function of salt concentration. The spectrum immediately after the pulse could be readily separated from the metal-dependent band which might grow in at longer times [58].

Spin-Pairing

A perplexing problem in both metal-ammonia and metal-amine solutions is the extent to which solvated electrons form paired spins. As indicated earlier, the intensity of the ESR singlet seems to be too low compared with the absorbance of the IR band to permit a one-to-one relationship between isolated solvated electrons (or cation-electron ion pairs) and the IR absorption. A similar situation obtains in ammonia [1] and has led to the picture of "loosely-bound" electrons and "ion-aggregates" [7, 9]. It would be most desirable to have quantitative susceptibility and absorbance data for the metal-amine and metal-ether systems in order to confirm the supposition that these systems also have such aggregates and that these are responsible for the IR absorption.

The relaxation time for an electron in an ion pair can become long enough to give a separate ESR signal provided the dielectric constant is low enough [59]. Indeed, the "monomer" signals observed so commonly in these solutions could be due to ion pairs between the solvated electron and the cation with varying contributions to the observed signal from partial occupancy of s-orbitals on the cation. One case in which the nature of the ion pair is better understood is that of lithium in ethylamine [37, 60—62] in which splitting by four equivalent nitrogen nuclei occurs but not by the lithium nucleus. Even the addition of diglyme [62] does not eliminate the nitrogen splitting. The interpretation of these results is based upon the slow exchange of four amine molecules in the first coordination sphere of Li^+. The ion pair with the solvated electron gives the observed pattern because of rapid rotation of the solvated lithium ion in the vicinity of the electron. The presence of a central narrow line in addition to the hyperfine pattern shows that at sufficiently low concentrations the relaxation time for the dissociation equilibrium

$$Li(NHR)_4^+ \cdot e_{solv}^- \rightleftarrows Li(NHR)_4^+ + e_{solv}^- \tag{3}$$

is long ($\gtrsim 10^{-7}$ sec) compared to the reciprocal of the hyperfine splitting. The narrow ESR singlet of e_{solv}^- also requires that direct spin-pairing processes such as

$$2\, e_{solv}^- \rightarrow e_2^= \tag{4}$$

be relatively slow ($\gtrsim 10^{-6}$ sec) [1].

Thermodynamic Considerations

Several qualitative trends in the concentrations of various species are evident for metal solutions. In a given solvent the solubility as a function of metal follows the sequence

$$Li > Cs > Rb > K > Na.$$

When the monomer can be detected by its ESR spectrum, the concentration for a given total metal concentration follows the order

$$Cs > Rb > K.$$

The Na monomer is rarely seen [61] and a monomer with hyperfine splitting by the lithium nucleus has not been observed. When both the metal-dependent absorption band* (of M^-) and the IR band (of e_{solv}^-) are present, the ratio of

* Although it will be demonstrated later, that the species responsible for the metal-dependent absorption band has the stoichiometry M^-, we will, for simplicity, refer to this species as M^- in what follows.

intensities of the band of M^- to that of e_{solv}^- follows the order

$$Na \gg K \approx Rb > Cs\,.$$

The band of Li^- has not been observed. Both the solubility and the ratio $(e_{solv}^-)/(M^-)$ decrease as the solvent polarity decreases.

Some of the trends mentioned above can be rationalized by thermodynamic arguments. Unfortunately, there are few quantitative data for comparison. The solubilities and conductances of alkali metals in ethylenediamine have been measured [25] as have optical spectra [63] and extinction coefficients [31]. The absorption spectrum of a near-saturated solution of potassium in ethylenediamine [63] indicates that about 25% of the metal is present as e_{solv}^- and $K^+ \cdot e_{solv}^-$ (plus perhaps loosely-bound aggregates). The remaining 75% is present as K^- and $K^+ \cdot K^-$. These estimates, together with the association constant and the extended Debye-Hückel equation, give $\Delta G^0 = 9.1$ kcal for the process:

$$K(s) \to K^+ + e_{solv}^- \text{ (hyp., ideal, c = 1 M)} \tag{5}$$

compared with a value of -2.1 kcal for the same process in ammonia [64]. Estimates of the limiting conductances of e_{solv}^-, K^+, and K^- in EDA lead to a predicted specific conductance at saturation of 0.20 ohm^{-1} cm^{-1} which compares favorably with the measured value of 0.27 ohm^{-1} cm^{-1}.

If we assume that the differences in free energy of solvation of the cations in EDA are the same as in NH_3 [65], so that $\Delta G_s^0 = 4.1$, 12.5, 8.6, and 8.1 kcal mole^{-1} for Li, Na, Rb, and Cs respectively, then the concentration product $(M^+)(e_{solv}^-)$ in the saturated solution can be estimated. Using the ion-pair association constant from conductance, it is then possible to estimate the contribution of e_{solv}^- and $M^+ \cdot e_{solv}^-$ to the total solubility as a function of the metal. The predicted values compared with the total solubilities are given in Table I.

Table I. Predicted contribution of e_{solv}^- and $M^+ \cdot e_{solv}^-$ to saturated solutions of metals in ethylenediamine

Metal	Predicted contribution (M.)	Total solubility [25] (M.)
Li	> 1 [a]	0.29
Na	0.00003	0.0024
K	(used as standard)	0.010
Rb	0.006	0.013
Cs	0.012	0.054

[a] Predicted concentration of e_{solv}^- is 0.03 M. This is too high to permit a valid calculation of the concentration of $Li^+ \cdot e^-$.

Clearly, the differences in free energy of solvation of the cation can be used to qualitatively interpret the solubility differences among the metals. One might be able to use electron affinities of the gaseous atoms and calculated solvation energies to make similar predictions for M^-, which becomes a major contributor in saturated solutions of Na, K, Rb, and Cs.

The observed variations in the concentration of monomers from one metal to another in such solvents as ethylamine can also be explained on thermodynamic grounds if one considers that ΔG^0 for the process:

$$M(g) \rightarrow M(solv) \tag{6}$$

is either small or else is nearly independent of metal. For example, if we *arbitrarily* choose $[K] = 6 \times 10^{-5}$ M in a saturated solution, then values of 10^{-15}, 10^{-7}, 5×10^{-4}, and 3×10^{-3} M would be predicted for Li, Na, Rb and Cs in their respective saturated solutions based upon the known free energies of atomization. Once again, the observed trends are explained. Of course, differences in solvation energies of the monomers would also affect these concentrations.

Undoubtedly, the decrease in solvation energy of the cation with decreased solvent polarity is largely responsible for the drastic reduction in solubility of the metals. The role of "crown" and "cryptate" in increasing the solubility lies primarily in the ability of these chelating agents to replace the first solvation layer. Since a large fraction of the total solvation energy arises from first-layer solvation, the replacement of relatively nonpolar molecules in the first-layer by the strongly chelating molecule greatly increases the total "solvation" energy and therefore enhances the solubility.

Alkali Anions

Historical Perspective

Having learned as freshmen that the alkali metals have valence states of 0 and $+1$, most chemists are skeptical when species such as Cs^- are mentioned. In fact, some cannot quite accept the existence of the solvated electron! Although the evidence for M^- in metal solutions has increased substantially in the last few years it has generated very little excitement among chemists in general compared, say, to the discovery of rare gas compounds. There are perhaps several reasons for this:

1. No salts containing the alkali anion have yet been isolated.
2. The history of confusion in our field over the identification of species has not inspired great confidence in the identification of new species.
3. Species of stoichiometry M^- have been repeatedly proposed for metal–ammonia solutions [3, 4, 6, 8] but their properties led to the suggestion that ion-clusters such as $e^- \cdot M^+ \cdot e^-$ were responsible for the observed behavior. Such clusters are not "genuine" anions and therefore elicit little enthusiasm.

If, indeed, identifiable alkali anions exist in metal solutions rather than just agglomerates of "old" species, then there is cause for excitement. The isolation of salts becomes a real possibility and the availability of new reducing agents varying from Na^- to Cs^- may be of utility in synthetic work. We have demonstrated [66] that solutions at least as concentated as 0.2 M in potassium in ethylamine can be prepared with the aid of "crown" (I). The optical characteristics indicate that the use of excess metal yields mainly the complexed cation and M^-. Concentrations as high as this might be of utility in synthesis. Homogeneous reductions by alkali metals in both ammonia and in amines have been used for some time. Perhaps the differences in reaction products and pathways can be interpreted in terms of reduction by either e^-_{solv} or by M^-.

Evidence for Anions

Optical Spectra. The most striking characteristic of metal–amine and metal–ether solutions which sets them apart from metal–ammonia solutions is the metal-dependent absorption band which occurs at higher energies than the IR band of e_{solv}^- and which shifts progressively to higher energies for solutions of Cs, Rb, K, Na in a given solvent. Figure 6 shows the absorption bands of e_{solv}^-, Cs^-, K^- and Na^- in THF at 25° C in the presence of either crown or cryptate. It seems likely that such large shifts in energy with cation result from more than just a weak cation-electron interaction. The dependence of the peak position on metal, solvent and temperature prompted Matalon, Golden, and Ottolenghi [16] to suggest that the species were indeed alkali anions, and that the transitions responsible for the absorption band were similar to the charge-transfer-to-solvent (CTTS) bands of I^- [67—69]. Some characteristics of a CTTS transition are:

1. pronounced dependence of the position of the absorption maximum upon solvent;
2. shift of the maximum to lower energies with an increase in temperature, with temperature coefficients of the order of $10—100\ cm^{-1}\ deg^{-1}$;
3. correlation between the shift of the peak position with solvent and the temperature coefficient. In fact, one model [67] predicts that the slope of a plot of $\bar{v}_{max}$ vs $d\bar{v}_{max}/dT$ will be equal to the absolute temperature;
4. correlation between the position of the absorption maximum and the size of the anion, with a shift to lower energies for larger ions. For a given solvent, it is predicted that the transition energy minus the electron affinity of the atom will be a linear function of $1/r$.

All of these characteristics have been observed [16] for the bands of M^-. In an extensive study of the solvent and temperature dependence, Lok, Tehan, and Dye [22] have shown excellent correlations between the peak positions of Na^- and K^- (Fig. 7). In addition the variation of the temperature coefficients with solvent is in accord with the prediction of the Smith-Symons model [67].

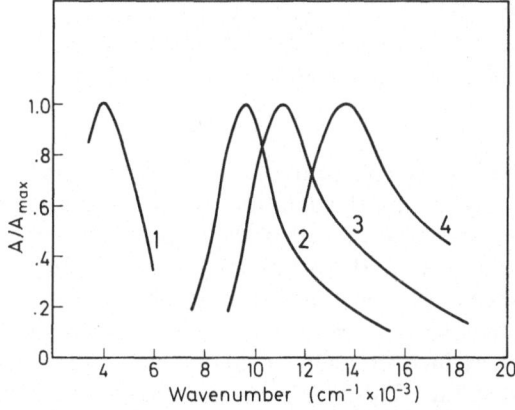

Fig. 6. Spectra at 25° C of e_{solv}^- (1), Cs^- (2), K^- (3) and Na^- (4) in THF in the presence of *I* or *II*

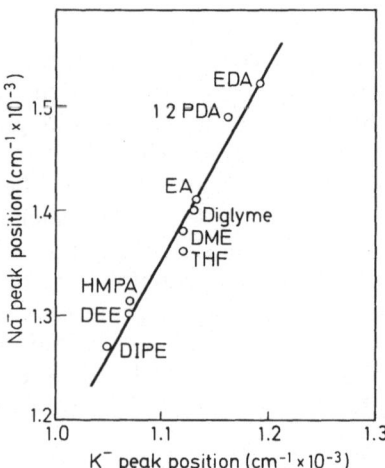

Fig. 7. The relation between the peak position of Na⁻ and of K⁻ in various solvents at 25° C

In contrast to the absorption band of e_{solv}^- (and aggregates?) in amines and ethers, which broadens with a decrease in temperature, and that of e_{solv}^- in NH_3 which shifts but is not broadened with a change in temperature, the absorption bands of K⁻ and Na⁻ *narrow* substantially with a decrease in temperature [22]. For example, the width at half-hight of the K⁻ band in ethylamine decreases from 3320 cm⁻¹ at 39° C to 2480 cm⁻¹ at − 24° C. Similarly, a decrease from 4150—3190 cm⁻¹ occurs between − 9° and − 71° C for the Na⁻ band in diethyl ether. Whether this is caused by effect of ion-pairing or by changes in solvent properties with temperature cannot be determined from the available data.

Although the comparison of spectra with the predictions of CTTS theory strongly suggests that alkali anions are responsible, it does not prove the stoichiometry M⁻. (After all, the optical properties of e_{solv}^- also meet most of these criteria.) However, the determination of the oscillator strength of Na⁻ in ethylenediamine by DeBacker and Dye [31] and its relation to that of e_{solv}^- in this solvent [58] leave little doubt about the stoichiometry. It had been established [25, 52, 63] that dilute solutions of Cs in EDA yield largely e_{solv}^- and that the reaction

$$2\,e_{solv}^- + Na^+ \rightarrow Na^- \tag{7}$$

lies far to the left whether the source of e_{solv}^- is a dilute cesium solution [31] or pulse radiolysis [52]. By mixing the solution of a sodium salt (NaBr or NaI) in EDA with a solution which contained a slight excess of cesium in EDA in a flow system equipped with a rapid-scan monochromator [70], it was possible to measure the absorbance of Na⁻ as a function of the concentration of added salt (after subtraction of the band e_{solv}^- remaining). The result obeyed the Beer-Lambert law and yielded an extinction coefficient at the maximum of $8.2 \pm 0.3 \times 10^4\,M^{-1}$ cm⁻¹ per mole of added Na⁺. This yields an oscillator strength of 1.9 ± 0.2 based upon one sodium nucleus per absorbing unit. If two sodium nuclei were present, this would become 3.8 ± 0.4 per absorbing unit. According to the f-sum rule [71, 72] we would expect that an oscillator strength of about 2 would require that at least two equivalent electrons be involved in the transition, which is satisfied by the species Na⁻. It is of interest to note, for comparison, that the oscillator strength of the $3^1P \leftarrow 3^1S$ transition of the gaseous magnesium atom is 1.745 [73]. It should be emphasized at this point that, because of the requirement of charge-balance, studies on pure sodium solutions would be expected to yield an extinction coefficient and oscillator strength per sodium nucleus which is only 1/2 that found by the addition of cesium solutions to a sodium salt.

The use of the extinction coefficient of Na⁻ permitted us [58] to determine that e_{solv}^- in EDA has an extinction coefficient of $2.0 \pm 0.3 \times 10^4$ and an oscillator strength of 0.88 ± 0.12. The oscillator strength agrees well with those of the hydrated and ammoniated electrons.

The determination of the extinction coefficient of e_{solv}^- in EDA was made in connection with a rate study by pulse radiolysis [58] of reaction (7). It was shown that the reaction was second-order in (e_{solv}^-) and depended upon (Na^+) only below about 10^{-2} M. The dependence upon (Na^+), (K^+) and (Cs^+) led us [58] to the conclusion that electron-pairing preceded the incorporation of Na⁺ into the final Na⁻ moiety. Since the decay of e_{solv}^- and the growth of Na⁻ were followed simultaneously, and showed the same rate, there were apparently no slow intermediate steps.

These results may be compared with earlier photochemical studies [32, 74—76] in which e_{solv}^- was produced by photolysis of M⁻ or its decomposition products. The data from the photolysis of metal solutions in diglyme, DME and tetraglyme [32] and of pyrenide radical anion in tetrahydrofuran [34] indicate the inter-mediate formation of species of stoichiometry M.

Glarum and Marshall [76] followed the ESR signal of e_{solv}^- in DME either under conditions of steady illumination or after a light flash. They found that the spectral response curves closely followed the absorption spectrum of M⁻ for K⁻, Rb⁻ and Cs⁻, thus showing that optical excitation of these species can lead to photodissociation. Observation of ESR *emission* in some cases showed that photodissociation can polarize the electron spins, presumably *via* an intermediate triplet state.

Conductance of M⁻. As shown by Dewald and Dye [25] for solutions of sodium in ethylenediamine and by Dewald and Browall [47] for methylamine solutions, the conductance behavior of Na⁻ is very different from that of e_{solv}^- (and its aggregates). The variation of the equivalent conductance with concentration for solutions of sodium in methylamine gives $\Lambda^0 = 158$ based upon the species Na⁺ and Na⁻ which gives $\lambda_{Na^-}^0 = 94$. (Walden's rule and the known conductance of Na⁺ in NH₃ were used.) Thus, the anion has a higher mobility than the cation, a common occurrence because of lesser solvation of anions. The higher viscosity of ethylenediamine reduces the limiting conductances to 22 and 32 for Na⁺ and Na⁻ respectively.

All of the conductance data are compatible with the presence of a "normal" cation and anion whose mobilities are similar to those of other ions. Although the association to ion pairs is relatively small, this could be caused by the large size of Na⁻ and/or the resistance to contact-pair formation. It would be interesting to determine whether anomalous conduction occurs at still higher concentrations *via* electron-pair "hopping".

Kinetics. The kinetics of formation of M^- has been discussed previously in connection with identification of the optical bands. These studies [58] indicated that the rate of dissociation of Na^- in EDA might be low ($\gtrsim 10^3 sec^{-1}$) so that the kinetics of reaction of M^- in appropriate solvents might not be limited by the reactivity of e_{solv}^- which is formed by dissociation. Indeed, the reactivity of Na^- towards water in EDA is much less than that of e_{solv}^- [77—79]. The rate limiting step for this reaction could well be the dissociation of Na^-, and the high order in (H_2O) [79] could be a solvent effect upon this dissociation. Another difference between the reactivity of e_{solv}^- and Na^- has been observed [79]. While the former adds rapidly to anthracene to form the anion radical, Np^-, the latter does not, but seems instead to form an intermediate charge transfer complex. Therefore, in its reactivity as well as its optical spectrum and conductivity, the species M^- has distinct characteristics which are better described in terms of a "true" anion than as a loose aggregate of two electrons and a cation.

Summary

A general scheme for the equilibria present in metal solutions is the following [21]:

$$2 M(s) \rightleftharpoons M^- + M^+ \tag{8}$$
$$\hookrightarrow M + e_{solv}^- \tag{9}$$
$$\hookrightarrow e_{solv}^- + M^+ . \tag{10}$$

Strongly solvating media and high dielectric constants tend to shift these equilibria to the right to give e_{solv}^- and M^+ with the species M^- and M present only in small concentrations if at all. Ammonia is the classic case of such behavior. Although the dielectric constant of HMPA is higher than that of NH_3, its lower solvating power for cations leads to the presence of some M^- [19]. Solvents such as EDA and methylamine give mixtures of M^-, M and e_{solv}^-, while the less polar amines and the ethers give largely M^- with only minor amounts of M and e_{solv}^-.

The cyclic polyethers complex the cations and tend to shift equilibria (8) and (10) to the right. This permits solubilization in such solvents as diethyl ether. The more powerful complexing agent, cryptate, (*II*), when present in amounts sufficient to complex all of the cations in solution can shift the equilibria in favor of e_{solv}^-. However, contact with metal causes more to dissolve, with the disappearance of e_{solv}^- and the appearance of M^-. Therefore, by controlling the ratio of complexing agent to total dissolved metal, it is possible to prepare solutions which contain largely M^+ (complexed) and either M^- ot e_{solv}^-. in "poor" solvents such as diethyl ether and THF we have not been able to convert M^- completely to e_{solv}^-. Interestingly, in the presence of either crown or cryptate the *ESR* pattern of the monomer, M, disappears [21].

Acknowledgements

The preparation of this review and a portion of the travel to this conference was supported by the U.S. Atomic Energy Commission under Contract No. AT (11–1) –958. I am grateful to past and present members of my research group who have performed much of the research described. Their names appear in the references. Special thanks go to Drs. V. A. Nicely and M. DeBacker for introducing the cation-complexing technique and to F. J. Tehan and M. T. Lok for their diligent use of these methods in the study of spectra. The samples of cryptate used in these studies were generously provided by Dr. C. H. Park of E. I. du Pont de Nemours Co. and by Prof. J. M. Lehn of the Institut de Chimie, Strasbourg. A grant of travel funds from the Dow Chemical Company which permitted me to attend this Conference is gratefully acknowledged.

References

1. Dye, J. L.: Pure Appl. Chem. **1970**, 1.
2. Ogg, R. A., Jr.: J. Chem. Phys. **14**, 295, 1141 (1946).
3. Bingel, W.: Ann. Physik **12**, 57 (1953).
4. Deigen, M. F.: Tr. Inst. Fiz. Akad. Nauk, Ukr. S.S.R. **5**, 119 (1954); — Zh. Eksp. Teor. Fiz. **26**, 300 (1954).
5. Becker, E., Lindquist, R. H., Alder, B. J.: J. Chem. Phys. **25**, 971 (1956).
6. Arnold, E., Patterson, A., Jr.: J. Chem. Phys. **41**, 3089, 3098 (1964).
7. Gold, M., Jolly, W. L., Pitzer, K. S.: J. Am. Chem. Soc. **84**, 2264 (1962).
8. Golden, S., Guttman, C., Tuttle, T. R., Jr.: J. Am. Chem. Soc. **87**, 135 (1965); — J. Chem. Phys. **44**, 3791 (1966).
9. Douthit, R. C., Dye, J. L.: J. Am. Chem. Soc. **82**, 4472 (1960).
10. Gold, M., Jolly, W. L.: Inorg. Chem. **1**, 818 (1962).
11. Burow, D. F., Lagowski, J. J.: Advan. Chem. Ser. No. 50, 125 (1965).
12. Tuttle, T. R., Jr., Rubinstein, G., Golden, S.: J. Phys. Chem. **75**, 3635 (1971).
13. McConnell, H. M., Holm, C. H.: J. Chem. Phys. **26**, 1517 (1957).
14. Acrivos, J. V., Pitzer, K. S.: J. Phys. Chem. **66**, 1693 (1962).
15. O'Reilly, D. E.: J. Chem. Phys. **41**, 3729 (1964).
16. Matalon, S., Golden, S., Ottolenghi, M.: J. Phys. Chem. **73**, 3098 (1969).
17. Cavigny, T., Normant, J., Normant, H.: Compt. Rend. Ser. C **258**, 3503 (1964).
18. Fraenkel, G., Ellis, S. H., Dix, D. T.: J. Am. Chem. Soc. **87**, 1406 (1965).
19. Brooks, J. M., Dewald, R. R.: J. Phys. Chem. **72**, 2650 (1968).
20. Dye, J. L., DeBacker, M. G., Nicely, V. A.: J. Am. Chem. Soc. **92**, 5226 (1970).
21. Dye, J. L., Lok, M. T., Tehan, F. J., Coolen, R. B., Papadakis, N., Ceraso, J. M., DeBacker, M.: Ber. Bunsen-Gesellsch. Phys. Chem. **75**, 659 (1971).
22. Lok, M. T., Tehan, F. J., Dye, J. L.: J. Phys. Chem. **76**, 2975 (1972).
23. Gibson, G. E., Phipps, T. E.: J. Am. Chem. Soc. **48**, 312 (1926).
24. Blades, H., Hodgins, J. W.: Can. J. Chem. **33**, 411 (1955).
25. Dewald, R. R., Dye, J. L.: J. Phys. Chem. **68**, 128 (1964).
26. Down, J. L., Lewis, J., Moore, B., Wilkinson, G.: J. Chem. Soc. **1959**, 3767.
27. Cafasso, F., Sundheim, B. R., J. Chem. Phys. **31**, 809 (1959).
28. Dainton, F. S., Wiles, D. M., Wright, A. N.: J. Chem. Soc. **1960**, 4283.
29. Catterall, R., Slater, J., Symons, M. C. R.: Pure Appl. Chem. **1970**, 329.
30. Hurley, I., Tuttle, T. R., Jr., Golden, S.: J. Chem. Phys. **48**, 2918 (1968).
31. DeBacker, M. G., Dye, J. L.: J. Phys. Chem. **75**, 3092 (1971).
32. Giling, L. J., Kloosterboer, J. G., Rettschnick, R. P. H., Van Voorst, J. D. W.: Chem. Phys. Letters **8**, 457 (1971); **8**, 462 (1971).
33. Gabor, G., Bar-Eli, K.: J. Phys. Chem. **75**, 286 (1971).
34. Fisher, M., Ramme, G., Claesson, S., Szwarc, M.: Chem. Phys. Letters **9**, 309 (1971).
35. Vos, K. D., Dye, J. L.: J. Chem. Phys. **38**, 2033 (1963).
36. Dalton, L. R., Rynbrandt, J. D., Hansen, E. M., Dye, J. L.: J. Chem. Phys. **44**, 3969 (1966).
37. Bar-Eli, K., Tuttle, T. R., Jr.: J. Chem. Phys. **40**, 2508 (1964).
38. Catterall, R., Hurley, I., Symons, M. C. R.: J. Chem. Soc. **1972**, 139.
39. Catterall, R., Slater, J., Symons, M. C. R.: J. Chem. Phys. **52**, 1003 (1970).

40. For other references to esr studies of metal solutions see Nicely, V. A., Dye, J. L.: J. Chem. Phys. **53**, 119 (1970).
41. Dalton, L. R.: M.S. Thesis, Michigan State University 1966.
42. Symons, M. C. R.: J. Chem. Phys. **13**, 99 (1959); – Quarterly Reviews, **30**, 1628 (1959).
43. Hohlstein, G., Wannagat, U.: Z. Anorg. Allg. Chem. **284**, 191 (1956); **288**, 193 (1956).
44. Kraus, C. A.: J. Am. Chem. Soc. **43**, 749 (1921).
45. Dewald, R. R., Roberts, J. H.: J. Phys. Chem. **72**. 4224 (1968); **73**, 2615 (1969). (See these papers for references to other conductance work in liquid ammonia).
46. Berns, D. S., Evers, E. C., Frank, P. W., Jr.: J. Am. Chem. Soc. **79**, 5118 (1957).
47. Dewald, R. R., Browall, K. W.: J. Phys. Chem. **74**, 129 (1970).
48. Fuoss, R. M.: J. Am. Chem. Soc. **80**, 5059 (1958).
49. Fuoss, R. M., Onsager, L.: Proc. Natl. Acad. Sci. U.S. **41**, 274, 1010 (1955); – J. Phys. Chem. **61**, 668 (1957).
50. Justice, J. C.: J. Chim. Phys. **65**, 353, 1708 (1968).
51. Compton, D. M. J., Bryant, J. F., Cesena, R. A., Gehman, B. L.: In: Pulse-Radiolysis, p. 43. New York: Academic Press 1965.
52. Dye, J. L., DeBacker, M. G., Dorfman, L. M.: J. Chem. Phys. **52**, 6251 (1970).
53. Dalton, L. R., Dye, J. L., Fielden, E. M., Hart, E. J.: J. Phys. Chem. **70**, 3358 (1966).
54. Dorfman, L. M., Jou, F. Y., Wageman, R.: Ber. Bunsen-Gesellsch. Phys. Chem. **75**, 681 (1971).
55. Pedersen, C. J.: J. Am. Chem. Soc. **89**, 7017 (1967); **92**, 386 (1970).
56. Dietrich, B., Lehn, J. M., Sauvage, J. P.: Tetrahedron Letters 2885 (1969); 2889 (1969); – J. Am. Chem. Soc. **92**, 2916 (1970).
57. Delahay, P.: J. Chem. Phys. **55**, 4188 (1971).
58. Dye, J. L., DeBacker, M. G., Eyre, J. A., Dorfman, L. M.: J. Phys. Chem. **76**, 839 (1972).
59. Dye, J. L.: Comment following paper by M. C. R. Symons, J. Phys. Chem. **71**, 172 (1967).
60. Catterall, R., Symons, M. C. R., Tipping, J. W.: Pure Appl. Chem. **1970**, 317.
61. Catterall, R., Symons, M. C. R., Tipping, J. W.: J. Chem. Soc. (London) A, **1967**, 1234.
62. Catterall, R., Hurley, I., Symons, M. C. R.: J. Chem. Soc., Dalton **1972**, 139.
63. Dewald, R. R., Dye, J. L.: J. Phys. Chem. **68**, 121 (1964).
64. Dye, J. L.: In: Lepoutre, G., Sienko, M. J. (Eds.): Metal-ammonia solutions, Colloque Weyl I, p. 137. New York: Benjamin 1964.
65. Jolly, W. L.: Chem. Rev. **50**, 351 (1952).
66. Tehan, F. J., Lok, M. T.: This laboratory, unpublished work.
67. Smith, M., Symons, M. C. R.: Trans. Faraday Soc. **54**, 338, 346 (1958).
68. Stein, G., Treinen, A.: Trans. Faraday Soc. **55**, 1086, 1091 (1959).
69. Burak, I., Treinen, A.: Trans. Faraday Soc. **59**, 1490 (1963).
70. Dye, J. L., Feldman, L. H.: Rev. Sci. Instr. **37**, 154 (1966).
71. Kuhn, W.: Z. Phys. **33**, 408 (1925).
72. Thomas, W.: Naturwissenschaften **13**, 627 (1925).
73. Bierman, L., Trefftz, E.: Z. Astrophys. **26**, 213 (1949).
74. Gaathon, A., Ottolenghi, M.: Israel J. Chem. **8**, 165 (1970).
75. Huppert, D., Bar-Eli, K. H.: J. Phys. Chem. **74**, 3285 (1970).
76. Glarum, S. H., Marshall, J. H.: J. Chem. Phys. **52**, 5555 (1970).
77. Feldman, L. H., Dewald, R. R., Dye, J. L.: Advan. Chem. Ser. No. 50 **1964**, 163.
78. Dye, J. L.: Accounts Chem. Res. **1**, 306 (1968).
79. DeBacker, M. G.: Ph. D. Thesis, Michigan State University (1970).

Discussion

L. M. DORFMAN   1. I should like to make a point about the spectra of e_{solv}^- which you determine in the ethers using solubilizing agents (crown or cryptate), and to seek some clarification from you regarding your reference to "the solvated electron" in these liquids. Now that we have completed a careful determination of the spectra of e_{solv}^- in various ethers, we have made comparisons with some of the spectra you have determined in crown or cryptate systems. We find significant differences, as I shall point out in my lecture on Friday. In our pulse radiolysis observations we are looking at the pure liquid ether, there being no alkali metal present. We find that your use of the cryptate shifts λ_{max} somewhat (as much as $1100\ cm^{-1}$ in one case), the maximum in the pure liquid being at the lower frequency. The shape is altered considerably, the spectra in cryptate solutions being somewhat broader ($W^{1/2}$ being $5500\ cm^{-1}$ compared with $3500\ cm^{-1}$ in one case). You would agree that the cryptate data and the data for e_{solv}^- in the pure liquid are not to be treated as a unified set of data?

The reason for the difference is hard to guess at, and may be a matter of interest. I shall make it clear on Friday that in some systems, the addition of a small amount of a second component can make a profound difference in the spectrum.

2. It may be of interest to point out that the bimolecular reaction of two cation-paired solvated electrons in ethylenediamine, e_{solv}^-, $M^+ + e_{solv}^-$, M^+, to which Dye has referred [Dye, DeBacker, Eyre, Dorfman: J. Phys. Chem. **76**, 839 (1972)] has an analogy and precedent in a similar reaction in water, in which the sodium ion does not play a similar role. The reaction:

$$e_{aq}^- + e_{aq}^- \xrightarrow{\ H_2O\ } H_2 + 2OH^-$$

was investigated over nine years ago, and its nature established by isotopic studies (Dorfman and Taub) and its rate constant determined.

J. L. DYE   1. I can think of three explanations for the differences between our results and those obtained for the isolated solvated electron by pulse radiolysis:

a) The poly-ethers might perturb the spectrum of e_{solv}^-. However, the concentrations are low ($\approx 10^{-3}\ M$) and I would expect no large effect of these ethers in ethereal solvents. Obviously, this can be checked by pulse radiolysis in the presence of crown or cryptate.

b) Formation of the ion pair or monomer, as suggested in the text of this paper. Once again, pulse radiolysis in the presence of salts could be used to test this hypothesis.

c) Formation of spin-paired species such as the ion clusters which have been proposed for the ammonia system.

L. M. DORFMAN   The question has been asked as to what generalization can be made about the type of liquid which will support electron solvation. While we tend to investigate liquids in which e_{solv}^- will be formed, there are many classes of compounds in which solvation will not occur. These are compounds containing an atom for functional group which will localize the electron in the form of an anion.

For example, we find that in halogenated compounds, e_{solv}^- is not formed, the electron probably being trapped as a halide ion. And it is probably not just a matter of a short lifetime for e_{solv}^-, because the dissociative attachment may occur before solvation would have taken place.

So it is a matter, among other things, of the magnitude of the electron affinity of atoms and groups in the molecule.

A. GAATHON   In a paper [Gaathon, A., Ottolenghi, M.: Isr. J. of Chem. **8**, 165 (1970)], two additional indications as to the nature of the species having metal-dependent absorption are given.

1. In steady photolysis, at $\lambda < 550$ nm of Na solution in mixture of propylamine and propylenediamine having signals due to both paramagnetic species, i.e., Na, e^-, increases of both signals are obtained, indicating that both species are formed from the "blue" species.

2. In comparing flash photolysis spectra obtained (photolysis at 350 mμ) with a $Na^+ RNH_2^-$ solution with spectra obtained from reversible photolysis of the "blue" line in the system Na in propylenediamine, there is a ratio of 2:1 between the optical densities of the IR species (e_{solv}^-) reacting to form the blue line. These results indicate that, as in the amide system, the reaction is

$$Na^+ + 2e^- \rightarrow Na^- \,.$$

The regeneration of the "blue" line is the reversible system $M + e^- \rightarrow M^-$.

J. L. DYE   These observations also support the assignment M^-. It should be noted that the results from the photolysis of RNH_2^- should be used with caution since photolysis can produce reactive species such as RNH^- which might compete with Na^+ for the solvated electron.

K. BAR-ELI   1. Historical comment: The credit for the first proof of M^- species should be given to Arnold and Patterson, and about the same time to Tuttle, Golden and Guttman.

2. Further proofs were obtained by us through laser photolysis (and salt effect on kinetics) of PDA and EDA solutions, and by the test for CTTS in mixture of solvents ($MeNH_2/EtNa_2$).

3. $Li/EtNH_2$ solutions were shown earlier to give ESR signals which conform with a species which contains 4 nitrogens around the atom.

J. L. DYE   1. I don't believe it is universally agreed that the existence of M^- in metal–ammonia solutions has been "proven". To my knowledge, the first suggestion of such a species was made by Bingel some ten years before the work of Arnold and Patterson and that of Tuttle, Golden and Guttman.

2. I agree that the work referred to, as well as the others referenced in the paper, all tend to support the assignment of the metal-dependent band to a species of stoichiometry M^-.

3. The ESR spectra of $Li/EtNH_2$ solutions are discussed in the paper but were not described in the oral presentation.

U. SCHINDEWOLF   1. If an M^- species really exists, the concentration of which should depend on concentration, transference number measurements should reveal it.

2. Conductance data of solvated electrons follow Walden's rule with respect to viscosity. Do you think, therefore, that solvated electrons follow a normal conduction mechanism, i.e., that they are dragged through the solution with the solvation shell? Walden's rule also implies that the mobility of a particle is in-

versely proportional to its radius. The huge solvated electrons in ammonia definitely do not follow Walden's rule in this respect.

3. Can cavity sizes be determined by density measurements in ethers and amines in which metal solubility can be increased up to about 1 molar by using crown or cryptate?

J. L. Dye 1. Transference number measurements would indeed be useful but would be experimentally difficult. The limiting conductance values indicate that $\lambda^0_{M^-}$ is slightly smaller than $\lambda^0_{Na^+}$ in ethylenediamine and in methylamine, if our interpretation of the species is correct.

2. Walden's rule implies only a viscosity-dependent migration of the species in question. Thus the proton in aqueous solution moves by an anomalous mechanism but still obeys Walden's rule fairly well in its temperature dependence. I might add that comparison of the conductivity of $K-NH_3$ and $K-ND_3$ solutions shows that the major difference in electron mobility arises from viscosity effects (Smith, G.: Ph. D. Thesis, Michigan State University, 1963).

3. Density measurements might, indeed, be feasible as suggested.

J. Jortner You have shown us several linear correlations for CTTS spectra, and I would like to inquire what is the meaning of these plots, and in particular whether you can assign a meaningful value to the slope. Kestner has shown us a linear empirical correlation between the maximum transition energy of the solvated electron and the static dielectric constant of the solvent, which is obviously misleading. Thus one has to exert some care in attaching too much significance to such linear correlations.

I would like to address myself to the theory of CTTS bands of anions in solutions. One can apply the same "molecular model" reviewed by Kestner for the solvated electron which includes both short-range and long-range interactions. The only modification required involves the introduction of a proper pseudopotential for the neutral atom.

J. L. Dye I share your reluctance to give a great deal of significance to comparison with CTTS theory. Our major aim here is to show a correlation with the spectral characteristics of other anions which show CTTS absorptions. It seems to me that the time is ripe for a re-examination of CTTS theory.

Extinction Coefficients of Solutions of Alkali Metals in Amines

S. Nehari and K. Bar-Eli

Abstract

The extinction coefficient of metal anion and solvated electron was measured in solutions of sodium and potassium in PDA and EDA. Measurements were carried out by a) measuring the absorbancies and concentration of metal and using the equilibrium $M^- \rightleftharpoons M^+ + 2e^-$; b) measuring the rate of increase of the bands with electrical current. Both methods give the result: $\varepsilon(M^-) = 42\,500 \ M^{-1} \ cm^{-1}$ $\varepsilon(e^-) = 2100 \ M^- \ cm^{-1}$. The equilibrium constant for the above reaction is calculated.

The extinction coefficients of the V and of the IR bands of solutions of sodium and potassium in ethylenediamine (EDA) and 1,2 propylenediamine (PDA) were measured. The V band was shown earlier to be associated with the metal anion. The proof is based on the following evidence: a) the linear dependence [1, 2] of $\tilde{v}_{max}$ on $\dfrac{d\tilde{v}}{dT}$ as expected for CTTS spectra, b) conductivity measurements, [3], c) laser photolysis of the V band and the kinetics of the recovery reaction [4, 5], d) this species was used by Golden et al. [6] to explain the magnetic susceptibility, Knight shift, conductivity and optical data of various authors [7]. The IR band is associated with the solvated electron [8]. Assuming therefore that the only equilibrium existing in these solutions is $M^- \rightleftharpoons M^+ + 2e^-$, one can obtain from the electroneutrality condition $[M^+] = [M^-] + [e^-]$ and from the conservation of matter $[M^+] + [M^-] = c$ (where c = total concentration of metal) the following relationship:

$$c = \frac{2OD_V}{\varepsilon_V} + \frac{OD_{IR}}{\varepsilon_{IR}}. \tag{1}$$

Measuring the two absorbancies, (OD_V and OD_{IR}), correcting for the overlap of the two bands (especially for potassium solutions) and determining the total concentration, we can solve the above equation (using the multiple regression technique) to obtain the two extinction coefficients.

A different method to measure the extinction coefficients is the following. An electric current is passed through a solution containing M^+, preferably a completely decomposed solution of the metal. Since the equilibrium between the various species is fast [4, 5], one obtains, assuming the mixing is very fast and the rate of decay is very slow, the following equation:

$$\frac{I\,1000}{FV} = \frac{d\,OD_V}{dt} \frac{2}{\varepsilon_V} + \frac{d\,OD_{IR}}{dt} \frac{1}{\varepsilon_{IR}} \tag{2}$$

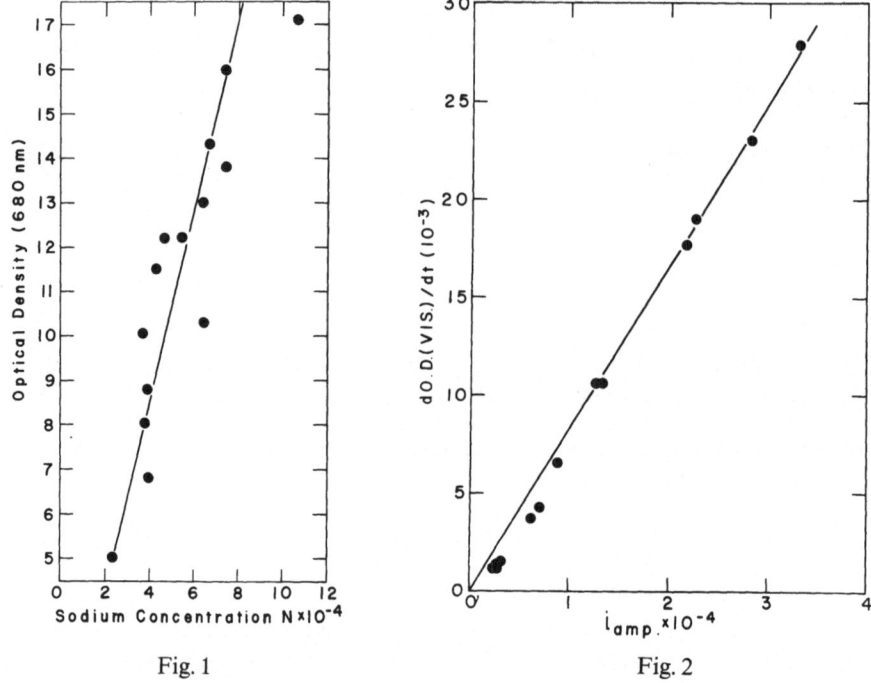

Fig. 1

Fig. 2

Fig. 1. Concentration vs. absorbancy of V band in solution of sodium in PDA (No Ir band). All OD values reduced to 1 cm path length

Fig. 2. Current vs. rate of increase of V band in decayed solution of sodium in PDA.

where I is the current in amperes, F is the Faraday, V is the volume of the sample, and $\dfrac{d\,OD}{dt}$ is the rate of increase of the band (measured at the start of the current to avoid any decay processes). Again using multiple regression we can obtain the two coefficients.

In Fig. 1 we present the linear plot of sodium concentration vs. absorbancy of the V band for a Na-PDA solution. Since there is no IR band in this case, the extinction coefficient is obtained directly from the slope. In Fig. 2 the plot of Eq. (2) for the same solution is shown. Figure 3 shows a plot of the equation

$$\frac{OD_{IR}}{OD_V} = \varepsilon_{IR}\,\frac{c}{OD_V} - \frac{2\varepsilon_{IR}}{\varepsilon_V} \tag{3}$$

for K-EDA solutions; and Fig. 4 shows a plot of $c - \dfrac{2\,OD_V}{\varepsilon_V}$ vs. OD_{IR} (assuming value of ε_V as in Na-PDA solutions) for the same solutions. Both of these plots were used as a visual aid for the scatter of the points rather than for computational purposes since the latter was done by handling Eq. (1) directly by multiple regression. The average results obtained are: $\varepsilon_V = 42\,500 \pm 1500\ \mathrm{M^{-1}\,cm^{-1}}$ and $\varepsilon_{IR} = 2100 \pm 1000\ \mathrm{M^{-1}\,cm^{-1}}$, the difference between the two methods being 10%. Table I shows pK values obtained using these extinction coefficients.

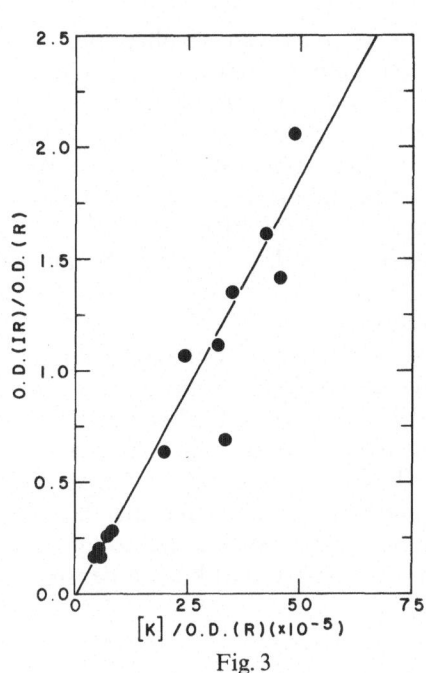 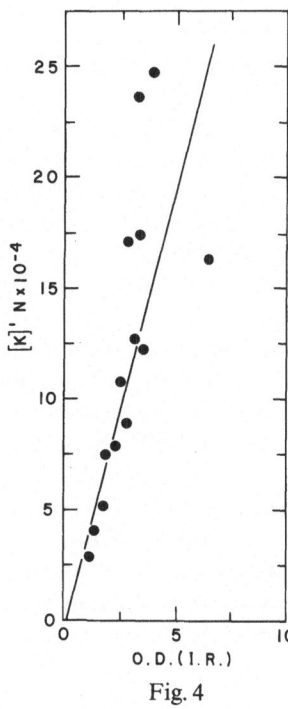

Fig. 3   Fig. 4

Fig. 3. Plot of Eq. (3) for potassium — EDA solution

Fig. 4. $[K]' = c - \dfrac{2\,OD_V}{\varepsilon_V}$ vs. OD_{IR} ($\varepsilon_V = 42\,500\ M^{-1}\,cm^{-1}$) for potassium — EDA solution

Table I

Solution	Na/PDA	Na/EDA	K/PDA	K/EDA
pK[a]	> 10[b]	7.5	6.4	4.9

[a] Uncertainties are less than half pK unit. Units of K are M^2.
[b] Calculated on basis of limit of observation of absorbancy.

These results should be compared with those of DeBacker and Dye [9], who found $\varepsilon_V = 82\,000\ M^{-1}\,cm^{-1}$ and a ratio [5] $\varepsilon_V/\varepsilon_{IR} \approx 4$. Also Huppert and Bar-Eli [4] found for the same ratio the value of ≈ 3. On the other hand, Windwer and Sundheim [10] estimated the value of $2.35 \times 10^4\ M^{-1}\,cm^{-1}$ for the V band, and showed that the V band is not associated with the ESR signal. In a similar manner we have shown that our result for ε_{IR} conforms with the ESR signal intensity. We cannot explain the discrepancy unless we invoke another, probably diamagnetic, unabsorbing species such as M_2. This conjecture will however defy any direct measurement of the extinction coefficients.

Experimental

Metals used were Koch-light 99.95% purity which were distilled into the vessel. Amines from Fluka AG. were dried over sodium (reflux for a few hours) and distilled twice from blue solutions to a sodium-containing reservoir, from which the final distillation to the reaction vessel took place. The optical cells were from 5—0.01 cm path length in order to increase the range. In some cases both a narrow and a wide cell were used when the intensities of the two bands varied widely. Optical data were taken on a Cary 14 Spectrophotometer at 680 nm for sodium V band, 870 nm for potassium V band, and 1280 nm for the IR band.

The concentrations were measured as follows:

1. Immediately after absorbancy measurement, an excess of ammonium bromide was added and the evolved hydrogen measured. 2. The solution was transferred to a side arm bulb, the solvent distilled back to the optical cell and transferred again to the bulb in order to wash all the metal into the bulb. The procedure was repeated a few times. The dry side arm bulb was then sealed off from the main vessel, a solution of dilute HCl was added to dissolve the metal, and metal content measured by atomic absorption. Each vessel contained 4 such side bulbs. Volume of solvent was found from its weight and known density. Agreement between the two methods was within 5%.

References

1. Matalon, S., Golden, S., Ottolenghi, M.: J. Phys. Chem. **73**, 3098 (1969).
2. Bar-Eli, K., Gabor, G.: J. Phys. Chem. **77**, 323 (1973).
3. Arnold, E., Patterson, A.: J. Chem. Phys. **41**, 3089, 3098 (1964).
4. Huppert, D., Bar-Eli, K.: J. Phys. Chem. **74**, 3285 (1970).
5. Dye, J. L., DeBacker, M. G., Eyre, J. A., Dorfman, L. M.: J. Phys. Chem. **76**, 839 (1972).
6. Golden, S., Guttman, C., Tuttle, T. R., Jr.: J. Chem. Phys. **44**, 3791 (1966).
7. (a) Hutchison, C. A., Jr., Pastor, R. C.: J. Chem. Phys. **21** 7959 (1953).
   (b) McConnell, H. M., Holm, C. A.: J. Chem. Phys. **26**, 1517 (1957).
   (c) Acrivos, J. V., Pitzer, K. S.: J. Phys. Chem. **66**, 1693 (1962).
   (d) O'Reilly, D. E.: J. Chem. Phys. **41**, 3729 (1964).
   (e) Kraus, C. A., Bray, W. C.: J. Am. Chem. Soc. **35**, 1315 (1913).
   (f) Kraus, C. A., Lucasse, W. W.: J. Am. Chem. Soc. **44**, 1941 (1922).
   (g) Gold, M., Jolly, W. L., Pitzer, K. S.: J. Am. Chem. Soc. **84**, 2264 (1962).
   (h) Douthit, R. C., Dye, J. L.: J. Am. Chem. Soc. **82**, 4472 (1960).
8. (a) Dorfman, L. M., Jou, F. Y., Wageman, R.: Ber. Bunsenges. Phys. Chem. **75**, 681 (1971).
   (b) Saito, E.: In: Lagowski, J. J., Sienko, M. J. (Eds.): Metal–ammonia solutions. Proceedings of Colloquium Weyl *II*, p. 483. London: Butterworths 1900.
9. DeBacker, M. G., Dye, J. L.: J. Phys. Chem. **75**, 3092 (1971).
10. Windwer, S., Sundheim, B. R.: J. Phys. Chem. **66**, 1254 (1962).

Discussion

J. L. DYE   The addition of salt does not prove that the reaction must go via $M + e^- \rightarrow M^-$ [see Dye, DeBacker, Eyre, Dorfman: J. Phys. Chem. **76**, 839 (1972)]. The formation of ion pairs, $M^+ \cdot e^-$ and the nearly diffusion-controlled pairing reactions:

$$e^- + e^- \rightarrow e_2^=$$

or

$$M^+ \cdot e^- + e^- \rightarrow e_2^= \cdot M^+ \quad \text{or} \quad e^- \cdot M^+ \cdot e^-$$

or

$$M^+ \cdot e^- + M^+ \cdot e^- \rightarrow e_2^= \cdot M_2^+ \quad \text{or} \quad (e^- \cdot M^+)_2$$

can make such a reaction independent of ionic strength.

K. BAR-ELI   It is agreed that the IR band is associated with the solvated electron. Its disappearance (after its creation by the flash) is second-order and is independent of ionic strength; therefore there must be a rate determining step in which an uncharged species reacts with either charged or another uncharged species formed before. The simplest mechanism which involves the minimum number of parameters and which conforms with the experimental data is $M + e^- \rightarrow M^-$.

J. L. DYE   You see a contribution from the IR band in sodium-ethylenediamine solutions which we do not see in the freeze-purified material. It is possible that your solvent contains ammonia produced by decomposition of the metal which is used for drying.

K. BAR-ELI   This is certainly a possibility. However, no IR contribution is seen in sodium-propylenediamine solutions, for which there is probably the same amount of ammonia produced by decomposition. Moreover, we do not observe any increase in the IR band in older solutions. And, lastly, the band appears to peak at 1280 nm which is more energetic than the solvated electron in ammonia.

Ultrafast Optical Processes

P. M. RENTZEPIS

Abstract

Molecular relaxation, radiationless transitions and most primary photochemical processes which have lifetimes of the order of 10^{-12} sec can be studied experimentally with picosecond pulses generated by mode-locked lasers. This paper reviews the application of picosecond pulses for the study of molecular relaxation. We describe several methods which we have utilized for the direct observation of radiative and radiationless transitions in liquids. Simultaneous time-frequency resolved emission and absorption spectra methods provide measurements of emission and absorption processes with a time resolution limited by the time width of the picosecond pulse. The techniques of picosecond spectroscopy were utilized for the study of the relaxation of a quasi-free electron to the localized electron state in water.

I. Introduction

The laser provided an improved source for conventional molecular spectra studies but it provided little opportunity for innovation in molecular relaxation measurements relative to the Q-switched lasers which generated pulse with peak power of approximately 5×10^8 W within about 10^{-8} sec. The development of mode-locked lasers capable, at present, of producing pulses of radiation within as little as $\sim 10^{-12}$ sec and power of more than 10^{10} W have superceded, if not eclipsed, the earlier techniques. These picosecond pulses have already found extensive application in the study of a host of nonlinear optical effects, plasma physics, molecular and electronic relaxation as well as engineering application in optical radar, etc. Our work in the construction of new picosecond spectroscopic techniques to obtain data not directly accessible or not accessible at all, forms the basis of this review.

Mode-Locking

The first mode-locking of a laser was achieved in 1963 with an acoustic-optic oscillator [1]. Thereafter solid state lasers were mode-locked by means of saturable absorbers such as polymethene dyes, [2—6] cryptocyanene or other dyes which have a broad absorption band at the laser wavelength and a repopulation time shorter than the optical round-trip time of the laser cavity. Experimentally one can easily obtain mode-locked laser pulse, as shown in Fig. 1a. The laser rod provides the amplifying medium, while the mirrors form the Fabry-Perot cavity. By inserting a bleachable dye which has a much broader absorption band than the laser line, standing waves are set up with discrete optical frequencies defined by the Fabry-Perot modes

$$v_n = \frac{nc}{2L}$$

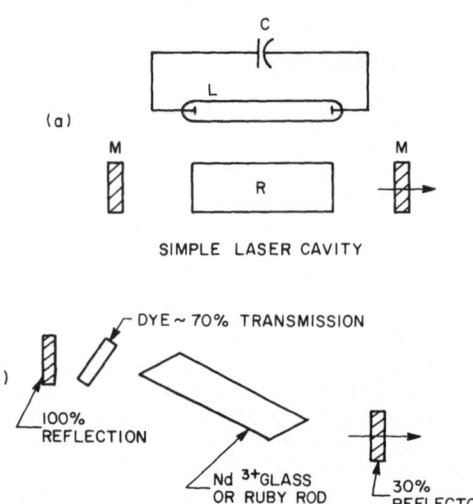

SIMPLE LASER CAVITY

MODE LOCKED LASER

a

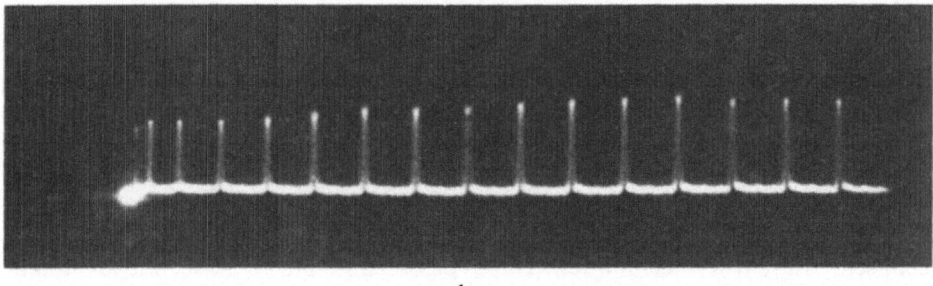

b

Fig. 1. a Schematic representation of a typical mode-locked Nd^{+3} glass laser. b Train of picosecond pulses detected by fast photodiode and displayed on 519 Tektronix oscilloscope

where L is the optical separation of the mirrors and c the light velocity. For a typical laser in which, there are $\sim 10^6$ modes oscillating within the cavity, the adjacent mode separation Δv is given by $\Delta v = v_n - v_{n-1} = c/2L$. In practice, however, the modes allowed have a spectral width Δv which is a function of the mirrors and of the effective aperture of the laser rod. The light beam passes through the bleachable dye and distorts the original pulse intensity distribution by increasing the peak intensities. In the cavity the beam passes back and forth many times, continuously sharpening the pulse and increasing the intensity, with the duration of the pulse theoretically limited by the width of the spectral band.

Computer calculations show that the pulse is sharpened after each successive pass, with the maximum achieved after about 15 passes. Although the saturable absorber provides a method for sharpening the pulses and separates them by $c/2L$, no unified model as yet provides sufficient information about the formation of the initial sharp pulse, which will be amplified in the laser cavity faster than the broad envelope. Normally the modes of the laser oscillate in a random fashion without any set phase relationship. The locking of the phase is provided by the bleachable dye which modulates the frequencies ($v_0 + \Delta v$), v_0 and ($v_0 - \Delta v$) (where $\Delta v = c/2L$) of the cavity interferometer and couples new side bands to the first three. With each additional path new side bands are coupled until all of the modes defined by the total bandwidth are coupled. Mode-locking is achieved experimentally by ascertaining that all of the optics (rod and Q-switch cell) are at the brewster angle to eliminate back-reflections. The pulse intensity can vary from high intensities in short pulse trains to low intensities in long trains by adjusting reflectivity of the mirror, the length of the cavity and the concentration of the Q-switch dye.

Mode-locked laser experiments have used both Nd^{+3} glass and ruby, and YAG:Nd^{+3}, to generate pulses in the 10^{-12} sec range. The Nd^{+3} glass is most interesting because of the 100 to 200 cm^{-1} bandwidth which provides the possibility of 10^{-13} sec pulses. In practice one usually generates pulses with a duration of 2 to 7 picoseconds utilizing Eastman-Kodak No. 4740 and No. 9860 dyes as the saturable absorber. In the case of the ruby, mode-locked pulses in the 40 psec range can be generated by using cryptocyanane (~ 25 psec), DDI (1,2-diethyl-2,2 dicarbocyanine iodide (~ 8 psec) and dicyanine A (~ 50 psec).

A typical train of mode-locked picosecond pulses is shown in Fig. 1b. The pulses are separated by 8 nsec, which corresponds to a 1.2 meter cavity. It is interesting to note that the pulse width is smaller than the resolution of either photodiodes or oscilloscopes of about 10^{-9} sec. One can obviously vary the separation of the pulses by changing the length of the cavity. In situations in which small separation is desired, one can utilize a front mirror with flat surfaces thus creating pulse separations equivalent to twice the thickness of the mirror [7].

It seems appropriate to point out that gas lasers have also been mode-locked, as a matter of fact, even before other lasers, with acoustic-optic crystals; however, due to the small spectral widths the pulse duration is limited to $\sim 10^{-10}$ sec. In the case of argon, mode-locking has also been achieved, and by frequency doubling, we have obtained pulses at ~ 2550 Å as short as ~ 200 psec [8] with a separation rate of 1 pulse per 3 nsec or longer. With the aid of fast electronics we are now able to monitor spectroscopic processes essentially continuously with a high signal-to-noise ratio. Lately also the CO_2 laser has been modified to generate picosecond pulses; [9, 10] however, because of its low frequency, its application is restricted to vibrational excitation of molecular systems [11, 77—79].

Single Picosecond Pulse

It has been repeatedly shown that the pulses contained within a pulse train vary in both intensity and duration. It is very important therefore to utilize, when possible, a single pulse when relaxation rates are in the range of a few picoseconds. The first high-intensity single pulse instrument was proposed by Vuylstoke [12]. He constructed a laser with mirrors having 100% reflectivity at low incident light intensity; however, near pulse intensity, the front mirror switched rapidly to zero reflectivity, dumping the pulse within the optical cavity round-trip time and a single pulse of very high power.

An alternative method for extracting single pulses from a mode-locked train utilizes a spark gap charged to about 10 kV [13]. The gap is adjustable so that one can select the most intense pulse to initiate a discharge. A high-voltage pulse, generated by the discharge, is used to increase the voltage on a partially energized polarizing pockels cell so that a 90° change of polarization is generated for about 5 nsec. A single pulse is thus transmitted which has a high intensity and short duration of ~ 2 psec. With the addition of a 25 cm long amplifier, a single picosecond pulse with ~ 1 joule of energy can be generated [13].

Measurement of Picosecond Pulses [7, 14, 15]

The pulse duration of nanosecond pulses is usually measured directly by means of a photomultiplier — (or photodiode) oscilloscope combination. High-speed photodiodes have been found with a resolving power of ~ 0.3 nsec. Because it is not possible to measure pulses in the 10^{-12} sec range using presently available electronic techniques, nonlinear optical methods have been developed which are capable of displaying the pulse duration and shape directly. Another, as yet less well-developed possibility is offered by interferometric methods which determine the autocorrelation function of the pulse amplitude; this is in turn related to the Fourier transform of the power density spectrum, as measured by a spectrometer.

It is obvious, however, from the relationship $\Delta \omega \, \Delta \tau = 2\pi$, that the linear techniques provide only a lower limit for the pulse duration except in the case where the complete spectral width is contributing to the mode-locked pulse. In practice it is generally observed that when a high background level is observed in the oscilloscope trace, the modes are only partially locked and the spectrum of the pulse appears to be less wide, indicating that the inverse bandwidth time-duration relationship is followed at least as a lower limit.

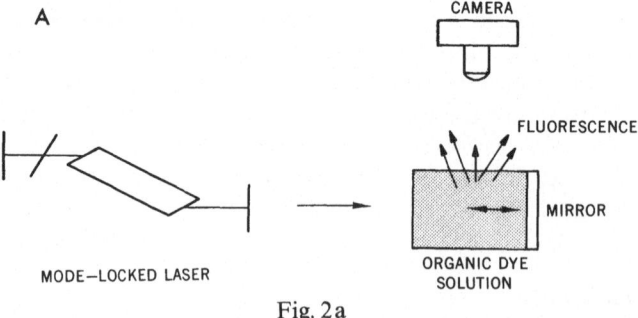

Fig. 2a

Fig. 2. Two-photon fluorescence technique (TPF). a The original apparatus for the direct measurement of picosecond pulses. b TPF technique using two picosecond pulses with different frequencies for improved contrast. c Two-photon photographic display. d Photograph of picosecond pulses by a) the original TPF technique, b) TPF using two different frequencies, c) same as above but removing the bromobenzene cell and dissolving the TPF dye in a dispersing solvent.

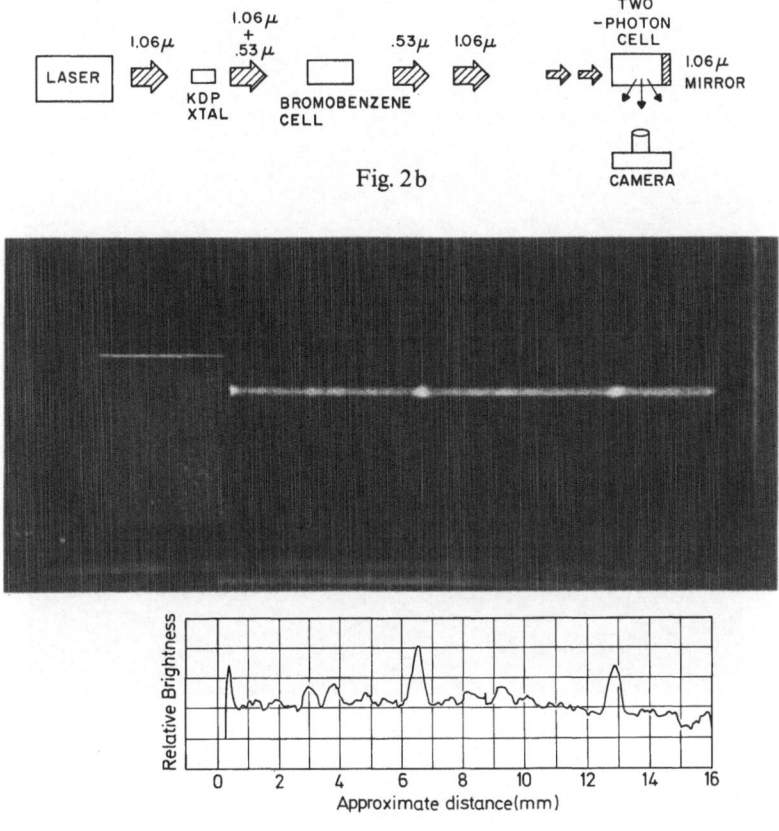

Fig. 2b

Fig. 2c

The method used most frequently for the measurement of picosecond-duration pulses is the two-photon fluorescence (TPF) method [7, 14, 15]. The simplicity, accuracy, and reliability of this method are the principal advantages over other methods such as the second harmonic. The TPF measures the correlation of two pulses as illustrated in Fig. 2a. The first reported experiment on TPF made use of the second harmonic, 5300 Å of a train of mode-locked light pulses of a Nd^{+3} glass laser. After filtering the fundamental and pump lights, the beam traversed a 20 cm cell containing a dibenzanthracene/benzene solution. Other fluorescent dyes have been used subsequently with this method and almost any dye is suitable if the solution does not absorb by one photon (it is not desirable, for example, to use methanol as the solvent with Nd^{+3} laser fundamental since it slightly absorbs at 1.06 µ), has a large two-photon absorption cross section, and a high quantum yield for emission. The pulses of the train were reflected by a mirror situated at the end of the cell. They then collided with the following members of the pulse train, inducing TPF at the point of overlap, Fig. 2b. The induced fluorescence even from a single pulse was sufficiently intense to be directly displayed photographically (camera with $f = 2.8$, polaroid film 3000 speed).

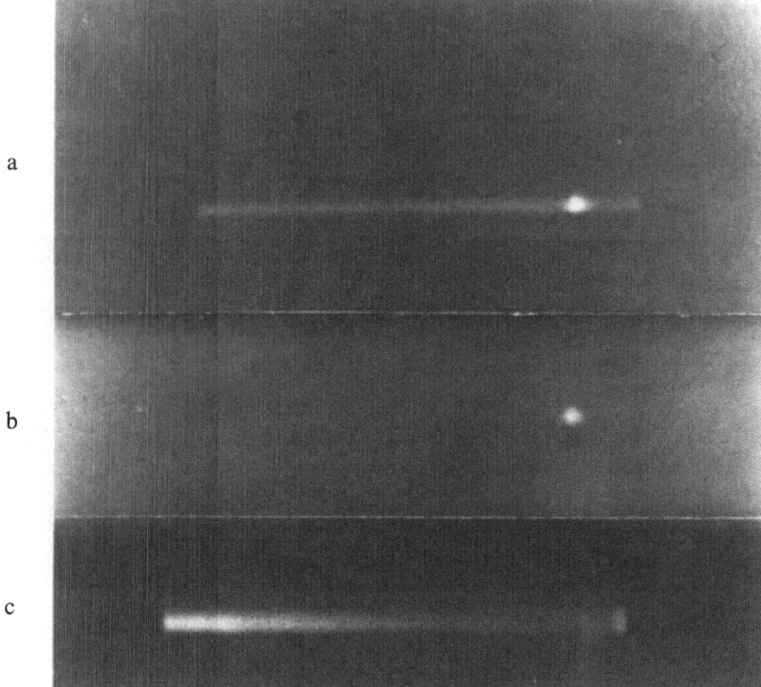

Fig. 2d

The pulse duration, t, is simply calculated by $t = \dfrac{dn}{\alpha c}$ where d is the length of the two-photon bright spot at half-maximum intensity above background, α, a constant depending on the shape of the pulse and n, the index of refraction of the solution. The separation of the pulses, t_2, displayed as the distance d_2 between bright spots, is given by $ct_2/2n$. A typical display of picosecond pulses by the TPF technique is shown in Fig. 2c. In this case the pulses are 2—3 psec and separated by 50 psec.

Figure 2c clearly shows not only the fluorescence spot at the point where the pulses overlap but a continuous background induced by each pulse alone as it traverses the two-photon solution.

Subsequently it was shown [16—20] that the theoretical maximum contrast ratio of the TPF spot to the background for a two-photon virtual process is 3:1 while this ratio for the free running (nonmode-locked) case is 1.5:1. The contrast ratios can be calculated assuming plane polarized waves of frequency w reflected by a mirror of reflectivity R (100% in this case). The standing wave field strength in front of the mirror is given by

$$E_{0m} = E_0[1 + R + 2R^{1/2} \cos(4\pi nd/a - \delta)]^{1/2}$$

where n is the refractive index, a is the wave length, δ is the phase shift due to imperfect reflection and d is the distance from the mirror. In the case of two-photon induced fluorescence, the intensity, I_n, is proportional to $(E_{0m})^4$. Taking a running average over a wave length,

$$I_n \sim E_0^4 (1 + 4R + R^2),$$

which is valid for any pulse length reflected onto itself by the mirror. The background fluorescence emitted by the individual pulses acting alone is $1 + R^2$.

From this it is obvious that the ratio between the fluorescence intensity of the point where the two-photon mode-locked pulse meets and the background is $3:1$ and in the free running or random case $1.5:1$.

The TPF method has the advantages of simplicity and accuracy; however, the low contrast ratio and inherent incapability of displaying the pulse symmetry are sufficient disadvantages to warrant the use of the simple and accurate three-photon method (3 PF) [21, 22], which circumvents the first of these disadvantages in a practical way and the second in principle.

Three-Photon Method

The three-photon method has all of the attractive features of the TPF techniques including the direct photographic recording of the pulse and the simplicity of operation. In addition, it has the advantages of (1) a contrast ratio of $10:1$ as opposed to 3 for TPF, (2) the potential for measurement of the phase-intensity relationship. The first makes it possible to observe experimentally processes which were previously masked by the intense background.

The experiment is represented schematically in Fig. 3a. The solution contains a dye which has its first allowed state at three times the frequency of the laser and which strongly fluoresces. In the first 3 PF experiments we utilized (BBOT)

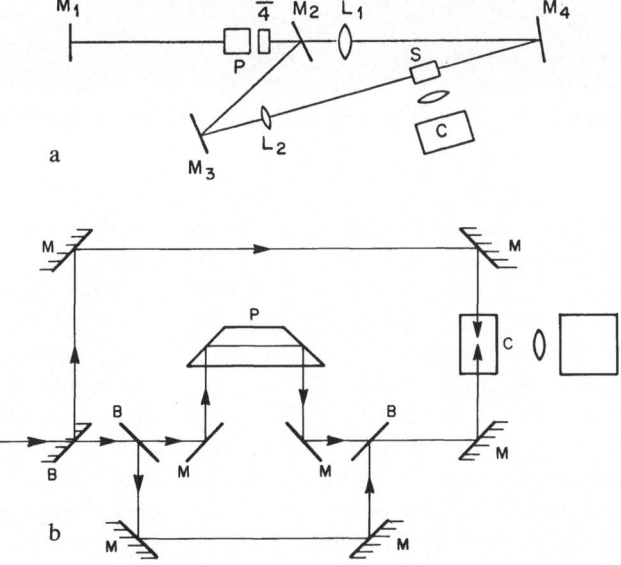

Fig. 3. Three-photon fluorescence method (3 PF). a Schematic representation for the measurement of the duration of a single picosecond pulse. b Method for the simultaneous measurement of asymmetry and duration of a picosecond pulse

2,5-bis-[5-terbutylbenzo xazolyl (2)] thiophene in methylcycloxane and 1,4-bis[2-(4-methyl-5-phenyloxyazolyl)] benzene (dimethyl POPOP) in methylcyclohexane. The absorption cross section of this dye at the three-photon wave length (3550 Å) of the Nd^{+3} glass laser is very high (ε 3590 Å $\approx$ 50000) and the quantum yield for emission is approximately unity.

The Nd^{+3} glass laser was mode-locked and Q-switched in the normal manner by Eastman Kodak No. 9860 or No. 9740 bleachable dyes. The pulse traversed the optical paths shown in Fig. 3a. To eliminate reflection back into the laser cavity a set of $\lambda/4$ plates and polarizers were inserted in the path. One could achieve similar results by a slight misalignment of the back mirror, thus reflecting the beam away from the laser cavity. The data of Fig. 4 are typical of well

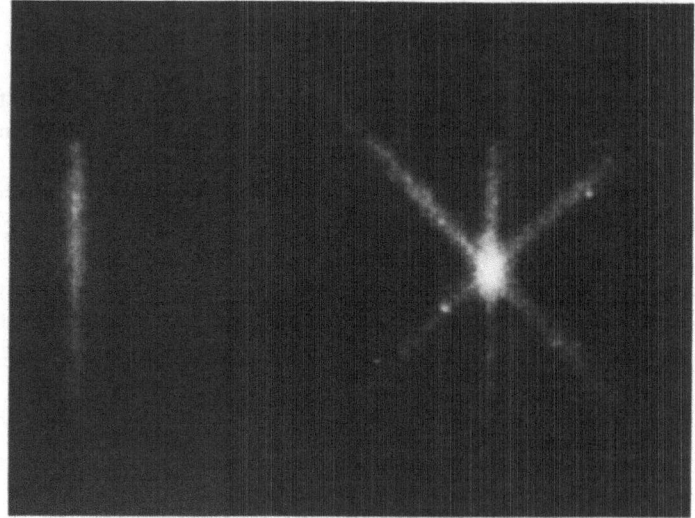

Fig. 4. Photographic display of picosecond duration and asymmetry of pulse(s) by 3 PF

mode-locked pulses with a contrast ratio of better than 9:1. The 3 PF method, due to its large range of contrast ratios, has also the potential of depicting the degree of mode-locking much more clearly. In fact, studies show that as the degree of mode-locking decreases, the pulses are broadened, the oscilloscope (519 Tektronix-photodiode combination) display has the appearance of noise between the pulse of the train and the 3 PF intensity ratio decreases. We observed that low contrast ratio with either the 2 PF or 3 PF methods was accompanied by a display of a "noisy" trace on the oscilloscope. Conversely, an oscillogram displaying a smooth signal without the background resulted in a relatively high ratio, i.e. 9:1. We believe these observations are of practical importance because one need not process the film and measure by a photodensitometer the contrast ratio of each photograph to verify that the laser does in fact generate mode-locked picosecond pulses. As we mentioned previously, one of the major advantages of the 3 PF method is that the third order correlation function relates the

3 PF intensity to the pulse symmetry [23]

$$G^{(3)}_{(\tau_1 \tau_2)} = \int I(t)\, I(t + \tau_1)\, I(t + \tau_2)\, dt$$

and

$$P^3(\tau) = \int [I^2(t)\, I(t + \tau_1) + I^2(t + \tau_2)]\, dt\,.$$

Analysis of such a third order correlation function may give the relative phases of the pulse components and their amplitude, thus unveiling the shape of the pulse.

The experimental arrangement of Fig. 3b shows the method exployed to achieve the variable pulse delay $(t_1\, t_2\, t_3)$ demanded by the $G^3_{(\tau_1 \tau_2)}$ function. The apparatus consists of beam splitters B which divide the original pulse into three equal parts. Each pulse is reflected by the mirrors M and traverses the optical path shown. The prism P is situated on a moving platform and provides the variable delay necessitated by G^3. The pulses recombine inside the cell c and the induced three-photon fluorescence is photographed by the camera situated normal to the path of the pulses.

We have previously observed that the pulse width within a train of picosecond pulses varies from ~ 2 psec to about 8 psec with the shorter pulses near the beginning of the train. Obviously it is difficult to observe the shape of a short pulse when one averages over the whole train; since the pulses are not continuously reproducible a single pulse method is much more reliable for determination of the pulse width and shape and also for experiments in picosecond spectroscopy and molecular relaxation. Recently we devised a method [22] which enables one to measure the pulse width and asymmetry with a single pulse. This method is a variation of the 3 PF technique diagramed in Fig. 3a. Refer again to Fig. 3b; the pulse is split into three components of equal amplitude while maintaining the original polarization. These now enter the dye cell along the x, y and z axes, overlap, and generate the fluorescence pattern shown in Fig. 4. If the pulse is very asymmetric the fluorescence pattern would be asymmetric. Appropriate computer simulation shows that in many cases a well mode-locked laser has essentially a gaussian shape and pulse width of ~ 2 psec.

We also mention here that several methods have been devised which essentially achieve pulse compression approaching the limit $\Delta\tau\, \Delta \approx 1$, and are capable of generating pulses with duration of $\sim 4 \times 10^{-13}$ sec [24—26].

II. Direct Measurement of Radiationless Transitions

In a general sense radiationless relaxation plays an important role in most photochemical processes. There are several theoretical treatments of radiationless transitions; i.e., Jortner and Berry [27], Bixon and Jortner [28, 29], Freed and Jortner [30], Nitzan et al. [31] and Robinson and Frosh [32], Lim and Bershon [33], and several reviews dealing with nuclear scattering which have a strong relevance to molecular systems [34]. The interpretation of these relaxation processes is based on the assumption that the system is a compound state. The system can be thought of as being composed of a set of zero-order states which, upon interaction, create the exact eigenstates of the Hamiltonian. As visualized by Jortner [35] the compound state can be subdivided into two zero-order states, a sparse and a very dense system. The relaxation process then proceeds into the continuum when there is a superposition of the zero-order discrete and dense states. Since the zero-order states do not portray a physical real-life situation we can consider the relaxation of a state as a function of its total width, γ, which is composed of several components, such as nonradiative width Δ and radiative width Γ, or $\gamma = \Gamma + \Delta$. In this spirit the width of a state is zero until coupling with the radiative field of a state takes place, as in the case of coupling between stationary states, where upon one observes a radiative transition with width Γ_i. Since nonradiative transitions are not allowed, Δ is essentially zero and $\gamma = \Gamma = \dfrac{1}{\tau_R}$, where τ_R is the radiative lifetime. Similarly, if the system is excited to a nonstationary state or, equivalently, the molecule is embedded in an inert matrix such as a nonreacting solvent, radiationless transitions are now allowed and the total width, γ, contains both components Γ and Δ. Δ can be constructed to include the collisional broadening and other deactivating processes, B,

$$\gamma = \Gamma_i + \Delta_i, \quad \text{and} \quad \Delta_i = \delta + a + B_i.$$

An an approximation, we will consider the excitation to a bound state which initially has zero width and in which a width of certain magnitude appears as the decay channels develop.

In this discussion of the relaxation processes in large molecules we will consider three types of decay channels — the mode of decay when: (a) the excited state is coupled to very large numbers ($\sim 10^8$) of levels resembling a quasi continuum, (b) a sparse number of levels, and (c) the intermediate case where the energy gap between the excited state and a lower state is a few thousand wave numbers.

In the theoretical treatment given by Bixon and Jortner [28] it is shown that within the statistical limit the process of relaxation of the excited state molecule would exhibit similar characteristics whether the molecule is isolated or in solution. We present here experimental evidence which shows that a large molecule with a large energy gap relaxes nonradiatively from an excited electronic state at the same rate both as a "isolated molecule" and when imbedded in an inert matrix, i.e. in solution.

The theoretical sparse level situation occurs in an actual experiment when a zero order state is strongly coupled to a small number of widely separated vibronic levels which belong to a lower electronic origin. Certain triatomic molecules [37] such as NO_2, SO_2, and CS_2 provide examples of a complex spectrum which may be due to the strong intramolecular interaction between the zero order excited state and coarsely spaced levels.

The intermediate coupling provides a very interesting case which just lately has been discussed theoretically [36, 38]. Some of the features predicted and partially exposed experimentally are 1. the nonexponential decay characteristics, [39, 40] 2. lengthening of the lifetime of the excited state [41, 42], 3. emission between highly excited states $S_2 \rightarrow S_1$, [43, 44] and between S_2 and the ground manifold [45], 4. deviations from the Stern-Volmer relationship [46, 47], and 5. quantum interference between levels in large molecules [40—42].

Molecules Corresponding to the Statistical Limit Azulene

The first experiments on radiationless relaxation [47, 48] of large molecules with picosecond pulses were performed late in 1966. At this time Jortner and his colleagues were concerned, independently, with the theoretical aspects of radiationless transitions of large molecules, and were focusing their attention on coupling with a dense manifold. For the first experiment ever performed on molecular relaxation with picosecond pulses [48], we selected azulene because it exhibits a fast relaxation rate from the first excited singlet state $^1B_1$ either by coupling to the ground state or to a low lying triplet [49, 50]. The energy gap of $14400\ cm^{-1}$, ω_1, from the ground state to the "origin" of the first $^1B_1$ state could be conveniently covered by the fundamental of the ruby laser while an upper vibronic level of the same excited state could also be populated by the second harmonic of a Nd^{+3} glass, at $18862\ cm^{-1}$, ω_2. The azulene was purified by sublimation and dissolved in spectroscopically pure methanol. The experimental arrangement is shown in Fig. 5. The fundamental $9431\ cm^{-1}$ beam of a mode-locked Nd^{+3} glass laser generated its second harmonic, $18862\ cm^{-1}$ in a $1 \times 1 \times 3$ cm phase matched KDP (potassium dihydrogen phosphate) crystal, with a conversion efficiency of $\sim 12\%$. The two pulses propagated simultaneously and entered the dispersion cell containing bromobenzene. The fundamental pulse, ω_1, propagates faster than ω_2 and the separation of the ω_1 and ω_2 pulses can be calculated accurately by the group velocity relationship; the wave number of the two pulses and refractive index of bromobenzene at ω_1 and ω_2 are known. In order to measure simultaneously the pulse width and relaxation time of azulene in solution for every picosecond train, and thus avoid the fluctuation in pulse intensity and duration that occurs between successive laser shots, part of each beam was reflected into a two-photon cell in the manner shown in Fig. 9. The fluorescence induced in each cell was photographed by an fl. 9 Polaroid camera. Specific care must be taken that the exposure of the film is within its linear range by using neutral density or other calibrated filters.

Pulse ω_1, enters the cell first. However, the photon energy is not sufficient to populate the first excited singlet state. Thus, pulse ω_1 will not induce fluorescence from the emitting second singlet as it travels through the azulene solution. At the end of the cell the dielectronic mirrors, Fig. 7, reflect ω_1 onto itself so it now

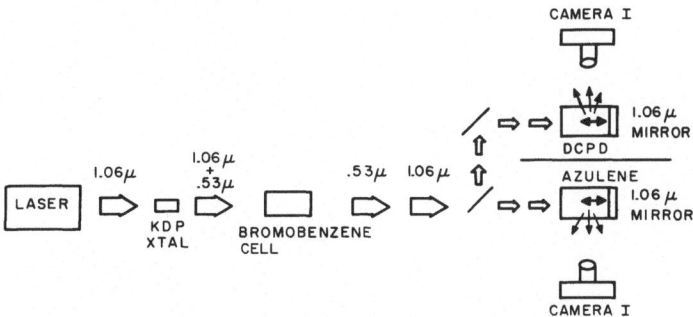

Fig. 5. Schematic version of the method for the direct observation of radiationless transition rate of azulene

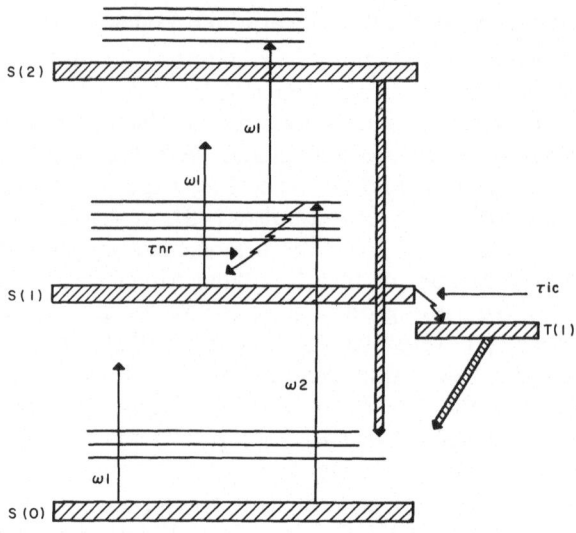

Fig. 6. Energy level diagram of azulene

travels in the opposite direction. Pulse ω_2 follows behind, at a predetermined distance which insures that it will overlap with the returning ω_1 pulse inside the azulene cell. Pulse ω_2 has a frequency of $18862\,cm^{-1}$ which is sufficient to populate an upper vibronic level of the azulene S_1. Azulene has been shown not to emit from the first singlet [55, 56], $(Q—10^{-6})$ [57], thus no emission is recorded. To avoid biphotonic excitation by ω_2 to the second excited singlet state of azulene, which does emit, the intensity of ω_2 is reduced by detuning the KDP crystal, or by appropriate filters. Following the course of events we see that w_2, Fig. 6, populates a vibronic level of S_1 situated at $18862\,cm^{-1}$ above the ground state. At the time and space within the cell where the two pulses w_1 and w_2 overlap, the combined energy of the pulses is the equivalent of $28293\,cm^{-1}$ which is near the origin of the S_2 state of azulene. At the position of overlap the frequency and photon density are sufficient to generate observable fluorescence from S_2 to S_0 during the time of the pulse overlap. When the pulses separate, the fluorescence

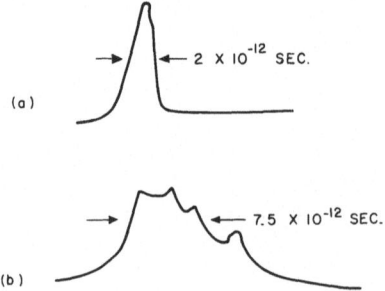

Fig. 7. Photodensitometer trace of a picosecond pulse and b lifetime of $18862\,cm^{-1}$ excited level of azulene

should also disappear, unless the level which was populated by w_2 has a finite lifetime which is longer than the width of the pulses, ~ 2 psec in this case. The w_1 pulse performs this "population interrogating" function as it travels against the direction of ω_2. It is evident that the distance which ω_1 traverses from the point of intersecting ω_2, and still encounters excited azulene corresponds to the direct measure of the relaxation time. The length of the fluorescence spot is simply transposed into time and the relaxation from the $18\,862$ cm^{-1} vibronic level is measured, after the appropriate deconvolution, as ~ 7 psec. This simple and accurate method made possible the first measurement of nonradiative molecular relaxation. This value of ~ 7 psec is in very good agreement with the independent theoretical studies based on the statistical limit model, of Jortner and Berry [27]. In a following section we shall discuss the relation of these experimental results to the theoretical predictions for the isolated molecule case.

Repopulation of the Azulene Ground State

The previous experiment revealed that relaxation from the $18\,862$ cm^{-1} vibronic level of S_1 occurs with a rate of 1.3×10^{11} sec^{-1}; however, the experiment does not a priori explain the mechanism of the relaxation. One cannot predict from this result whether the coupling takes place with the ground state or with the first excited triplet. To elucidate the mechanism we measured directly the relaxation rate to the ground state of azulene in solution [48] by exciting with $14\,400$ cm^{-1} picosecond pulses from a mode-locked ruby laser. In the apparatus which we originally designed, shown in Fig. 8 [48, 49], the picosecond pulses are partly reflected on photodiodes p_1 and p_3. The ratio of the intensities of $p_1 : p_3$ gives directly the number of molecules excited. The pulse traverses the

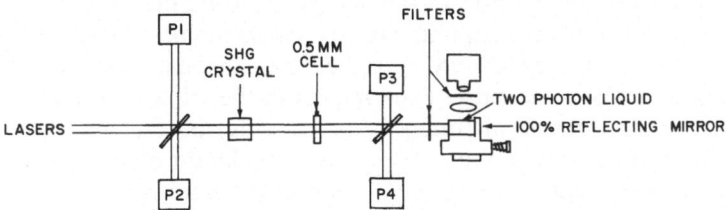

Fig. 8. The apparatus used for the rate of repopulation of the ground state of azulene

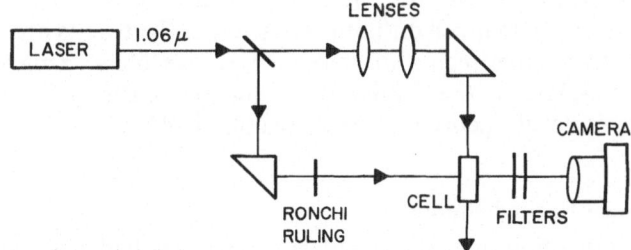

Fig. 9. The cross beam method which enables one to measure in one shot essentially all the parameters of Fig. 8

azulene cell and then enters into the "two-photon liquid" located at the end of the apparatus. Here the length of the pulse is measured with the two-photon method by the fluorescence spot recorded on the film in the above-described manner. The reflected pulse is attenuated by the filters shown in the figure and its intensity is measured again, before and after passage through the azulene cell, by p_4 and p_2. The returning pulse is attenuated to the extent that it is 95% absorbed when azulene is in the ground state. The rate of repopulation is directly measured by this apparatus in the following manner. The first pulse passes through the azulene and upon reflection by the mirror in the "two-photon cell", returns into the azulene solution where it meets with the second pulse which is travelling in the opposite direction. The ratio of the intensities at $p_4 : p_2$ measures the concentration in the ground state. If the two pulses are synchronized and intersect each other within the 0.5 mm azulene cell, the interrogating pulse measures the population of the ground state at the time of excitation, $t = 0$. To measure directly the repopulation after a few picoseconds, the mirror situated on the sliding micrometer platform at the right-hand end of the apparatus was translated so as to change the time elapsed between the excitation pulse, and the arrival of the interrogating pulse. With this continuous translation one obtains a histogram of the repopulation to the ground state from S_1. The repopulation time is found to be $\sim 8 \times 10^{-12}$ sec. This result strongly supports the view that the $^1A_1$ state preferentially couples with the ground rather than with a low lying triplet state.

The method just described, which we originated for the study of azulene and polymethane dyes, has been subsequently utilized by several research workers in the studies of Q-switching dyes [50], rotational relaxation [51, 52], and energy transfer in rhodamine 6 G [53].

To confirm that the dissipation of the energy from the first excited singlet state manifold occurs via direct coupling with the dense manifold of the ground state ($\sim 10^8$ levels/cm^{-1} at ~ 14000 cm^{-1}), rather than through the low lying triplet, we performed similar absorption experiments in the triplet state which indicate that the electronic transition (intersystem crossing) of azulene, $S_1 \rightarrow T_1$ is approximately 10 times slower than the $S_1 \rightarrow S_0$ rate. Drent et al. [54] reported the intersystem crossing time as 6 psec. Although their data are in excellent agreement with ours, they analyzed their data by considering the exciting pulse as having a delta width compared to the fluorescence spot of azulene. If the analysis of the data is performed by a deconvolution of the pulse, one obtains a 60×10^{-12} sec lifetime, vs. 6×10^{-12} sec as in Ref. 54. If the analysis is performed in the same manner on both sets of data, the results are in very good agreement. Further experiments in solution and gas phase azulene have supported direct coupling to the ground state and the statistical coupling model for the azulene molecule excited in the S_1 manifold [44, 45].

Emission Between Excited States [55]

Fluorescence between excited states of large molecules had not been previously observed although these radiative decay processes are frequently observed in diatomic molecules. The difficulty in observing emission from higher excited states of large molecules rests mainly in the very efficient electronic relaxation. The lifetime of an excited state is a function of its radiative $\tau_r(j \to i)$ and non-radiative $\tau_{nr}(j)$ widths,

$$[\tau(j)]^{-1} = [\sum_{i<j} (j \to i)]^{-1} + [\tau_{nr}(j)]^{-1}$$

In most cases the nonradiative decay $\tau_{nr}(j)$ of an upper electronic state, i.e. the second excited singlet $(j = 2)$, is fast, $\tau_{nr}(2) \ll \tau_2(2 \to 1)$. The quantum yield $y_{(2 \to 1)} = \tau_{nr}(^2 t_{r(2 \to 1)}) \ll 1$, and therefore very difficult to observe experimentally.

The anomalous nature of azulene [55—57] renders this molecule suitable for the experimental observation of emission between the first two excited singlet states because of the long lifetime of the $^1A_1$ excited state and the low symmetry of the molecule. The azulene solution was excited by the second harmonic of a mode-locked ruby laser and the emission observed with a $S-1$ type photomultiplier. The appropriate filters, beam splitting and polarizing components were inserted for the elimination of pump light from the detecting instruments. The characteristics of the emission are:

1. a strong band at ~ 7600 Å and a weaker one at ~ 7400 Å [58];
2. deuterated azulene enhances the quantum yield by $\sim 20\%$ compared to normal azulene. This observation is consistent with the deuterium effect observed for the $S_2 \to S_0$ fluorescence and contrary to the effect observed for the emission between the first singlet and the ground state [59];
3. the quantum yield was found to be $< 10^{-6}$;
4. very recently [60, 45], azulene in the gas phase has been excited by Q-switched ruby laser pulses, (20 nsec, 100 MW peak power). The characteristics of the $S_2 \to S_1$ emission from gaseous azulene coincide with the solution experiments except that the quantum yield for $S_2 \to S_1$ is larger by ~ 100.

This discrepancy can be due either to energy deactivation provided by the solvent or, since all other parameters are in good agreement i.e. spectra, lifetimes, isotope effect, to the possibility of stimulated emission $S_2 \to S_0$ induced by the

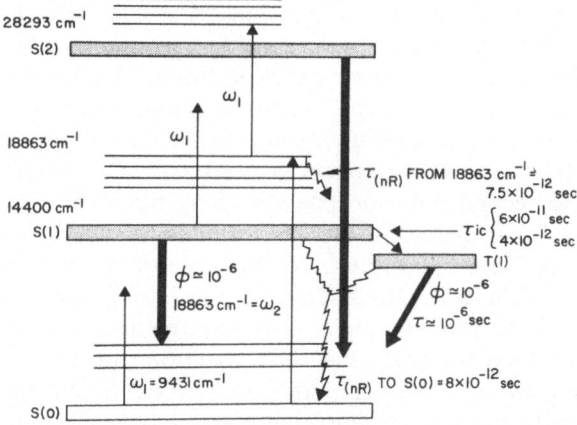

Fig. 10. Compilation of data for azulene by picosecond spectroscopy

high power picosecond pulses. In the latter case the apparent decrease in $S_2 \rightarrow S_1$ emission will result in the low quantum yield. This possibility is currently under study [61].

The observation of the emission from the first singlet state to the ground state has been observed in solution and, recently, in the gas phase, by methods similar to those described for observation of the $S_2 \rightarrow S_1$ transition. The gas phase and solution data are in good agreement and further validate the argument that azulene predominantly by direct coupling with upper vibronic levels of the ground state. It should be noticed also that the theoretical predictions of Jortner, Berry and Freed, with regard to the relaxation of a molecule belonging to the statistical model, are confirmed by the experimental observation that azulene in both solution and gas phase relaxes with the same rate, and exhibits to a great extent the predicted spectroscopic properties of a statistical case. The data for the azulene molecule in solution are displayed in Fig. 10.

III. Time and Frequency Resolved Spectra

The methods by which we were first able to measure picosecond relaxation rates by absorption techniques, especially by the mirror translation method, have the following disadvantages:
1. variation in the intensity of the pulses within a train;
2. wide variation in the widths of the pulses in a single train;
3. being time-consuming, requiring at least a laser shot per experimental point;
4. wide variation of pulse width and intensity from shot to shot;
and, in addition, wave length resolution is totally absent. In the past few years we have utilized completely different experimental methods for picosecond spectroscopy and relaxation measurements, which permit simultaneous time and frequency resolution with a single picosecond pulse [62].

The time-frequency resolving method has been utilized for both absorption and emission measurements of relaxation processes [62−64]. The apparatus [62], which we believe to be the first such apparatus designed for simultaneous time-frequency resolution is shown in Fig. 13.

In this particular case a picosecond light pulse from a 1.06μ (9431 cm^{-1}) Nd^{+3} glass laser generates the second harmonic for use as an interrogating pulse, although stimulated Raman or laser emission from a dye has been used for this purpose. The fundamental (9431 cm^{-1}) frequency pulse traverses the path shown in Fig. 13 and enters the optical shutter cell [64]. The shutter is a 0.1×20 mm CS_2 cell positioned between two polarizers crossed at $45°$ to the field of the 1.06μ pulse. As this picosecond pulse propagates along the 20 mm length of the CS_2 cell, it induces a birefringence in the CS_2, similar to that induced by the high electrical pulses of Kerr cell shutters. In the case of the picosecond pulse, the shutter transmits light through different points along the 20 mm cell of the shutter in step with the propagation of the opening pulse. The reorientation time for CS_2 is about 1.8 psec. Thus, the time resolution at each point within the CS_2 is restricted, in this "kerr liquid", to 1.8 psec or by the length of the pulse, which usually has a longer duration than 1.8 psec. Therefore, a spot of the shutter is "open", i.e. transmits light through the cross polarizers, only during

the orientation time and is then closed. Hence, the light recorded by the spectrometer is dependent upon the time and point at which it arrives in the CS_2 cell.
The second harmonic pulse is reflected by the mirror shown in Fig. 11 and
induces the excitation, in this case, of rhodamine 6 G. The emitted radiation
passes through the optical shutter during the time, and at the point, where the
shutter is open for 2 psec. The path of the fundamental pulse along the length
of the CS_2 cell provides the time-resolving mechanism. The CS_2 cell is imaged
along the length of the slit of the spectrometer. The excitation pulse, corresponding to $t = 0$ with regard to fluorescence from the molecule under investigation,
intersects the fundamental at the upper part of the CS_2 cell. Subsequent emission
is transmitted through the shutter at various lower points of the CS_2 cell
depending upon the time at which emission originates.

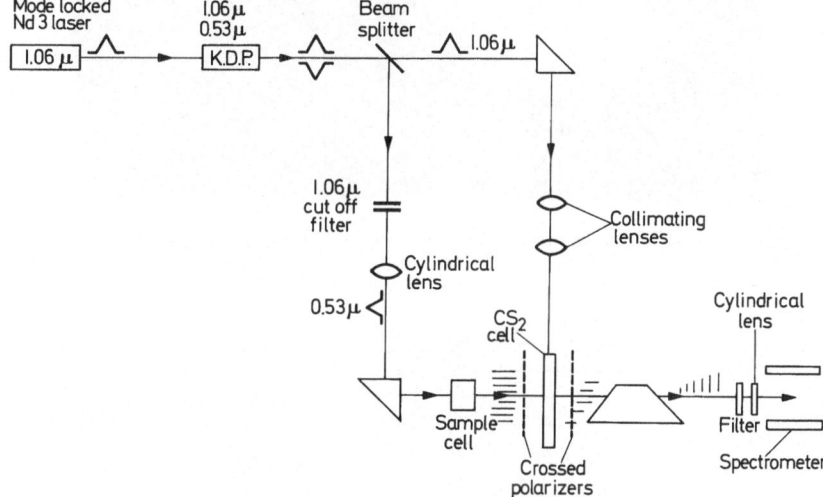

Fig. 11. Method for time-frequency resolution of picosecond spectra

The light transmitted through the shutter is now time-resolved, with time increasing from top to bottom in the CS_2 cell, since the shutter cell is imaged
along the length of the slit of the spectrometer with $t = 0$ on the top of the slit.
The resulting spectrum now has two coordinates — one, the normal frequency
display (v), the other the time in picoseconds (t) [63]. This apparatus was first
used for studies of vibrational relaxation in rhodamine 6 G.
Rhodamine 6 G was excited by 0.53 μ picosecond pulse, resulting in stimulated
emission from the rhodamine solution. The exciting 0.53 μ pulse was synchronized so that it arrived simultaneously with the shutter-opening pulse and was
transmitted at the upper point of the CS_2 cell corresponding to $t = 0$. The time-
frequency plate, Fig. 12, shows the 0.53 μ pulse at $t = 0$. Later, a very small
amount of spontaneous emission from the upper vibrational levels of the first
singlet [65] and, after a period of ~6 psec, the onset of the stimulated emission
was observed with a duration of ~10 psec. The spectral resolution achieved by

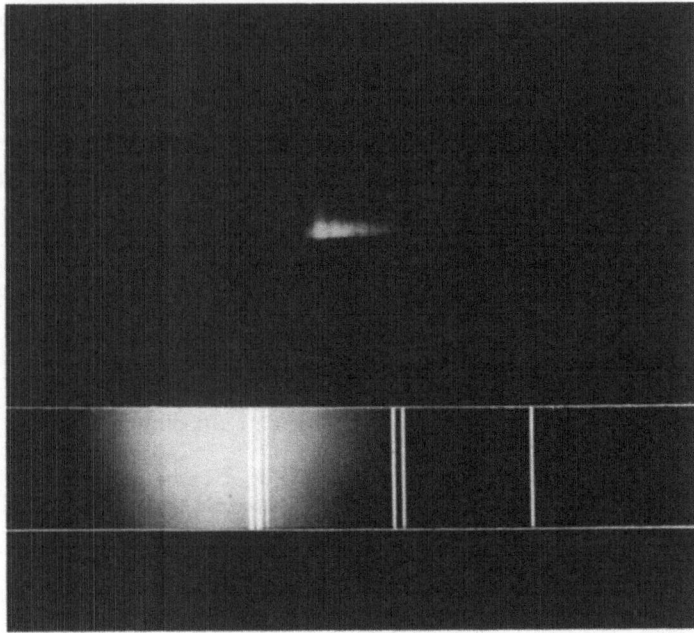

Fig. 12. Simultaneous time-frequency resolved spectrum of rhodamine 6 G

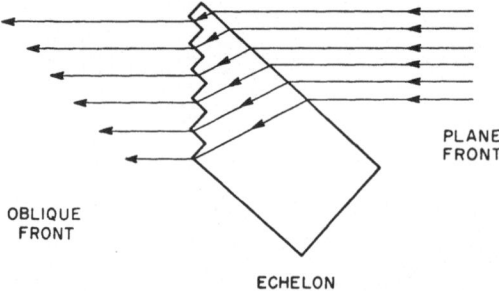

Fig. 13. The echelon, a single pulse passing through or reflected by the echelon is segmented into several pulses having any digitized separation equal intensity and the duration of the original pulse

this method is strictly spectrometer limited. The rhodamine 6 G quantum yield for emission is almost unity, and since the stimulated emission occurs by a four level mechanism, the 6 psec time for the onset of stimulated emission represents the rate of vibrational relaxation within the electronically excited state.

The observed relaxation is due to either 1) intramolecular energy distribution by anharmonic coupling of the optically active modes with vibrational states resulting in the redistribution of the excess energy to intramolecular vibrational modes or, 2) dissipation of the energy by coupling to lattice modes. The vibrational relaxation is related to the width of the absorption bands by the Fourier transform and can be related to the solvent-solute, "supermolecule" interaction consisting of the solute and the neighboring solvent molecules.

By the same method, the stimulated Raman emission [63] from the solvent was measured to be delayed by less than 3 psec, in agreement with theoretical postulations that molecular vibrations are only transiently excited by picosecond pulses. A further step in the development of high resolution time-frequency spectroscopy is the utilization of an echelon [65—67], which in practice generates picosecond pulses of identical duration and intensity yet with selected constant separation (Fig. 13). This simple yet very reliable method is shown in Fig. 14.

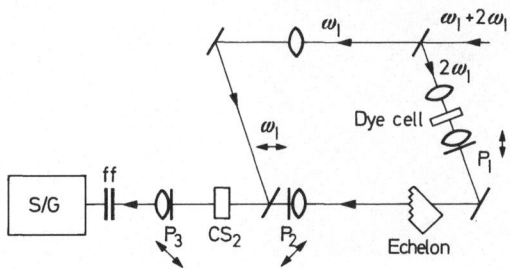

Fig. 14. Schematic of the echelon apparatus. This type with obvious variations has been used for both emission and absorption picosecond spectroscopy

A train of pulses traverses the pockels cell where a single pulse is extracted. This single pulse passes through the "stepped delay-echelon", which consists of a quartz block cut into several equal steps, or a stack of plates. As the pulse travels through, or is reflected by the echelon, it experiences variable delays corresponding to the length of that particular part of the echelon. The relative delay is simply a function of the refractive index, n, of the echelon medium, the thickness of the step d, and the angle of incident Θ,

$$t = (d/c)\,[n^2 - \sin^2 \Theta^{1/2} - \cos \Theta]$$

After the echelon, the output now consists of a train of pulses of equal intensity, identical duration, and separated selectively by one or more picoseconds. This set of pulses has been utilized either as the absorption or interrogation pulse placed before the sample cell, or as the time-resolving element in emission by locating the echelon after the sample cell. Two typical experimental arrangements which have been applied in the study of relaxation processes are shown in Figs. 14 and 16. One application, for example, was to measure accurately the reorientation mechanism of molecules such as CS_2, which control the time resolution of the experiment, if the pulse duration is ultrashort.

To measure directly the reorientation time of CS_2, we placed the CS_2 cell between crossed polarizers as in a normal "shutter" and utilized the apparatus shown in Fig. 14. A single pulse is extracted from the train of picosecond pulses and split into two parts, one with 90 % of the intensity and the other with

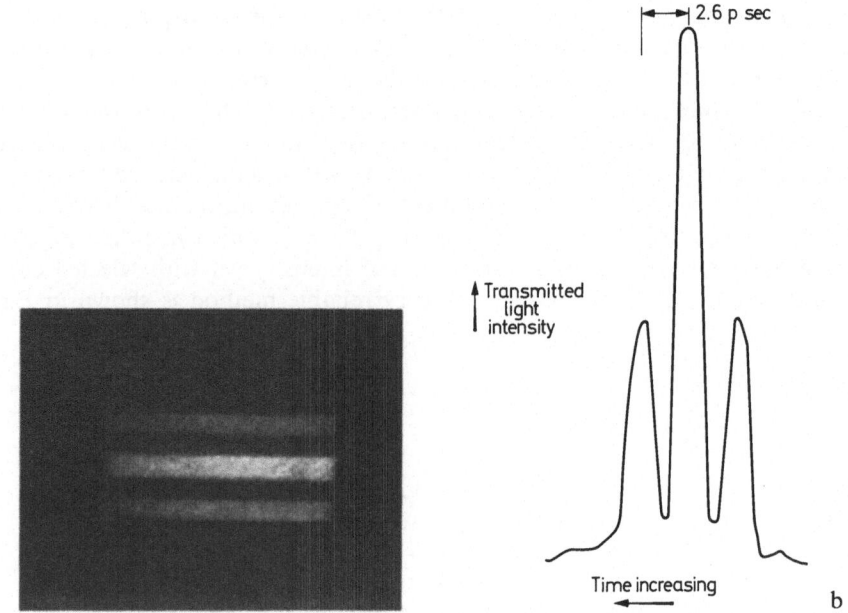

Fig. 15. a Photograph of the reorientational relaxation of CS_2 by the echelon method, used as in absorption. b Photodensitometer trace of the echelon segments

10%. The intense pulse acts as the shutter-opening pulse while the weaker one passes through the echelon. The echelon output, which now consists of a set of picosecond pulses, enters the CS_2 cell collinearly and simultaneously with the intense pulse. The transmitted segments of the echelon are then displayed on a photographic plate. Since each segment represents 1.2 psec, the number and intensity of the transmitted segments give directly the reorientation time of CS_2.

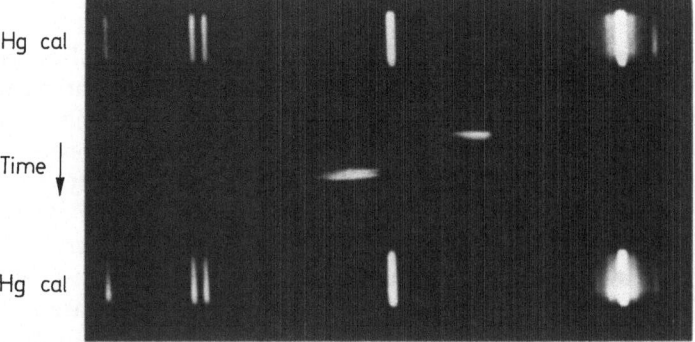

Fig. 16. The echelon method in emission. Stimulated emission from rhodamine 6 G in ethanol. Vertical echelon segments — 3 psec per segment. Center broad emission at 545 nm. Structures segment at 7 psec and 558 nm

Obviously, by minor variation of the apparatus, one can easily obtain the re-orientation relaxation rate of other molecules in the excited and ground electronic states. The resulting absorption of the echelon-generated pulses, from which we measured the total opening time of the shutter as 3.6 psec, is shown in Fig. 15a, b. This time includes both the CS_2 reorientation.

Several other chemical processes have been studied by us which are not described in this chapter. They include: (1) the decay characteristics of an excited electronic state of a large molecule characterized by a small electronic energy gap — these experiments measure the radiative decay of the second excited singlet state of 3, 4 benzpyrene [42] and of naphthalene [41] in the gaseous phase; (2) the decay of the S, state of benzophemone [67—70], which in solution exhibits ultrafast intersystem crossing and in the gas phase, neveals the features of intermediate level structure; (3) the energy transfer mechanism within single long chain molecules, such as polymers [71] and biological molecules [72, 73]; (4) the relaxation mechanism of dyes such as rhodamine 6 G, BBOT and polymethanes [74]; (5) cis-trans isomerization [51] of molecules such as rhodopsin; and (6) dynamic reorientation of large molecules [63]. Other current studies include the cage effect [76], and several other fast processes which we believe will provide information concerning the mechanism and characteristics of ultrafast molecular relaxation via picosecond spectroscopy.

IV. Relaxation of Excess Electrons in a Polar Solvent

The wealth of physical information concerning the physical properties of excess electron states in liquids [77—82] pertains mainly to "long-time" experiments where the features of the energetically stable electron are monitored. When an excess electron is introduced into the conduction band of a liquid by an adiabatic process, an experiment can be performed so that the liquid does not rearrange during the electron injection process, whereupon the electron is initially produced in the quasi free state. Two types of adiabatic injection experiments can be performed: (a) photoemission from a cathode [83] or emission from a field emission tip [84] immersed in a nonpolar liquid; (b) photoionization of an impurity state in a nonpolar [85] or in a polar liquid, which involves excitation to the bottom of the conduction band. The localized electron itself can act as an "impurity" which can be photoionized [86]. Thus, in a particular liquid where the localized excess electron state is energetically favored, the quasi free electron state can be produced by adiabatic injection or by photoionization and will subsequently relax to the localized state. This relaxation process will be accompanied by configurational changes in the liquid. The dynamics of electron localization can be described as a radiationless transition [81] and, originating from the breakdown of the Born-Oppenheimer approximation, as induced by nonadiabatic coupling between the quasi-free and the ground localized states. In polar liquids, Walker [87] recently studied the optical bleaching of solvated electrons in water by a Q-switched laser (pulse length 20 msec) and by extrapolation concluded that the quasi-free-bound relaxation rate is shorter than 6 psec.

We have observed the quasi-free-bound electron relaxation in a polar solvent utilizing the techniques of ultrafast picosecond spectroscopy [88]. In view of the

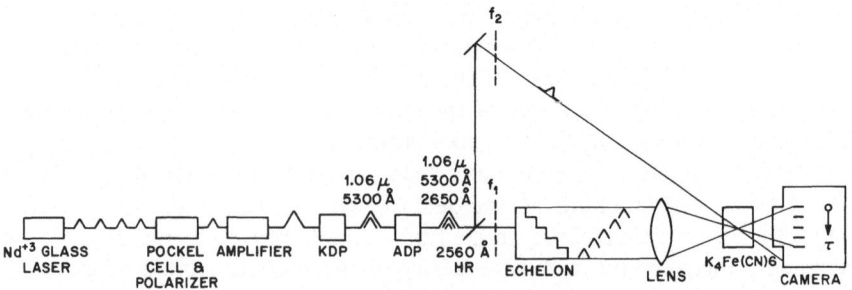

Fig. 17. Schematic representation of the experimental arrangement for the study of the relaxation of the solvated electron. The absorption of the hydrated electron was monitored at 1.06 μ or at 5300 Å

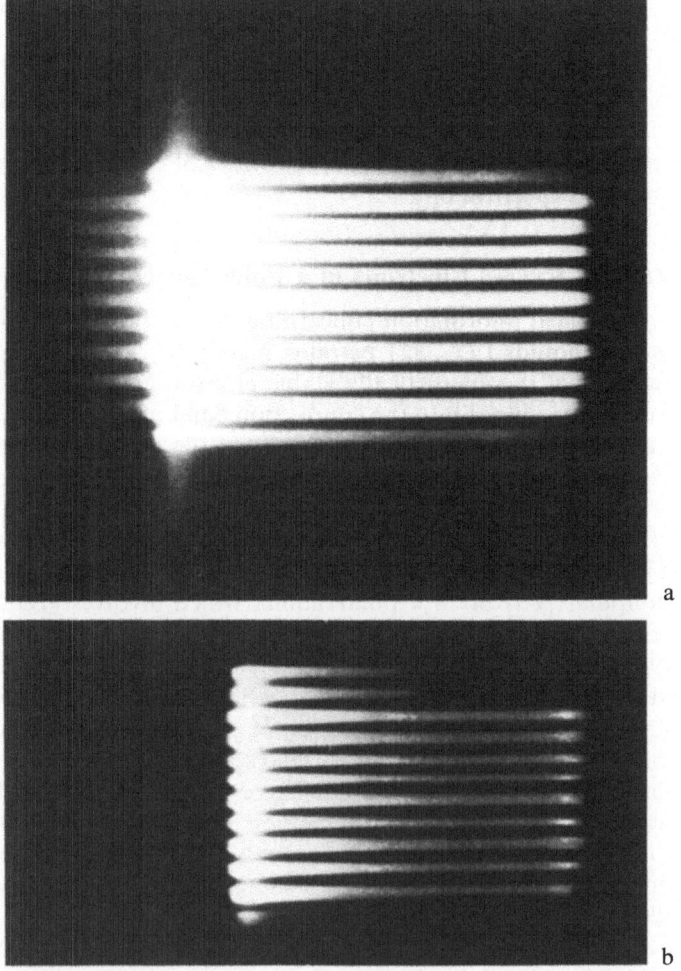

Fig. 18. Onset of the optical absorption of the hydrated electron interrogated at 1.06 μ. Each echelon segment is 0.81 mm in width, corresponding to the resolution of ∼2 psec. a No photoionization. b Photoionization at 2650 Å

current interest in the properties of the hydrated electron we have chosen to study the dynamics of electron localization in water. The quasi-free electron was generated by photoionization of the ferrocyanide anion excited at 2650 Å (i.e. the fourth harmonic of neodymium-glass laser) while the formation of the localized hydrated electron was interrogated by its absorption at $1.06\,\mu$ (the fundamental wave length of the neodymium-glass laser), at 5300 Å (the second harmonic of neodymium-glass laser) and in the spectral region 4500—9000 Å utilizing a continuum picosecond pulse [89].

The experiment was performed with a mode-locked Nd^{+3}-glass laser which generated a set of picosecond pulses. A single pulse was extracted, amplified and after passing through an angle phase-matched KDP and a temperature-matched ADP, the fundamental wave length, $1.06\,\mu$, was converted to 5300 Å by the KDP and this in turn to 2650 Å. The single pulse had a width of $\sim 2\,$psec, measured by the three-photon fluorescence method. The schematic representation of Fig. 17 portrays the optical path of the exciting 2650 Å light and the $1.06\,\mu$ interrogating echelon train. Third echelon segment ($1.06\,\mu$) and the exciting 2650 Å light arrive simultaneously in the $K_4\,Fe(CN)_6/H_2O$ solution after filtering out the light of all other wave lengths, and imaging the echelon on the camera. It was observed (see Fig. 18) that the absorption of the $1.06\,\mu$ pulse occurred at the fourth echelon segment within $\sim 2\,$psec, indicating that the initial rate of formation of the hydrated electron is $\sim 5 \times 10^{11}\,sec^{-1}$. Interrogation of the absorption of the hydrated electron at 5300 Å revealed that the optical absorption sets in within $\sim 4\,$psec. Subsequent experiments utilizing a continuum picosecond pulse [83] (see Fig. 19, p. 492) show that the total hydrated band evolution developes in time from the low frequencies to high frequencies within $\sim 4\,$psec.

References

1. Hargrove, L. E., Fork, R. L., Pollack, M. A.: Appl. Phys. Letters **5**, 4 (1964).
2. Mocker, H. W., Collins, R. J.: Appl. Phys. Letters **9**, 270 (1965).
3. Stetser, D. A., DeMaria, A. J.: Appl. Phys. Letters **9**, 118 (1966).
4. Bloembergen, N.: Comm. Solid State Phys. **1**, 37 (1968).
5. Armstrong, J. A.: Appl. Phys. Letters **10**, 16 (1967).
6. Rentzepis, P. M.: Chem. Phys. Letters **2**, 117 (1968).
7. Giordmaine, J. A., Rentzepis, P. M., Shapiro, S. L., Wecht, K. W.: Appl. Phys. Letters **11**, 216 (1967).
8. Busch, G. E., Rentzepis, P. M.: Picosecond molecular relaxation with a continuous mode-locked laser. Chem. Phys. Letters, to be published.
9. Patel, C. K. N.: Appl. Phys. Letters **18**, 25 (1971).
10. Wood, O. R., Abrams, R. L., Bridges, T. J.: Appl. Phys. Letters **17**, 376 (1970).
11. See for example, Shaw, E. D., Patel, C. K. N.: Appl. Phys. Letters **18**, 215 and **18**, 274 (1971).
12. Vuylstoke, A. A.: J. Appl. Phys. **34**, 1615 (1963).
13. DeMaria, A. J.: Proc. IEEE **57**, 1 (1969).
14. Rentzepis, P. M., Duguay, M. A.: Appl. Phys. Letters **11**, 218 (1967).
15. Weber, H. P.: Phys. Letters **27**, 321 (1968).
16. Klauder, J. R., Duguay, M. A., Giodmaine, J. A., Shapiro, S. L.: Appl. Phys. Letters **13**, 174 (1968).
17. Harrach, R. J.: Phys. Letters **28**, 393 (1968).
18. Weber, H. P., Dandliker, R.: IEEE J. Quantum El **4**, 1009 (1968).

19. Drexhage, K. H.: Appl. Phys. Letters **14**, 318 (1969).
20. Kuznetosova, T. I.: Soviet Phys. JETP **28**, 1303 (1969).
21. Rentzepis, P. M., Mitschele, C. J., Saxman, A. C.: Appl. Phys. Letters **17**, 122 (1970).
22. Rentzepis, P. M., Douglass, D. C.: Measurement of pulse shape by the three photon fluorescence method, to be published.
23. Blount, E. I., Klauder, J. R.: J. Appl. Phys. **40**, 2874 (1966).
24. Treacy, E. B.: Appl. Phys Letters **17**, 14 (1970).
25. Shapiro, S. L., Duguay, M. A.: Physics Letters **28 A**, 698 (1969).
26. Bradley, D. J., New, C. H. C., Laughey, S. J.: Phys. Letters **30 A**, 78 (1969).
27. Jortner, J., Berry, R. S.: J. Chem. Phys. **48**, 2757 (1968).
28. Bixon, M., Jortner, J.: J. Chem. Phys. **48**, 715 (1968).
29. Bixon, M., Jortner, J.: J. Chem. Phys. **50**, 4061 (1969).
30. Freed, K., Jortner, J.: J. Chem. Phys. **50**, 2916 (1969).
31. Nitzan, A., Jortner, J., Rentzepis, P. M.: Molecular Physics **22**, 583 (1972).
32. Robinson, G. W., Frosh, R. P.: J. Chem. Phys. **38**, 1187 (1964).
33. Lin, S. H., Bersohn, R.: J. Chem. Phys. **48**, 2723 (1968).
34. Block, B., Feschbach, H.: Ann. Phys. (N.Y.) **23**, 47 (1963).
35. Jortner, J.: J. Chim. Phys. Paris **9**, 1970 (1969).
36. Nitzan, A., Jortner, J., Rentzepis, P. M.: Proc. Roy. Soc. Lond. **A 327**, 367 (1972).
37. Douglas, A. E.: J. Chem. Phys. **45**, 1007 (1966).
38. Nitzan, A., Jortner, J.: To be published.
39. Busch, G. E., Rentzepis, P. M., Jortner, J.: J. Chem. Phys. Jan. issue (1972).
40. Busch, G. E., Rentzepis, P. M., Jortner, J.: Chem. Phys. Letters (1971).
41. Wannier, P. G., Rentzepis, P. M., Jortner, J.: Chem. Phys. Letters **10**, 102 (1971).
42. Wannier, P. G., Rentzepis, P. M., Jortner, J.: Chem. Phys. Letters **10**, 193 (1971).
43. Rentzepis, P. M., Jortner, J., Jones, R. P.: **4**, 599 (1970).
44. Huppert, D., Jortner, J., Rentzepis, P. M.: J. Chem. Phys. to be published (1972).
45. Hubbert, D., Jortner, J., Rentzepis, P. M.: Chem. Phys. Letters, to be published (1972).
46. Nitzan, A., Jortner, J., Drent, E., Kommandeur, J.: Chem. Phys. Letters, in press.
47. Drent, E.: Ph. D. Thesis, Univ. of Groningen, The Netherland (1971).
48. Rentzepis, P. M.: Chem. Phys. Letters **2**, 117 (1968).
49. Rentzepis, P. M.: Chem. Phys. Letters **3**, 717 (1969).
50. Scarlet, R. I., Figueria, J. F., Mahr, H.: Appl. Phys. Letters **13**, 71 (1968).
51. Rentzepis, P. M.: Photochem. and Photobio. **8**, 579 (1968).
52. Eisenthal, K. B., Drexhage, K. H.: J. Chem. Phys. **51**, 5720 (1969).
53. Rehm, D., Eisenthal, K. B.: Chem. Phys. Letters **9**, 387 (1971).
54. Drent, E., Van Der Deije, G. M., Zandstra, P. J.: Chem. Phys. Letters **2**, 526 (1968).
55. Beer, M., Longuet-Higgins, H. C.: J. Chem. Phys. **23**, 1390 (1955).
56. Viswanath, G., Kasha, M.: J. Chem. Phys. **24**, 574 (1955).
57. Rentzepis, P. M.: Science **169**, 239 (1970).
58. Rentzepis, P. M., Mitschele, C. J.: Anal. Chemistry **42**, 20 (1970).
59. Malley, M. M., Rentzepis, P. M.: J. of Luminescence **1, 2** 448 (1970).
60. Hubbert, D.: Ph. D. Thesis, Univ. of Tel-Aviv, Israel (1972).
61. Busch, G., Jortner, J., Rentzepis, P. M.: unpublished data.
62. Rentzepis, P. M., Topp, M. R., Jones, R. P., Jortner, J.: Phys. Rev. Letters **25**, 1742 (1970).
63. Rentzepis, P. M., Topp, M. R.: Annal. N. Y. Ac. Sci. **33**, 284 (1971).
64. Topp, M. R., Rentzepis, P. M.: Chem. Phys. Letters **4**, 1 (1971).
65. Topp, M. R., Rentzepis, P. M., Jones, R. P.: J. Appl. Phys. **42**, 3451.
66. Topp, M. R., Rentzepis, P. M.: J. Chem. Phys. in press (1972).
67. Rentzepis, P. M., Busch, G. E.: Molecular Photochem. in press (1972).
68. Busch, G. E., Rentzepis, P. M.: to be published.
69. Nitzan, A., Jortner, J., Rentzepis, P. M.: Chem. Phys. Letters **8**, 445 (1971).
70. Rentzepis, P. M.: Photochem. and Photobio., to be published (1972).
71. Jortner, J., Rentzepis, P. M.: to be published.
72. Rentzepis, P. M., Jortner, J.: To be published.
73. Malley, M. M., Rentzepis, P. M.: Chem. Phys. Letters **3**, 534 (1969); — **7**, 57 (1970).
74. Topp, M. R., Rentzepis, P. M.: Phys. Rev. **3 A**, 538 (1970).

75. In collaboration with Drs. Lamola, A. A., and Busch, G. E.
76. In collaboration with Prof. Szwarc, M., and Dr. Busch, G. E.
77. Metal–ammonia solution, Lepoutre, G., Sienko, M. J., Eds. New York: Benjamin 1964.
78. Solvated electron advances in chemistry, series No. 50, Americal chemical society, Washington, D.C. (1964).
79. Proceedings of colloque Weyl II (Cornell 1969). Lagowski, J. J., Sienko, M. J., Eds. London: Butterworth 1970.
80. Hart, E. J., Anbar, M.: The hydrated electron. New York: Wiley 1970.
81. Jortner, J.: Berichte der Bunsen-Gesellschaft für physikalische Chemie, 75, 696 (1971).
82. Copeland, D., Kestner, N. R., Jortner, J.: J. Chem. Phys. 53, 1189 (1970).
83. Woolf, M. A., Rayfield, G. W.: Phys. Rev. Letters 15, 235 (1935).
84. Halpern, B., Gommer, R.: J. Chem. Phys. 43, 1069 (1968).
85. Raz, B., Jortner, J.: Proc. Roy. Soc. A 317, 113 (1970).
86. Northby, J. A., Sanders, T. M.: Phys. Rev. Letters 18, 1184 (1967).
87. Walker, J.: Chem. Phys. (1970).
88. Rentzepis, P. M., Jones, R. P., Jortner, J.: Chem. Phys. Letters 5, 480 (1972).
89. Rentzepis, P. M., Jones, R. D., Jortner, J.: Dynamics of excess electrons in Polar solvents. J. Chem. Phys. in Press.

Discussion

L. M. DORFMAN   It should be noted that these new data for the dynamics of solvation, obtained in these elegant experiments or Rentzepis, represent the second picosecond observation of this process, the first having been published by Hunt and his colleagues. It appears to me that the two sets of observations may be at variance. Hunt, using a stroboscopic pulse radiolysis technique which utilizes the 10 psec substructure of electron pulses of a linac, has observed the spectrum of e_{solv}^- in water and some alcohols, and has concluded that in water the same spectrum observed by Hart and Boag on a μ sec scale is fully formed in about 10 psecs. Similarly, in ethanol, in 10 psecs he sees the same band we observed on a μ sec scale some years ago. This would seem at variance with Rentzepis' observation that "in the green, the band is beginning to form in about 7 psecs". [Editor's note: The correct number reported by Rentzepis in his paper is 4 psec.] This represents a considerable difference since at least 3 or 4 half-lives would be required for complete formation, whereas Rentzepis is apparently seeing part of the first half-life. What is the reason for the difference?

G. CZAPSKI   Dorfman raised the apparent disagreement in hydration times of the electron as observed in picosecond pulse radiolysis and on the present flash photolysis studies. The pulse radiolysis studies indicate shorter hydration times [Bronskill, M. J., Wolf, R. K., Hunt, J. W.: J. Chem. Phys. 53, 4201 (1970)]. I would like to suggest a possible explanation for shorter solvation times in pulse radiolysis experiments.

In radiolysis, when we are concerned with processes occurring at times as short at $10^{-11} - 10^{-12}$ sec after the deposition of the radiation, one should consider the "temperature" in the reaction zone. Mozumder [Mozumder, A : Adv. Rad. Chem. Vol. I, Ed. Burton, M., Magee, J. L.: Interscience N.Y. 1969] calculated that on the average, in the spur, 30 eV are converted into heat in a volume with a radius of 20 Å in less than 10^{-12} sec. This would correspond to a "temperature increase" of 30° C in this region. (If a larger fraction of this energy is deposited initially only in the central core of the spur ($r \sim 15$ Å), the "temperature increase" may be as high as 70° C.)

Mozumder calculated the thermal equilibration time as being about 6 psec. Any process occurring at times shorter than 10^{-11} sec may be influenced by the pre-equilibrium conditions. This effect could account for the faster solvation of electrons than would be expected at room temperature. The solvation times of electrons in water as well as in various alcohols, was found to be below 10^{-11} sec in the pulse radiolysis technique, although the expected value for some of these alcohols, at room temperature, exceeded the observed solvation time by up to an order of magnitude [Mozumder, A.: J. Chem. Phys. **50**, 3153 (1969)]. This observation can be explained as being due to faster process at times prior to the achievement of thermal equilibrium at the "solvation zone".

M. S. MATHESON   First, I want to amplify the comment just made by Dorfman. In Hunt's stroboscopic pulse radiolysis experiments on the solvation of the electron, he sets a time resolution of about 20 psec, since it takes the analyzing Cerenkov light 20 psec longer to transverse the 2 cm cell than it takes the electron. That is, electrons and photons entering the cell simultaneously are dephased by 20 psec after transversing 2 cms of water. Nevertheless, he concludes from deconvoluting his absorption vs. time curves that the electron is solvated in water in less than 10 psec, indeed, he believes in much less than 10. Again, as Dorfman has said, Rentzepis' result is difficult to reconcile with Hunt's result. The fact that Hunt produces electrons with a range of energies while Rentzepis produces monoenergetic electrons in his experiment, may possibly provide an explanation. That is, a monoenergetic electron of about 1 eV is thermalized in a few picoseconds and then begins to solvate, the solvation process being rapid. The extra five picoseconds for the green absorption to appear is also a measure of the solvation time. Would Dr Rentzepis please comment on these suggestions?

P. M. RENTZEPIS   Our experiments demonstrate that the hydrated electron is formed in an equilibrium configuration in about 4 psec. Thus after 4 psec the absorption at 5300 Å is developed. We have not yet elucidated the details of the kinetic analysis concerning this ultrafast relaxation process.

U. EVEN   What is the meaning of "local temperature" in a pulse radiolysis experiment?

J. JORTNER   One cannot, of course, define a thermodynamic property, but rather must refer to local distribution of vibrationally, rotationally and translationally excited molecules.

L. M. DORFMAN   I would like to add that a relaxation for e^-_{solv} from the infrared to the equilibrium spectrum is not unexpected. Platzman pointed out in 1953 that the solvation process would involve the infrared frequencies (lower trapping energy) first. Partial electron solvation can occur without involving the complete dipole relaxation. The relaxation process is indeed slowed down by decreasing the temperature. Baxendale has observed, by nanosecond pulse radiolysis, that in low temperature alcohol glasses, part of a band is first seen in the infrared, which shifts over tens of nsecs to the absorption band finally seen at 700 nm and longer wave lengths at 1 nsec.

J. JORTNER   The direct observation of the time evolution of the solvated electron band demonstrates the roles of short and long range interactions for electron localization in polar solvents. The continuum polar modes relax fast and provide a sufficiently deep well to localize the electron, which then absorbs around 1 μ. Subsequently the first coordination layer becomes oriented leading to the development of the final spectrum of the hydrated electron.

Quasi-Free Electrons in Polar Liquids

PAUL DELAHAY

Abstract

Information on solvated and quasi-free electrons obtained from photoelectron emission (PEE) spectroscopy is discussed. Topics covered: production of quasi-free electrons in solution by optical transition; resolution of the absorption spectrum of solvated electrons in various solvents; rate of loss of kinetic energy of quasi-free electrons in solution from *PEE* spectra; energy distribution curve for electrons emitted by solutions in vacuum.

I. Introduction

The study of quasi-free electrons is closely allied to solvated electrons in this paper, and such an association has indeed historical roots. Thus, the formation of quasi-free electrons by optical excitation of solvated electrons (in liquid ammonia) was proposed long ago by Ogg [1], the loss of kinetic energy of quasi-free electrons moving in polar liquids was analyzed by Platzman [2] in a paper in which the existence of the hydrated electron was predicted. These two references pertain to two major, but poorly understood aspects of quasi-free electrons in polar liquids. PEE spectroscopy provides a way of studying these two aspects and offers perhaps unassailable proof that quasi-free electrons can be produced by optical excitation of a solute in polar liquids. The final state of the emission process indeed corresponds to free electrons in a gas [3] (in some cases at sufficiently low pressure to be considered a vacuum).

The following, recently proposed model [4] for PEE by solutions embodies the two major aspects just mentioned:

The process of PEE by solutions is decomposed (in order to render it tractable) into three steps: a) production of quasi-free electrons in the bulk of the solution by optical excitation; b) their transport to the solution-gas (vacuum) interface; c) their transfer across the interface into the gas (vacuum). Such three steps were initially proposed by Spicer [5] for PEE by semiconductors and metals, but the treatment of each step differs for solutions and solids.

Two kinds of experimental information are available in PEE spectroscopy to test theory: a) dependence of the PEE quantum yield upon photon energy; b) distribution of the kinetic energy of electrons emitted in vacuum. Data on a) can be gathered for a variety of solvents and can be corrected, if necessary, for back scattering of electrons in the gas phase when the solvent vapor pressure is too high to allow the neglect of this effect. Determination of b), namely energy distribution curves (EDC), requires the use of solvents with sufficiently low vapor pressure [e.g., water-free glycerol, hexamethyl phosphoric triamide (HMPA; probably satisfactory as judged from preliminary work mentioned in III-B)].

Methodology has been developed for determination of PEE quantum yields of solutions (from PEE currents) for aqueous solutions of emitters [6], emitters in H₂O–NaOH glasses [7], solvated electron solutions (in HMPA [8] and liquid ammonia [9, 10]), organic anion radicals [11] (naphthalene, anthracene). The Millikan drop method was also recently applied to this problem [12, 13]. Experimental determination of EDC's from retarding potentials was recently developed [14] for PEE by solutions by transposition of Spicer's extensive work on solids [15]. Application of the Millikan drop method was also reported for this purpose [16].

II. Production of Quasi-Free Electrons in Solution by Optical Transition

The production of quasi-free electrons in solution by optical excitation is of interest to PEE spectroscopy and other areas, e.g., photoconduction [17], absorption spectroscopy, etc. Several processes by which quasi-free electrons might be produced in solution could be conceived by transposing to solutions the extensive information available on this problem for solids. Only rather limited information is available at present for solutions, and we shall retain here only direct transition to the continuum in solution. Other processes are not excluded and may indeed be significant for various emitters.

The main problem is to obtain the dependence of the probability for direct transition to the continuum on photon energy. A simple solution to this problem was recently proposed [4] for solvated electrons by transposition of the case of atomic hydrogen. The main underlying assumption consists of the use of a simple coulombic potential, just as for a Landau type of polaron. It is then possible to derive an equation for the bound-continuum absorption band if a further assumption is made about the distribution of configurations. A Gaussian distribution was used, and reasons in favor of this choice are given in Ref. [18]. Experimental absorption bands are then regarded as composed of one or several bound-bound (bb) bands and one bound-continuum (bc) band. Only the bc band is significant at sufficiently high photon energies, and the foregoing analysis can be tested experimentally. To a good approximation, the experimental absorbancy in the high-energy tail of experimental absorption spectra of solvated (or trapped) electrons should be proportional to the photon energy to the power $-8/3$. This is indeed the case (Fig. 1) for several solvents [4, 18] but not for others (liquid ammonia, water). The contribution arising from electronic polarization of the medium is neglected in this approach, as was recently noted [18], which may account for the failure of the $E^{-8/3}$ plot for liquid ammonia and water.

Once an equation is accepted for bc bands, it is possible to resolve absorption spectra of solvated electrons in various solvents. This was recently done [18], and encouraging results were obtained for solvents having a low β-factor, where β is the difference between the reciprocals of the optical and static dielectric constants. Resolution with a fitting as good as experimental errors was obtained for trapped electrons in 3-methyl pentane at 77 °K (Fig. 2) and solvated electrons in tetrahydrofuran with only one Gaussian bb band and one bc band. Resolution for solvents with higher β's, such as HMPA, and especially liquid ammonia and water,

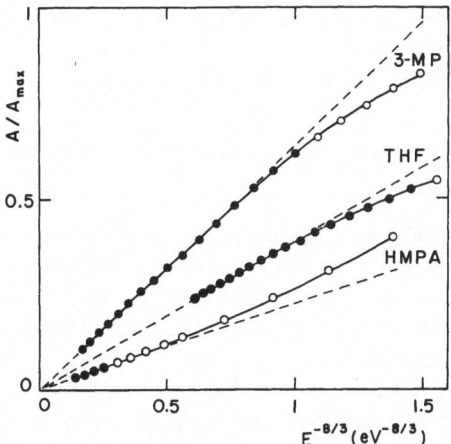

Fig. 1. Plot of normalized experimental absorbancy against photon energy to the power $-8/3$ for trapped electrons in 3-methyl pentane (at 77° K) and solvated electrons in tetrahydrofuran and hexamethyl phosphoric triamide (both at room temperature) according to ref. 18. Solid circles correspond to range of energy in which departure from linearity should not exceed -3%

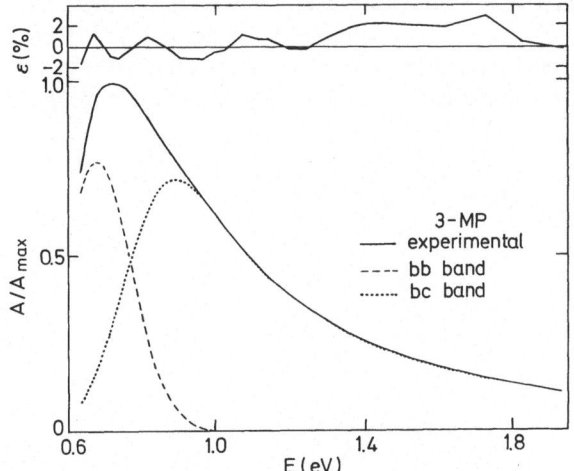

Fig. 2. Resolution of the absorption spectrum of trapped electrons in 3-methyl pentane at 77° K according to Ref. [18]

is less convincing because of the necessity of introducing more than one bb band and the poorer approximation provided by the equation for the bc band (for ammonia and water). Interesting correlations with the β-factor of the solvent also follow: major contribution of the bc band for solvents with low β's versus small contribution of this band for high β's; monotonic increase of the energy at the bottom of the continuum with β for all solvents studied. Details are given in Ref. [18].

This resolution method is somewhat empirical but not to the extent of mere data fitting. It suggests that solvated (or trapped) electrons in solvents with low β may be more easily amenable to detailed model calculations [19] than the usually treated cases of ammonia and water. Improvement in the equation for the bc band is desirable, especially for solvents with high β's.

III. Loss of Kinetic Energy of Quasi-Free Electrons in Solution

A. From Quantum Yield

The PEE quantum yield at a given photon energy is proportional to the probability of transition to the continuum. If the functional dependence of this probability on photon energy is known (or assumed!), one can normalize PEE quantum yields to a constant transition probability. Such normalized quantum yields seem to vary linearly with the photon energy over a relatively wide range for solvated electrons in HMPA [4]. One can show that this result implies, to a first approximation, that the range of quasi-free electrons in solutions varies linearly with photon energy. The range depends on the rate dW/dt at which quasi-free electrons lose their kinetic energy during their motion prior to thermalization. Analysis as a simple random walk problem (neglecting coulombic interaction) leads to the conclusion that dW/dt is inversely proportional to the speed of quasi-free electrons. Moreover, dW/dt for quasi-free electrons in liquid ammonia, as computed from Häsing's data [9] for liquid ammonia, is of the order of magnitude (e.g., 10^{13} eV sec^{-1} for 1 eV-electrons) estimated by Platzman [2]. Details are given in Ref. [4]. This approach could be refined considerably, but the results just summarized already show that PEE does provide a way of estimating dW/dt.

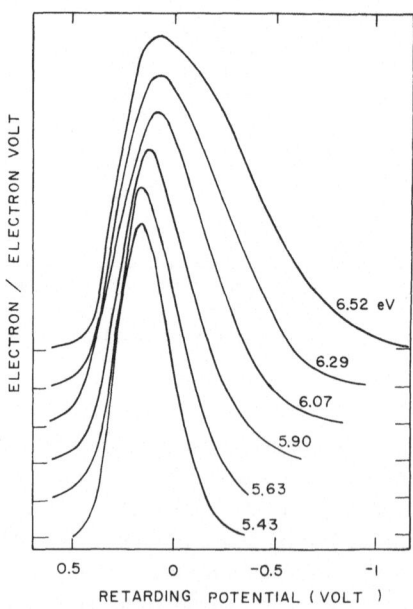

Fig. 3. Energy distribution curves for 0.44 M lithium ferrocyanide in water-free glycerol at room temperature for different photon energies according to Ref. [14]. Curves are normalized to the maximum

One serious problem must be mentioned, namely the necessity of knowing the emitter concentration in solution. The answer is direct for simple systems such as ferrocyanide in aqueous solution (even so, there are multiple equilibria), but then the production of quasi-free electrons by optical excitation is not amenable to quantitative analysis at this time. Conversely, this process is hopefully better understood for solvated electrons in some solvents, but the composition of such solutions is not known quantitatively despite much arduous work over many years. Dilute solutions ($< 10^{-3}$ mole liter^{-1}) of alkali metals in liquid ammonia are now being investigated [10] since knowledge about the composition of such solutions is more comprehensive than for other solvents.

B. From Energy Distribution Curves

The distribution of kinetic energy of electrons emitted in vacuum by a solution depends on the loss of kinetic energy by quasi-free electrons during their transport to the solution-vacuum interface. Experimental EDC for solutions are asymmetric bell-shaped curves with an elongated tail toward high energies. The width at half-peak increases with photon energy for ferrocyanide in glycerol (Fig. 3). Qualitative analysis shows that this effect is compatible with a mechanism involving loss of kinetic energy by quasi-free electrons during their motion. A preliminary quantitative analysis was developed to extract the function $dW/dt = f(W)$ from EDC's. This work is too recent to warrant further discussion at this stage, but the approach indeed appears very promising.

Acknowledgments

The work reported here was supported by the National Science Foundation and the Office of Naval Research. The author is also indebted to the John Simon Guggenheim Memorial Foundation for a fellowship during his sabbatical leave.

References

1. Ogg, R. A., Jr.: J. Am. Chem. Soc. **68**, 155 (1946).
2. Platzman, R. L.: Basic Mechanisms in Radiobiology. II. Physical and Chemical Aspects (National Research Council, Washington, D. C., 1953), Publication 305, pp. 22—50.
3. We do not consider here subsequent molecule-electron collisions and possible attachment.
4. Delahay, P.: J. Chem. Phys. **55**, 4188 (1971).
5. See, for instance, Sommer, A. H., Spicer, W. E.: In: Lavach, S. (Ed.): Photoelectronic Materials and Devices, pp. 175—221. New York: Van Nostrand 1965.
6. Nemec, L., Delahay, P.: J. Chem. Phys. **57**, 2135 (1972). See this paper for earlier references.
7. Bomchil, G., Delahay, P., Levin, I.: J. Chem. Phys. **56**, 5194 (1972).
8. Baron, B., Delahay, P., Lugo, R.: J. Chem. Phys. **55**, 4180 (1971).
9. Häsing, J.: Ann. Physik **37**, 509 (1940).
10. Aulich, H., Baron, B., Delahay, P., Lugo, R.: J. Chem. Phys. (in press).
11. Aulich, H., Baron, B., Delahay, P.: J. Chem. Phys. **58**, 603 (1973).
12. Ballard, R. E., Griffiths, G. A.: J. Chem. Soc. A**1971**, 1960.
13. Kinder, J.: Photoemission von Elektronen aus Elektronendonatoren in Elektrolytlösung. Dissertation, Tehnical University, Munich, 1971. Dissertation kindly supplied by its author.
14. Nemec, L., Baron, B., Delahay, P.: Chem. Phys. Letters **16**, 278 (1972).

15. Berglund, C. N., Spicer, W. E.: Phys. Rev. **136**, A1030, A1044 (1964); Krolikowski, W. F., Spicer, W. E.: Phys. Rev. **185**, 882 (1969). See series of papers by Spicer and co-workers on this topic.
16. Ballard, R. E., Griffiths, G. A.: Chem. Commun. 1472 (1971).
17. See, as an example Takeda, S. S., Houser, N. E., Jarnagin, R. C.: J. Chem. Phys. **54**, 3195 (1971).
18. Lugo, R., Delahay, P.: J. Chem. Phys. **57**, 2122 (1972).
19. E.g., Copeland, D. A., Kestner, N. R., Jortner, J.: J. Chem. Phys. **53**, 1189 (1970).

Discussion

N. R. KESTNER   Your use of hydrogen atom transition probabilities for the electron to the continuum may be reasonable in some of your systems since even in the ammonia and water system the higher excited states and the 2p levels are well explained by the simple Landau-Jortner polaron model. In other words when our more detailed models are solved for the higher excited states (not the 2s state) the results are similar to the older models. Near the center of the "cavity" the calculations are closer to a particle in a box but far from the cavity they are coulombic. Thus in cases with shallow traps and non-zero β values as in many of your examples the higher excited states and transition probabilities are probably similar to the hydrogenic models.

P. DELAHAY   I am glad to hear your favorable reaction to the use of the $E^{-8/3}$ relationship for bound-continuum transitions.

N. R. KESTNER   Related to my earlier comments does the $E^{-8/3}$ plot look poorer for water than for ammonia? Our model calculations would suggest this.

P. DELAHAY   It is hard to tell whether deviations from a linear $E^{-8/3}$ plot are greater for water than ammonia. Absorbancy data in the high-energy tail are available for water but hardly for ammonia (emphasis on maximum and width of band, possibly experimental difficulties). The spectrum for NH_3 in Jackman and Keenan [J. Inorg. Nucl. Chem. **30**, 2047 (1968)] shows only the tail and is calibrated in absorbancy. Hence, A/A_{max} is not known, and the intercept of the $E^{-8/3}$ plot along the ordinate cannot reliably be estimated in terms of A/A_{max} (as was done for water in Ref. [18] listed in my paper). Otherwise, deviation from linearity seems quite comparable for water and ammonia.

J. JORTNER   I would like to make three comments concerning these interesting results:

1. It is still an open question whether the absorption cross sections for bound continuum transitions can be adequately presented by the hydrogenic $E^{-8/3}$ power law. We should bear in mind that the potential exerted by the first coordination layer is reminiscent of a square well. Neil Kestner mentioned yesterday that the order of the 2p and 2s levels in a polar solvent is reversed, as is the case for a square well. Recently, Gaathon found that variational calculations for the ground state of the localized electron in polar solvents yield a lower energy utilizing spherical Bessel and Hankel variational wave functions. Thus the cross section for

the continuum absorption may deviate markedly from the hydrogenic model even in polar solvents.

2. The bound continuum transitions are inhomogeneously broadened by the contribution of several solvent configurations and by phonon broadening as is the case for bound-bound transitions.

3. I would like to offer an explanation for the large cross sections for bound continuum transitions in 3 MP in contrast to the low oscillator strength for these transitions in the case of ammonia. In the case of a polar solvent the potential well is deep (~ -6 eV) at small distances and coulombic at large distances and most of the intensity corresponds to bound-bound transitions. In the case of a nonpolar system the long range coulomb part is absent and the potential well resembles a spherical well. This relatively shallow spherical well sustains only a finite number of bound states. When the 2p state is close to the onset of the continuum the intensity of the 1s→p and the 1s→continuum transition will be comparable.

P. DELAHAY 1. Firstly, we agree, I am sure, that we are dealing with approximations, either in your 1959 model or in your recent one developed with Kestner. The question then is: How crude is the approximation provided by the $E^{-8/3}$ relationship for bound-continuum transitions in the high-energy tail? Evaluation based on bound-bound transitions should be overly pessimistic because the square-well type is then predominant. A better answer, it seems to me, would be provided by the calculation of transition probability for bound-continuum transitions for a potential composed of a square well with a coulombic part at larger distances. Comparison with the purely coulombic well should then answer the question. One then could ask: How realistic is the improved well?
The question you raised can also be answered, in part at least, by checking the experimental validity of the $E^{-8/3}$ plot. This can be misleading, as noted in Ref. [18] of my paper, but the fit is good for solvents with low beta factor. The main point of our approach, I should perhaps reiterate, is to stress the significance of bound-continuum transitions, at least for solvents of low beta. Such transitions have been considered, and rightly so, as unimportant for ammonia and water.

2. The $E^{-8/3}$ plots hold only at sufficiently high energies (several band widths above continuum) at which configuration distribution no longer matters much (cf. Ref. [18] of my paper). Your point, however, is entirely pertinent at lower energies, and the equation we used for bound-continuum is affected accordingly. However, see excellent fit for THF and 3MP.

3. I agree. This point is taken up briefly in Ref. [18], now in press.

M. MATHESON Have experiments varying the concentration of absorber permitted the determination of the range in solvent of electrons of given energy?

P. DELAHAY Experiments at a single concentration of solvated electrons allow determination of the range (see example in Ref. [4] of my paper). Calculations for different concentrations are, of course, highly desirable, and this is one reason why we are now investigating photoelectron emission by solvated electrons in ammonia. However, unravelling the chemistry of the system is arduous for less well-documented solvents (e.g., HMPA).

B. RAZ   Have you attempted to obtain information from the energy dependence of the photoemission yield curves near the threshold?

P. DELAHAY   No, we did not. It could be done by considering configuration distribution, but analysis at higher energies is easier.

Spectra of the Ammoniated Electron in the Presence of Amide

E. Saito

Abstract

In the determination of the near-infrared spectrum of the solvated electron in liquid ammonia, an anomaly, evidenced by a larger transparence of the solution, has been observed at wavelengths corresponding to the bands of ammonia in the 1500—2300 nm region. This anomaly showed up in the studies of the electron formed by the chemical equilibrium

$$NH_2^- + 1/2\, H_2 \rightleftharpoons e^- + NH_3$$

that is, in the presence of relatively high concentrations of KNH_2 and large hydrogen pressures. Under these circumstances it was difficult to attribute the cause of this anomaly to the solutes. Such solutions possess the property of stabilizing electrons created by photolysis or radiolysis, and we have used this characteristic to study the influence of the electron alone on this anomaly. Effectively, this anomaly increases with the concentration of the solvated electron which suggests the formation of a complex between e^-, KNH_2 and a certain number of ammonia molecules strongly linked so that they do not participate in the vibrational modes corresponding to the quoted wavelengths.

Introduction

The first part of this study was aimed at measuring the very low concentrations of the solvated electron arising from the equilibrium discovered by Jolly [1]

$$NH_2^- + 1/2\, H_2 \rightleftharpoons e_{am}^- + NH_3 \,. \tag{1}$$

Using a Beckman DK-1A double-beam spectrophotometer, we used cells with relatively long optical path lengths — 1.2, 2.5 or even 5.0 cm. With the same amide solution in the sample and reference cells, we could obtain a relatively horizontal base line as a function of wavelength. When hydrogen was introduced and sufficient time allowed for the equilibrium to be established, we could record the solvated electron spectrum. At 1520 nm, this spectrum presented a dip which meant a greater transparency. Also, above 1800 nm, there was a rapid decrease in absorption to values lower than those corresponding to the normal spectrum of the solvated electron.

With such path lengths, we could suspect changes in the physical properties of the solution. It is well known that there are drastic changes in the refractive index when scanning through an absorption peak of the solvent. The density of the solution is increased when hydrogen pressure is applied. Also as explained in the experimental section there is a teflon O-ring between the silica cell window and a hollow brass nut. The higher pressures in the interior of the cell crush this O-ring, thereby increasing the optical path length of the sample cell. Such are the variables which may be taken into account to explain the dip.

In the second part of this study we used the property possessed by amide solutions in the presence of hydrogen to stabilize electrons formed by photolysis or γ-radiolysis [2]. For these experiments, described in another paper [3], all cells were of silica. With the same amide solution in the sample and reference cells, we obtained a base line with some perturbations arising from slight differences in optical paths. Hydrogen originating from the attack of ammonia on the alkali metal was also present. When the sample cell was irradiated or photolyzed, we obtained stable electrons. In this system all the variables remained constant except the electron concentration and here again the dips occurred at 1520 nm as well as at higher wavelengths. Thus this anomaly requires the existence of the solvated electron, and in the discussion we present a tentative explanation.

Experimental

The cells used for the study of the chemical equilibrium (1) were of stainless steel with silica windows 10 mm thick, 22 mm in diameter. The cells were made to withstand pressures up to 200 atm. To avoid leakage of the solution a polyethylene O-ring was placed between the inner surface of the window and the steel compartment and a teflon O-ring between the outer surface and a brass screw. This brass screw has a hole of 12 mm in diameter to let the light beam through. By tightening or loosening the screw, the path lengths could be modified slightly. Silica cylinders with optically polished end faces of adequate lengths introduced in the cell allowed the path lengths to be reduced to 1 mm or less. Thus, we could observe the effect of differences of path lengths on the bands of ammonia.

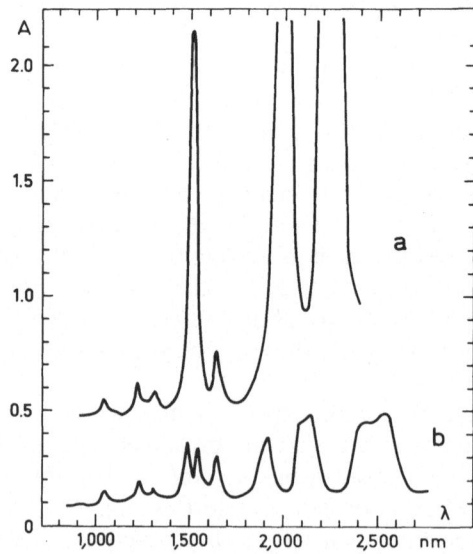

Fig. 1. Spectrum of NH_3. *a* Optical path length 0.100 cm, Ref. = air *b* Optical path length sample = 0.55 cm, Ref. = 0.50 cm

The steel cell had a side arm in which pure K metal or KNH_2 prepared beforehand could be introduced while flushing with dry argon. This side arm served also to condense ammonia. Hydrogen could be introduced up to pressures of 200 atm via a steel valve.

The cells in silica used in our photolysis or radiolysis studies are described in detail in the next paper. For these studies, the silica optical cell and irradiation tube could be thoroughly cleaned and heated under vacuum to eliminate traces of water.

Results

The spectrum of pure ammonia in the 1000—2500 nm region is given in Fig. 1a for a path length of 0.100 cm. The important bands have the following assignments given by Barrow and Lagowski [4]:

nm	cm^{-1}	Assignment
1530	6523	$2v_1$
1640	6090	$v_2 + v_3 + v_4$
2000	5000	$v_2 + v_4$
2240	4470	$v_2 + v_3$
2280	4386	$v_1 + v_2$

The main bands have the following values $v_2 = 3285\ cm^{-1}$, $v_2 = 1035 - 1066\ cm^{-1}$, $v_3 = 3375\ cm^{-1}$ and $v_4 = 1632\ cm^{-1}$.

In Fig. 2 we give a typical spectrum of amide solutions with pure ammonia as reference. A supplementary peak occurs at 1600 nm which is roughly proportional to the amide concentration and is probably the harmonic $2v$ of $v_{1\,NH_2^-}$ found by Corset [5] at 3186 cm^{-1}.

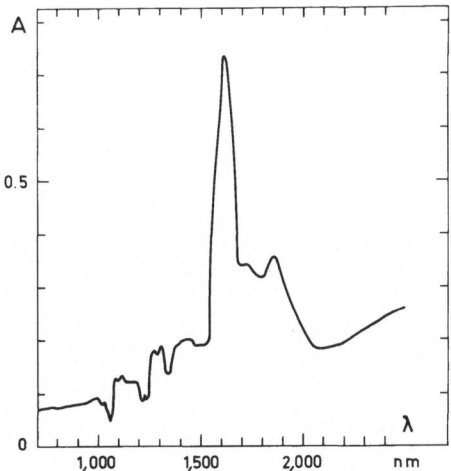

Fig. 2. Spectrum of amide solution with pure ammonia as reference Conc. $KNH_2 = 0.58$ mole l^{-1}, H_2 pressure $= 0.83$ atm e_{am}^- is present at a concentration $\sim 10^{-6}$ mole l^{-1} due to the thermal equilibrium

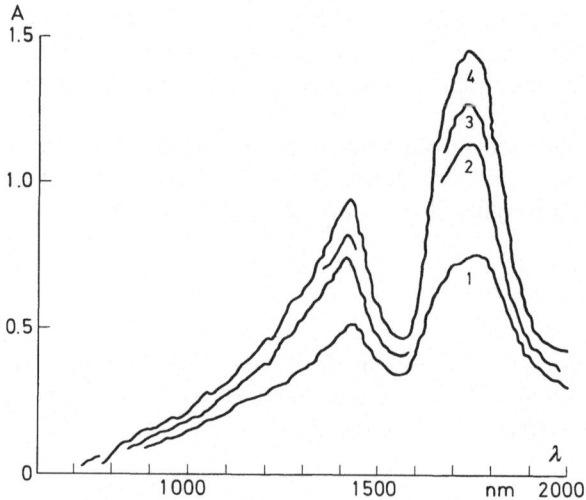

Fig. 3. Spectra of e_{am}^- at various hydrogen pressures. Optical path length = 5.1 cm, $t° = 25°$ C. Conc. $KNH_2 = 0.22$ mole l^{-1}, hydrogen pressures: *1* 9.4 atm, *2* 25 atm, *3* 29 atm, *4* 50 atm

The spectra of solvated electrons obtained from chemical equilibrium (1) at various pressures of hydrogen are shown in Fig. 3. There is a dip at 1520 nm and the longer wavelength part of the electron band is perturbed, the slope being much steeper than that for the normal solvated electron absorption curve.

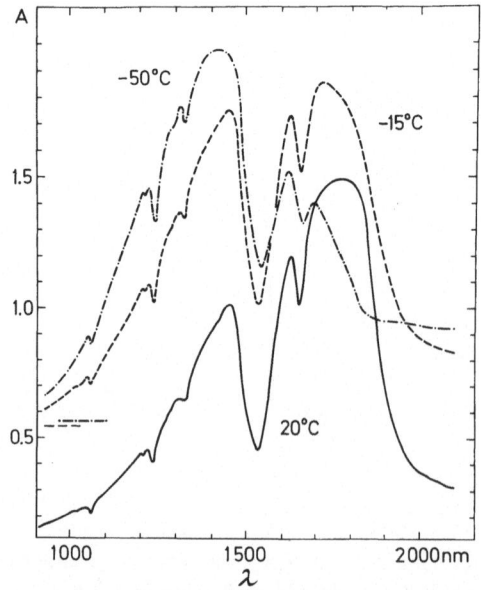

Fig. 4. Spectra of e_{am}^- produced by photolysis recorded at different temperatures Conc. $KNH_2 = 0.58$ mole l^{-1}. Optical path length = 2.5 cm

The increase in density of the solution when it is compressed by high pressures of hydrogen may be considered equivalent to longer path lengths. To observe the effect on the ammonia spectrum of differences in path lengths, between sample and reference cells, we used the possibility of modifying the path lengths described in the experimental section. The spectra were recorded for relatively short path lengths of both reference and sample cells and after each incremental increase of the sample cell optical path. When the difference between the reference and sample cells increases, the 2000 and 2300 nm bands begin to show dips at the top of the peaks. When the difference attains 0.5 mm for a path length of 5 mm, the 1520 band also shows the dip, as one can see in Fig. 1b.

In Fig. 4 we present another set of spectra obtained in silica cells by radiolysis or photolysis of the amide solution in the presence of a constant pressure of hydrogen. The shift of the maximum of the solvated electron band with temperature is seen in this figure. We notice that the dips are always at the corresponding wavelengths of the ammonia peaks. We see also that the e_{am}^- band at 20° C has the steep slope on the longer wavelength side due to the dip of the 2000 nm band. At $-50°$ C, this electron absorption curve is sufficiently shifted to have the normal shape towards the longer wavelengths, but its maximum is just on the anomaly and it is difficult to estimate its exact height.

Discussion

The four main peaks of liquid ammonia shown in Fig. 1 belong to the harmonics or combinations of the fundamental frequencies of the ammonia molecule. In their work on the internal reflection spectroscopy of alkali metal–ammonia solutions, Burrow and Lagowski [4] came to the conclusion that since there was no apparent modification of these bands in the presence of alkali metal from concentrations of 10^{-2} mole 1^{-1} up to bronze solutions, the vibrational modes are relatively insensitive to the presence of charged species.

In our experiments, as long as we were working with the steel cells for which the optical path lengths and the density of the solution were variable due to the increasing hydrogen pressure, we were tempted to dismiss these transparent dips as arising from the changes in index of refraction. The experiments done with sample cells longer than the reference cell by known increments showed effective "dips" on these bands.

But when by photolysis, we were able to produce stable electrons in the all-silica cells without modifying any other parameter, and still observe the dips, it became difficult to invoke changes in refractive index or density with just 10^{-5} mole 1^{-1} of the solvated electron.

Spectra of amide solutions relative to pure ammonia do not show a dip; on the contrary there is an additional peak at 1600 nm. Hydrogen gas does not perturb the ammonia spectrum, so if we observe the transparent dips on the solvated electron band, their existence seems to require the presence of amide, hydrogen and e_{am}^-. A tentative explanation is that a complex $e^- \cdot n(NH_3) \cdot m(NH_2^-)$ is formed in which the n ammonia molecules do not participate in the v_1 vibrations or have their combinations of fundamental frequencies disrupted, leading to less absorption of this solution with reference to an equivalent solution without the solvated electron.

References

1. Kirschke, E. J., Jolly, W. L.: Inorg. Chem. **6**, 855 (1967).
2. Belloni, J., Fradin de la Renaudiere, J.: Nature Phys. Sci. **232**, 173 (1971).
3. Belloni, J., Saito, E.: Colloquium Weyl III.
4. Burrow, D. F., Lagowski, J. J.: J. Phys. Chem. **72**, 169 (1968).
5. Corset, J.: Thesis. Bordeaux 1967.

Discussion

J. H. ROBERTS   The experimentally observed spectrum resembles a superposition of the absorption due to the solvated electron and a negative difference spectrum of the solvent. Several factors may contribute to the negative absorption of the solvent, but the experimental conditions exclude all but two. The change in concentration of ammonia molecules due to the volume expansion would account for only a very small part of the observed negative absorption. The other factor is the decrease in absorption of the near infrared solvent bands due to the structure enhancement observed in metal ammonia solutions (see papers by Roberts, J. H., Lagowski, J. J., p. 39, and by Rusch, P. F., Lagowski, J. J., p. 169). The observed splittings in the bands may in part be due to the presence of amide ion. It would be interesting to see the same type of difference spectrum for a solution of sodium in ammonia.

E. SAITO   I have explained that indeed the dip arose from a negative spectrum of ammonia (Fig. 4) and it is surprising that by the formation at concentrations as low as 10^{-5} mole l^{-1} of e_{am}^- we can observe the dip. At such a concentration the proportion of e^- to NH_3 molecules is only $1—3.5 \times 10^6$ and should not make any change in density due to volume expansion.

Moreover, in the earlier experiments with the steel cell in which the e_{am}^- is produced by the thermal equilibrium (1), the increase of hydrogen pressure in the sample cell leads to higher densities relative to that in the reference cell, so this again rules out the volume expansion explanation.

J. JORTNER   This experiment should be conducted in ND_3 where the interference of the infrared solvent bands is eliminated.

E. SAITO   An experiment in ND_3 should show effectively if there is any influence of the amide ion on the e_{am}^- spectrum — the dips due to the solvent spectrum would be displaced to longer wavelengths beyond the maximum of the band.

U. SCHINDEWOLF   Investigating the absorption spectra of solvated electrons formed by the "Jolly reaction" $(1/2\, H_2 + NH_2^- \rightarrow e_{NH_3}^-)$, we also observed the anomaly in the spectral range of the light absorption of ammonia around $1.5\, \mu$. In the published spectra, we dotted this range, because we did not think it real, being an artifact only.

E. SAITO   Yes, I have noticed your dotted lines. Because you were using a single beam instrument, at the 1520 nm peak, you were comparing high values of optical densities and I supposed you were not sure of the negative difference.

Raman Spectroscopy of Metal-Ammonia Solutions

B. L. SMITH and W. H. KOEHLER

Abstract

A laser Raman spectrometer has been constructed which readily accepts a specially designed dewar assembly. After the resolution and polarization performance characteristics of the spectrometer were determined, polarization studies on liquid ammonia were undertaken. Depolarization ratios for the fundamental modes of ammonia are reported, and a doublet has been observed for the v_2 band. Sodium-ammonia solutions in the concentration range 0—50×10^{-4} M were studied. No scattering center attributed to the solvated electron was found, nor were the positions of the bands altered by the presence of the solute in this concentration region. The intensity of the scattered radiation was found to decrease with increasing metal concentration and has been attributed to the absorbance of these solutions. The Raman effect in F-centers is discussed and compared to metal–ammonia solutions.

Introduction

The nature of the species present in a very dilute metal–ammonia solution continues to be a matter of speculation. Numerous models have been proposed, but as yet no single theory is clearly superior. A discussion of the various models will not be attempted in this work because numerous review articles are available [1]. Copeland, Kestner, and Jortner [2] have proposed a model for localized excess electron states in ammonia, and this model has a particular bearing on the present work. According to Copeland *et al.*, very dilute solutions may be described in terms of unassociated solvated cations and electrons. The electron is envisioned as existing in a cavity created by some number of pre-ferentially oriented ammonia molecules.

The proposed model predicts a totally symmetric vibration in the ground state with a frequency given by

$$v = \frac{1}{2\pi}\left[\frac{K}{\mu}\right]^{1/2},$$

where K is $\frac{1}{2}(\partial^2 E_t / \partial R^2)_{R=R_0}$, and $\mu = N m_{\mathrm{NH}_3}$. E_t is the total energy of the ground state, R is the cavity radius, N is the number of solvent molecules in the primary solvation shell, and m_{NH_3} is the reduced mass of the ammonia molecule. Depending on the value of N and V_0, the electronic energy of the quasi free electron state, the totally symmetric vibration is predicted to be between 25 cm^{-1} and 60 cm^{-1}. These data are summarized in Table I.

Rusch [3], applying the treatment of Klick and Schulman [4] to metal–ammonia solutions, has suggested that the symmetric "breathing" mode may be in the 400—700 cm^{-1} region. These results were obtained using the formulation for

Table I. Properties of the totally symmetric vibration

Ref.	N	V_0 (eV)	$\bar{\nu}$ (cm^{-1})
Copeland, Kestner, and Jortner [2]	4	2.0	25.0
		1.5	37.7
		1.0	44.1
		0.5	52.3
		0.0	62.4
		-0.5	62.0
	6	1.0	28.7
		0.5	36.5
		0.0	35.9
		-0.5	41.0
	8	0.5	31.8
		0.0	31.4
		-0.5	35.1
Rusch [3]		$\nu_g = \dfrac{2kT}{h}\left[\dfrac{W_{h/2}\,(\text{low temp. limit})}{W_{h/2}\,(\text{high temp. limit})}\right]^2$	
		$\nu_g = 400 - 700 \text{ cm}^{-1}$	

the frequency of the "breathing" mode which is given by

$$\nu_g = \frac{2kT}{h}\left[\frac{W_{h/2}(\text{L.T.})}{W_{h/2}(\text{H.T.})}\right]^2,$$

where ν_g is the frequency of the ground state vibration, $W_{h/2}$ is the width at half-height of the absorption band, and L.T. and H.T. refer to the low and high temperature limits, respectively.

Regardless of the model used to explain the observed properties of these solutions, the effect of the metal on the solvent structure is of primary importance. Several investigators have addressed themselves to this problem, but at present the results are incomplete and appear contradictory. Beckman and Pitzer [5], using external reflection techniques, studied the intermediate and concentrated solutions in the infrared. Burow and Lagowski [6], using internal reflection techniques, addressed themselves to much the same problem. Rusch [7], using transmission techniques, studied the effect of metal concentration on the 3300 cm^{-1} envelope of ammonia. The results of the aforementioned investigations are summarized in Table II.

In addition to the effect of metal on the solvent structure, the assignment of the Raman bands of liquid ammonia is contradictory and still speculative. It is not the purpose of this work to review all the vibrational data reported on ammonia; however, the Raman data available on liquid ammonia is summarized in Table III.

Experimental

A laser Raman spectrometer was constructed from three basic components: 1. Control Laser Corp. Ar$^+$ laser, 2. Spex 1401 double monochromator, and 3.

Table II. Infrared spectroscopy of metal–ammonia solutions

Ref.	Assignment	Solvent	Concentration (MPM > 0.5)	2 × 10⁻³ m Na	1.7 × 10⁻² m Na	1.3 × 10⁻¹ m Na	1.4 m Na	10.3 m Na	1.0 × 10⁻² m Li	1.0 × 10⁻² m K	1.0 × 10⁻¹ m Li	1.0 × 10⁻¹ m K	5 × 10⁻³ [Li]	5 × 10⁻³ [K]	5 × 10⁻² [Li]	5 × 10⁻² [K]
Beckman and Pitzer [5]	ν_2	—	1030—1050 cm⁻¹													
	ν_1	—	3190													
	ν_3	—	3370													
Burow and Lagowski [6]	$2\nu_4$	3220 cm⁻¹		3220 cm⁻¹	3220 cm⁻¹	3230 cm⁻¹	3240 cm⁻¹	3220 cm⁻¹	3170 cm⁻¹	3230 cm⁻¹	3220 cm⁻¹	3240 cm⁻¹				
	ν_1	3280		3280	3280	3230	—	3300	3280	3280	3280	3280				
	ν_3	3410		3410	3410	3410	3410	3410	3380	3420	3410	3400				
Rusch and Lagowski [7]	$2\nu_4$	3155 cm⁻¹											3158 cm⁻¹	3156 cm⁻¹	3153 cm⁻¹	3152 cm⁻¹
	ν_1	3286											3265	3268	3250	3252
	ν_3	3453											3429	3426	3409	3413

Table III. Raman spectroscopy of liquid ammonia

Date	Observed bands and assignments					$t°\,C$	Ref.
1929			3210( )	3310( )	3380( )	−40	Daure [8]
1930	1070(ν_2)		3216( )	3304(ν_1)	3380(ν_3)	+25	Bhagavantam [9]
1930	—	1594(ν_4)	3208( )	3296(ν_1)	3388(ν_3)		Dadieu et al. [10]
1936	1070( )	1580( )	3210( )	3300( )	3380( )	−40	Costeanu [11]
1953	1031 / 1054 / 1077(ν_2)	1624(ν_4)	3212( )	3303(ν_1)	3384(ν_3)	+25	Kinumaki et al. [12]
1954	—	—	3218(2ν_4)	3300(ν_1)	3373(ν_3)	−34 to −70	Plint et al. [13]
1968	—	—	3215(2ν_4)	3301(ν_1)	3384(ν_3)	+25	Seillier et al. [14]
1970	1070(ν_2)	1641(ν_4)	3206(ν_1)	3296(2ν_4)	3363(ν_3)	−55	Birchall et al. [15]
1970	1061(ν_2)	1645(ν_4)	3218( )	3303( )	3386(ν_3)	+40 to −80	Bettignies et al. [16]
1972	1060±5	1634±5	3215±2	3301±2	3380±2	−67	Present work

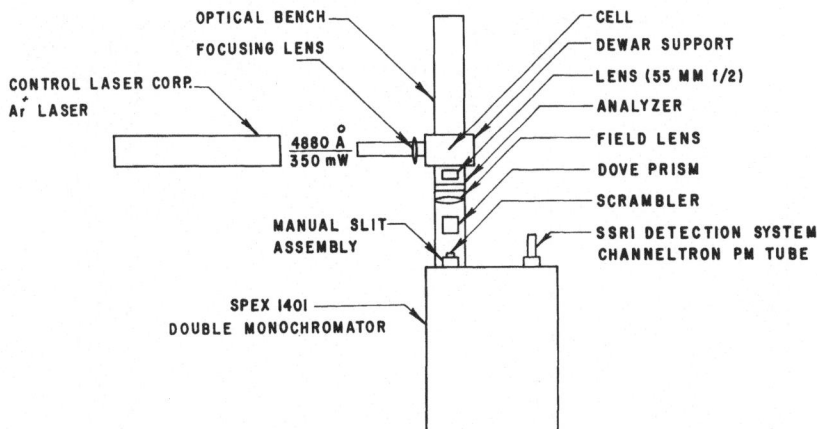

Fig. 1. Laser raman spectrometer

Solid State Research photon-counting detection system. The spectrometer utilized the conventional 90° geometry and is shown schematically in Fig. 1.

Radiation of appropriate frequency and intensity was focused on the sample cell (Fig. 4) and the scattered radiation collected at right angles. Both the 4880 Å line (50–350 mW) and the 4579 Å line (maximum power 100 mW) were used in this work. The beam was focused to a diameter of approximately 0.2 mm at the sample cell.

The scattered radiation was collected and collimated with a conventional camera lens (55 mm $f/2$), passed through an analyzer, and focused on the entrance slit with a lens compatible with the f-number of the monochromator. Prior to the slit assembly, a dove prism was used to rotate the image 90° (this rendered the image compatible with the vertical entrance slit), and a calcite wedge scrambler was inserted to remove the polarization dependence of the monochromator.

An SSRI photon-counting detection system was employed with the Spex 1401 monochromator. This detection system incorporates a Bendix Channeltron photomultiplier tube which has a dark count of 10 counts/second at room temperature.

The Raman cell itself was little more than a U-tube. The extrusion permitted the closest approach of the scattering center to the collection lens within the confines of the dewar. This design also enabled the focal point of the collection lens to be set at the scattering center which resulted in good collimation. The design permitted only one pass of the laser which resulted in lower intensities but more quantitative polarization measurements.

The solutions were prepared and studied in a modified version of the dewar assembly described by Quinn and Lagowski [17]. The dewar and insert are shown in Figs. 2 and 3. Prior to solution preparation, the assembly was subjected to a cleaning procedure previously described [18]. To prepare a solution, a quantity of ammonia was distilled off sodium and condensed in the insert. The volume was determined from the calibrated pipette, and was known to within ± 1%. A known weight of purified sodium was added using the winch assembly

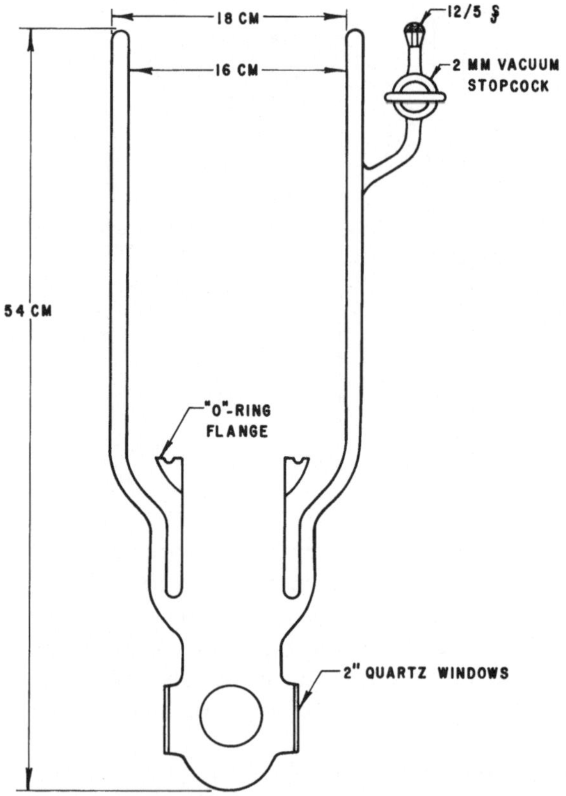

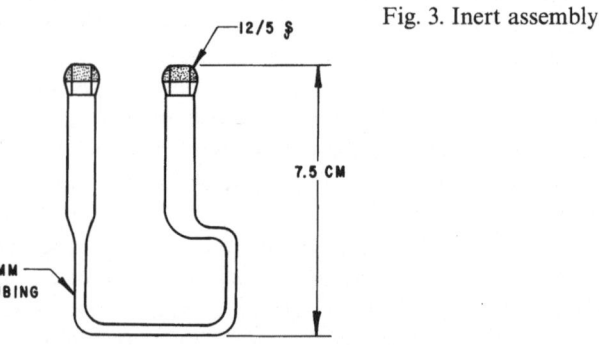

Fig. 2. Dewar assembly

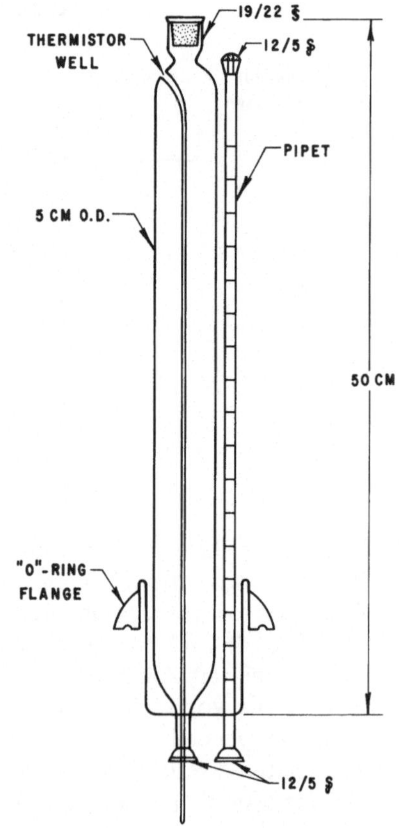

Fig. 3. Inert assembly

Fig. 4. Raman cell

shown in Fig. 5. Temperature of the solutions was monitored with a thermistor which extended into the optical cell.

In general, spectra were recorded without the analyzer in the fore optics and then with the analyzer set first parallel and then perpendicular. The spectra were recorded at constant slit width (usually $300\,\mu$) which corresponded to spectral band widths in the order of 4–$6\,cm^{-1}$. The spectrum was generally scanned at $10\,cm^{-1}$/min for the first several hundred wave numbers, then increased to $50\,cm^{-1}$/min.

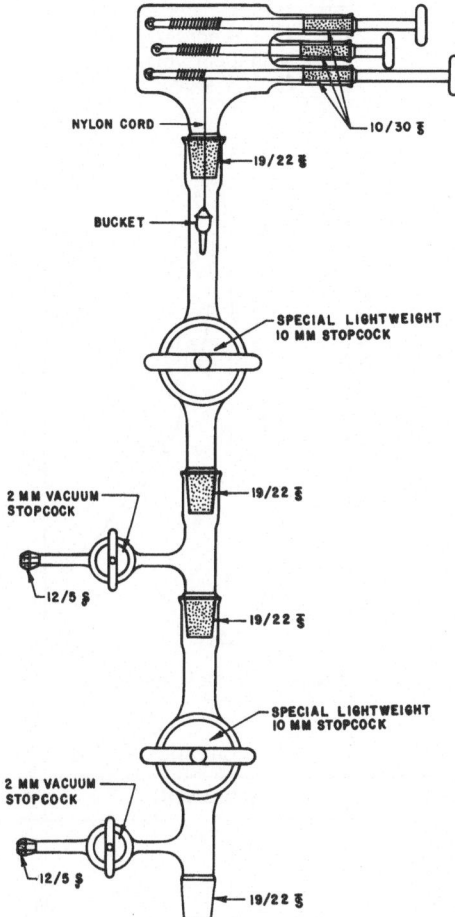

Fig. 5. Metal handling assembly

Results and Discussion

Prior to initiating the metal–ammonia studies, the resolution and polarization performance of the spectrometer were determined. Figure 6 shows the 459 cm^{-1} line of carbon tetrachloride. Three bands are evident with peaks estimated at 456.1 cm^{-1}, 458.8 cm^{-1}, and 460.6 cm^{-1}. These three bands arise from the isotopic distribution of chlorine and have been reported by Herzberg [19] at 455.1 cm^{-1}, 458.4 cm^{-1}, and 461.5 cm^{-1}.

No attempt was made in the present study to eliminate all of the errors generally associated with quantitative polarization measurements. The design of the spectrometer and cell assembly was predicated on obtaining Raman spectra of metal–ammonia solutions rather than absolute depolarization ratios. Consideration was given to such factors as monochromator polarization efficiency, image magnification, and cell design, but when necessary these parameters were compromised. For a detailed discussion concerning the effect of these parameters

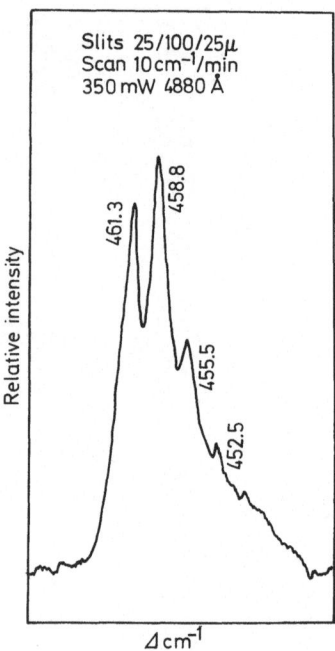

Fig. 6. 459 cm^{-1} line of CCl$_4$

Table IV. Depolarization ratios of CCl$_4$

Line	p(observed)	p(predicted)
225 cm^{-1}	0.79	0.75
315 cm^{-1}	0.78	0.75
459 cm^{-1}	0.01	0.0

on polarization measurements, the reader is referred to several excellent papers on the subject [20, 21].

Carbon tetrachloride was used to determine the accuracy of the polarization measurements reported in this work. The results are presented in Table IV.

Inspection of the results reveals that the depolarization ratios are higher than the predicted values or than the experimental values previously reported [22–24]. Although not absolute, the measurements were generally in error by no more than 10%.

The Raman spectrum of liquid ammonia was determined, and salient features of a typical spectrum are shown in Figs. 7, 8, 9, and 10. With reference to Fig. 10, a broad structureless band centered around 200 cm^{-1} is evident and is attributed to unresolved rotational structure. Figures 7 and 8 show bands at 3380 cm^{-1}, 3301 cm^{-1}, 3215 cm^{-1}, 1634 cm^{-1}, and 1060 cm^{-1}. The bands at 1060 cm^{-1} and 1634 cm^{-1} have been assigned as v_2 and v_4 respectively. The 3380 cm^{-1} band has been assigned as v_3;, however, the assignments of the 3215 cm^{-1} and 3301 cm^{-1} bands are less certain. The polarization studies pre-

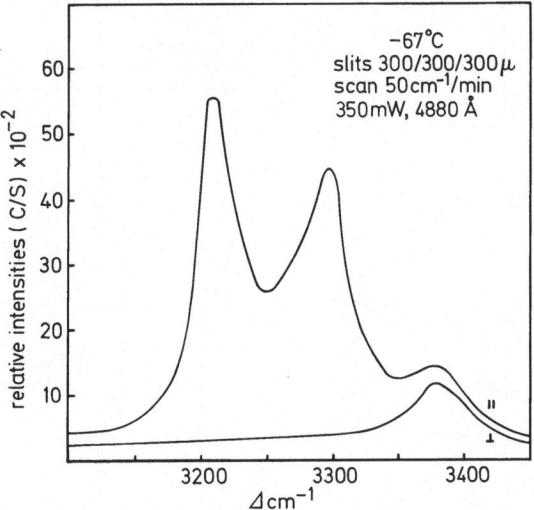

Fig. 7. $(v_1, v_3, 2v_4)$ bands of liquid ammonia

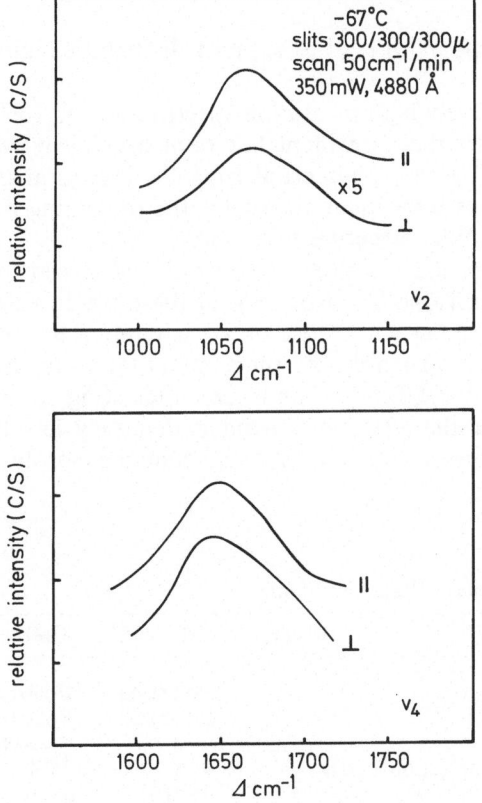

Fig. 8. v_2 and v_4 bands of liquid ammonia

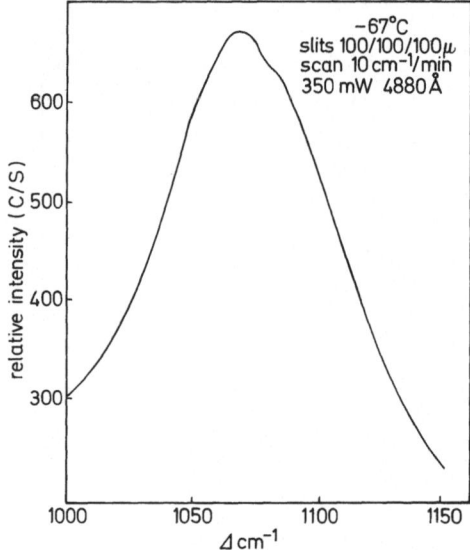

Fig. 9. v_2 band of liquid ammonia

sented in this work (*vide infra*) tend to support the band assignments of Birchall *et al.* [15].

Figure 9 shows a relatively high resolution spectrum of the 1060 cm^{-1} (v_2) band. The unresolved doublet that is evident has been previously reported [12] and arises from inversion. Based on numerous runs, the best estimate of the splitting is 10–15 cm^{-1}. Attempts were made to resolve the remaining bands in the spectrum, but no other doublet structure was found.

Typical polarization spectra are shown in Figs. 7 and 8, and the depolarization ratios are presented in Table V. Inspection of the table reveals that the results are in qualitative agreement with those previously reported. The depolarization ratios for v_2 and v_4 are probably the most quantitative to date. The ϱ values for the 3215 cm^{-1} and 3300 cm^{-1} bands are interesting in that they are not equal. The ϱ value for the 3215 cm^{-1} band is definitely less than that for the 3300 cm^{-1} band; however, both are so small that the absolute values are uncertain.

Table V. Polarization studies of liquid ammonia

Ref.	1060 cm^{-1}	1640 cm^{-1}	3215 cm^{-1}	3300 cm^{-1}	3380 cm^{-1}
[8]	—	—	Polarized	Polarized	Depolarized
[9]	—	—	Polarized	Polarized	Depolarized
[14]	—	—	$\varrho = 0.08$	$\varrho = 0.08$	$\varrho = 0.4$
[15]	50% Polarization	0%	>95%	>95%	0
Present work	$\varrho = 0.15$	$\varrho = 0.5$	$\varrho < 0.01$	$\varrho = 0.01$	$\varrho = 0.5$

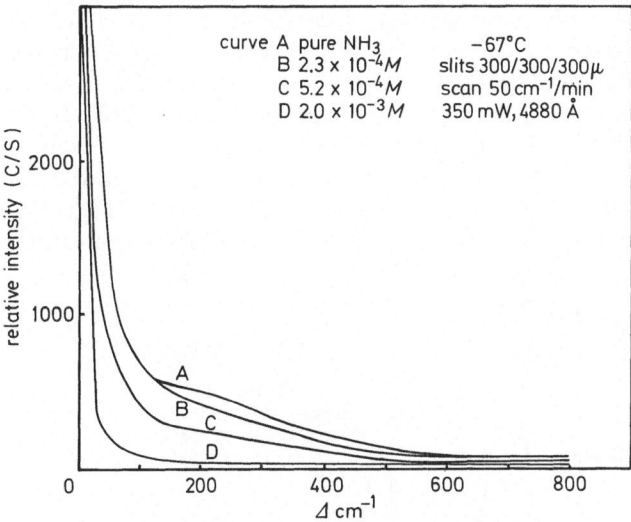

Fig. 10. Raman spectra of sodium–ammonia solutions

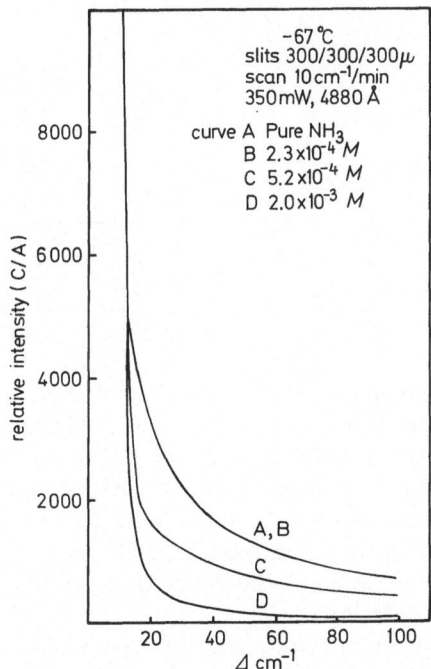

Fig. 11. Raman spectrum of sodium–ammonia solutions

Typical Raman spectra of sodium-liquid ammonia solutions are shown in Figs. 10, 11, and 12. The broad band ($\bar{v} = 200 \text{ cm}^{-1}$) observed in pure ammonia was not observed in any of the metal solutions. This is to be expected if the band is un-resolved rotational structure because dissolution of metal must result in a solvent ordering process with a subsequent decrease in rotational freedom. Careful attention was given to the region $0-1000 \, cm^{-1}$ because of the predicted "breathing" mode vibration; however, no new scattering center was observed. In fact, the spectrum of pure ammonia was indistinguishable from that of a $2.3 \times 10^{-4} \, M$ solution in the region $0-100 \, cm^{-1}$. The only essential difference between the spectrum of pure ammonia and any solution in the concentration region $(2 - 50 \times 10^{-4} \, M)$ was in intensity. As is evident in Figs. 10, 11 and 12, the scattered intensity decreases with increasing metal concentration.

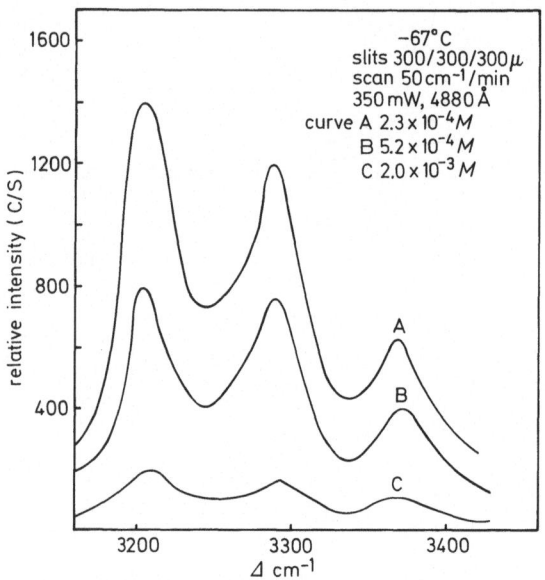

Fig. 12. $(v_1, v_3, 2v_4)$ bands of ammonia in sodium–ammonia solutions

Figure 13 is a plot of band maxima as a function of metal concentration and reveals that in the concentration range $0 - 5 \times 10^{-3} \, M$ there are no shifts in band positions. Although the data are not shown, the $1060 \, cm^{-1}$ and $1640 \, cm^{-1}$ bands were also unaffected by metal up to $5 \times 10^{-4} \, M$. At higher concentrations, the intensity of the v_2 and v_4 bands was too low to be observed. It is perhaps important to note that three concentrations $(2.3, 5.2, \text{ and } 20 \times 10^{-4} \, M)$ represent sequential addition of metal, whereas the other three concentrations represent three different solution preparations.

These results are in disagreement with the infrared work of Rusch [7] who reported a shift in the positions of v_1 and v_3 with increasing metal concentration (Table II). Some caution must be exercised in comparing the two sets of data because the band positions reported by Rusch were obtained from computer-resolved infrared spectra and the band positions reported in this work are based on unresolved Raman data; however, the spectra presented in Fig. 12 are of sufficiently high resolution that very little uncertainty exists in band positions.

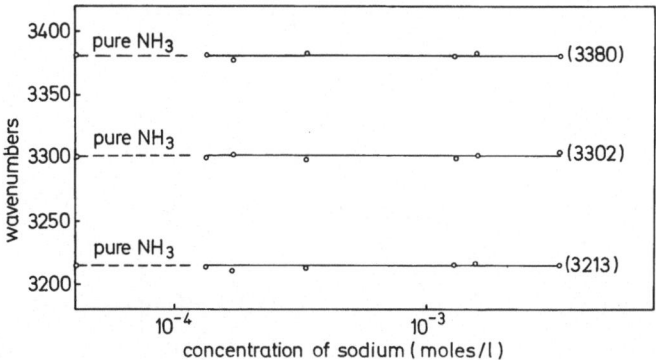

Fig. 13. Band position as a function of sodium concentration

The decrease in scattered intensity with increasing metal concentration is readily explained by considering the absorption spectrum of the solutions; the spectrum is asymmetric and tails into the visible region of spectrum. Using the optical data of Gold and Jolly [25], a plot was made of (A/l) as a function of concentration at 5830 Å. Absorbance/path length, A/l, was used rather than absorbance to eliminate the path length dependency; 5830 Å corresponds to $\Delta v = 3300 \text{ cm}^{-1}$. Figure 14 shows at least a qualitative relationship between absorbance and Raman scattering intensity. Figure 15 attempts to illustrate the problem. This illustration is not to be construed as quantitative. Using 4880 Å excitation, $\Delta v = 3300 \text{ cm}^{-1}$ corresponds to 5830 Å, and it is obvious that the absorbance at this wavelength increases with concentration. There is also appreciable absorption at 4880 Å so that two effects are operative which tend to diminish the scattered intensity. First the excitation line energy is reduced due to absorption and secondly the scattered radiation is subject to absorption. At concentrations greater than 10^{-2} M, the 4880 Å line was completely absorbed and severe heating resulted. The spectrum of the 5.5×10^{-3} M solution was obtained using 4580 Å excitation. Although the power of this line was less than the 4880 Å line, the decreased absorption at 4580 Å, the v^4 dependence, and the

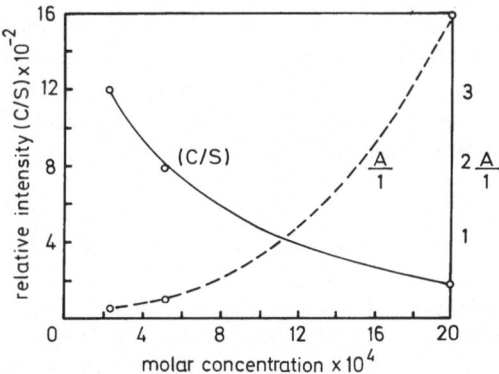

Fig. 14. Raman emission intensity and absorbance of sodium–ammonia solutions at 5830 Å

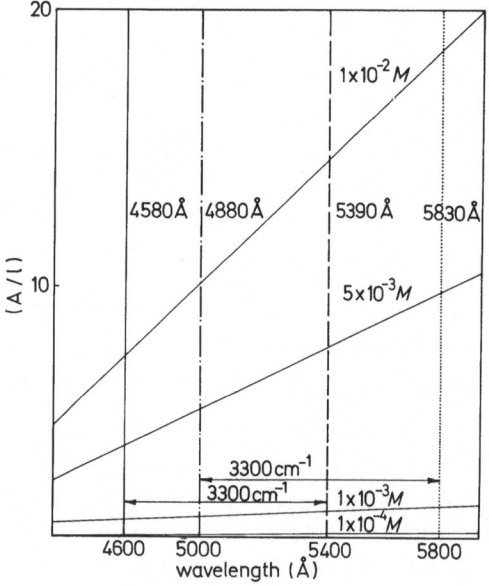

Fig. 15. Absorbance and raman excitation/emission

decreased absorption at 5390 ($\Delta \bar{v} = 3300 \text{ cm}^{-1}$) resulted in a meaningful spectrum. Attempts to study concentrations greater than 5.5×10^{-3} M with the 4580 Å line were unsuccessful due to absorption of both the excitation and emission frequencies. Investiations are being undertaken to determine the feasibility of using the anti-Stokes lines at higher concentrations.

Unfortunately it is impossible to state absolutely whether the predicted scattering center is or is not present. Arguments can be raised that the concentration of these centers was too low to be detected. Increasing the concentration is of course the wrong approach because of the corresponding loss in intensity and the certainty of association in the solutions. The most obvious approach would seem to be multi-pass cell. Several such cells have been designed and tested, but at this time some engineering problems remain.

Since an analogy is often made between F-centers and metal–ammonia solutions and since much of the theoretical treatment of these solutions has as its origin F-center theory, it is interesting to extend the treatment of the Raman effect in F-centers to metal–ammonia solutions. Kleinman [26] has calculated the Raman polarizability tensor using the expression

$$\alpha_R = f \frac{e^2}{m(\omega_{GF}^2 - \omega_0^2)} \left(\frac{\beta - 1}{\beta} \right) \left[\frac{2\hbar/M\omega_G}{(X_G - X_F)^2} \right]^{1/2},$$

where α_R is the polarizability tensor, f the oscillator strength, ω_{GF} the angular frequency of the electronic absorption, ω_0 the angular frequency of the excitation energy, M the reduced mass, ω_G the angular frequency of the lattice vibration, X_G and X_F the positions of the energy minimum of the ground and excited states, respectively. β is the "absorption-emission discrepancy" as shown in the relationship

$$(\psi_{G1} M_{GF}(X)^2 \psi_{G1}) = \beta^2(\psi_{F1} M_{GF}(X)^2 \psi_{F1}),$$

when ψ_{G1} and ψ_{F1} represent the lowest vibrational states associated with the ground electronic state and the first excited electronic state, and $M_{GF}(X)$ is the matrix element of the dipole-moment operator between electronic states G and F.

Kleinman calculated $\alpha_R = 9 \times 10^{-25}$ cm^3 for an F-center using He–Ne excitation. This value is very close to the value of $\alpha_R = 2 \times 10^{-25}$ cm^3 reported [27] for the strong lines of carbon tetrachloride.

Worlock and Porto [28] reported observing Raman scattering from F-centers in KCl and NaCl. The Raman line was broad (~ 200 cm^{-1}) and centered at $\Delta v = 200$ cm^{-1}. In these experiments the concentration of scattering centers was approximately 10^{16} to 10^{17} centers/cm^3. Radhakrishna and Sehgal [29] reported similar spectra of F-centers in KCl and KBr at densities of 10^{18} centers/cm^3.

Kleinman's equation was applied to metal–ammonia solutions using the following data:

$$f = 0.77$$

$$\omega_{GF} = 1.26 \times 10^{15} \ (1.5 \ \mu)$$

$$\omega_0 = 3.68 \times 10^{15} \ (4880 \ \text{Å})$$

$$\beta \ \text{varies from 1 to 10}$$

$$M = 1.7 \times 10^{-22} \ \text{g} \ (6 \ NH_3 \ \text{molecules})$$

$$\omega_G = 6.4 \times 10^{12} \ (35 \ \text{cm}^{-1})$$

$$\left. \begin{array}{l} X_G = 2.2 \ \text{Å} \\ X_F = 2.6 \ \text{Å} \end{array} \right\} \quad \text{for} \quad N = 6, \quad V_0 = 0.0$$

and the following result was obtained

$$\alpha_R = 6 \times 10^{-24} \left(\frac{\beta - 1}{\beta} \right).$$

If $\beta = 1$, then $\alpha_R = 0$, but as β becomes large α_R approaches 6×10^{-24} as a limit. Unfortunately little is known about the excited state and consequently β is indeterminate, but unless β is very small α_R is of the same magnitude as that obtained for F-centers. It should also be pointed out that concentrations in the range 5×10^{-4} M to 5×10^{-3} M correspond to 3×10^{17} to 3×10^{18} electrons/cm^3 (assuming no association).

In view of these results it is perhaps surprising that no Raman effect was observed in the metal–ammonia solutions; however, caution must be exercised in carrying the analogy too far. Several factors could be operative which could reduce the scattering below detectable limits. The interactions in this system may be much less localized than in F-centers which would produce broad and continuous Raman shifts. Certainly the difference in lifetimes of the centers must be considered because F-centers are very longlived compared to the lifetime of a given cavity (as deduced from nuclear magnetic resonance relaxation times) [30]. The relative ease of reforming cavities coupled with the low cavity concentration may reduce the concentration of cavities at any instant below detectable limits. The high radiation density is another factor whose effect is difficult to predict. Localized heating might possibly reduce the concentration of cavities through density effects or even destroy cavities. In this context it should be mentioned that spectra were determined with and without focusing the laser and at very low laser power outputs. There was no discernible difference (other than intensity) between these spectra and those observed at high power outputs. There is also the possibility that a cavity as has been described does not exist. Although no scattering center was found, the conclusion must be reached that sodium metal does not perturb the solvent structure in a detectable manner in the concentration region studied.

Acknowledgement

We gratefully acknowledge the financial support of The Robert A. Welch Foundation. We are indebted to the University of Dallas for the loan of the Spex 1401.

References

1. Metal–ammonia solutions, Colloque Weyl II. Lagowski, J. J., Sienko, M. J., (Eds.). London: Butterworths 1970.
2. Copeland, D. A., Kestner, N. R., Jortner, J.: J. Chem. Phys. **53**, 1189 (1970).
3. Rusch, P. F.: Université Catholique de Lille, France, Private communication.
4. Klick, C. C., Schulman, J. H.: Solid state physics, Vol. 5, p. 97. New York: Academic Press 1957.
5. Beckman, T. A., Pitzer, K. S.: J. Phys. Chem. **65**, 1527 (1961).
6. Burow, D. F., Lagowski, J. J.: J. Phys. Chem. **72**, 169 (1968).
7. Rusch, P. F.: Ph. D. Dissertation. The University of Texas at Austin, 1971.
8. Daure, P.: Ann. Phys. **12**, 375 (1929).
9. Bhagavantam, S.: Indian J. Phys. **5**, 54 (1930).
10. Dadieu, A., Kohlrausch, K. W.: Naturwissensch. **18**, 154 (1930).
11. Costeanu, M. G.: Compt. Rend. **207**, 285 (1938).
12. Kinumaki, S., Aida, K.: Sc. Repts. Res. Inst. Tokohu Univ. **6**, 186 (1954).
13. Plint, C. A., Small, R. M. B., Welsh, H. L.: Can. J. Phys. **32**, 653 (1954).
14. Seillier, G., Ceccaldi, M., Leicknam, J. P.: Method. Phys. Anal. (GAM) **4**, 388 (1968).
15. Birchall, T., Drummond, I.: J. Chem. Soc. (A) **1970**, 1859.
16. Bettignies, B., Wallart, F.: Compt. Rend. **271**, 640 (1970).
17. Quinn, R., Lagowski, J. J.: J. Phys. Chem. **73**, 2326 (1969).
18. Burow, D. F., Lagowski, J. J.: Solvated electron. Advances in Chemistry Series 50. Washington, D. C.: Am. Chem. Soc. 1965.
19. Herzberg, G.: Infrared and Raman spectra. New York: Van Nostrand 1966.
20. Claasen, H. H., Selig, H., Shanier, J.: Appl. Spectry. **23**, 8 (1969).
21. Allemand, C. D.: Appl. Spectry. **24**, 348 (1970).
22. Murphy, W. F., Evans, M. V., Bender, P.: J. Chem. Phys. **47**, 1836 (1967).
23. Slomba, A. F., Hinman, C. D., Siegler, E. H.: Proc. Conf. Anal. Chem. Appl. Spectry. Pittsburg 1965.
24. Douglas, A. E., Rank, D. H.: J. Opt. Soc. Am. **38**, 281 (1948).
25. Gold, M., Jolly, W. L.: Inorg. Chem. **1**, 818 (1962).
26. Kleinman, D. A.: Phys. Rev. **134**, A 423 (1964).
27. Brandmüller, J., Moser, H.: Einführung in die Ramanspektroskopie. Darmstadt: Dr. Dietrich Steinkopff 1962.
28. Worlock, J. M., Porto, S. P. S.: Phys. Rev. Letters **15**, 697 (1965).
29. Radhakrishna, S., Sehgal, H. K.: Phys. Letters **29** A, 286 (1969).
30. Catterall, R.: In: Lagowski, J. J., Sienko, M. J. (Eds.): Metal-ammonia solutions. Colloque Weyl II. London: Butterworths 1970.

Raman Spectra of Dilute Metal–Ammonia Solutions

M. G. DeBacker, P. F. Rusch, B. DeBettignies, and G. Lepoutre

Abstract

Studies of the Raman spectrum of dilute potassium-ammonia solutions are reported. No new bands appeared, and no difference between the spectrum of pure ammonia and that of the solution could be detected. Possible reasons for the failure to observe any new bands are discussed.

The most striking feature of the optical absorption spectrum of dilute and moderately concentrated metal–ammonia solutions is the presence of a single absorption band in the near-infrared region with a maximum at about 0.8 eV. This band is very intense ($\varepsilon = 4.5 \times 10^4$ liter mole^{-1} cm^{-1}), broad (half-width = 0.4 eV) and noticeably asymmetric towards higher energy. It is generally accepted [1, 2] that this band results from an optical transition of the solvated electron. The cavity model [3] is able to explain the absorption band in terms of a transition of the solvated electron from a 1s (ground) to a 2p (excited) state.

In a recent paper [4], Copeland, Kestner, and Jortner provide a theoretical description of the solvated electron in terms of a configuration coordinate (cc) model. The coordinate chosen is a measure of the radial displacement of n solvent molecules about the center of the solvated electron density. These calculations yield total energy curves characteristic of stable systems for both the ground and excited states. From the total energy curve for the 1s state it is possible to calculate the vibrational energy levels of the cavity by making the following assumptions. First, the ammonia molecules forming the cavity are essentially hard spheres, and second, the n cavity molecules act as a harmonic oscillator such that the potential energy curve in the region of the minimum can be approximated by a parabola. Thus from the second derivative of the curve evaluated at the minimum, the frequency of vibration can be calculated as a function of n and of the potential energy V_0 which represents the energy of a quasi-free electron in the medium. The calculated frequencies are in the range 25—75 cm^{-1}, and Copeland, Kestner, and Jortner suggest that this vibration can be observed by Raman scattering. By analogy with similar calculations on electron impurity centers in the solid state, this vibration is called a "symmetric breathing mode".

In the solid state cc calculations for the F-center [5], the symmetric breathing mode is found to be of rather low energy and can be used to explain quantitatively the observed broadening of the electron absorption band. On the basis of these calculations it was predicted [6] that this mode of vibration of the cc (i.e. the lattice surrounding the F-center) should be observable by Raman scattering. Although there was a Raman spectrum observed with F-centers in NaCl and KCl, the results showed "no evidence for a strong sharp localized mode of vibration" [7].

The absorption spectrum of the solvated electron has been investigated by many researchers [8, 9] but the spectrum of the solvent has, until recently [10], not received much attention. The infrared (IR) spectrum of the solvent has been observed [11] for solutions of potassium and lithium in ammonia. In this investigation it was observed that the presence of the solvated electron perturbed the spectrum in the $3\,\mu$ region much more strongly than ordinary salts in ammonia. A study of the Raman spectrum of dilute potassium-ammonia solutions was initiated in order to look for new bands or for the perturbation of the ammonia vibration bands. It is the purpose of this paper to present some of the preliminary results.

Experimental

Potassium solutions with metal concentrations ranging from 10^{-3}—10^{-5} M were prepared in cylindrical 6 mm OD Pyrex tubing. These cells could withstand the pressure of ammonia at room temperature. The stability of the solutions at room temperature was excellent; no decomposition problems occurred during the runs.

The Raman spectrometer used was a CODERG PH 1. This instrument has a double monochromator equipped with gratings ruled at 1800 lines/mm. Entrance and exit slits were adjustable and could give resolutions from 0.1—$10\,\mathrm{cm}^{-1}$. The detector was a photomultiplier of the S 20 type with a spectral range from 3000—$8000\,\text{Å}$, and a maximum sensitivity at $4200\,\text{Å}$. The light source was a tunable Krypton laser (CRL model 52). The total power output of this laser is of the order of 1 watt. The light beam was focused on the cell containing the solution and passed through it twice. The scattered light was received in a direction perpendicular to the incident light. The cell holder was equipped with a cryostat cooled by the controlled evaporation of liquid nitrogen. The temperature could be regulated with an accuracy of $\pm 2°\,\mathrm{C}$.

All of our study was done by comparing the spectra of pure ammonia and of a metal solution taken under exactly the same conditions. It is good to keep in mind that liquid ammonia is a poor scatterer of light. The laser exciting line has to be chosen carefully in order to use the photomultiplier where its sensitivity is good. To detect the weakest of the bands of pure ammonia, slit-widths of the order of $10\,\mathrm{cm}^{-1}$ were needed. The concentration of the solutions had to be adjusted in order to have enough light reaching the detector.

Results

The spectral region from 0—$4000\,\mathrm{cm}^{-1}$ was carefully investigated. The region between 0 and $100\,\mathrm{cm}^{-1}$ was studied with special attention since it was where Jortner *et al.* predicted the appearance of a new band. A careful alignment of the optical system and the selection of narrow slits allowed us to work as close as $10\,\mathrm{cm}^{-1}$ from the Rayleigh line. This region was studied with seven exciting lines from the Krypton laser (from 4619—$6764\,\text{Å}$). No new bands appeared, and we could not detect any difference between the spectrum of pure NH_3 and that of the K–NH_3 solution. The region of the vibration bands of NH_3 was studied with the 4762 Å line and settings such the vibration bands of ammonia were well defined. It was impossible to detect any new band.

As shown in Fig. 1, the use of the 4762 Å (2.605 eV) exciting line places the Raman spectrum of the fundamental solvent vibrations outside of the region of intense solvated electron absorption. This should allow the use of solutions of high concentration. The first feature noted was a marked decrease of the intensity of the solvent spectrum; much of this is expected on the basis of absorption of scattered light by the blue band.

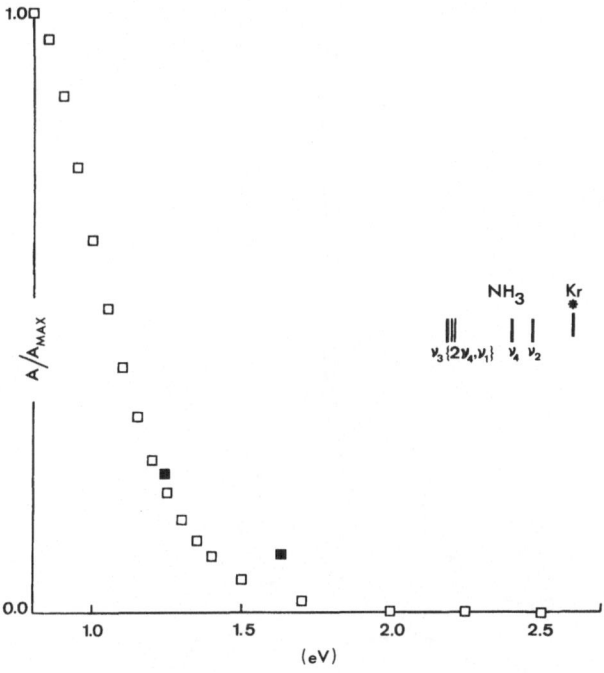

Fig. 1. Position of the Kr exciting line (4762 Å) and the solvent fundamental vibrations relative to the solvated electron absorption peak (data from Ref. [9])

In the region of 1000 cm^{-1} we found the v_2 and v_4 fundamental vibrations of ammonia. These bands were weak even in pure NH_3 and the signal-to-noise ratio was of the order of 10. In one experiment we detected a slight shift of the v_2 band to lower energies but this feature could not be reproduced. Within experimental error, the normalized bands of the pure ammonia and of the metal-ammonia solutions were identical. The lowering of the v_2 and v_4 bands was the same for both bands when going from pure solvent to the metal solution.

In the region of 3000 cm^{-1}, three peaks were observed. Typical spectra of pure NH_3 and a 10^{-3} M K–NH_3 solution at +23° C are shown in Fig. 2. The least intense peak of this group (3384 cm^{-1}) has been assigned the asymmetric N–H stretching vibration v_3 [12, 13]. The two remaining peaks (3305 and 3215 cm^{-1}) are attributed to the asymmetric H–N–H bending overtone ($2v_4$) and to the symmetric N–H stretching vibration (v_1). The overtone occurs with enhanced intensity because of Fermi resonance with the fundamental vibration v_1. An unambiguous assignment of these two, more intense peaks is not possible for several reasons. First, both peaks show a depolarization ratio approaching zero, which indicates that the vibrations each have a common symmetry element. Secondly, the large number of combination and overtone peaks observed in the near-infrared region of the spectrum of liquid NH_3 and $M–NH_3$ solutions is not observed in the Raman spectrum. Thus it is impossible to deduce the assignments by forming sums (and differences) of the fundamental frequencies. An assignment of the pure solvent

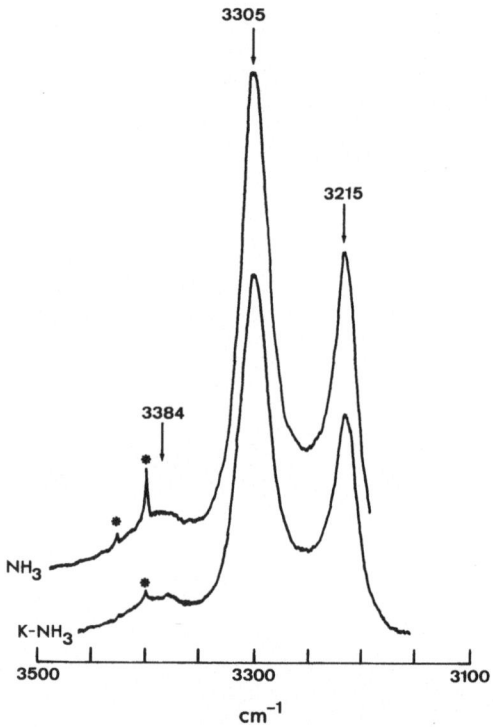

Fig. 2. Raman spectra of the solvent fundamental vibrations in the $3\,\mu$ region for pure NH_3 and 10^{-3} M K–NH_3 at $+20°$ C. Starred peaks are laser emission lines

Table I. Observed laser Raman frequencies of pure liquid ammonia

$t(°C)$	v_2	v_4	$2v_4$	v_1	v_3	Ref.
-63	1060	1634	3296	3206	3375	[14]
-55	1070	1641	3296	3206	3363	[15]
$+25$	1066	1638	3301	3215	3379	[15]
$+23$	1056	1634	3305[a]	3215[a]	3384	This work
$+20$	1055	1645	3218[a]	3303[a]	3386	[16]
$+25$			3215[a]	3301[a]	3384	[17]

[a] Doublet assigned to v_1 and $2v_4$ in Fermi resonance.

bands is given in Table I where the data from this investigation are compared to other laser Raman data.

Three features of the $3000 \, cm^{-1}$ region spectrum merit special attention. First, there appear to be no new peaks. Secondly, within the resolution of the observation there is no shift of the peaks in the $K-NH_3$ solution relative to pure solvent at the same temperature. Thirdly, the intensity of the peaks is decreased in the $K-NH_3$ solution as compared to pure solvent. All three features are in apparent contradiction to the observed infrared spectrum [11] in the same region. The lack of new peaks or shift of the solvent peaks can be shown by normalizing both spectra of Fig. 2; both spectra can be completely superimposed.

Discussion

In spite of the fact that no new bands were observed, our results cannot exclude the existence of a Raman active symmetric breathing mode of the solvated electron cavity. There are several reasons for this: 1. The exciting line used should have an energy lower than the first electronic transition. At present, there are no Raman spectrometers capable of operating in the near-infrared region. 2. The concentration of scattering centers may be too low to observe the transitions. In most Raman experiments, the concentration of the solution must be at least 10^{-2} M to give rise to a signal. Raman spectra of F-centers have been reported [7] at much lower concentrations than those used in this study, but the sample was cooled to liquid nitrogen temperature to narrow the electron absorption band. 3. The use of a single cc may not be a valid description of the cavity. It has been stated [5] that even though the single cc may be a useful device, it may not correspond to a real coordinate of the system. Such a statement is particularly important when discussing the liquid state (i.e. solvated electrons). The cc description of an electron impurity center in the solid state makes use of the fact that the nearest neighbors of the excess electron are ions in positions which are more or less fixed in space. In the case of a solvated electron the neighbors are the solvent molecules which have normal modes of vibration of their own. Furthermore, neither the positions of the neighbors nor the orientation of the cavity is fixed in space.

Both the Raman and the IR data on metal–ammonia solutions suggest that there is an interaction between the solvated electron and the solvent. The fact that the solvent vibrational peak occurs at the same position in both pure solvent and in metal solutions is somewhat surprising. From the infrared absorption data, one would expect that the totally symmetric vibration (v_1 and v_2) would shift with increasing metal concentration. It might even be succesfully argued that either the positions or half-widths or both would change because of the effects of site symmetry of the solvent molecule in the absorption of the cavity.

The decrease in intensity is probably due to the absorption of the scattered radiation by the high-energy tail of the solvated electron absorption peak. A quantitative comparison of the intensities in this preliminary investigation is not possible. There is also a lack of data for the extinction coefficient of the solvated electron at these energies and temperature.

References

1. Metal–ammonia solutions. Lepoutre, G., Sienko, M. J., (Eds.). New York: Benjamin 1964.
2. Metal–ammonia solutions. Lagowski, J. J., Sienko, M. J. (Eds.). London: Butterworths 1970.
3. Jortner, J.: J. Chem. Phys. **30**, 839 (1959).
4. Copeland, D. A., Kestner, N. R., Jortner, J.: J. Chem. Phys. **53**, 1189 (1970).
5. Lax, M.: J. Chem. Phys. **20**, 1752 (1952).
6. Kleinman, D. A.: Phys. Rev. **134**, A423 (1964).
7. Worlock, J. M., Porto, S. P. S.: Phys. Rev. Letters **15**, 697 (1965).
8. Lepoutre, G., Corset, J.: In Ref. [1], pp. 186.
9. Gold, M., Jolly, W. L.: Inorg. Chem. **1**, 818 (1962).
10. Rusch, P. F.: Ph. D. Thesis. The University of Texas at Austin, 1971.
11. Rusch, P. F., Lagowski, J. J.: These proceedings, p. 168.
12. Herzberg, G.: Molecular spectra and molecular structure, Vol. II. Infrared and Raman spectra of polyatomic molecules. New York: Van Nostrand 1945.
13. De Bettignies, B., Wallart, F.: Compt. Rend. **271**, 640 (1970).
14. Smith, B. L., Koehler, W. H.: J. Phys. Chem. (to be published).
15. Birchall, T., Drummond, I.: J. Chem. Soc. A, **1970**, 1859.
16. De Bettignies, B.: These de 3ᵉ Cycle. Université des Sciences et Techniques, Lille, 1971.
17. Ceccaldi, M., Leickman, J. P.: CEA-R-3586 (1968). Commissariat à l'énergie Atomique, France.

General Discussion of Papers by Smith/Koehler and DeBacker/Rusch/DeBettignies/Lepoutre

J. JORTNER   The fact that the totally symmetric radial mode of the solvated electron is not amenable to experimental observation by Raman scattering indicates that the current theoretical models are somewhat oversimplified. In particular, other angular modes may contribute substantially, as reflected by the theoretical underestimate of the electronic line width when only the radial mode is included. Furthermore, the role of rapid solvent exchange in the first coordination layer may lead to a smearing of the Raman line.

It was stated by some solid state theoreticians that the solvated electron just provides an analogue of the F-center in an ionic crystal. The interesting Raman data reported by Koehler indicate some major differences between these systems.

W. H. KOEHLER   I would like to emphasize that the calculations of α_R by Kleinman were applied to these solutions and gave values of the same order of magnitude. Also, our concentrations were the same as those in F-center systems where scattering was observed. If we are to pursue the application of F-center theory, we must then seek an explanation for the absence of scattering in these solutions. — Could not solvent exchange give rise to a variation in cavity size and contribute to band broadening of the blue absorption band?

J. JORTNER   Yes.

Metal–Ammonia Solutions VIII. Infrared Spectroscopy of the Solvent

P. F. Rusch and J. J. Lagowski

Abstract

The infrared absorption spectra of liquid ammonia and of lithium- and potassium-ammonia solutions at low temperatures have been obtained in the 3μ region. A spectrophotometric cell employing sapphire windows sealed to pyrex was used in the investigation. The envelope of strongly overlapping bands in the $3\,\mu$ region has been mathematically resolved into three components which are assigned to $2v_4$, v_1, and v_3 in order of increasing energy. The N-H stretching frequency of the solvent was observed to shift to lower energy with increasing concentration of both lithium and potassium. The shift is independent of the metal in solution over the concentration range from 0.005 M to 0.05 M. Results are interpreted in terms of new solvent-containing species in the presence of solvated electrons. Solvent molecules in these species are polarized by the solvated electron and show weakened N-H bonds in the presence of increased electron density.

Introduction

Solutions of alkali metals in liquid ammonia have been the subject of many experimental and theoretical investigations [1, 2] since liquid ammonia appears to be uniquely suited for stabilizing the solvated electron.

By necessity every theoretical description of these systems requires that the solvent be intimately involved in the process of stabilizing the solvated electron. In these descriptions the role of the solvent has been inferred by analogy with similar systems. For the most part these investigations, both theoretical and experimental, have been concerned with the properties of the solvated electron. Only a small amount of experimental work has addressed itself to a description of the nature of the solvent. In view of this emphasis on the solvated electron, it appears that what is needed at this point is an investigation of the other half of this system, the solvent.

Previous investigations have used proton magnetic resonance and optical reflection techniques to determine the nature of the solvent in metal–ammonia solutions. Proton magnetic resonance studies of metal ammonia solutions [3] yield a different type of information about the solvent properties than do optical investigations. Optical reflection studies in the fundamental region of the infrared have proved to be only moderately successful. Beckman and Pitzer [4] reported specular reflection, in the region from $1—20\mu$, of sodium-ammonia solutions over a wide range of concentrations. Reflection bands attributed to the solvent were observed in the $3\,\mu$ region. The positions of these reflection bands can not be directly related to absorption bands since the optical constants of the solutions were not obtainable from the experimental procedure used. Burow and Lagowski [5] used a silicon prism to obtain internal reflection spectra of sodium, lithium and potassium-ammonia solutions over a wide range of concentrations. Severe noise difficulties prevented the unambiguous assignment of the solvent bands. The purpose of the present study was to investigate the vibrational spectrum of the solvent in the fundamental region of the infrared using a transmission technique.

Experimental Section

In this investigation it was desired to obtain the IR transmission spectra of metal–ammonia solutions over as wide a range of spectral frequencies as possible. Spectral data in the near-infrared region can be used to estimate the metal concentration from the absorption band [6] near 1.5μ. Decomposition of the solutions can be determined by observation of the amide ion charge-transfer-to-solvent (CTTS) absorption [7] at 0.335μ. From previous investigations of liquid ammonia [8, 9] it is known that the N–H stretching vibrations have fundamental frequencies in the region around 3 μ while the bending fundamental vibrations occur in the regions around 6μ and 11 μ.

Design of a suitable optical cell required the use of a window material which rigidly satisfied the following criteria:

1. Optical transparency throughout a broad spectral region, particularly in the IR;
2. chemical inertness towards metal–ammonia solutions;
3. thermal stability over the liquid range of ammonia ($-70\,°C$ to $-33\,°C$);
4. ability to make a vacuum-tight seal to prevent decomposition of the solutions.

Obtaining the infrared spectrum of a normal liquid or solution at ambient temperatures requires that only the first two criteria be satisfied. For this reason most references on techniques of IR spectroscopy of liquids suggest the use of alkali metal halide windows and amalgamated cells. For studies of aqueous solutions alkaline earth fluoride optics are used. The alkali metal halides are soluble in liquid ammonia, and the thermal stability of alkaline earth fluorides is not acceptable.

Synthetic sapphire meets all of the design criteria for use as a window material. It is transparent in the region from 0.2 μ to more than 4.5 μ and is commercially available in the form of optical flats graded sealed to Pyrex glass. These graded seals have exceptional thermal stability, and sapphire is inert to metal–ammonia solutions. An optical cell similar to the one used by Gold and Jolly [10] was constructed. Spacing and alignment of the cell windows were achieved by use of a glass lathe during construction. Using ammonia gas [11] and metal–ammonia solutions of known concentration, the cell path length was determined to be about 0.2 mm, which is adequate for the IR study of liquids.

The sapphire optical cell was attached by means of ball-and-socket joints to a solution preparation vessel. This assembly was similar to those reported previously [6] and was contained in the evacuatable dewar [12] assembly described by Quinn [13]. Optical windows on the dewar assembly were of sodium chloride of sufficient thickness to withstand one atmosphere differential pressure. Solution preparation techniques used were those described earlier [6]. The entire dewar solution-vessel assembly was placed in a specially constructed support stand which was designed to permit reproducible positioning in both the Cary-14 and Beckman IR-7 spectrometers.

The optical cell assembly containing a freshly prepared metal–ammonia solution was transported to the Cary-14 to determine the UV-VIS-NIR spectrum and then to the Beckman IR-7 for determination of the IR spectrum. After the IR spectrum was recorded the cell was returned to the Cary-14 for observation of selected wavelengths to check for decomposition and solvated electron concentration. Initial IR spectral determination of the pure solvent demonstrated the need for a compromise to achieve the best instrument settings. Fundamental vibrations of ammonia in the 3μ region are intense and severely tax the ability of the instrument to perform well in the double beam mode. Although the black-body energy curve of the source is highest in this region, the grating used in the double monochromator system is operating in the fourth order. Furthermore the double beam mode of operation was not useful because of the difficulties of using an optical null instrument to measure the spectra of low temperature samples [14]. The inherent intensities of the bands of interest dictated the use of the 0—10% T scale which further required that the single beam mode of operation be used. The instrument settings selected proved to be the best compromise among scan speed, noise level, and resolution and gave reproducible spectra with a resolution of about $10\ \mathrm{cm}^{-1}$ in the region of the solvent absorptions.

Results and Discussion

From the IR spectrum of gaseous ammonia it is known [15] that the molecule has C_{3v} symmetry. A molecule with this symmetry has four normal modes of vibration and all the normal modes are both IR and Raman active [15]. The most striking feature of the gas phase IR spectrum [16] of ammonia is that two of the normal modes occur very close together in the region near 3μ. These modes are labelled v_1 and v_3; they are of symmetry types A_1 and E respectively and are attributed to N–H stretching. Normal modes involving H–N–H angle deformation are labelled v_2 and v_4 for the type A_1 and E modes, respectively. The intensities of the bending modes are much greater than the stretching modes, and their normal frequencies are beyond the region of sapphire transparency.

Liquid ammonia. The spectrum of liquid ammonia in the 0.2 mm sapphire cell is shown in Fig. 1. Prominent features include the intense solvent cutoff at 0.225μ and the series of intense, narrow bands in the near-infrared region. In the fundamental region of the infrared near 3μ there is an envelope of overlapping bands. Previous investigators [8, 9] have attempted to resolve the components of this envelope using the method of symmetrical contours. Although this is an acceptable method of curve resolution it suffers from the disadvantage that only one symmetric contour may be developed at a time. The envelope appears to have three components. Two of these components are the N–H stretching fundamentals v_1 and v_3 while the third component is attributed to an unusually intense bending overtone which appears because of Fermi Resonance [15]. The infrared spectrum of the pure solvent showed little spectral change as a function of temperature.

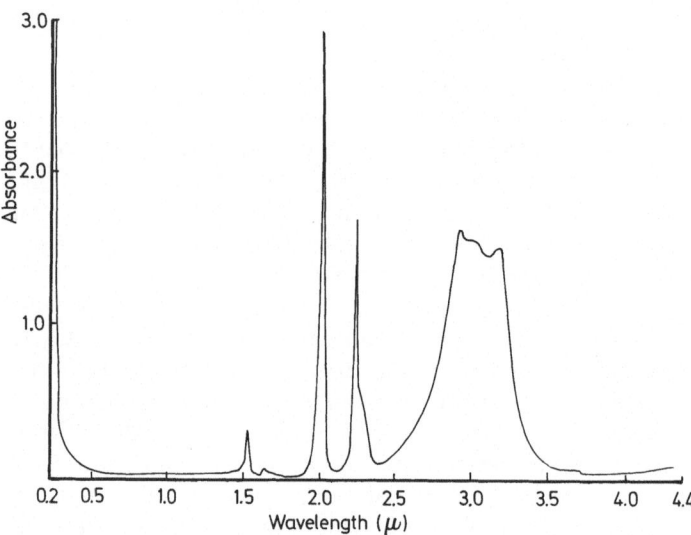

Fig. 1. Spectrum of liquid ammonia at − 70° C

Throughout this investigation a nonlinear least squares method was employed to perform the curve resolution in the 3 μ region. This method has been incorporated into a computer program [17] which was executed using a CDC-6600 computer system [12]. All the band positions reported in the 3 μ region were obtained in this manner and can be considered accurate within the resolution of the Beckman IR-7 spectrometer. All resolutions were performed using three component bands described by either Lorentzian or Gaussian-Lorentzian combination [18]. There are some valid objections to the use of computerized methods for resolving overlapping spectral bands. In general, the best least-squares fit of the data is obtained with an infinite number of bands. At the outset this may appear to be a severe limitation. In reality, however, any envelope of overlapping bands has certain prominent features which limit the number of components to be resolved. In the absence of any well-defined structure in the envelope there may be pertinent theories which help predict the number of bands. With judicious use, the method of mathematical resolution of overlapping spectral bands provides a standard method of analysis which can be kept relatively free from human prejudice in the results.

Resolution of the envelope of the three overlapping bands in the 3 μ region of the pure solvent spectrum gave the following positions of the bands: 3151 cm^{-1} $(2v_4)$, 3286 cm^{-1} (v_1), and 3453 cm^{-1} (v_3). The assignments are consistent with past experimental data and were verified in the following way. Using the simple valence force field (VFF) equations for a molecule of C_{3v} symmetry [15] the positions of v_3 and v_4 $(= 2v_4/2)$ were used to calculate the force constants for bond stretch and bond angle deformation. These force constants were then used to calculate v_1 and v_2 which was not observed directly. With the normal frequencies of all of the normal vibrations, the positions of overtone and combination bands were calculated. The results of this last calculation led to an assignment of all observed bands which was in agreement with all past data [8, 9, 15].

Metal–ammonia solutions. As the concentration of alkali metal is increased in liquid ammonia, the solutions exhibit properties more like metals than like normal ionic solutions. Because of this metallic behavior, the solutions reflect radiation [4, 19] very strongly.

Thus the solutions suitable for observation of transmission spectra are in the dilute to intermediate range of concentration. From previous investigations [6] it appears that there is no change in the solvated electron spectrum as a function of the alkali metal in solution. The spectral properties of the amide ion, on the other hand, show some dependence on the cation radius for the lighter alkali metals [20]. The position of the amide ion CTTS absorption changes most as the cation radius changes from that of Li^+ to that of K^+ while for larger cations there is no change.

Based on previous data for the optical properties of metal–ammonia solutions as a function of concentration, metal, and temperature, the range of experiments was limited to the study of lithium and potassium solutions at $-70\,°C$. The metal concentration was in the range 0.005—0.05 M. Attempts to observe the infrared spectrum of solutions more concentrated than 0.05 M failed due to lack of transmitted radiation even in the 0.2 mm cell.

The spectrum of a 0.005 M Li/NH_3 solution is shown in Fig. 2. Even though the intense solvated electron absorption band is present at about $1.5\,\mu$, several overtone and combination bands of the pure solvent are easily observed. There is no evidence of the amide ion CTTS absorption at $0.335\,\mu$. Resolution of the overlapping bands in the $3\,\mu$ region yields the three vibrational frequencies for $2v_4$, v_1 and v_3. A simplified VFF calculation was performed to calculate the frequency of v_2. Using this calculated value and the positions of the resolved bands, the positions

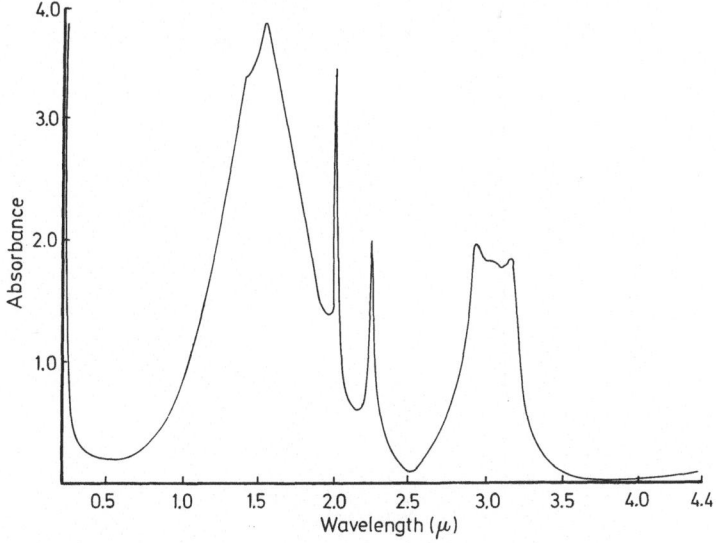

Fig. 2. Spectrum of 0.005 M Li/NH_3 solution at $-70°$ C

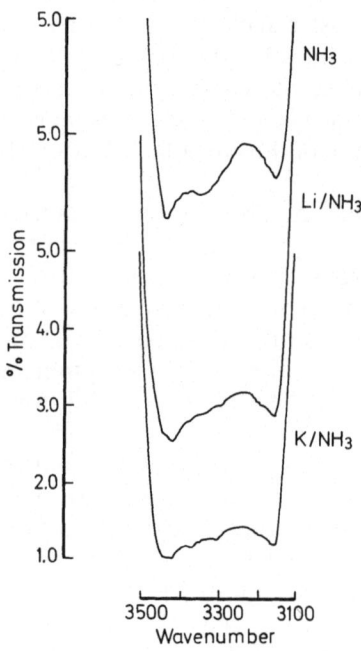

Fig. 3. Infrared spectra of 0.005 M Li and K ammonia solutions at −70° C

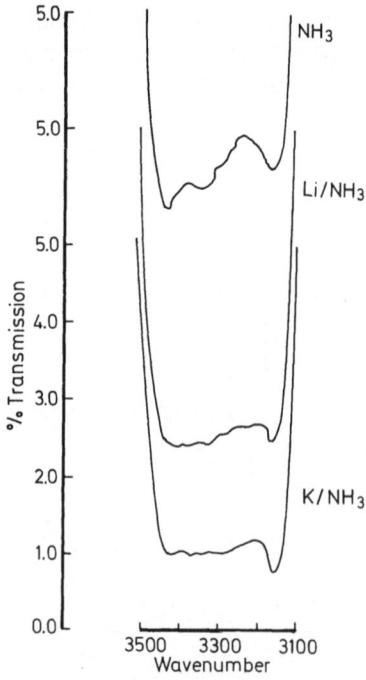

Fig. 4. Infrared spectra of 0.05 M Li and K ammonia solutions at −70° C

Table I. Typical resolved band positions from the 3 μ envelope

Solution	cm⁻¹					
	$2\nu_4$		ν_1		ν_3	
NH_3	3155	± 1.4	3281	± 3.0	3449	± 3.6
0.005 M Li/NH₃	3158	± 1.4	3265	± 2.7	3429	± 2.5
0.005 M K/NH₃	3156	± 1.4	3268	± 2.8	3426	± 2.5
0.05 M Li/NH₃	3153	± 2.0	3250	± 4.0	3409	± 5.0
0.05 M K/NH₃	3152	± 1.4	3252	± 3.0	3413	± 4.3

of the overtone and combination bands were determined. The assignment of the fundamental frequencies of the solvent in 0.005 M lithium and potassium solutions is identical to that of the pure solvent. In all other metal–ammonia spectra the bands in the 3 μ region were assigned to $2\nu_4$, ν_1 and ν_3 in order of increasing energy.

Detailed infrared spectra of dilute (0.005 M) lithium- and potassium-ammonia solutions are compared (Fig. 3) to the pure solvent at the same temperature. Similar spectra are shown in Fig. 4 for the more concentrated (0.05 M) lithium- and potassium-ammonia solutions. Each of the IR spectra was resolved into three component peaks assigned to $2\nu_4$, ν_1, and ν_3 in order of increasing energy. Typical resolved band positions are given in Table I where the error in resolved peak position is also listed. Resolved band positions reported in Table I have been plotted as a function of metal concentration (Figs. 5—7). The position of both the symmetric (ν_1) and asymmetric (ν_3) stretching fundamentals is observed to shift to

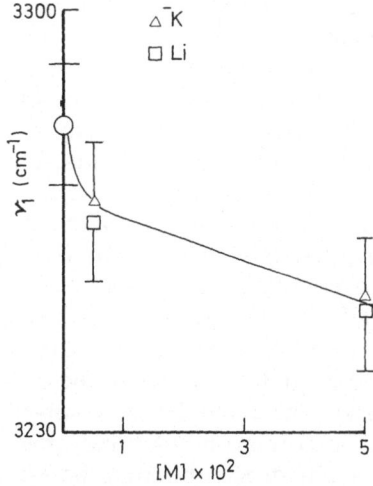

Fig. 5. Resolved ν_1 position as a function of metal concentration

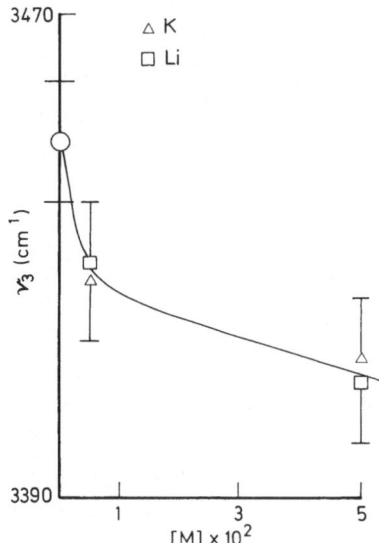

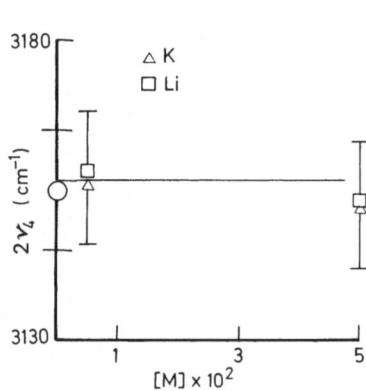

Fig. 6. Resolved v_3 position as a function of metal concentration

Fig. 7. Resolved $2v_4$ position as a function of metal concentration

lower energy as a function of increasing metal concentration. These shifts are independent of the metal in solution. The asymmetric bending overtone ($2v_4$) remains relatively unchanged as a function of metal concentration.

Any lowering of the N–H fundamental stretching vibration (v_1) is indicative of a lower force constant for the vibration, thus inferring a weakened bond. Formation of hydrogen bonds [21] may account for this lowering of the N–H stretching vibration. The normal hydrogen bonding in the pure solvent should, however, be destroyed by the addition of an alkali metal [22]. In metal–ammonia solutions many of the solvent molecules will be involved in solvation layers about the charged species, and these molecules may exhibit weakened N–H bands.

The effect on the N–H vibrational modes when ammonia becomes associated with ions and/or forms hydrogen bonds can be ascertained by observing the spectrum of ammonia molecules in solutions of salts. The infrared spectrum of concentrated salt solutions (>5 M) in liquid ammonia has been observed by Corset, Huong and Lascombe [23]. In the 3μ region of the spectrum they observed a series of overlapping bands with two prominent peaks which were assigned to $2v_4$ and v_3. These investigators showed that the frequency of the v_3 (asymmetric N–H stretching fundamental) vibration was shifted to lower energy as a linear function of the quantity Z/X^2 where Z is the cation charge and X is the sum of the cation crystal radius and the radius of the ammonia molecule in the pure liquid. In solutions of alkali metal iodides the v_3 solvent vibration was shifted more than in solutions of the corresponding alkali metal nitrates. This anion effect is attributed to increased hydrogen bonding between the ammonia molecules in cation solvation layers and the iodide ion. From these infrared data on concentrated salt solutions in liquid ammonia it can be concluded that two effects may be operating to weaken the solvent N–H bond. Coordination of the ammonia lone pair of electrons and/or formation of additional hydrogen bonds is possible. These two effects operating either separately or collectively result in a lowering of the N–H stretching frequency.

The presence of an alkali metal in liquid ammonia has an effect on the solvent which is quite different from that of the corresponding alkali metal salt. First, in metal–ammonia solutions, the shift of the v_3 frequency of the solvent is independent of the cation (Fig. 6). In salt-ammonia solutions there is a weak cation dependence of this shift attributed to ion-dipole interaction with the solvent. Second, the magnitude of the shift as a function of concentration is greater for metal–ammonia solutions then for salt-ammonia solutions. A 10.25 M solution of KI in ammonia shifts the v_3 frequency to a value 18 cm^{-1} lower than the pure solvent under similar conditions of pressure and temperature [23]. A 0.05 M solution of potassium metal in ammonia produces a 36 cm^{-1} shift of v_3 relative to the pure solvent under the same conditions. The magnitude of the v_3 shift per unit concentration is approximately 400 times greater for potassium metal than for KI. Since the shift of the v_3 solvent vibration is independent of the metal in solution (comparing Li and K), it is implied that the presence of the solvated electron in liquid ammonia accounts for the observed shift.

The solvated electron is a well-established species in metal–ammonia solutions. It is axiomatic that the solvent must contribute to the stabilization of the solvated electron in liquid ammonia. The solvated electron species in dilute metal–ammonia solutions seems to be well described by the cavity model [24]. In this model the solvated electron is stabilized in a cavity by the orientation of 4 to 6 solvent molecules oriented so that the protons are closest to the center of the cavity. Solvent molecules in a cavity have characteristics which can account for the observed infrared spectrum of the solvent in metal–ammonia solutions. First, these molecules would be expected to have weakened N–H bonds because of the increased electron density in the cavity. This would lead to a lowering of the N–H stretching frequencies compared to the pure solvent. Secondly, the solvent molecules in cavities are free from cation interactions because the cavities appear to be isolated from the corresponding cations [13]. Thirdly, the solvated electron in a cavity polarizes the solvent molecules of the cavity so that they have perturbed dipole moments [24], a part which will be shown to be very important.

As the concentration of alkali metal is increased in liquid ammonia it appears that some association occurs to form new species. A simple description of a typical aggregate species is given by the expanded metal model [22]. The basic unit of this model is the monomer composed of about 6 solvent molecules with the ionized electron stabilized in an expanded metal orbital. The solvent molecules in the monomer are oriented so that the nitrogens are closest to the cation and the electron orbital is in the region of the protons. Solvent molecules in this environment can also account for the observed infrared spectrum of the solvent in metal–ammonia solutions. The increased electron density about the protons and the coordination of the lone pair of ammonia electrons to the cation result in a weakened N–H bond for the solvent molecules in the monomer.

The fundamental stretching vibrations of ammonia in metal–ammonia solutions display shifts, relative to the pure solvent, which are characteristic of a system in which hydrogen-bonded species are being formed as a function of increasing metal concentration. The formation of hydrogen bonded species is accompanied by a shift to lower energy for the fundamental vibration involving symmetrical N–H bond stretching [21]. Often a new band is observed as a low-energy component of the symmetric bond stretching fundamental. The new component is attributed to the hydrogen bonded species which contains molecules with weakened bonds. Fundamental vibrations of the molecule which involve asymmetric bond stretching are also shifted to lower energy but usually to a smaller extent than the symmetric bond stretching fundamental. Bending fundamental vibrations often are not shifted by the formation of a hydrogen bonded species.

The concentration of solvent molecules in cavities is 4—6 times the concentration of the alkali metal present, depending on the number of solvent molecules per cavity [24]. For an alkali metal concentration of 0.05 M the concentration of associated solvent molecules is 0.2—0.3 M while the concentration of the bulk solvent is more than 200 times greater. Concentration of the bulk molecules can be calculated from the density of the pure solvent corrected for the concentration of the associated molecules in cavities. In the case of a 0.005 M alkali metal solution in liquid ammonia the bulk solvent molecules are more than 2000 times more concentrated than the associated molecules. Based on the effect of concentration alone, the theoretical second component of the absorption would be too weak to produce the shift observed for a dilute metal–ammonia solution. Assuming that Beer's Law is obeyed by both the bulk and the associated solvent molecules for a given vibration leads one to a consideration of the extinction coefficient (absorptivity) of the two different types of molecules. It is reasonable to assume that the bulk solvent molecules in dilute metal–ammonia solutions will have the same absorptivity as those in the pure solvent. The polarized state of the associated molecules [24] suggests that they may have a much larger absorptivity than the bulk molecules. Selection rules for infrared transitions require that there be a change in the dipole moment of the molecule for a given fundamental vibration. Solvent molecules in cavities may well have larger dipole moments as a result of polarization by the solvated electron and therefore also have larger absorptivity. With a large absorptivity even a very low concentration of molecules would have a significant absorption spectrum. The absorptivity of the associated molecules must be from 50—100 times greater than that of the bulk solvent molecules to produce spectral components of sufficient intensity to shift the resolved band position.

Conclusion

By analogy with systems in which hydrogen bonding is known to occur, it is possible to account for the shift of the resolved v_1 and v_3 band positions in metal–ammonia solutions. In dilute metal–ammonia solutions the formation of cavities effectively removes molecules from the bulk solvent. The solvent molecules in the cavities have fundamental bond stretching vibrations which are of lower energy than the bulk solvent molecules. Experimentally these lower energy vibrations appear as additional components in the spectrum. The sum of the two components, one for bulk molecules and one for molecules in cavities, would then account for the resolved band position shift to lower energy.

Acknowledgements

We gratefully acknowledge the financial assistance of the Robert A. Welch Foundation and the National Science Foundation.

References

1. Metal–ammonia solutions physicochemical properties. Lepoutre, G., Sienko, M. J. (Eds.). New York: Benjamin 1964.
2. Metal–ammonia solutions. Lagowski, J. J., Sienko, M. J. (Eds.). London: Butterworths 1970.
3. Hughes, T. R., Jr.: In: Lepoutre, G., Sienko, M. J. (Eds.): Metal–ammonia solutions physicochemical properties, pp. 211—14. New York: Benjamin 1964. — Hughes, T. R., Jr.: J. Chem. Phys. **38**, 202—9 (1963).
4. Beckman, T. A., Pitzer, K. S.: J. Phys. Chem. **65**, 1527—32 (1961).
5. Burow, D. F., Lagowski, J. J.: J. Phys. Chem. **72**, 169—75 (1968).
6. Burow, D. F., Lagowski, J. J.: Advan. chem. ser. No. 50, p. 125. Am. Chem. Soc. Washington, D. C. 1965.
7. Cuthrell, R. E., Lagowski, J. J.: J. Phys. Chem. **71**, 1298—1301 (1967).
8. Demidenkova, I. V., Scherba, L. D.: Izv. Akad. Nauk. SS SR Ser. Fiz. **22**, 1122—4 (1958).
9. Plint, C. A., Small, R. M. B., Welsh, H. L.: Can. J. Phys. **32**, 653—61 (1954).
10. Gold, M., Jolly, W. L.: Inorg. Chem. **1**, 818—27 (1962).
11. Gunther, F. A., Barkley, J. H., Kolbezen, M. J., Blinn, R. C., Staggs, E. A.: Anal. Chem. **28**, 1985—9 (1956).
12. For detailed drawings, specifications and procedures, see Rusch, P. F.: Ph. D. Dissertation, The University of Texas at Austin, 1971.
13. Quinn, R. K., Lagowski, J. J.: J. Phys. Chem. **72**, 1314—8 (1968).
14. Potts, W. J., Jr.: Chemical infrared spectroscopy, Vol. I, Techniques. New York: John Wiley and Sons 1963.
15. Herzberg, G.: Molecular spectra and molecular structure. Part II: Infrared and Raman spectra of polyatomic molecules. New York: Van Nostrand 1945.
16. Pierson, R. H., Fletcher, A. N., Gantz, E. S.: Anal. Chem. **28**, 1218—38 (1956).
17. Program RESOL was written by Dr. D. D. Tunnicliff, Shell Development Company, Emeryville, California, who kindly supplied the initial version.
18. Young, R. P., Jones, R. N.: Chem. Rev. **71**, 219—28 (1971).
19. Koehler, W. H., Lagowski, J. J.: J. Phys. Chem. **73**, 2329—35 (1969).
20. Caruso, J. A., Takemoto, J. H., Lagowski, J. J.: Spectr. Letters **1**, 311—16 (1968).
21. The Hydrogen Bond. Pimentel, G., McClellan, A.: San Francisco: W. H. Freeman and Co. 1960.
22. Becker, E., Lindquist, R. H., Alder, B. J.: J. Chem. Phys. **25**, 971—5 (1956).
23. Corset, J., Huong, P. V., Lascombe, J.: Spectrochim. Acta **24A**, 1385—96 (1968).
24. Copeland, D. A., Kestner, N. R., Jortner, J.: J. Chem. Phys. **53**, 1189—1216 (1970).

Discussion

U. E. EVEN   The solvent IR bands are considerably shifted in the presence of the solvated electron, while the Raman lines are not affected. How do you explain this discrepancy?

P. F. RUSCH   This is an inevitable question. The obvious difference is that the Raman spectrum is induced by polarizability changes while the IR spectrum originates from the change of the dipole moment with nuclear coordinates.

J. JORTNER   The different sources of intensity for the Raman and the IR spectra can result in different effects of the solvated electron on the absolute intensities; however, they cannot explain the different frequency shifts observed in IR and Raman spectroscopy.

J. H. ROBERTS   These results are in agreement with the observation of Roberts and Lagowski (these proceedings) and with Saito (these proceedings). The fact that the conclusions drawn from the studies of the fundamentals of ammonia in Li and K solutions (this paper), the near-infrared of the solvent bands of Na/ND_3 solutions (Roberts and Lagowski) and the near-infrared difference spectrum of solutions produced by a reversal of the equilibrium with amide reported by Saito are all in agreement in very strong evidence in support of the suggestion that the formation of the solvated electron causes an enhancement of the solvent structure. Moreover, this effect is observed at concentrations several orders of magnitude lower than similar effects in salt solutions.

J. JORTNER   You have demonstrated that intermolecular interactions between the solvent and the localized excess electrons affect the intramolecular vibrational frequencies of the solvent molecule. It would be interesting to study cooperative vibrational excitations, where two molecules are simultaneously excited by one photon. These cooperative excitations are induced by the intermolecular coupling between the solvent molecule, which may be grossly modified in the presence of solvated electrons.

Metal–Ammonia Solutions: Transition Range

Gerard Lepoutre

Abstract

This paper considers the nonmetal-metal transition and the liquid-liquid phase separation in metal–ammonia solutions in the intermediate concentration range.

Introduction

It is well known that metal–ammonia solutions display a non metal-metal (NMM) transition, and, in the same range of concentrations, a liquid-liquid phase transition (LLP) at lower temperatures. This paper is an investigation of the solutions in the concentration range where these two properties occur. Characteristic experimental properties are reviewed and interpreted at two different levels: a phenomenological description is first obtained, and then an attempt is made to describe the solutions in terms of the structures of their components and the interactions between them.

I. Characteristic Experimental Properties

Some pertinent data are reviewed concerning the NMM and LLP transitions.

A. Liquid-Liquid Phase Transition

The general behavior of the phase transition has been reviewed elsewhere [1, 2] and is reproduced in Fig. 1.

An idea of the extent of the "transition" range of concentrations can be obtained from Fig. 1: the sodium phase diagram extends from 1 to 10 MPM (moles per cent metal in NH_3), or about 0.4 to 3.5 M (moles metal per liter).

Three features of the LLP transition deserve specific comment: the effects of charge and mass of the cation on the critical coordinates, the approach of phase separation from higher temperatures and the shape of the consolute curve.

Effects of charge and mass. As seen in Fig. 1, all monovalent metals have their critical concentration at about 4 MPM. The critical concentration for calcium solutions is closer to 2 MPM. These critical concentrations are exceptionally low.

Critical temperatures vary sharply with two factors, charge and mass; they are much higher for calcium than for monovalent cations. The critical temperature is lower for larger masses in the series of alkali cations. In fact, it would be below the freezing point for Cs. There is an inversion for Li, whose critical temperature

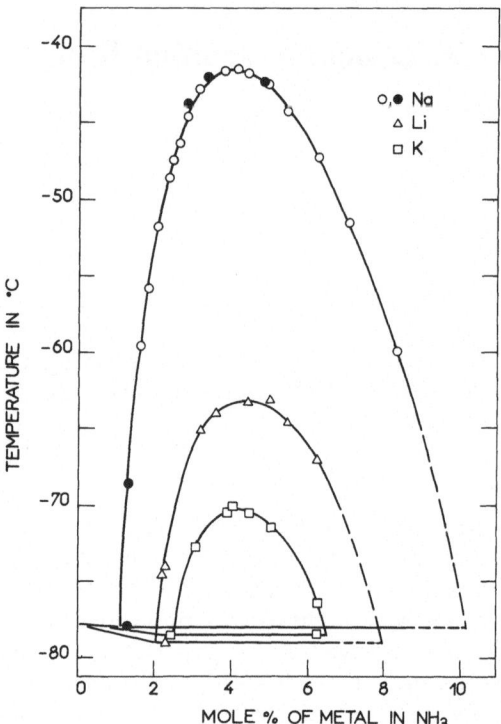

Fig. 1. Liquid-liquid phase separation (after M. J. Sienko, Colloque Weyl I)

is lower than that of Na. This will be commented upon later. In general, the critical temperature is higher for smaller cations and for higher charge.

Approach from higher temperatures. The mean ionic thermodynamic activity of sodium in ammonia has been calculated for all concentrations at 240 °K from vapor pressures and e.m.f. measurements [3]. The concentration dependence is shown in Fig. 2. The curve is almost horizontal between 1.5 and 7.5 MPM with an inflection point appearing at 4 MPM. A horizontal tangent at the critical temperature, 231.5 °K is to be expected. A horizontal gap would appear at lower temperatures. A very strong indication of the demixtion appears therefore in the activities of the solute at temperatures as high as ten degrees over the critical temperature. It is even observable at more than thirty degrees above the critical temperature.

Other properties are disturbed in a more immediate vicinity of the phase separation, as seen in Fig. 3 for the temperature dependence of the thermoelectric power. The slope is modified within five degrees above the critical temperature [4].

Shape of the consolute curve. It has been shown [5] that the consolute curve, which is parabolic at lower temperatures, becomes cubic with 1.8° C of the critical temperature. Other critical parameters have recently been obtained by Thompson, through e.m.f.'s of the cell [13].

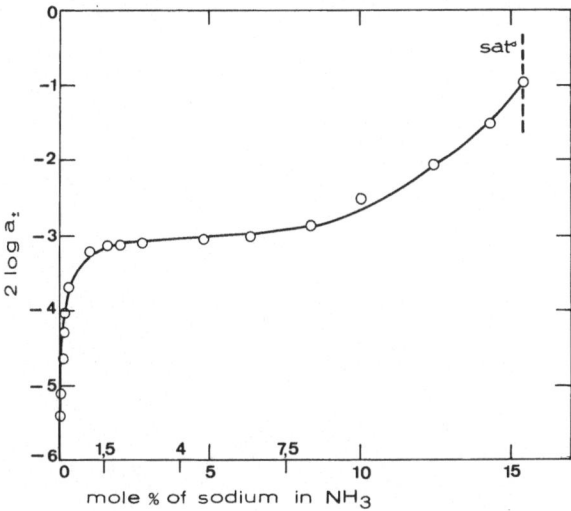

Fig. 2. Mean ionic activity of Na in NH_3

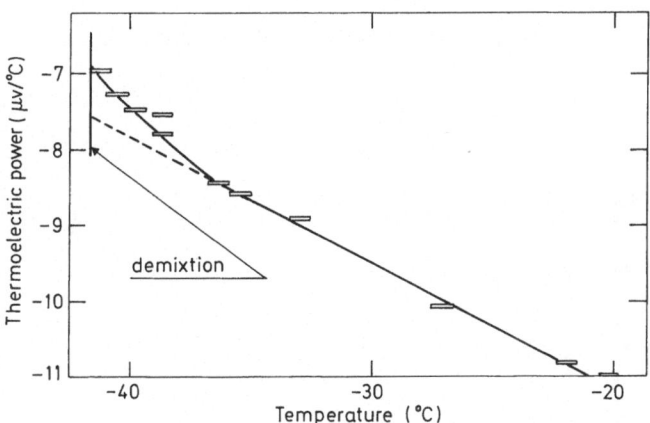

Fig. 3. Thermoelectric power near critical point

B. Non-Metal — Metal Transition

The effect of a NMM transition can be observed in transport properties i.e. thermo-electric power, electrical conductivity, Hall effect, and in equilibrium properties (Knight shifts, magnetic susceptibilities).

Transport properties. Thermoelectric powers (Fig. 4) drop from nonmetallic values in the dilute range to metallic values in the concentrated range. An anomalous behavior is observed between 1.5 and 3 MPM (for sodium and potassium): there is a leveling followed by a sharp decrease, and the sign of the temperature coefficient is reversed in this range of concentration [6].

Conductivities increase by four orders of magnitude along the transition range. Their temperature coefficients are positive, with a maximum at 2 MPM (Fig. 5) followed by a rapid decrease [7].

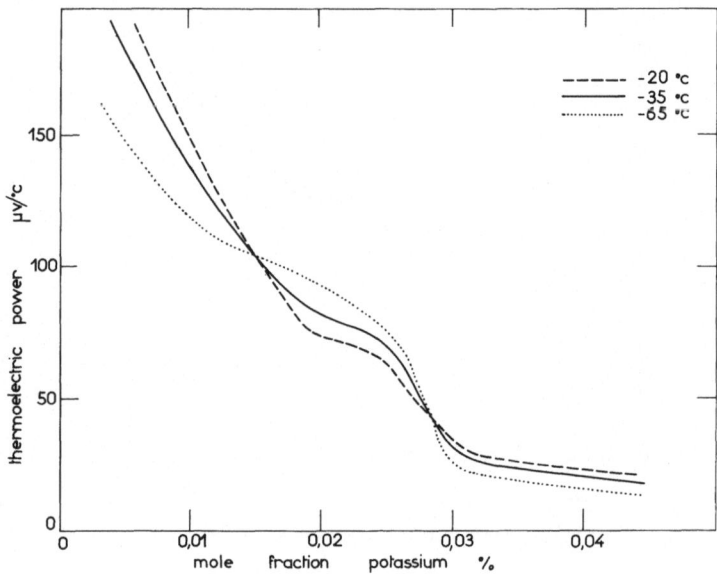

Fig. 4. Thermoelectric power of K in NH_3

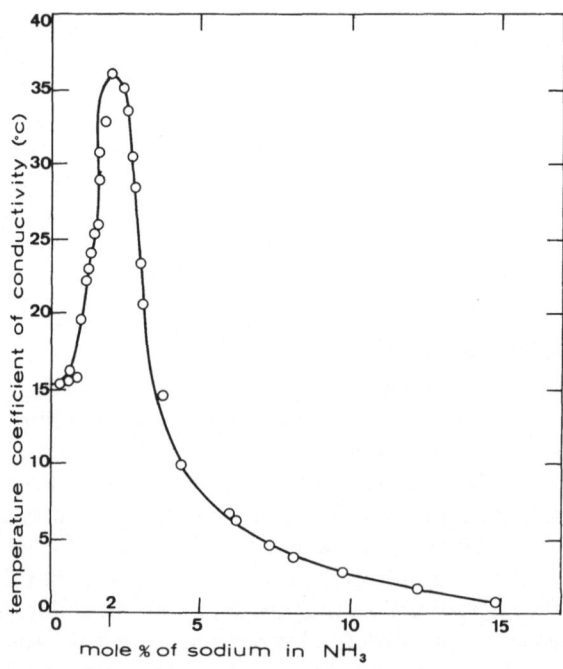

Fig. 5. Temperature coefficient of conductivity, Na in NH_3

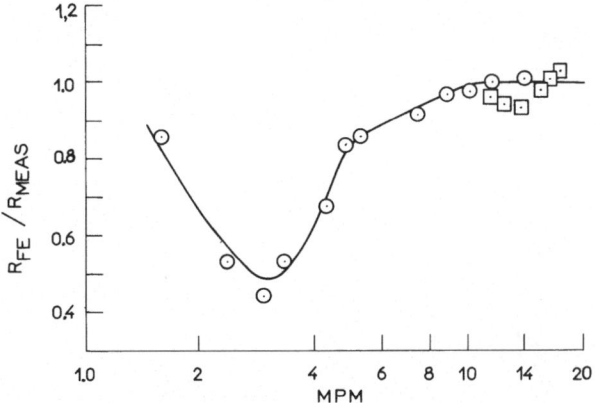

Fig. 6. Ratio of Hall coefficients for free electrons and for experimental material, Li in NH_3. [After Cohen, M. H., Thompson, J. C.: Adv. Phys. **17** (70), 857 (1968)]

Hall coefficients (Fig. 6) rise at 3 MPM to twice the value corresponding to a free electron gas. At higher concentrations, they decrease rapidly towards the free electron value, which is reached at $8 - 10$ MPM [8].

In summary, striking features of the transport properties occur at concentrations lower than critical, between 1.5 and 3 MPM; at higher concentrations, there is a fast, smooth trend towards metallic behavior, without any peculiarity at 4 MPM (critical concentration).

Equilibrium properties. Thermodynamic activities have been displayed (Fig. 2) with a plateau from 1.5 to 7.5 MPM and an inflection point near 4 MPM.

Magnetic susceptibilities increase with dilutions below 0.8 MPM and increase with concentrations above 2 MPM [9] (Fig. 7). A minimum with vanishing or

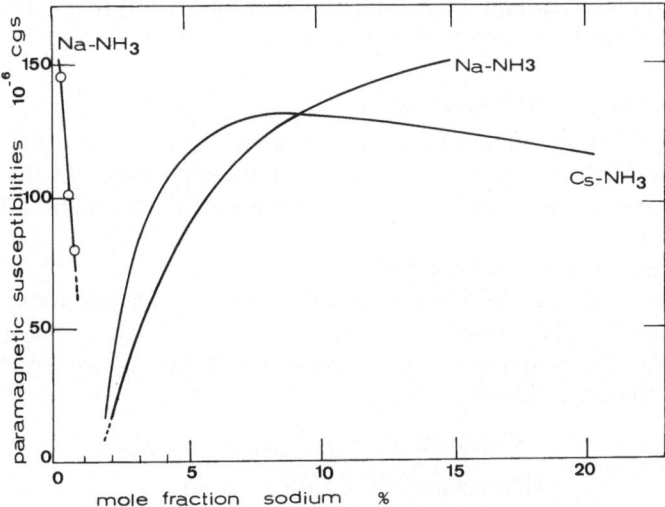

Fig. 7. Paramagnetic susceptibilities, Na and Cs in NH_3

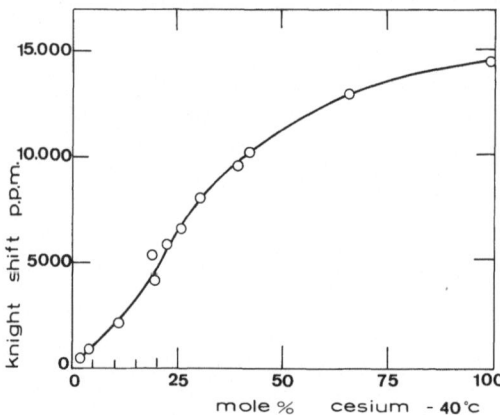

Fig. 8. Knight shift, Cs in NH_3

negative values of the susceptibility must occur in the vicinity of 1 MPM, where experimental data for Na are unfortunately lacking. The temperature coefficient is positive, with a maximum in the vicinity of 1 MPM, followed by a sharp decrease.

Knight shifts follow closely the magnetic susceptibilities in the transition range of concentrations [9] (Fig. 8).

It will be useful to remember that paramagnetic susceptibilities and Knight shifts are very low in the vicinity of 1 MPM; both increase rapidly throughout the transition range.

II. Phenomenological Description

Using the previous data, a description of the solutions will now be given, with the help of thermodynamics and of a small number of assumptions; this will be done in terms of electrons and in terms of chemical species.

A. Electrons and the NMM Transition [9]

Transport properties are functions of electronic mobilities, which increase very fast along the NMM transition. To account for the high values of the Hall coefficient in the transition range, it has been necessary [9] to make the following two assumptions:

a) the mobilities of the cations are negligible

b) there are two different mechanisms of transport for the electrons; they are labelled "l" and "d"; corresponding mobilities are μ_l and μ_d; n_l and n_d are the densities of electrons, with $n_l + n_d = n$, total density of the valence electrons.

It is then easy to derive three equations:

$$\text{Conductivity} = f_1(\mu_l, \mu_d, n_l, n) \tag{1}$$

$$\text{Hall coefficient} : f_2(\mu_l, \mu_d, n_l, n) \tag{2}$$

$$\text{Thermoelectric power} = f_3(\mu_l, \mu_d, n_l, n, S_l^*, S_d^*) \tag{3}$$

With reasonable extrapolations for the entropies of transfer S_1^* and S_d^*, the resolution of the three equations at each total concentration n gives μ_1, μ_d and n_1 (or n_d) as functions of the total concentration. The results are shown in Figs. 9 and 10.

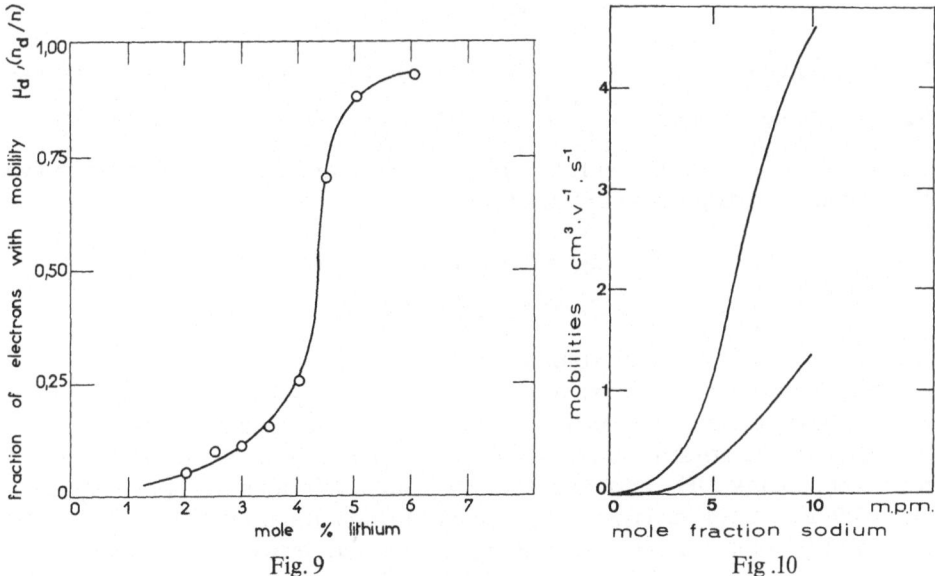

Fig. 9

Fig. 10

Fig. 9. Density of carrier d(n_d/n), Li in NH_3

Fig. 10. Mobilities of carriers l and d, Li in NH_3

It can be seen from Fig. 9 that the proportion of carriers "d" increases from 0 to 100% in the transition range, with a sharp rise at about 4 MPM, the critical concentration.

It can be seen from Fig. 10 that the mobility "d" is much higher than the mobility "l". It should be noted that the smooth curves in Figs. 9 and 10 account for the peculiar behavior of the thermoelectric power between 1.5 and 3 MPM (Fig. 3).

If the mobility $\mu = 1$ cm²/volt, sec is taken as the threshold between localized and delocalized electrons, it is seen that electrons d are delocalized at 4 MPM, while l electrons remain localized until they disappear.

Since the mobilities μ_d rapidly become much higher than μ_1 while the proportion n_d/n of carriers d is also increasing, it is easily understood that the peculiar features of the transport properties appear at rather low concentrations, i.e. between 1.5 and 3 MPM. At higher concentrations, a major part of the transport is assumed by mechanism "d", and the contribution of "l" vanishes rapidly. Static properties as activities are more likely monitored by n_1 and n_d, which vary rapidly around 4 MPM. It seems safe to state that 4 MPM is at the "center" of the NMM transition.

B. Chemical Species and the LLP Transition [4]

For a liquid mixture of two components a positive deviation from Raoult's law (excess vapor pressure) indicates the possibility of a liquid-liquid phase separation. If the deviation from Raoult's law follows the simple law:

$$RT \log \gamma_1 = A_{12} \bar{V}_1 \varphi_2^2 \qquad (4)$$

(with $y_1 = p_1/p_1^0$, A_{12} = coefficient of solvent-solute interaction; φ_2 = volumic fraction of the solute $= \dfrac{x_2 \bar{V}_2}{x_1 \bar{V}_1 + x_2 \bar{V}_2}$; x_1, x_2 = mole fractions of the solvent and solute $\bar{V}_1$, $\bar{V}_2$ = molar volumes of the solvent and solute) it can be shown that the critical parameters are given by:

$$\begin{aligned} x_{2c} &= f(\bar{V}_{1c}, \bar{V}_{2c}) \\ T_c &= A_{12} f(\bar{V}_{1c}, \bar{V}_{2c}, x_{2c}) \end{aligned} \qquad (5)$$

(with $\bar{V}_{1c}$, $\bar{V}_{2c}$, partial molar volumes at the critical point).
For metal–ammonia solutions, it is impossible to apply Eq. (5) with the components ammonia (NH_3) and metal (M). Their respective volumes cannot yield the low value of the critical concentration. It is possible, however, to apply Eq. (5) with components NH_3 and $M(NH_3)_n$. The mole fractions for such a mixture are easily expressed in terms of the analytical mole fractions (for NH_3 and M) and of the number n of ammonia molecules transferred to the solute. The volume $\bar{V}_1$ is taken as the molar volume of pure ammonia (25 cm^3/mole). The volume $\bar{V}_2$ is taken as the sum of the apparent volume of $M(V_{Na} = 65$ cm^3) and of the volumes of n ammonia molecules.
Equation (5) has then only one unknown, n, and is solved for sodium with $n = 6.5$. Using this value in Eq. (3), it is found that $A_{12} = 6.45$ cal/cm^3.
But values of n and A_{12} can be computed in an independent manner through the vapor pressures and Eq. (1), The value of n is adjusted so that A_{12} does not vary with concentration, and the values thus found are

$$n = 6.5 \text{ to } 7, \qquad A_{12} = 6.4 \pm 0.5 \text{ cm}^3/\text{mole}.$$

The validity of the model is therefore confirmed. It is also understood why the critical concentration is so low: it is simply the result of the large difference between V_2 and V_1, which is due to the nature of the solute, $Na(NH_3)_{6.5}$; such a thing could not happen if the solute were simply Na.
The coordination numbers of the other monovalent cations are about the same as those for sodium, because their critical concentrations are similar. This fact is confirmed, although without high precision, by vapor pressures. Calcium must be much more solvated, because its critical concentration is lower, which is consistent with its double charge; n is found to be of the order of 20.
Critical temperatures depend on A_{12}. It is interesting to find that the values of A_{12} computed from vapor pressures are consistent with the inversion which puts lithium between sodium and potassium on the phase diagrams (Fig. 1).

C. Summary of Part 2

A model with two electronic mobilities has been introduced for describing the NMM transition, and the two mobilities and the densities of carriers for each mechanism have been computed. Nothing has been said until now about the nature of these two mobilities. No link has been suggested between the transport properties and the LLP transition.

A model with two components, NH_3 and $M(NH_3)_n$, has been introduced for describing the LLP transition; a coordination number and an interaction coefficient have been computed. Nothing has been said until now about the nature of $M(NH_3)_n$ or about the nature of the interactions between the two components. No link has been suggested between the two component system and the two electronic mobilities, between the LLP and the NMM transitions.

Two different descriptions have been obtained at the phenomenological level. Is it possible to merge them into a single model?

III. Tentative Microscopic Description

A model should explain both the LLP and the NMM transitions. It should therefore relate in some way the two components NH_3 and $M(NH_3)_n$, to the two conduction processes, "l" and "d". Component $M(NH_3)_n$ consists of solvated cations and of electrons. It has been possible to treat these two subcomponents as a single component because their concentrations are always equal (electroneutrality). Electrons are therefore implicitly present within the component $M(NH_3)_n$.

The questions concerning structures and interactions can then be analyzed along the following lines:

— structure of the component $M(NH_3)_n$, i.e. interactions between cations and electrons within $M(NH_3)_n$.

— interactions of the component $M(NH_3)_n$, i.e. cations and electrons, with the solvent NH_3.

Hopefully, this analysis will yield some understanding of the origins of the LLP and NMM transitions and of their relations.

A. Analysis of Structures and Interactions

a) Component $M(NH_3)_n$. On the concentrated side of the transition range, electrons are, for the most part, delocalized on the liquid lattice of solvated cations (Knight shifts). This quasi–metal must have a quasi–metallic lattice energy.

On the dilute side, most electrons are localized and paired (low susceptibility). These pairs of localized electrons are in preferential association with pairs of solvated cations (association constants for this double pairing

$$2[M(NH_3)_n^+ - e^-(NH_3)_m] \rightleftarrows [M(NH_3)_n^+ \cdot e^-(NH_2)_m]_2 ,$$

have been computed at lower concentrations [10]).

The component $M(NH_3)_n$ therefore undergoes a radical structural change along the transition range. This change is analogous to the vaporization of a metal:

$$2x \, Na_{(lattice)} \rightarrow x \, Na_{2(gas)} .$$

An important difference is that the electrons, instead of being localized on the cations as in Na_2(gas), are localized in the solvent in $[M(NH_3)_n^+ \cdot e^-(NH_3)_m]_2$, as will be described below.

b) *Interactions of* $M(NH_3)_n$ *with* NH_3. At dilute concentrations, two phenomena of opposite effects occur: ammonia molecules are in attractive interaction with electrons to give the solvated electrons. The component $M(NH_3)_n$, however, consists mainly of dimers $(M(NH_3)^+ \cdot e^-(NH_3)_m]_2$ i.e. pairs of solvated electrons associated with pairs of solvated cations. Although this includes the attractive interaction with ammonia which gives the solvated electrons, it implies also that $M(NH_3)_n$ prefers to pair with itself than to mix the rest of the ammonia. The concentration of the solute is less than the analytical concentration. The solvent may therefore exhibit a positive deviation from Raoult's law.

At high concentrations, bulk ammonia molecules are not well accommodated within the liquid lattice of solvated cations. As shown by vapor pressures, there is again a preferrential association of $M(NH_3)_n$ with other $M(NH_3)_n$ (lattice energy) and of NH_3 with NH_3, rather than with each other; the former coefficient A_{12} is characteristic of these preferential associations.

In summary, component NH_3 is poorly accommodated in component $M(NH_3)_n$; the chemical potential of NH_3 is therefore always higher than in an ideal mixture. Component $M(NH_3)_n$ tends to associate with itself in two different ways: in the concentrated range, $M(NH_3)_n$ is associated with itself by delocalization of the electrons on the lattice of solvated cations $M(NH_3)_n^+$; in the dilute range, localized electrons are associated with solvated cations, and paired, according to $[M(NH_3)_n^+, e^-(NH_3)_m]_2$.

B. Percolation, as a First Link between the Two Transitions

At this point, it is interesting to note that the transitions are fairly well described by the percolation of hard spheres of $M(NH_3)_n$, as for alkali metals vapors.

In the case of pure alkali metals it has been shown that just above the critical liquid gas temperature T_2, a NMM transition occurs in the vicinity of the critical density n_c (see annex I and references given there). Moreover, the ratio of the densities at the critical point and at the melting point, n_c/n_m (equal to the ratio of the packing fractions at the same points), is the same for the various alkalis and is equal to 0.23. It has been inferred that the NMM transition occurs through percolation, when, around the critical density, the packing fraction becomes sufficient for such a percolation to occur.

For Li–NH_3, solutions, a pure component $Li(NH_3)_4$ is known in the solid state, with a melting point at 90 °K. It may be assumed as before that for component $Li(NH_3)_4$ the liquid–liquid critical point is equivalent to a liquid-gas critical point. Densities n_m and n_c can then be computed by ignoring the component NH_3, and the ratio n_c/n_m is again found equal to 0.23, as for pure alkali metals. It appears therefore that a simple model of percolation of hard spheres is suitable as a first link between the LLP and the NMM transitions. In the transition range, a fraction of the electrons is conducting along the narrow percolation paths created by the solvated cations, $M(NH_3)_n^+$. It should be added that the other fraction may still be conducting within the ammonia, which we have ignored but which can still solvate electrons.

At lower temperatures, the phenomenon of percolation paths is replaced by that of concentrated aggregates which separate from the rest of the ammonia and the solvated electrons.

C. Electrons and the LLP Transition

Below the critical temperature, there are two liquids in equilibrium. The coefficient A_{12} is characteristic of the unwillingness of $M(NH_3)_n$ (component 2) and NH_3 (component 1) to form a solution. What is the contribution of electrons to A_{12}?

If k_{11}, k_{22}, k_{12} represent the binary attractions between the components, A_{12} can be expressed as

$$A_{12} = 2k_{12} - k_{11} - k_{22}$$

Positive values of A_{12} give phase separation with T_c proportional to A_{12}. The individual k's are positive for repulsion, negative for attraction.

Delocalization of electrons in component 2 yields a high negative value of k_{22} due to the metallic cohesive energy. This can be partly compensated by the tendency of electrons to be solvated in ammonia ($k_{12} < 0$). It has been seen, however, that this effect is itself balanced by a dimerization of the solvated electrons associated with cations. The coefficient k_{22} appears therefore to be an important factor in A_{12}. The lack of good solubility is linked to the cohesive energy of electrons and cations in the liquid metal $M(NH_3)_n$.

At higher temperatures, there is a continuous transition. The thermal energy may expand the lattice until there is enough NH_3 to solvate some electrons in shallow traps; there is then, at each concentration, a continuous series of electronic energies.

The chemical potentials (Fermi energy) were evidently equal in the two separate liquid phases at lower temperatures. At higher temperatures, they still display a plateau along the transition range, as observed for the vapor pressures of the solvent and the activities of the solute (Fig. 2). The thermal energy performs the mixing, but imperfectly; the transition is continuous, but far from ideal.

Closer to the critical temperature, the thermoelectric power has shown a peculiar behavior (Fig. 3), probably characteristic of local fluctuations. Two different electronic states would then coexist in two different kinds of fluctuating microregions.

Interaction coefficients A_{12} reflect a lack of solublility of NH_3 with $M(NH_3)_n$, the latter being a liquid metal. It should be possible to interpret the magnitudes of A_{12} for the various cations in terms of cohesive energies for the corresponding liquid metals $M(NH_3)_n$; they decrease, as the critical temperatures, in the order Ca^{++}, Na^+, Li^+, K^+, Cs^+.

D. Chemical Species Responsible for the Electronic States along the NMM Transition

With respect to concentration, the NMM transition with its two transport processes may be described as a transition from electrons localized in NH_3 to electrons delocalized on $M(NH_3)_n$.

Two features have to be kept in mind: 1) the chemical potentials (Fermi energy) remain almost constant in the transition range, well above the critical temperature; 2) starting at 1 MPM, electrons which were solvated and paired have a rising paramagnetic susceptibility.

It will then be assumed that below 1 MPM there is a gap between two localized states. The lower state is filled, the higher is empty. In the lower, the electron is solvated in an ammonia cavity; the higher is due to the solvated cations, around which the electrons might orbit. As concentration increases, the band of solvated electrons broadens by overlap of the ammonia cavities. It is shifted to higher energies by lack of solvating ammonia. Where the lower and higher bands begin to overlap a pseudogap is generated. Occupied states appear in the pseudogap at and above 1 MPM. Electrons in these states have higher mobilities but are still localized; they have a paramagnetic susceptibility.

At 5 MPM the g-factor of Mott becomes equal to 1/3 [9, 11] (g is the ratio of the densities of states at the Fermi level for the actual material and for a gas of free electrons; it may be computed from Knight shifts, conductivities, susceptibilities.) This is the threshold of delocalization. It has been seen also that the mobility μ_d reaches the value of 1 cm^2/volt $\cdot$ sec which is recognized as the threshold of delocalization.

At 10 MPM, the former lower band has disappeared; the cationic band contains all of the valence electrons.

According to these views, there is a continuous series of electronic energies in the transition range. Lower energies, which are reminiscent of the solvated electrons, characterize interactions between electrons and NH$_3$, are responsible for the mobilities λ_l, and can be visualized as localized states in a band. Higher energies characterize interactions between electrons and the M(NH$_3$)$_n^+$ lattice, and are responsible for the mobility μ_d. The higher mobilities μ_d, however, are still monitored by the diffusion of electrons on both NH$_3$ centers and M(NH$_3$)$_n$ centers [12].

The present description is tentative and needs discussion.

Conclusion

Much work, some of which is in progress, remains to be done.
Experimentally:

— systematic data should be collected, so that the parameters μ_l, μ_d, n_l, A_{12}, n (coordination number) may be known for cations of different charge and mass.

— X-ray and neutron diffraction and diffusion measurements are needed. At higher temperatures, are there two different distances between the cations (those forming a band and those associated with pairs of solvated electrons)? Near the critical temperature, are there clusters?
Theoretically:

— phenomenological relations, already derived, should yield the temperature coefficients of the above parameters.

— primarily, the values of these parameters should be recomputed from the model of a M(NH$_3$)$_n$ metal expanded by NH$_3$.

The present work, quantiative at the phenomenological level, is only tentative and qualitative at the microscopic level. It is hoped that half of the way has been covered and that it will be easier for theoreticians to interpret the phenomenological parameters than the plain experimental data.

Note:

A priori calculations have been attempted by Teoh [13] and by Lelieur [9].

Teoh calculates the liquid-liquid critical temperatures by applying an equation of state of the Van den Waals type to a mixture of solvated ions and solvated electrons. He has to postulate a value for the fraction of electrons solvated at the critical concentration. His method would deserve further criticism.

Lelieur assumes that there is a critical volume V_c which is "metallic" if it contains two electrons or more, "non-metallic" if it contains less than two electrons. With a gaussian distribution of small fluctuations, he can compute n_d/n, fraction of delocalized electrons, as a function of total concentration; V_c is taken as a parameter. The increase of n_d/n with concentration is less sharp than the increase computed from experimental data (Fig. 10).

Further calculations should include the species NH_3 and $M(NH_3)_n$.

References

1. Sienko, M. J.: In: Metal–ammonia solutions, Colloque Weyl I, G. Lepoutre, M. J. Sienko, Eds., p. 23. New York: Benjamin 1964.
2. Thompson, J. C., Teoh, H., Antoniewicz, P. R.: J. Phys. Chem. **75**, (3) 399 (1971).
3. Dye, J. L., Lepoutre, G., Marshall, P. R., Pajot, P.: see Ref. I, p. 92.
4. Damay, P.: This colloquium and Thèse, Université Paris VI, May 1972.
5. Chieux, P., Sienko, M. J.: J. Chem. Phys. **53** (2) 566 (1970).
6. Damay, P., Depoorter, M., Chieux, P., Lepoutre, G.: In: Metal–ammonia solutions, Colloque Weyl II, J. J. Lagowski, M. J. Sienko, Eds., p. 233. London: Butterworths 1970.
7. Dewald, R. R.: See Ref. 6, p. 497.
8. Nasby, R. D., Thompson, J. C.: J. Chem. Phys. **49**, 969 (1968).
9. Lelieur, J. P.: This Colloquium, and, Thèse, Université d'Orsay, France n° 884 A, 1972.
10. Demortier, A., DeBacker, M., Lepoutre, G.: J. Chem. Phys. **69** (3), 380 (1972).
11. Acrivos, J. V., Mott, N. F.: Phil. Mag. **240** (187), 19, (1971).
12. Teoh, H.: Ph. D. Dissertation, Univ. of Texas (Austin) 1970. (Univ. Microf. 70—18297).
13. Ichikawa, K., Thompson, J. C.: This Colloquium, p. 230.

Appendix

A Percolation Process for Metal to Nonmetal Transition in Metal–Ammonia Systems

P. Chieux and G. Lepoutre

R. G. Ross [1] has noted that the ratios of liquid densities at the critical point and at the melting point, $\varrho_{cp}/\varrho_{mp}$, are approximately constant at about 0.23 for the alkali metals. This value can be combined with the packing fraction of 0.45 used by Ashcroft and Lekner [2] to fit a hard sphere model to the experimental liquid structure factor at the melting point of the alkali metals; a common packing fraction of hard spheres of about 0.11 is then obtained at the critical point. As has been shown by Scher and Zallen [3], bulk conduction occurs in a random mixture of conducting and insulating hard spheres when the packing fraction of conducting spheres is greater than about 0.15. A simple percolation process for hard spheres is therefore suggested by R. G. Ross for the metal-insulator transition which occurs in the alkali metal vapors at about their critical density. It will be shown that the metal-nonmetal transition in metal–ammonia systems can be described in the same terms.

For lithium ammonia solutions, the solid ammine $Li(NH_3)_4$ is known to exist with a melting point at $-185°$ C and a liquid density $\varrho_{cp} \simeq 0.53$ gr/cm^3 (by extrapolation of Lo's data). By analogy with the pure alkali metals it may be assumed that the $Li(NH_3)_4$ entity is not affected by dilution; the solution at the liquid–liquid critical point is then a mixture of $Li(NH_3)_4$ and pure NH_3 molecules. The concentration of $Li(NH_3)_4$ entities at the critical consolute point is known since it is nominally the same as the lithium solution concentration, 0.00165 moles/cm^3. The concentration of the $Li(NH_3)_4$ entities at the melting point of the compound is 0.00707 moles/cm^3. The packing fraction is then $0.00165/0.00707 = 0.23$, which is the same value as for pure lithium metal [1]. A percolation process of hard spheres of $Li(NH_3)_4$ may therefore be suggested to describe the metal to nonmetal transition in this solution.

It would be interesting to investigate in the same way the alkaline earth-NH_3 solutions where solid ammines such as $Ca(NH_3)_6$ are known to exist; however a lack of accurate density values prevents a complete analysis.

A comparison can be made between the percolation process and Mott's criterion for nonmetal to metal transition $n^{1/3} a_H \geqq 0.25$. The packing fraction quoted above would give $n^{1/3} a_H \geqq 0.30$. Values of a can then be computed in the case of M–NH_3 solutions; and a dielectric constant ε can then be introduced since $a_H = a_0$ (Bohr radius of the hydrogen atom). Using $n^{1/3} a_H \geqq 0.30$, the following values of ε are calculated:

$$Na-NH_3 \quad at \quad -41° C \quad \varepsilon = 5.99$$
$$Li-NH_3 \quad at \quad -63° C \quad \varepsilon = 5.78$$
$$K-NH_3 \quad at \quad -70° C \quad \varepsilon = 5.78$$

This is constant consistent with the computation of ε by the Simpson formula [5].

The temperature effect is of the right order of magnitude; however if we can trust the present values for static and dielectric constant for pure NH_3, the temperature effect is of the wrong sign.

References

1. Ross, R. G.: Phys. Letters **34 A** (3), 183—184 (1971).
2. Ashcroft, N. W., Lekner, J.: Phys. Rev. **145** (3), 83—90 (1966).
3. Scher, H., Zallen, R.: J. Chem. Phys. **53**, 3759 (1970).
4. Mott, N. F.: Rev. Mod. Phys. **40** 677 (1968).
5. Sienko, M. J.: In Colloque Weyl I.

Thermodynamics and Critical Phenomena in Metal–Ammonia Solutions

P. Damay

Abstract

The vapor pressure and the liquid–liquid phase separation of metal–ammonia solutions are related using thermodynamic data. An effective solvation number of 6 is found for potassium in ammonia solutions. Metal–ammonia solutions are assumed to be a mixture of this compound and free ammonia.

Introduction

The vapor pressure and the liquid–liquid phase separation of metal–ammonia solutions are related from a thermodynamic point of view. By a hypothesis on the structure of the solution, a relationship may be derived between these two phenomena. The hypothesis is that the solute is not the pure metal but rather a compound $M(NH_3)_n$. With this model we explain the peculiarities of the vapor pressure-concentration curve. Knowing the excess free energy for the solvent we calculate the critical temperature and the critical concentration with excellent precision. Two other thermodynamic quantities, the partial molar heat capacity computed from compressibility data and the effect of salt upon the critical temperature, reinforce the validity of the model.

I. An Analysis of the Vapor Pressure-Concentration Curve

We use the data of Marshall [1] for the sodium–ammonia system at $-35°$ C. The vapor pressure becomes lower with higher concentrations but it is more interesting to consider the activity coefficient for the solvent. This quantity is shown in Fig. 1. The ammonia activity coefficient, γ_1, is greater than unity at low concentrations; it goes through a maximum and becomes lower than unity at high concentrations. This peculiar behavior suggests that there are two competing phenomena. One of them leads to a negative deviation from Raoult's law ($\gamma_1 < 1$), the other to a positive deviation ($\gamma_1 > 1$).

A negative deviation from Raoult's law is generally interpreted by solvation. Hypothesizing that a number n of ammonia molecules are strongly fixed to the metal we consider two kinds of ammonia molecules: the molecules of free solvent and the molecules strongly fixed to the metal. The latter do not contribute to the vapor pressure. If the number, n, is greater than 4, the ammonia activity coefficient is higher than unity throughout the range of concentrations for the solution $NH_3–M(NH_3)_n$.

We must now account for the positive deviation from Raoult's law. If we consider that the departure from Raoult's law is the result of pair interaction, the excess

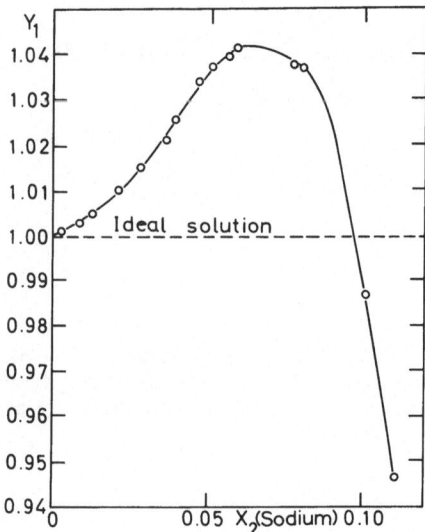

Fig. 1. Activity coefficient of ammonia in the Na–NH$_3$ solutions at $-35°$ C

free energy may be written

$$G^{ex} = k x_1 x_2 \tag{1}$$

where x_1 and x_2 are the molar fractions and k is a coefficient of interaction.

When the two components are very different in size, it is suggested, for example by Hildebrand [2], that it is better to use volume fraction instead of molar fraction.

Equation (1) becomes

$$G^{ex} = k' \varphi_1 \varphi_2 \tag{2}$$

where k' is a coefficient of interaction and φ_1, φ_2 are the volume fractions.

By differentiating with respect to x_1, it is possible to find the excess chemical potential of the solvent,

$$\mu_1^{ex} = k' \varphi_2^2 = A_{12} V_1 \varphi_2^2 \tag{3}$$

if $k' = A_{12} V_1$.

Such relationships are observed in potassium iodide–ammonia solutions in the intermediate range of concentrations [3] and in the ammonia–trimethylamine system [4].

In applying the same relationships to the ammonia–M(NH$_3$)n system, have to find two parameters, n, the solvation number and A_{12}, the coefficient of interaction. If the pair interaction hypothesis is valid, A_{12} must be concentration independent.

The volume V_1 is defined as the pure ammonia molar volume and the volume V_2 is the apparent molar volume of the metal plus n times the volume of pure ammonia. Thus

$$V_1 = 24.90 \text{ cm}^3 \quad \text{at} \quad -35° \text{ C}$$

$$V_2 = (65 + 25n) \text{ cm}^3 \quad \text{for} \quad \text{Na(NH}_3)_n.$$

The number n is taken to be an adjustable parameter. If, for a given n, the coefficient A_{12} is concentration independent, we shall say that, at the same time, the number n is reasonable and the pair interaction hypothesis is plausible.

If low values of n are used, the coefficient A_{12} decreases continuously with concentration. When n is high, A_{12} increases continuously. For $n = 6.5$ to 7 this coefficient is nearly constant in the $0.015 < x_2 < 0.09$ concentration range. At lower concentrations, the solvation model is no longer valid because the structure of the solution is different in dilute solutions. At high concentrations, the number of free ammonia molecules becomes too small to have a significant precision. Thus the hypotheses of the existence of $M(NH_3)_n$ and of pair interaction give a satisfactory explanation for the peculiar behavior of sodium–ammonia vapor pressure in the intermediate range of concentrations.

Results for solutions of others metals are not as good because the vapor pressure data are less precise.

II. Critical Phenomena

If the chemical potential of the solvent is known, it is possible to predict the critical phenomena by applying the stability criterion. Liquid–liquid phase separation can exist only if the deviation from Raoult's law is positive. The chemical potential for the solvent is written:

$$\mu_1 = RT \ln a_1 = f(T, p) + RT \ln x_1 + RT \ln \gamma_1 . \tag{4}$$

By using relation (3), it becomes:

$$\mu_1 = f(T, p) + RT \ln x_1 + A_{12} V_1 \varphi_2^2 . \tag{5}$$

The solution is not thermodynamically stable if:

$$\left(\frac{\partial \mu_1}{\partial x_1}\right)_{T,p} < 0 \quad \text{or} \quad \left(\frac{\partial \mu_1}{\partial x_2}\right)_{T,p} > 0 \quad \text{because} \quad dx_1 = -dx_2 .$$

By differentiating:

$$\left(\frac{\partial \mu_1}{\partial x_2}\right)_{T,p} = -\frac{RT}{x_1} + \frac{2 A_{12} x_2 V_1^2 V_2^2}{(x_1 V_1 + x_2 V_2)^3} . \tag{6}$$

At the critical point:

$$\left(\frac{\partial \mu_1}{\partial x_2}\right)_{T,p} = \left(\frac{\partial^2 \mu_1}{\partial x_2^2}\right)_{T,p} = 0 . \tag{7}$$

So that:

$$T_c = \frac{2}{R} \left(\frac{\varphi_1 \varphi_2 V_1 V_2 A_{12}}{x_1 V_1 + x_2 V_2}\right)_c . \tag{8}$$

All of the quantities in brackets are critical quantities. By differentiating a second time

$$(x_1)_c = (1 - x_2)_c = \left[\frac{(V_1^2 + V_2^2 - V_1 V_2)^{1/2} - V_1}{V_2 - V_1}\right]_c . \tag{9}$$

The critical concentration depends only on the volumes of the components, and the critical temperature depends on the volumes and on the coefficient A_{12}.

For sodium–ammonia solutions the calculated critical concentration is 0.0412 and the calculated critical temperature is 231.3 °K with $n = 6.5$. The experimental values are 0.041 for the concentration and 231.53 °K for the critical temperature [5].

The results for the other alkali metal solutions are good for the critical concentration but not as good for the critical temperature. (Errors of 24 °K for lithium and 53 °K for potassium were obtained.) In the latter two cases, determination of A_{12} was not as good because the vapor pressure data are much poorer and are not known near the critical point.

It is possible to calculate n and A_{12} from the critical constants. Thus the solvation number, n, is determined using two independent methods, the vapor pressure and the critical concentration. The coefficient A_{12} is also calculated using two independent methods, the vapor pressure and the critical temperature.

The value of n and A_{12} found for several alkali metal—ammonia solutions are reported in Table I.

Table I. Determination of the solvation number, n, and of the interaction coefficient by different methods

Metal	Vapor pressure		Critical phenomena		Heat Capacity
	n	A_{12} cal/cm^3	n	A_{12} cal/cm^3	n
Sodium	6.5 to 7	6.42	6.5	6.45	
Lithium	5 to 9		6.5	6.24	
Potassium	5 to 9		6	5.87	6
Calcium			20		

So the two phenomena, vapor pressure and liquid–liquid phase separation are well interpreted by the same model of solvation and pair interaction between the compound $M(NH_3)_n$ and the free ammonia.

We now apply the same idea to two other quantities — heat capacity and the effect of upon critical temperature.

III. Heat Capacities

The heat capacities at constant pressure, C_p, and at constant volume, C_V, have been computed for potassium–ammonia solutions from the adiabatic compressibility data of Bowen [6] and from the isothermal compressibility data of Böddeker and Vogelsgesang [7].

Two well-known relations have been used:

$$\frac{\beta_T}{\beta_S} = \frac{C_p}{C_V} \tag{10}$$

and

$$\beta_T - \beta_S = \frac{T}{C_p} \alpha^2 V \tag{11}$$

where β_T is the isothermal compressibility,
$\quad\quad \beta_S$ is the adiabatic compressibility,
$\quad\quad T$ is the temperature,
$\quad\quad \alpha$ is the thermal expansion coefficient,
and $\quad V$ is the volume of one mole of mixture.

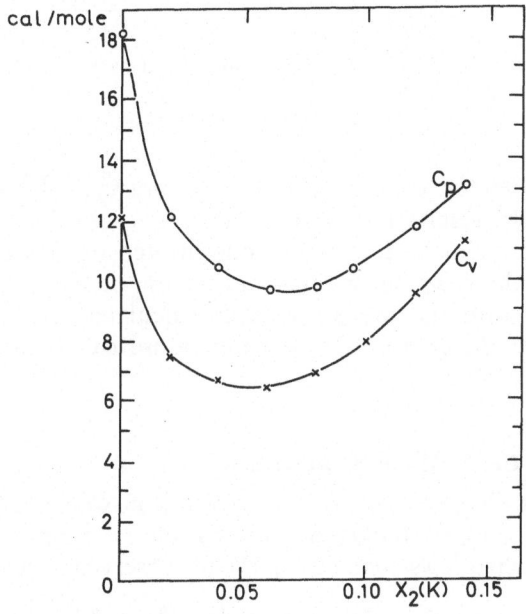

Fig. 2. Heat capacities of K–NH$_3$ solutions at -35° C calculated from compressibility data

The results are given in Fig. 2.
From these calculated values the partial molar heat capacities have been determined. The partial molar heat capacity at constant pressure for ammonia goes from 18.2 to 0 calories per mole from dilute solutions to saturated solutions. The constant volume heat capacity goes from 12.1 to -3 calories per mole.
Assuming the existence of a M(NH$_3$)$_n$ compound we assume that the heat capacity value for free ammonia is the value of pure ammonia and that the partial molar heat capacity of bound ammonia has the value of the saturated solution throughout the range of concentrations.
Thus

$$\bar{C}_p(\text{NH}_3) = nx_2 C_p(\text{Bound NH}_3) + (1 - nx_2) C_p(\text{free NH}_3)$$

and

$$\bar{C}_V(\text{NH}_3) = nx_2 C_V(\text{Bound NH}_3) + (1 - nx_2) C_V(\text{free NH}_3).$$

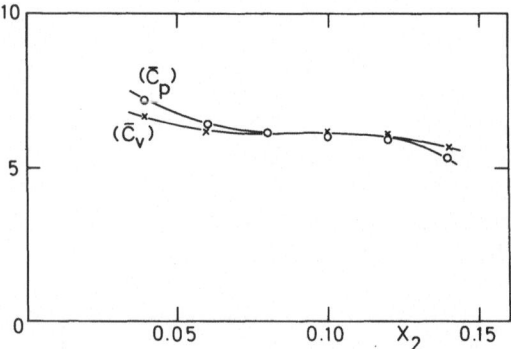

Fig. 3. Solvation number for K–NH$_3$ solutions calculated from partial molar heat capacity of ammonia

It is possible to determine the solvation number, n, from $\bar{C}_p(\mathrm{NH}_3)$ and $\bar{C}_V(\mathrm{NH}_3)$ at all concentrations. The results are given in Fig. 3. It is seen that the value of n is close to 6 in the intermediate and in the concentrated ranges. The solvation number 6 found from the molar heat capacities for the potassium in ammonia is in good agreement with the preceding value calculated by the two other methods. The effect of salt on the critical temperature can be analyzed from the same point of view.

IV. Effect of Salt on the Critical Temperature

Prigogine and Defay [8] have derived a relation which gives the variation of the critical temperature in a liquid binary mixture by adding a small amount, x_3, of a third component when laws such as Eq. (1) are observed. They find:

$$q = \left(\frac{\partial T_c}{\partial x_3}\right)_{x_3 \to 0} = -\frac{1}{2R}\frac{(k_{12} - k_{23} + k_{13})(k_{12} - k_{12} + k_{23})}{k_{12}}. \tag{12}$$

This equation assumes that the volumes of the three components are the same.

In metal–ammonia solutions the volumes of the three components are very different. We have derived a generalized relation taking into account the differences in volumes [10]. Let the subscript 1 represent free ammonia, 2 the compound $M(\mathrm{NH}_3)_n$ and 3 the salt added. We find:

$$q = \left(\frac{\partial T_c}{\partial x_3}\right)_{x_3 \to 0}$$
$$= -\frac{1}{2R}\left[\frac{4A_{12}^2 V_1 V_2 \varphi_1 \varphi_2 + 2A_{12}V_3(\varphi_1 V_2 - \varphi_2 V_1)[A_{12} - A_{23} + A_{12}(\varphi_1 - \varphi_2)]}{A_{12}(x_1 V_1 + x_2 V_2)}\right.$$
$$\left. -\frac{V_3^2\{(A_{13} - A_{23})^2 + A_{12}^2(\varphi_1 - \varphi_2)^2 + 2(A_{13} - A_{23})(\varphi_1 - \varphi_2)A_{12}\}}{A_{12}(x_1 V_1 + x_2 V_2)}\right] \tag{13}$$

where the A_{ij} are the interaction coefficients between the i and j components and V_i are the different volumes.

The calculation is possible for the system $K-KI-NH_3$ because the six parameters are known in this case:

A_{13} has been determined previously by our measurement of vapor pressure.

A_{12}, V_1 and V_2 have been calculated in the first part of this work.

A_{23} may be determined by the solubility of the salt in the concentrated potassium–ammonia solutions.

V_3 is calculated from the density data of the $K-KI-NH_3$ system [9].

The calculated value of q is $2100\,°K \pm 400$ and the experimental value is $1960\,°K$ [10]. The agreement between the calculated and the experimental values is good.

Conclusions

Four independent thermodynamic quantities are analyzed in a model of solvation for the solutions of alkali metals in liquid ammonia. In this model the existence of a compound $M(NH_3)_n$ in the solution is assumed. The interactions between this compound and free ammonia are binary interactions. The solvation number, n, and the interaction coefficients have been determined.

References

 1. Marshall, P. R.: J. Chem. Eng. Data 7, 399 (1962).
 2. Hildebrand, J. H.: Solubility of non electrolytes, p. 150. New York: Reinhold Publishing Corporation 1950.
 3. Damay, P., Lepoutre, G.: J. Chim. Phys. 66, 809 (1969).
 4. Kottarathil, T.: Unpublished results.
 5. Kraus, C. A., Lucasse, W. W.: J. Am. Chem. Soc. 44, 1949 (1922).
 6. Bowen, D. E.: In: Metal–ammonia solutions, p. 355. Lagowski, J. J., Sienko, M. J., Eds. London: Butterworths 1970.
 7. Boddeker, V. W., Vogelgesang, R.: Ber. Bunsenges. Phys. Chem. 75, 638 (1971).
 8. Prigogine, I., Defay, R.: Thermodynamique chimique. Liège: Desoer 1950.
 9. Johnson, W. C., Martens, R. I.: J. Am. Chem. Soc. 58, 15 (1936).
10. Damay, P.: Thèse, Paris 1972.

Transport Properties of Lithium–Ammonia Solutions in the Intermediate Concentration Range

J. P. LELIEUR, P. DAMAY, and G. LEPOUTRE

Abstract

Transport properties (Hall effect, thermoelectric power and electrical conductivity) of Li–NH$_3$ solutions in the intermediate range of concentration are analyzed. It is shown that these properties can be phenomenologically interpreted in a two-carrier model. The concentration of the more mobile carriers is found to increase abruptly in a very narrow range of concentrations, close to the critical concentration of demixtion.

Introduction

Some transport properties of metal–ammonia solutions are well known, especially in the intermediate concentration range where a nonmetal–metal transition is displayed and where these properties show peculiar trends. For instance, the ratio of the free electron Hall coefficient to the experimental Hall coefficient of lithium–ammonia solutions has a minimum at about 3 MPM.; the thermoelectric power shows a plateau in the same range of concentrations and the electrical conductivity varies by about three orders of magnitude. The purpose of this paper is to show that this minimum observed in the Hall ratio is a consequence of the nature of the dilute metal–ammonia solutions, and that a two-carrier model can be easily applied in the intermediate concentration range to the three previously mentioned transport properties to shed some light on the mechanism of the observed nonmetal–metal transition.

Hall Effect

Nasby and Thompson [1] expressed their results of the Hall effect experiments in Li–NH$_3$ solutions by the ratio α defined by

$$\alpha = \frac{R_{\text{F.E.}}}{R_{\text{H}}} \tag{1}$$

where R_{H} is the experimental Hall coefficient and $R_{\text{F.E.}}$ is the corresponding free electron gas Hall coefficient. The values of α versus metal concentration are plotted in Fig. 1. For concentrations greater than about 10 MPM., α is equal to 1, which means that solutions are free electron-like, relative to the Hall effect experiments. Below 10 MPM., α is smaller than 1, and it can be inferred that the free electron concentration is smaller than the true electronic concentration.

The minimum of α at about 3 MPM. is surprising. For instance, nothing like that is found in low density mercury [2]. Recently [3, 4], this minimum in α has been explained; an outline is given below.

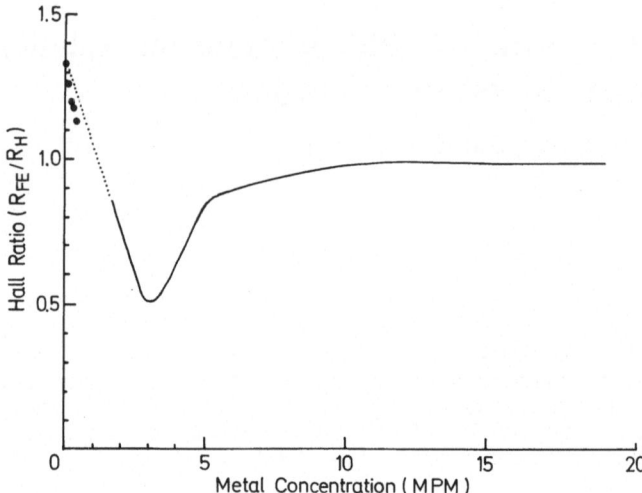

Fig. 1. Hall effect ratio $\alpha = \dfrac{R_{\text{F.E.}}}{R_{\text{H}}}$ for Li–NH$_3$ (full curve) and calculated values for dilute solutions (circles)

Three different carriers are known to exist at various concentrations and may contribute to the Hall effect. They are:

1. solvated electrons (charge e_1, mobility μ_1, concentration n_1), characteristic of dilute solutions;
2. mobile electrons (e_d, μ_d, n_d), characteristic of concentrated solutions;
3. solvated cations (e_+, μ_+, n_+).

We note that:

$$n_+ = n; \quad e_+ = e; \quad \text{so that} \quad n_d = n - n_1 \quad \text{and} \quad e_l = e_d = -e$$

(n is the valence electron concentration).

For carriers (e_i, μ_i, n_i), the Hall effect theory for weak magnetic field leads to the expression of the Hall coefficient [5]:

$$R_{\text{H}} = \frac{\sum\limits_i \mu_i^2 n_i e_i}{\left[\sum\limits_i \mu_i n_i e_i\right]^2}. \tag{1}$$

In the present case, we obtain:

$$R_{\text{H}} = \frac{1}{e} \frac{\mu_+^2 n - \mu_1^2 n_1 - \mu_d^2 (n - n_1)}{[\mu_+ n - \mu_1 n_1 - \mu_d (n - n_1)]^2}. \tag{2}$$

In the free electron gas, $R_{\text{F.E.}} = -\dfrac{1}{n_e}$ (because $\mu_+ = \mu_1 = n_1 = 0$), so we define:

$$\alpha = \frac{R_{\text{F.E.}}}{R_{\text{H}}} = -\frac{[\mu_+ n - \mu_1 n_1 - \mu_d (n - n_1)]^2}{n[\mu_+^2 n - \mu_1^2 n_1 - \mu_d^2 (n - n_1)]}. \tag{3}$$

For dilute metal–ammonia solutions (concentrations less than, say, 0.1 MPM.) it is normal to say that all valence electrons are solvated, so $n_1 = n_+ = n$ (except possibly association effects discussed later), and Eq. (3) gives:

$$\alpha = - \frac{|\mu_+| + |\mu_1|}{|\mu_+| - |\mu_1|}. \tag{4}$$

Introducing transference numbers t_+ and t_- defined by

$$t_+ = \frac{|\mu_+|}{|\mu_+| + |\mu_1|} \qquad t_- = \frac{|\mu_1|}{|\mu_+| + |\mu_1|} \tag{5}$$

the following value of α is obtained:

$$\alpha = \frac{1}{t_- - t_+}. \tag{6}$$

For extremely dilute metal–ammonia solutions, it is well known [6] that $t_- = 7/8$ and $t_+ = 1/8$, so $\alpha = 4/3$. We see that this value (point A, Fig. 1) is obtained in the extrapolation of experimental results (full curve). Transference numbers are not available for dilute Li–NH$_3$ solutions. We used transference numbers of Na–NH$_3$ solutions [6] to obtain values of α in the dilute concentration range. These values are close to the extrapolation of experimental results. It is seen that the Hall effect could eventually be used in electrolytic solutions to determine transference numbers, but this method would probably not be a sensitive one.

At about 1 MPM., the electric current is almost entirely due to electrons [6], so that in this concentration range $t_- \to 1$ and $t_+ \to 0$. In this case Eq. (6), which supposes that all electrons are solvated and localized, gives $\alpha \to 1$. Extrapolation of experimental results shows that α becomes less than 1 above about 1 MPM. That means that Eq. (6) is no longer valid, and that the Hall effect in this range cannot be explained only by electrolytic processes. Another process must necessarily intervene, and we have to suppose that $n_1 < n$. We now show that this is the condition for α to be smaller than 1. Neglecting cations ($\mu_+ = 0$), Eq. (3) can be written:

$$\alpha = \frac{\mu_1^2 n^2 + (\mu_d - \mu_1)^2 \, n_d^2 + 2\mu_1(\mu_d - \mu_1) \, n n_d}{\mu_1^2 n^2 + (\mu_d^2 - \mu_1^2) \, n n_d}. \tag{7}$$

Neglecting the second order term in n_d^2, it can readily be seen that α is smaller than 1 if $\mu_1 < \mu_d$.

This implies there are two electronic populations n_1 and n_d. Thus we can say that below 10 MPM., an additional electronic species different from the free electron appears, and above about 1 MPM., an electronic species different from the solvated electron has to be introduced. The reason why α becomes less than 1 above 1 MPM. cannot be association, since in this concentration range the mean distance between cations is less than the Bjerrum distance and the very notion of association becomes meaningless [3, 4]. Therefore in the 1—10 MPM. concentration range, one can perform calculations assuming that two electronic carriers exist (n_1, μ_1 and n_d, μ_d) and neglecting solvated cations. Eq. (3) can be

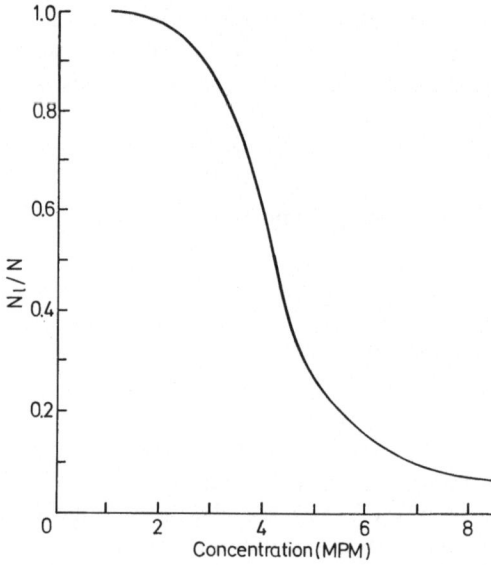

Fig. 2. n_1/n values deduced from the Hall effect assuming $\mu_1/\mu_d = $ constant

written:

$$\alpha = \frac{\left[\dfrac{\mu_1}{\mu_d}\dfrac{n_1}{n} + 1 - \dfrac{n_1}{n}\right]^2}{\dfrac{\mu_1^2}{\mu_d^2}\dfrac{n_1}{n} + 1 - \dfrac{n_1}{n}}.\tag{8}$$

The electronic concentration n_1/n can be deduced from experimental values of α by making an assumption for the value of the ratio μ_1/μ_d, taking $\mu_1/\mu_d \simeq 0.18$ in order to fit the minimum of α at 3 MPM., and assuming that μ_1/μ_d is constant in the $1 - 10$ MPM. concentration range. We found $\dfrac{n_1}{n}$ (Fig. 2) to be equal to 1 at 1 MPM., subsequently to decrease rapidly between 3 and 6 MPM. and to become very small at 10 MPM.

Using the same two-carrier model for electrical conductivity, we have:

$$\sigma = \mu_d n e\left(\frac{n_1}{n}\frac{\mu_1}{\mu_d} + 1 - \frac{n_1}{n}\right).\tag{9}$$

The expressions for α in Eq. (8) and for σ in Eq. (9) are not sufficient to determine μ_1, μ_d, n_1/n without hypothesis. A third experimental transport property is necessary, i.e. thermoelectric power.

Thermoelectric Power

The thermoelectric power (TEP) of metal–ammonia solutions is now well known [7—11]. Large negative values in very dilute solutions are indicative of electrolytic processes [7]. In the concentrated range, TEP has values typical of

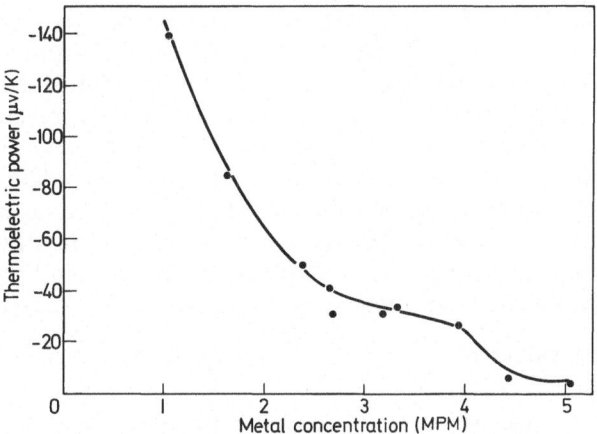

Fig. 3. Experimental thermoelectric power for Li–NH$_3$ solutions (◯) at $-35°$ C

metals (a few $\mu V/C$), and TEP values for free electron gas have been calculated with electron concentrations deduced from density or from the Hall effect. Very poor agreement between experimental and calculated values was obtained [4]. A "bump" has been experimentally obtained in the 1—3 MPM. concentration range (Fig. 3) [11].

Assuming two electronic carriers (n_l, μ_l and n_d, μ_d) between 1 and 10 MPM., we can write the TEP (S^*) of the solution [12]:

$$S^* = S_l^* \frac{\sigma_l}{\sigma_l + \sigma_d} + S_d^* \frac{\sigma_d}{\sigma_l + \sigma_d} \qquad (10)$$

where $\sigma_l = n_l \mu_l e$ and $\sigma_d = n_d \mu_d e$. S_l^* is the TEP of "localized" electrons, if they were alone in solution at their concentration n_l, S_d^* is the equivalent quantity for the other electronic group. Equation (10) can be written

$$S^* = S_l^* \frac{\dfrac{n_l}{n} \dfrac{\mu_l}{\mu_d}}{\dfrac{n_l}{n} \dfrac{\mu_l}{\mu_d} + 1 - \dfrac{n_l}{n}} + S_d^* \frac{1 - \dfrac{n_l}{n}}{\dfrac{n_l}{n} \dfrac{\mu_l}{\mu_d} + 1 - \dfrac{n_l}{n}} . \qquad (11)$$

Previously [3, 4] we computed S^* for K–NH$_3$ solutions (because data for Li–NH$_3$ were lacking), using $\dfrac{\mu_l}{\mu_d} \simeq 0.18$, the n_l/n values shown in Fig. 2 and reasonable values for S_l^* and S_d^*. We found that the general behavior of S^* is fairly well explained and even the observed "bump" qualitatively explained. Therefore it seems that this two-carrier model is valuable. Using simultaneously the experimental data for electrical conductivity, Hall effect and thermoelectric power and the corresponding equations, it was possible to determine n_l/n, μ_l, μ_d. Unfortunately α and σ were relative to Li–NH$_3$, and S^* to K–NH$_3$ solutions. Experimental values for TEP of Li–NH$_3$ solutions were recently obtained [11], (Fig. 3), and the present paper deals only with Li–NH$_3$ solutions.

Expressions of the Hall effect [Eq. (8)], electrical conductivity [Eq. (9)] and thermoelectric power [Eq. (11)] can therefore be solved without hypothesis, except for S_1^* and S_d^* values. For S_d^*, the procedure used for K–NH$_3$ solutions [3, 4] is impossible because TEP of Li–NH$_3$ has a sign reversal close to saturation, which is not predicted by the free electron theory. So taking experimental S^* values at 20 and 6.5 MPM., we linearly extrapolate these values towards more dilute solutions to get the needed S_d^* values. To determine S_1^* we supposed that between 1 and 1.5 MPM., experimental S^* values are due to the localized electrons (n_1, μ_1), and we extrapolated the curve between 1 and 1.5 MPM. towards more concentrated solutions to get the needed S_1^* values. Numerical computations give values of $\dfrac{n_d}{n} = 1 - \dfrac{n_1}{n}$, shown in Fig. 4. Values for μ_1, μ_d

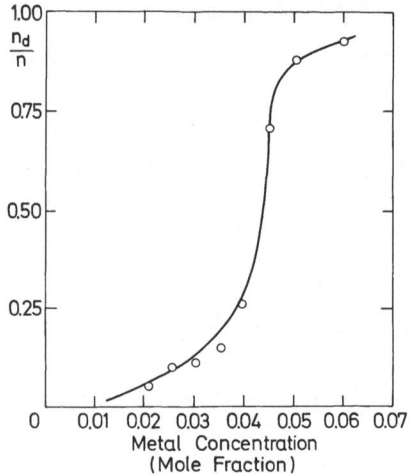

Fig. 4. n_d/n for Li–NH$_3$ solutions at -35°C deduced from Eqs. (8), (9) and (11)

and μ_1/μ_d have also been obtained. On the curve of Fig. 4, n_d/n goes from 25% at 4 MPM to 75% at 4.6 MPM. So the concentration change in the more mobile electronic group is very fast and we note that it is located exactly at the critical concentration of the liquid-liquid equilibrium. μ_1 and μ_d increase with concentration. At 1 MPM., when the second electronic carrier (n_d, μ_d) has to be introduced to explain the behavior of the Hall data, μ_d is found to be not very different from μ_1. The two mobilities increase rapidly, but μ_d increases much more rapidly than μ_1, so that $\mu_d \simeq 1$ cm^2/V sec at about 5 MPM. while μ_1 reaches the same mobility only at 10 MPM. where the number of that species is close to zero.

Discussion

The above model is phenomenological. We showed that the interpretation of the Hall effect data in Li–NH$_3$ solutions needs a two-carrier model. This model is reasonably consistent with electrical conductivity and thermoelectric power;

the general behavior of Fig. 4, and the values obtained for μ_l and μ_d throw light on the modification of electronic processes in the metal–nonmetal transition. For concentrations just below 1 MPM., electrons are solvated (and localized). They are essentially paired as shown by their magnetic susceptibility [3]. At about 1 MPM., a second electronic carrier (n_d, μ_d) appears, as required by the Hall data. As the metal concentration increases, n_d and μ_d increase. We propose that when the second carrier appears at about 1 MPM., the electronic states associated with it are in a pseudogap arising from the weak overlap of two bands. The lower band is a narrow band of localized states in cavities (solvated electrons); the upper band arises from cations solvated by a few ammonia molecules. As metal concentration increases, the overlap of the bands and the density of states in the pseudogap increase. The pseudogap is then more and more populated by n_d states when metal concentration increases. Above about 10 MPM., all the electrons are delocalized in the band due to solvated cations. From magnetic susceptibility measurements, it has been shown [3] that the Mott g-ratio $(g = N(E_F)/N(E_F)_{F.E.})$ is greater than about $1/3$ for concentrations greater than about 5 MPM. According to Mott [13], electronic states are then delocalized at the Fermi level. At 5 MPM., our model gives $\mu_d \simeq 1 \text{ cm}^2/\text{V} \cdot \text{sec}$ [3, 4] and the experimental electrical conductivity is about $300 \, \Omega^{-1} \text{cm}^{-1}$ [3], which is the minimum value predicted by Mott [13] for the electrical conductivity in extended electronic states. Therefore, electronic states at the Fermi level in the pseudogap are thought to be delocalized at about 5 MPM.

Such calculations as the above should be performed for Na–NH$_3$ and K–NH$_3$ solutions. In each case, however, we have no adequate values for R_H. It is expected the typical variations of α in Li–NH$_3$ (Fig. 1) would be found.

References

1. Nasby,R.D., Thompson,J.C.: J. Chem. Phys. **49**, 969 (1968).
2. Even,U., Jortner,J.: Phys. Rev. Letters **128**, 31 (1972).
3. Lelieur,J.P.: Thèse doctorat d'Etat. Orsay 1972.
4. Lelieur,J.P., Lepoutre,G., Thompson,J.C.: Phil. Mag. **26** (5), 1205 (1972).
5. Belissent,M.C., et al.: J. Chim. Phys. **68**, 355 (1971).
6. Dye,J.: Proc. Colloque Weyl I, p. 136. New-York: Benjamin 1964.
7. Dewald,J.F., Lepoutre,G.: J. Am. Chem. Soc. **76**, 3369 (1954); — **78**, 2953 (1956); — **78**, 2956 (1956).
8. Damay,P., Depoorter,M., Chieux,P., Lepoutre,G.: Proc. Colloque Weyl II, p. 233. London: Butterworths 1970.
9. Damay,P., Chieux,P., Lepoutre,G.: Ber. Bunsenges. Phys. Chem. **75**, 647 (1971).
10. Depoorter,M.: Thèse troisième cycle. Lille 1970.
11. Damay,P.: Thèse doctorat d'état. Paris 1972.
12. Wilson,A.H.: Theory of metals, second edition, p. 205. Cambridge: University Press.
13. Mott,N.F.: Advan. Phys. **16**, 49 (1967); — Phil. Mag. **19**, 835 (1969).

General Discussion of Papers by Lepoutre, Damay and Lelieur/Damay/Lepoutre

J. J. LAGOWSKI   I should like to point out one possible difficulty in Lepoutre's calculation for the distribution of electrons among the species in the proposed equilibria in 0.04 M metal–ammonia solutions. Such calculations based on the first equilibrium $[M^+(NH_3)_n + e^-(NH_3)_m \rightleftarrows (M^+(NH_3)_n, e^-(NH_3)_m)]$ probably represent the distribution of electrons between the species involved with reasonable accuracy. The dimerization process written by Prof. Lepoutre as $[2(M^+(NH_3)_n, e^-(NH_3)_m) \rightleftarrows (M^+(NH_3)_n, e^-(NH_3)_m)_2]$ which occurs as the concentration of metal increases, is expected for relatively higher concentrations of charged species in a solvent with a low dielectric constant; it is, for example, observed in the conductivity of concentrated solutions of electrolytes in liquid ammonia. The equilibrium constant for the association of monomers has been estimated from magnetic susceptibility measurements since the species involved differ in their spin. If the arguments developed in my plenary lecture concerning the markedly greater solvation of electrons as compared to cations are correct, we should expect that relatively small increments of metal would rapidly decrease the concentration of free ammonia to the point where association of monomers to dimers is the important process. Thus the number of ammonia molecules associated with the electron in the monomer and in the dimer (given as m in these equations) would not be expected to be the same. Indeed, the difference between solvation numbers for the electron in these species should change as a function of concentration. Consequently, the value of the equilibrium constant for the dimerization process from magnetic data may reflect the distribution of spin between the species, but it has little significance in terms of the detailed description of species in a chemical sense.

K. G. BREITSCHWERDT   In your association equilibria of solvated ions and solvated electrons you mentioned that in a 0.04 M solution you expect about 11% of the ions and solvated electrons to be in a dipolar form. Our measurements of the dielectric properties indicate however that in this concentration range the permittivity is essentially that of pure ammonia and that not more than approximately one per cent of the dipolar species can be present.

J. JORTNER   Since the percolation concept was invoked in Lepoutre's interesting lecture, I shall attempt to clarify this point. In an inhomogeneous system we can consider allowed and forbidden regions for electron transport. In the particular case of metal–ammonia solutions we can assume that large aggregates are formed due to the association of a large number of metal ion solvated electron quartets. These aggregates provide the allowed volume. Kirkpatrick [Phys. Rev. Letters 27, 1722 (1971)] has marked out the conductivity for the classical percolation model. The most important conclusions are that for a fraction, C, of allowed volume, $C > 0.4$ the conductivity is linear in C, while for $C_c < C < 0.4$ the conductivity tends to zero as $(C - C_c)^{8/5}$ where $C_c = 0.22$ representing the critical percolation probability.

The model proposed by Lelieur for the transport properties involves two types of carriers in a homogeneous medium. It is difficult to understand how free and localized electrons can coexist in a homogeneous system. The way out of this dilemma is to invoke the percolation theory. However, if we accept the in-homogeneous model for intermediate range metal–ammonia solutions, the equations for the transport properties have to be modified.

J. JORTNER   I would like to comment on the relation between the critical thermodynamic density and the "critical" density for the metal–nonmetal transition. In the case of subcritical and supercritical mercury the "critical" density for metal–nonmetal transition is $\varrho_{\text{MNM}} = 9 \text{ g cm}^{-3}$ while the thermo-dynamic critical density is $\varrho_{\text{CT}} = 4.5 \text{ g cm}^{-3}$. Thus in this one-component system, where metallic properties originate from the overlap between s and p bands, there is no direct relation between ϱ_{MNM} and ϱ_{CT}.

A. GARROWAY   For the $Na(NH_3)$ system, what value of "n" do you find is consistent with the vapor pressure data above 9 MPM? Do you have results for the $Li(NH_3)$ solutions?

P. DAMAY   At concentrations higher than 9 MPM, the number of free ammonia is too little to obtain a precise value of "n" from the vapor pressure data analysis. Nevertheless, the solvation number, n, seems to decrease as concen-tration increases. The same behavior is shown from heat capacities above $x_2 = 0.12$. It seems that the solvation number, n, remains constant in the inter-mediate range but may decrease when there is a lack of ammonia.

P. SCHETTLER   Does the expression that was presented for ΔG^{ex} suggest that long-range or short-range forces are at work?

P. DAMAY   Short-range or van der Waals forces.

NMR Measurements of Self-Diffusion
in Lithium–Ammonia and Sodium–Ammonia Solutions*

A. N. GARROWAY and R. M. COTTS

Abstract

Self-diffusion coefficients of ^{7}Li, ^{23}Na and ^{1}H have been measured in 1—20 MPM solutions of lithium–ammonia at 223° and 233 °K and in 2—15.5 MPM solutions of sodium–ammonia at 233 °K by the NMR spin echo pulsed magnetic gradient technique. In addition, the temperature dependences of the self-diffusion coefficients were measured for seven representative concentrations of lithium–ammonia below 240 °K. The data indicate that the lithium and sodium ions are solvated by four ammonia molecules over the time scale of molecular diffusion. The concentration dependence of the self-diffusion coefficients of the metal ion complexes and of the free ammonia molecules is found to be consistent, evaluated by the Stokes-Einstein relation, with the available viscosity data. The melting point self-diffusion coefficient and the temperature dependence of 20 MPM lithium–ammonia are fit to the Ascarelli-Paskin and Cohen-Turnbull diffusion models for a reasonable choice of packing fraction.

I. Introduction

In the metallic regime one of the most striking characteristics of metal–ammonia solutions is the rapid increase [1, 2] of electrical conductivity with increasing metal concentration. On increasing from 8—20 MPM, the electrical conductivity of the lithium–ammonia solution increases by a factor of almost ten and reaches a value of $15 \times 10^3 \, (\Omega \cdot cm)^{-1}$, a value comparable to that of liquid mercury. Ashcroft and Russakoff [3] have successfully explained this concentration dependence by employing a model in which the lithium ions are solvated by λ ammonia molecules; they chose $\lambda = 4$. They concluded that the $Li(NH_3)_4^+$ complex is a weak scatterer of electrons, whereas the free ammonia molecules are strong scatterers. As the lithium concentration is increased above 8 MPM, the number of free ammonia molecules is rapidly decreased, and it is primarily this depletion which accounts for the strong concentration dependence. This solvation model has also been applied [4, 5] to the analysis of the concentration dependence of the compressibility of lithium–, sodium–, potassium–, cesium–, and calcium–ammonia solutions.

There is as yet no direct evidence for the existence of the $Li(NH_3)_4^+$ complex in the liquid phase. However, the compound lithium tetramine does exist [6] below 89 °K and such solvation might be expected to persist at higher temperatures and perhaps at lower concentrations. If the lithium ion were indeed solvated by four ammonia molecules and if the complex $Li(NH_3)_4^+$ diffuses as an entity, then in the saturated solution (20 MPM) the self-diffusion coefficients of the lithium and ammonia should be equal. Furthermore, if this

* Work supported in part by the National Science Foundation (Grant GP-27531) and the Advanced Research Projects Agency, through the Cornell Materials Science Center.

solvation persists in concentrations below 20 MPM, then it would be expected that the larger, more massive $Li(NH_3)_4^+$ species should diffuse more slowly than the free ammonia molecules. Conversely, if the lithium ion were not solvated, then the lithium ion should be more mobile than the ammonia molecules. The self-diffusion coefficients of the ^7Li and of the ammonia protons may be measured independently in the lithium–ammonia solution by the nuclear magnetic resonance (NMR) spin echo pulsed field gradient technique as a test of the solvation hypothesis.

Although the sodium–ammonia solution is similar to lithium–ammonia, no compound of sodium–ammonia is known to exist. Measurement of the self-diffusion coefficients of the ^{23}Na and the ammonia protons in sodium-ammonia also provides a test of the solvation hypothesis and allows comparison with the behavior of lithium–ammonia.

Diffusion measurements in metal–ammonia solutions are of interest for other reasons. Not only is the conductivity of the saturated lithium–ammonia solution comparable to the conductivity of other liquid metals, but the Hall coefficient, thermopower, and Knight shifts are characteristic [7] of liquid metals. If the concentrated lithium–ammonia solution behaves as a simple metal, then it may provide a test of molecular transport theories in a significantly lower temperature regime than more common liquid metals: in 20 MPM lithium–ammonia diffusion may be studied over a factor of three in absolute temperature at pressures well below one atmosphere. Secondly, the mechanical properties of the metal–ammonia solutions change markedly with concentration. On going from pure ammonia to saturated lithium–ammonia the density of the solution decreases by 30%. The viscosity [8] decreases by almost 50% from its value for pure ammonia on the addition of 15.4 MPM of lithium. Also, the compressibility [9] of 20 MPM lithium–ammonia is about twice as large as in pure ammonia. It would then be expected that the self-diffusion coefficients should show a strong concentration dependence and might corroborate explanations which have evolved for the density, viscosity and compressibility results.

For these motivations, diffusion measurements were performed in 1—20 MPM lithium–ammonia and in 2—15.5 MPM sodium–ammonia solutions. (Signal-to-noise problems precluded diffusion measurements on the metal nuclei at concentrations much below 1 MPM.)

II. Experimental Technique

The Pulsed Field Gradient Spin Echo Method

In its simplest form the NMR spin echo method for measuring self-diffusion coefficients employs a $\pi/2$ rf pulse followed after a time τ by a π pulse. A spin echo appears at 2τ provided that 2τ is not much longer than the homogeneous T_2 of the nuclear system. If a magnetic gradient is applied after the $\pi/2$ and π pulses, then a particular spin will constructively interfere with the echo at 2τ provided that the phase gained by precession in the local static magnetic field seen by the spin is equal in the intervals $(0, \tau)$ and $(\tau, 2\tau)$. If the magnetic histories are unequal for many spins, due to diffusion along the magnetic gradient, then the echo amplitude will be reduced from its value in the absence of diffusion.

While diffusion may be measured in a steady (dc) gradient, it may be advantageous to apply a pulsed gradient after the $\pi/2$ pulse and then an identical pulsed gradient after the π pulse. For a $\pi/2 - \tau - \pi$ sequence the attenuation of the spin echo for pulsed gradients of amplitude G and duration δ, separated by a time τ and delayed from the $\pi/2$ pulse by t_1 is [10],

$$M(2\tau) = M_0' \exp - 2\tau/T_2 \exp - \{\gamma^2 D G^2 \delta^2 (\tau - \delta/3) - 2/3 \gamma^2 D g^2 \tau^3$$
$$- 2\gamma^2 D \delta^2 \boldsymbol{G} \cdot \boldsymbol{g} (\tau^2 - t_1^2 - \delta t_1 - 1/3 \delta^2)\} , \tag{1}$$

where M_0' represents the equilibrium magnetization, T_2 the homogeneous transverse relaxation, γ the nuclear gyromagnetic ratio, D the self-diffusion coefficient, and g the background gradient of the laboratory magnet.

For the large pulsed gradients used in these measurements, the background term (in g^2) and the crossterm (in $\boldsymbol{G} \cdot \boldsymbol{g}$) are small compared to the first term and may be ignored. To avoid measuring T_2, all data were taken for a fixed τ. Corrected [11] for the finite rise (t_r) and fall (t_f) times of the gradient pulses, Eq. (1) becomes

$$M(2\tau) = M_0 \exp - \gamma^2 D G^2 \delta^2 (\tau - \delta/3 + A(2\tau/\delta - 1)) \tag{2}$$

where $A = t_r - t_f$ and $M_0 = M_0' \exp - 2\tau/T_2$. The echo signal was recorded by a Fabri-Tek signal averager for 8—15 values of (G, δ) and the echo amplitude data, weighted by the signal-to-noise ratio of the averaged signal, were then computer fit via a least squares routine to determine D, the self-diffusion coefficient.

Sample Geometry

In the concentrated (above 8 MPM) metal–ammonia solutions, an applied rf field will induce eddy currents which attenuate and phase shift the applied fields over a distance of the skin depth. For the 20 MPM lithium–ammonia solution this skin depth is about 95 μ at 18.5 MHz, the nuclear Larmor frequency selected for the lithium–ammonia measurements. Thus for a cylindrical sample of radius much greater than 95 μ, the magnetic resonance is restricted to a thin annulus at the surface of the liquid and the effective NMR filling factor is significantly reduced, lowering the signal-to-noise ratio from its value in an insulating sample.

In NMR experiments in metals, the standard solution to this problem is to prepare a dispersion of particles of dimensions smaller than the skin depth. The particles are then coated with an oxide layer or physically separated by an insulating medium such as oil. However, because of the chemical instability of the metal–ammonia solutions in the presence of foreign material, a drastic increase in the surface-to-volume ratio might lead to severe chemical degradation. Furthermore, use of small particles requires a correction for restricted diffusion if the spins diffuse a distance comparable to the particle dimensions in the course of the spin echo experiment. While these corrections are known [12], they require detailed information about the distribution of particle sizes in the dispersion. To avoid these complications, all diffusion measurements were performed on "bulk" cylindrical samples, 4 mm i.d. Pyrex tubes filled 2.5 cm with the metal–ammonia solutions.

The problem of measuring self-diffusion coefficients in conducting liquids is treated briefly elsewhere [13, 14]. In a liquid in which the spins diffuse only a small fraction of the skin depth within the time of the measurement (τ), the functional dependence of the echo amplitude on G, τ, and δ is identical to Eq. (2). In solutions of high conductivity, however, only spins near the surface of the sample contribute to the observed echo, and these spins have a high probability of striking the sample wall during the measurement; hence the measured self-diffusion coefficient will be reduced from its value in an infinite, non-conducting medium. The measured self-diffusion coefficient of ^7Li in 20 MPM lithium–ammonia at 240 °K is estimated [13] to be 8% lower than the unrestricted value. For less concentrated solutions the effects of restricted diffusion are correspondingly reduced. Since these are small corrections and since the diffusion data at high concentrations are imprecise, no corrections have been made to the measured self-diffusion coefficients for restricted medium effects.

Sample Preparation

Matheson (UHP) ammonia, scrubbed of water in a sodium–ammonia solution, was distilled onto the alkali metals which had been cut and weighed in a helium-filled dry box. The gaseous ammonia was metered out by a mercury manometer and a calibrated volume. The stated compositions of the prepared solutions are accurate to about 1.5%, e.g. 20 ± 0.3 MPM.

None of the samples prepared for diffusion measurements showed visible signs of deterioration; these had been stored at 77 °K for months and warmed to 90—240 °K for hours during measurement. The sodium–ammonia used to dry the ammonia showed little evolution of hydrogen gas over periods of weeks even though it remained at dry ice–alcohol bath temperature. As a check on reproducibility, the self-diffusion coefficients of a 15.1 MPM lithium–ammonia solution were remeasured after 6 months storage and the data agree to within statistical uncertainty. Fifteen samples were destructively analyzed for degradation by measuring the quantity of hydrogen liberated on breaking the sample tubes under vacuum at 77 °K. The degradation observed, expressed as the fraction of metal which had been converted into metal amide, ranged from 10^{-4} to 2×10^{-2}, with most samples showing deterioration of about 5×10^{-3}. The few samples older than one year exhibited the greatest degradation.

There is no evidence that the level of sample deterioration affected the diffusion measurement. Certainly the actual metal concentration of the solution was slightly reduced by the deterioration; however, this reduction was always less than or, at worst, equal to the precision to which the samples were originally prepared.

Spectrometer and Gradient Pulser

A medium power, phase sensitive, pulsed NMR spectrometer was used. Extremely low level signals were averaged over 64—512 pulse sequence repetitions by a Fabri-Tek signal averager. The homebuilt gradient pulser [11] produced a maximum gradient of approximately 200 G/cm. The pulsed currents and gradient coil constant were measured independently. As a test of the calibration, the self-diffusion coefficient of distilled water was measured to be $1.90 \pm 0.12 \times 10^{-5} \mathrm{cm}^2/$ sec at $17.7 \pm 0.1°$ C. Using the measured temperature dependence [15] of $0.063 \times 10^{-5} \mathrm{cm}^2/\mathrm{sec} °K$, this corresponds to $2.36 \pm 0.12 \times 10^{-5} \mathrm{cm}^2/\mathrm{sec}$ at $25°$ C and compares favorably with other values in the literature [16].

III. Results

Before presenting the diffusion results, it must be established that the self-diffusion coefficient measured for the ammonia protons is actually the self-diffusion co-efficient for the ammonia molecules. That is, the distance moved by a proton via interammonia exchange must be small compared to the distance traversed by an ammonia molecule during its lifetime against intermolecular proton exchange. Satisfaction of this requirement may be shown explicitly and independ-ent measurement of the ^{14}N self-diffusion coefficient is not required.

In liquid ammonia the interaction of the ^{14}N spin $(I_N = 1)$ with the proton spin $(I_p = 1/2)$ produces a term in the proton spin Hamiltonian [17]

$$J h\, I_N \cdot I_p \tag{3}$$

where h is Planck's constant. This perturbation produces a triplet splitting of the nuclear Zeeman levels and, in the absence of rapid exchange, a well-resolved triplet is observed in the proton NMR spectrum. The coupling constant J was first determined by Ogg and Ray [18] to be $J = 46$ Hz. This interaction produces an additional mechanism for the longitudinal and transverse relaxation of the proton [19]:

$$
\begin{aligned}
(T_1^{-1}) &= (T_1^{-1})_{DD} + (T_1^{-1})_{SR} + 2/3(2\pi J)^2\, I_N(I_N + 1)\, \tau_J/(1 + \omega^2\tau_J^2) \\
(T_2^{-1}) &= (T_2^{-1})_{DD} + (T_2^{-1})_{SR} + 1/3(2\pi J)^2\, I_N(I_N + 1)\, (\tau_J + \tau_J/(1 + \omega^2\tau_J^2))
\end{aligned} \tag{4}
$$

where ω is the proton Larmor angular frequency.

For ammonia the spin-rotation interactions $(T_1^{-1})_{SR}$ and $(T_2^{-1})_{SR}$ are relatively weak [20] below 250 °K. The dipolar interactions $(T_1^{-1})_{DD}$ and $(T_2^{-1})_{DD}$ are equal since liquid ammonia is in the extreme motional narrowing regime. The correlation time of the spin-spin interaction is τ_J. Mechanisms which modulate the spin-spin interaction are intermolecular proton exchange (the proton may sample nitrogens of different spin orientations) and nitrogen (^{14}N) longitudinal relaxation. Swift, Marks, and Pinkowitz [17] have shown from the ratio of the triplet peak heights that the dominant mechanism is the dipolar and quadru-polar relaxation of the nitrogen spins in pure liquid ammonia and also in dilute potassium–ammonia solutions.

However, by assuming that the modulation of the spin-spin interaction is due entirely to intermolecular proton exchange, a lower bound on the exchange cor-relation time may be calculated. O'Reilly et al. [19] have measured $T_1 = 10$ sec and $T_2 = 50$ msec for anhydrous ammonia at 240 °K. Using $J = 46$ Hz, then $\tau_J = 3.6 \times 10^{-4}$ sec. Therefore, the correlation time for proton exchange must be greater than 3.6×10^{-4} sec in pure ammonia. In lithium– and sodium–am-monia solutions this spin-spin interaction will certainly occur although the coupling constant J may differ from 46 Hz. Hence the inequality of the proton

T_1 and T_2 still allows one to calculate a lower bound to the proton exchange time.

Relaxation times were not explicitly measured in these experiments. However, T_1 and T_2 were estimated to be of the order of 3—10 sec and 5—10 ms, respectively. The inequality of these relaxation times indicates that intermolecular proton exchange lifetimes are greater than 10^{-4} sec in these metal–ammonia solutions. Since this is many orders of magnitude greater than molecular

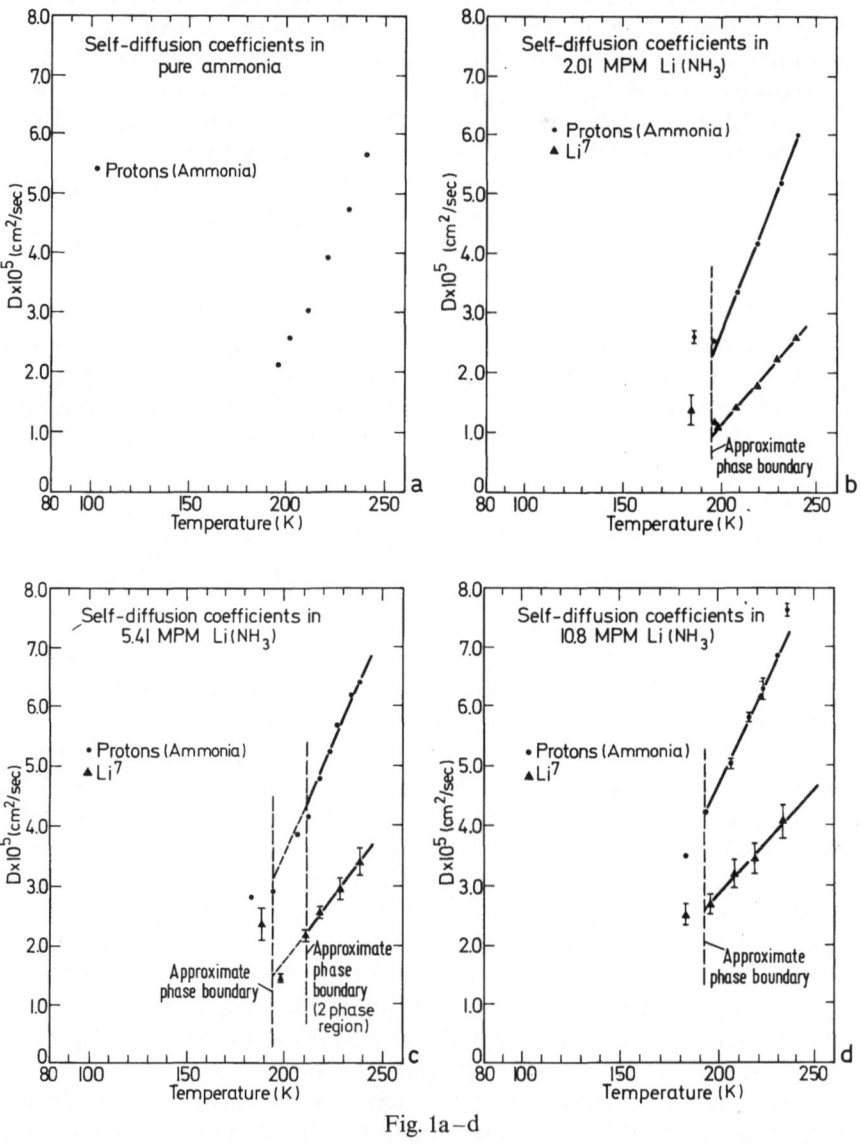

Fig. 1a—d

Fig. 1a—f. Self-discussion coefficients in lithium–ammonia solutions as a function of temperature. The solid curves are drawn for comparison only and have no theoretical significance

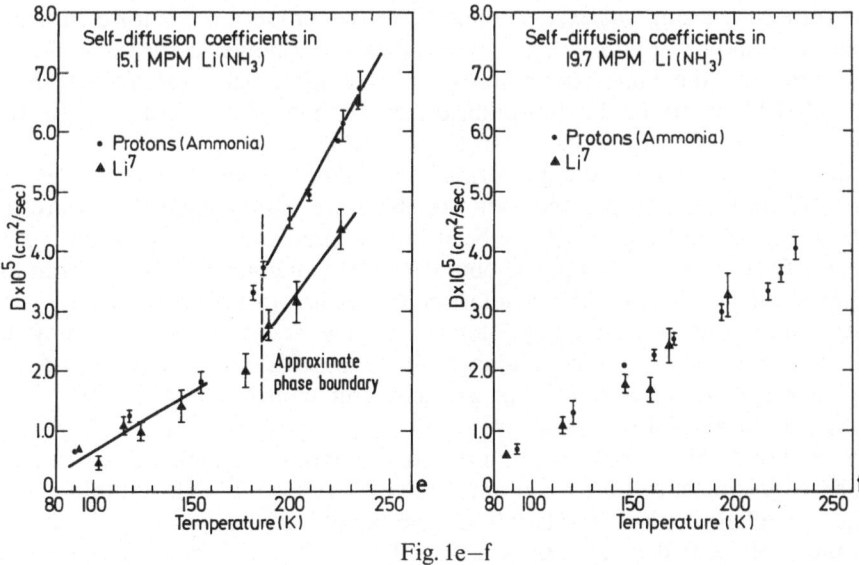

Fig. 1e–f

ammonia jump times, the self-diffusion coefficient measured for the protons is equal to the coefficient for the ammonia molecules.

The self-diffusion coefficients of lithium and ammonia were measured as a function of temperature for the 2.01, 5.41, 10.8, 15.1, 19.7, and 20.0 MPM lithium–ammonia samples; the self-diffusion coefficient in pure ammonia was also measured. These results are shown in Figs. 1 and 2. Below saturation these solutions

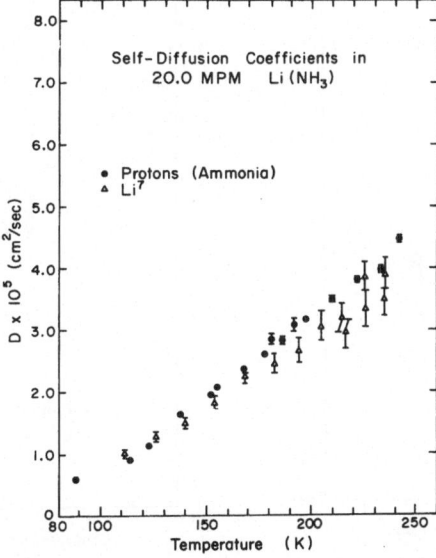

Fig. 2. Self-diffusion coefficients in a 20.0 MPM lithium–ammonia solution as a function of temperature

concentrate by freezing out excess ammonia as the temperature is lowered below the phase boundary. Figures 1b, c, d, e indicate a change in the self-diffusion coefficients near the phase boundaries. In particular, below about 160 °K in the 15.1 MPM solution, the self-diffusion coefficients of the lithium and ammonia are essentially equal, just as in the 20 MPM solution.

Diffusion measurements were performed for other concentrations at the isotherms 223 and 233 °K, and the data are shown in Fig. 3. Also shown are the results of Figs. 1 and 2 which have been interpolated to 223 and 233 °K by an Arrhenius fit; data below the phase boundary were, of course, excluded from the fit. The results for the 19.7 MPM solution are similar to those for the 20 MPM sample and are omitted from Fig. 3 for clarity. The error bars indicate only the statistical uncertainty of the least squares fit. All measurements may be systematically in error by 4.5%, since the gradient coil constant and currents were measured to 2.1 and 0.15%, respectively.

Figure 3 shows that both the lithium and ammonia self-diffusion coefficients increase with concentration to about 12 MPM. Above 12 MPM lithium self-diffusion coefficient then levels off or perhaps decreases slightly while the ammonia coefficient decreases by a factor of two until it is essentially equal to the lithium self-diffusion coefficient at 20 MPM. The ammonia self-diffusion coefficient for a 22 MPM sample is plotted at the 22 MPM abscissa. However, since lithium–ammonia saturates at about 20 MPM, this sample consists of a saturated lithium-ammonia solution and a small amount of lithium metal, and the measured self-diffusion coefficient is consistent with the result for the 20 MPM solution.

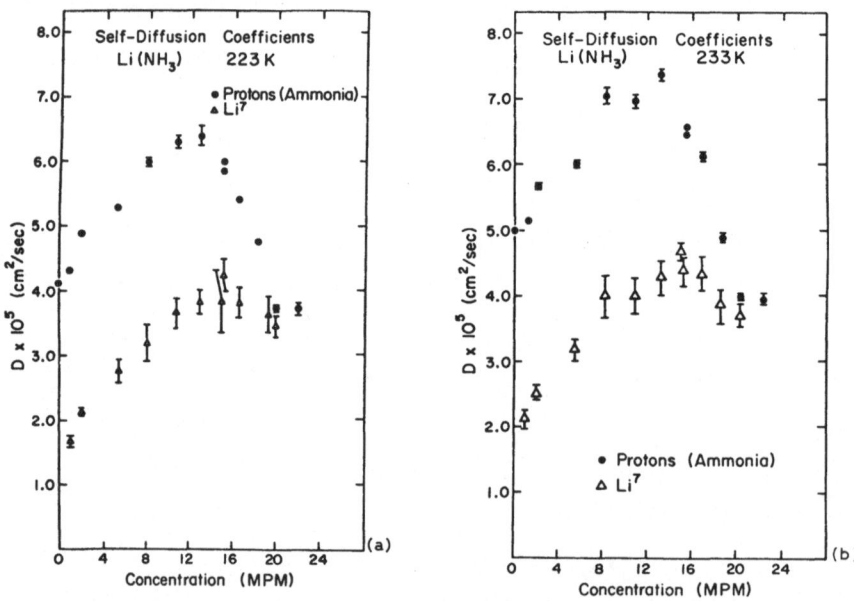

Fig. 3. Self-diffusion coefficients in lithium–ammonia as a function of concentration. This figure includes the data of Figs. 1 and 2 as well as data for additional samples

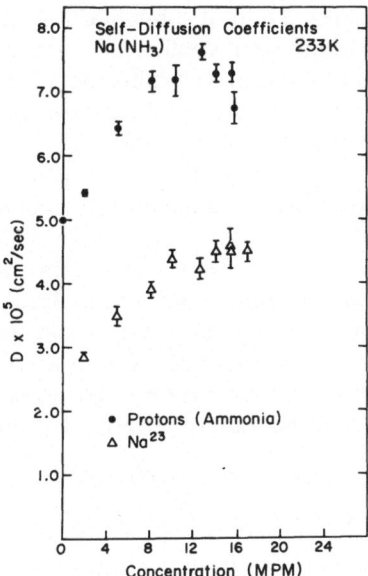

Fig. 4. Self-diffusion coefficients in sodium–ammonia as a function of concentration

For the sodium–ammonia solution, diffusion data were taken only at 233 °K for 2.0, 5.0, 8.0, 12.5, 14.0 and 15.5 MPM samples and are displayed in Fig. 4. Comparing Figs. 3 and 4 the concentration dependence and indeed the actual values of the metal and ammonia self-diffusion coefficients are seen to be very similar for both the lithium and sodium systems. The equality of the sodium self-diffusion coefficients in the 15.5 and 17.0 MPM solutions suggests that saturation has occurred, at approximately 15.5 MPM at 233 °K.

IV. Discussion

The elementary Stokes-Einstein theory of diffusion in liquids, while a macroscopic theory, has been applied with some success to molecular fluids and may be used to decide if the metal–ammonia diffusion results are consistent with a solvation model. In the Stokes-Einstein theory particles of radius "a" are considered to diffuse in a homogeneous fluid of viscosity v. The self-diffusion coefficient is [21]

$$D = \frac{1}{6\pi a}\frac{kT}{v},\tag{5}$$

where k is Boltzmann's constant and T is the absolute temperature.

Since the ionic radius of ^{7}Li (0.60 Å) [22] is much smaller than the ammonia van der Waals radius (1.54 Å) [22], Eq. (5) predicts that the observed self-diffusion coefficient of the lithium should be much greater than for ammonia. Furthermore, since the viscosities of lithium–ammonia and sodium–ammonia solutions are approximately equal at equivalent concentrations, the expected self-diffusion coefficient of lithium should be greater than that of sodium (ionic radius 0.95 Å)

[22]. The diffusion results (Figs. 3 and 4) show that for all concentrations except 20 MPM the ammonia self-diffusion coefficient exceeds that of lithium and further that the sodium and lithium self-diffusion coefficients are essentially equal at equivalent concentrations. Hence the diffusing species do not appear to be free metal ions and free ammonia molecules: the smaller lithium self-diffusion coefficient indicates that the lithium is impeded in its motion.

The lithium–ammonia results may be understood by considering the fluid to be a homogeneous mixture of lithium ions solvated by four ammonia molecules and free ammonia molecules. The lithium ion is assumed to be solvated during a diffusive step; that is, the cage of four ammonia molecules is dragged along as the lithium changes position. (If the time scale of solvation were shorter, then the lithium ion would appear unsolvated in its diffusional motion, in contradiction to the data of Fig. 3.) For a concentration x, expressed in MPM, the fraction of free ammonia molecules and the fraction of bound ammonia molecules is

$$F_f = (100 - 5x)/(100 - x),$$
$$F_b = 4x/(100 - x). \tag{6}$$

If the exchange between the free and bound species were slow compared to the time of the diffusion measurement (7 ms), then two separate diffusion coefficients should be observed and the proton echo given by the superposition

$$M_0 \{ F_b \exp - \gamma^2 D_b G^2 \delta^2 (\tau - \delta/3 + A(2\tau/\delta - 1))$$
$$+ F_f \exp - \gamma^2 D_f G^2 \delta^2 (\tau - \delta/3 + A(2\tau/\delta - 1)) \} \tag{7}$$

where D_b and D_f represent the bound and free self-diffusion coefficients. The Stokes-Einstein relation would predict that $D_f \simeq 2D_b$. At 10.8 MPM, 52% of the ammonia is free and the remainder is bound. However, for the 10.8 MPM sample there is no evidence from the echo amplitude data for the superposition of two exponential functions associated with two values of D. Therefore the exchange of free and bound ammonia molecules is rapid on the millisecond time scale, and over this interval a particular ammonia molecule will spend a fraction of time F_b attached to lithium ions and a fraction F_f as a free entity. The measured self-diffusion coefficient then represents the following average:

$$\langle D \rangle = F_b D_b + F_f D_f. \tag{8}$$

Under this four-to-one solvation model, at 20 MPM all the ammonia molecules are associated with lithium complexes and thus the lithium and (averaged) ammonia self-diffusion coefficients should be equal. Figure 2 shows that the coefficients are essentially equal from 90°—240 °K. The slight discrepancy is not understood. If the solvation involves exactly a four-to-one ratio, then the data are consistent with a 19.7 MPM solution. In fact, since the samples were prepared to only 1.5% of their stated composition, the stoichiometry of the 20 MPM sample is 20.0 ± 0.3 MPM. (The fraction of lithium metal which had degraded into lithium amide, measured by the technique outlined in Section II, was 3×10^{-3}, too small to account completely for the observed discrepancy.) On the other hand, if the composition were exactly 20.0 MPM, then the effective solvation number is 3.8—3.9. It is possible that the solvation number is temperature dependent. Unfortunately, the scatter of the diffusion data precludes a check for such temperature dependence.

The equality of the self-diffusion coefficients at saturation (20 MPM) is the strongest indication that the lithium ions are solvated by four ammonia molecules. At lower concentrations the solvation number cannot be unambiguously determined, although the ratio of the metal–to–ammonia self-diffusion coefficients definitely implies some solvation.

The sodium–ammonia solution does not reach a concentration at which the sodium and ammonia self-diffusion coefficients are equal and the solvation number cannot be exactly prescribed. Since at 233 °K in the 15.5 MPM solution the self-diffusion coefficient of sodium is smaller than that of ammonia, the solvation number must be less than $(100 - 15.5)/15.5 = 5.45$. The reassuring similarity of the lithium–ammonia and sodium–ammonia diffusion data suggests that the sodium ion is also solvated by four ammonia molecules.

By assuming that the solvation number is concentration independent, the self-diffusion coefficient of the free ammonia molecules may be unfolded from the experimental data by Eq. (8); the free ammonia results so derived are shown in Fig. 5 for lithium–ammonia at 223 °K and for both lithium– and sodium–ammonia at 233 °K. Again, the similarity of the data indicates that the solvation numbers are not very different for the two systems.

For both the lithium– and sodium–ammonia solutions, the Stokes-Einstein radii $(a = kT/6\pi Dv)$ may be calculated from the measured self-diffusion coefficients and the viscosity information. Viscosity data are available [8] for lithium–ammonia at 233 °K and higher temperatures and were extrapolated to 223 °K. Sodium–ammonia viscosity data are available [23] at 233 °K. The calculated radii of the metal complexes and of the free ammonia molecules are

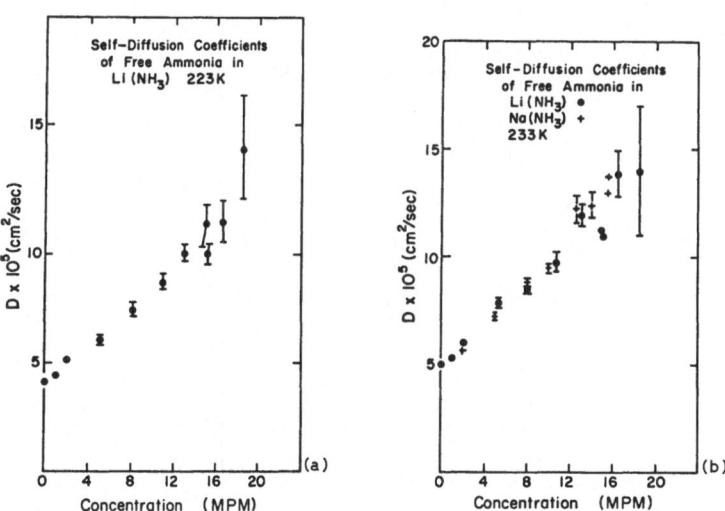

Fig. 5. Self-diffusion coefficients of the free ammonia molecules in lithium– and sodium–ammonia as a function of concentration. A solvation number of four has been assumed for both systems

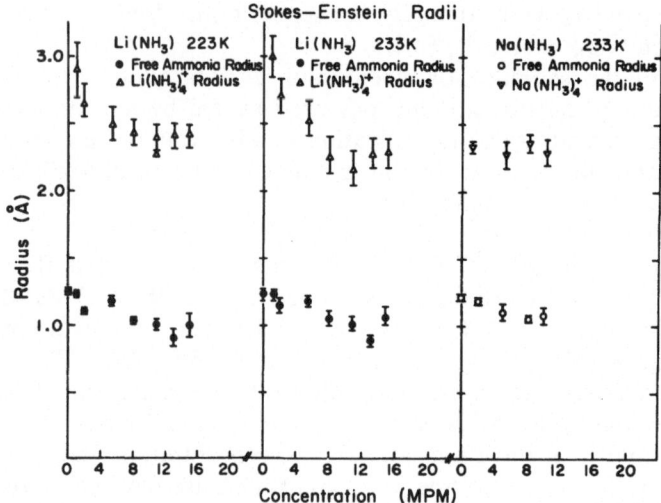

Fig. 6. Stokes-Einstein radii for lithium– and sodium–ammonia as a function of concentration. The radii were calculated from the measured self-diffusion coefficients and the available viscosity data

displayed in Fig. 6; error bars represent only the statistical uncertainty in D and do not reflect systematic errors in D or uncertainties in the viscosity data. The calculated radii for the $Li(NH_3)_4^+$ and $Na(NH_3)_4^+$ complexes are essentially equal and approximately twice as large as the free ammonia radius. The 20% decrease at about 4 MPM is not understood. McCall and Douglass [24] have

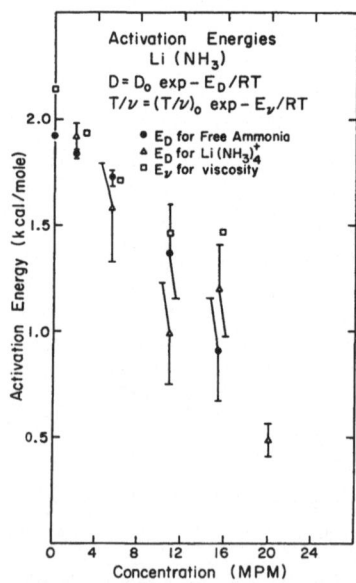

Fig. 7. Activation energies of the self-diffusion coefficients and of viscosity in lithium–ammonia

noted a similar but not universal decrease for aqueous solutions of electrolytes at molarities of about 2 M.

To stress the simple Stokes-Einstein relation even further, the temperature dependence of the self-diffusion coefficients and viscosity may be compared provided the characteristic radii are temperature independent. For the limited subset of samples for which temperature dependences were measured, the self-diffusion coefficients of the metal complexes and the free ammonia molecules were fit to the Arrhenius relation $D = D_0 \exp - E_D/RT$. The results, shown in Fig. 7, are somewhat imprecise due to the paucity of data. Viscosity activation energies, derived from a fit to $v/T = (v/T)_0 \exp - E_v/RT$ are also shown. If the Stokes-Einstein relation is obeyed, then $E_D = E_v$. From Fig. 6 the concentration dependences of E_v and E_D for both free ammonia and the metal complex are similar and numerical agreement is fair.

Unfortunately, viscosity data are not available beyond 15 MPM, where the lithium self-diffusion coefficient shows a leveling off or even a decrease with increasing concentration (Fig. 3). It may be speculated that the viscosity data would also mirror this behavior.

The Saturated Lithium—Ammonia Solution

The equality of the lithium and (averaged) ammonia self-diffusion coefficients suggests the 20 MPM solution may be treated as a single-component fluid. It is therefore appealing to treat the solution in the context of a simple hard sphere model and to predict the self-diffusion coefficient at melting and its temperature dependence by diffusion models which have been successful for other liquid metals.

The Ascarelli-Paskin model for diffusion in a dense gas of hard spheres predicts [25]

$$D = 0.73 \left(\frac{\eta_m}{\eta}\right)\left(\frac{r}{2}\right)\left(\frac{\pi kT}{m}\right)^{1/2} \frac{1}{z_m(T_m \varrho/T \varrho_m) - 1}, \tag{9}$$

where η is the packing fraction, r the hard sphere radius, m the mass of the diffusing species, and z the hard sphere compressibility. The subsript "m" denotes the melting point value. The packing fraction is given by

$$\eta = \frac{4\pi}{3} r^3 \varrho, \tag{10}$$

where ϱ is the number density of hard spheres. Vadovic and Colver [26—28] recommend using the Carnahan-Starling expression for z [29]:

$$z = (1 + \eta + \eta^2 - \eta^3)/(1 - \eta)^3. \tag{11}$$

Ashcroft and Lekner [30] found that the structure factor of liquid metals may be well approximated by taking $\eta = 0.45$ at melting. The temperature dependence (logarithmic derivative) at melting of the self-diffusion coefficient is calculated to be, for $\eta_m = 0.45$,

$$\frac{T}{D}\left(\frac{dD}{dT}\right)_m = 1.86 + 1.70 \, \alpha_m T_m, \tag{12}$$

where α_m is the thermal volume expansion coefficient at melting and T_m is the melting temperature, 89 °K in the case of 20 MPM lithium–ammonia. Since $\alpha_m T_m \ll 1$ for normal metals, Eq. (12) predicts that the temperature dependence of the self-diffusion coefficient is essentially equal for all metals, within the confines of this model.

For the 20 MPM solution, extrapolating Lo's density data [31] for ϱ_m and α_m, Eqs. (11) and (12) predict $D_m = 2.2 \times 10^{-5}$ cm²/sec and $T/D\, dD/dT)_m = 1.92$. The experimentally determined values are: $D_m = 0.59 \pm 0.5 \times 10^{-5}$ cm²/sec and $T/D\, dD/dT)_m = 1.97 \pm 0.05$. Agreement between the experimental values and the predictions of the Ascarelli-Paskin model (for $\eta_m = 0.45$) is poor for D_m and perhaps fortuitous for the temperature dependence.

A value of η_m may be calculated from the hard sphere model for the isothermal compressibility K_T [32]:

$$K_T = \frac{v_M}{kT} \frac{(1-\eta)^4}{(1+2\eta)^2}, \tag{13}$$

where v_M is the volume per molecule. While this model ignores [4] the electronic contribution to the compressibility, it has been employed [4, 5] with some success for concentrated metal–ammonia solutions. Since the isothermal compressibility is approximately equal to the adiabatic compressibility for metals, fitting the adiabatic compressibility obtained from sound velocity measurement [33] near melting to Eq. (13) yields $\eta_m = 0.61$. Using this value of η_m in the Ascarelli-Paskin model, $D_m = 0.73 \times 10^{-5}$ cm²/sec and $T/D\, dD/dT)_m = 1.72$. Thus a somewhat better fit occurs for $\eta_m = 0.61$, a large but not unphysical value.

The diffusion model of Cohen and Turnbull [34] may also be applied. Hard spheres of radius "r" are considered to move into holes whose size must exceed some critical volume v^*. Fluctuations in the free volume v_f available to each sphere open up holes for a diffusive step. The Cohen-Turnbull result is

$$D = \frac{1}{6}(2r)\left(\frac{kT}{m}\right)^{1/2} \exp -\Gamma v^*/v_f, \tag{14}$$

where Γ is a correction factor of the order of unity for the overlap of the free volume. The temperature dependence is then

$$\frac{T}{D}\frac{dD}{dT} = \frac{1}{2} + \frac{T}{r}\frac{dr}{dT} + \left(\frac{\Gamma v^*}{v_f}\right)\frac{T}{v_f}\frac{dv_f}{dT}. \tag{15}$$

Various forms are available [34, 35] for the functional relation of v_f upon v_M, the mean volume per molecule. In particular, v_f may be related to v_M by the hard sphere compressibility [13]

$$\frac{v_M}{v_f}\frac{dv_f}{dv_M} = z. \tag{16}$$

Using Eqs. (15), (16) and (11) and setting $\eta_m = 0.45$, the temperature dependence is

$$\frac{T}{D}\left(\frac{dD}{dT}\right)_m = 0.332 + \left(9.385\left(\frac{\Gamma v^*}{v_f}\right)_m\right) + 0.215\,\alpha_m T_m. \tag{17}$$

The measured value of D_m may be used to determine $(\Gamma v^*/v_f)_m = 2.8$ by Eq. (14) and hence Eq. (17) predicts that $T/D \, dD/dT)_m = 1.37$, a value lower than the experimental value of 1.97 ± 0.05. By varying η_m and hence z_m, a fit to both D_m and $T/D \, dD/dT)_m$ may be forced; the Cohen-Turnbull predictions agree with experiment if η_m is selected to be 0.52.

Both the Ascarelli-Paskin and Cohen-Turnbull models require melting point packing fractions greater than 0.45 to agree with the experimental results. This may indicate that the rather knobby structure of the tetrahedral array of ammonia molecules about the lithium ion impedes diffusion and hence the effective radius and therefore the packing fraction of the $Li(NH_3)_4^+$ complex are increased over the value expected for a more spherical species.

Saxton and Sherby [36] have noted that the measured self-diffusion coefficients of pure metals may be fit to the relation $D = D_0 \exp - NT_m/T$, where N ranges from 2.5 to 4.0 for different metals. The 20 MPM lithium–ammonia diffusion data gives $N = 2.75 \pm 0.05$, where T_m is taken to be 89 °K. Thus the temperature dependence of D for the $Li(NH_3)_4^+$ complex in 20 MPM lithium–ammonia is consistent with those of other liquid metals.

V. Conclusions

The self-diffusion coefficients of ^7Li and ^1H in lithium–ammonia and ^{23}Na and ^1H in sodium–ammonia have been measured in the liquid phase over the range of 1 MPM — saturation for lithium–ammonia — and 2 MPM — saturation for sodium–ammonia. In the liquid the lifetime of an ammonia molecule against intermolecular proton exchange is long on the time scale of molecular collisions and the measured self-diffusion coefficient of the proton is equal to that of the ammonia molecule. The observation that the measured ammonia self-diffusion coefficient is greater than the lithium self-diffusion coefficient, except at 20 MPM where they are essentially equal, suggests strongly that the lithium ion is solvated by four ammonia molecules. The similarity of the sodium–ammonia and lithium ammonia-diffusion data indicates that the solvation number is also four in the case of sodium–ammonia. Since exchange among free and bound ammonia molecules is rapid on the time scale of the experiment, the self-diffusion coefficient represents an average over the two states, and the self-diffusion coefficient of the free ammonia molecules was unfolded from the experimental diffusion data on the assumption that the solvation number is concentration independent. The Stokes-Einstein radii calculated for the metal complexes and for the free ammonia molecules from the available viscosity information are essentially concentration independent in both lithium– and sodium–ammonia, indicating that the solvation number is not strongly concentration dependent. Fair agreement is obtained for lithium–ammonia on comparing the activation energies of the self-diffusion coefficients with those of the Arrhenius fit to T/v.

Two models of diffusion, the Cohen-Turnbull and Ascarelli-Paskin models, were applied to the 20 MPM lithium–ammonia solution since it may be regarded as a single-component system. Agreement with model predictions is achieved for melting point packing fractions greater than 0.45.

Acknowledgements

Professor N. W. Ashcroft had originally suggested that NMR diffusion measurements might disclose solvation effects in concentrated lithium–ammonia solutions and we have benefited from further discussions with him. This work has also profited from the perspectives of Professor M. J. Sienko.

References

1. Morgan, J. A., Schroeder, R. L., Thompson, J. C.: J. Chem. Phys. **43**, 4494 (1965).
2. Kyser, D. S., Thompson, J. C.: J. Chem. Phys. **43**, 3910 (1965).
3. Ashcroft, N. W., Russakoff, G.: Phys. Rev. A **1**, 39 (1970).
4. Schroeder, R. L., Thompson, J. C.: Phys. Rev. **179**, 124 (1969).
5. Thompson, J. C.: Phys. Rev. A **4**, 802 (1971).
6. Mammano, N., Sienko, M. J.: J. Am. Chem. Soc. **90**, 6322 (1968).
7. Cohen, M. H., Thompson, J. C.: Advan. Phys. **17**, 857 (1969).
8. Demortier, A., Labry, P., Lepoutre, G.: J. Chim. Phys. **68**, 498 (1971).
9. Maybury, R. H., Coulter, L. V.: J. Chem. Phys. **19**, 1326 (1951).
10. Stejskal, E. D., Tanner, J. E.: J. Chem. Phys. **42**, 288 (1965).
11. Murday, J. S.: Ph. D. Thesis. Cornell University 1970, unpublished.
12. Robertson, B.: Phys. Rev. **151**, 273 (1966).
13. Garroway, A. N.: Ph. D. Thesis. Cornell University 1972, unpublished.
14. Garroway, A. N., Cotts, R. M.: Phys. Rev., to be published.
15. Simpson, J. H., Carr, H. Y.: Phys. Rev. **111**, 1201 (1958).
16. Pruppacher, H. R.: J. Chem. Phys. **56**, 101 (1972).
17. Swift, T. J., Marks, S. B., Pinkowitz, R. A.: In: Colloque Weyl II. London: Butterworth 1970.
18. Ogg, R. A., Ray, J. D.: J. Chem. Phys. **36**, 1515 (1957).
19. O'Reilly, D. E., Peterson, E. M., Lammert, S. R.: J. Chem. Phys. **52**, 1700 (1970).
20. Powles, J. G., Rhodes, M., Strange, J. H.: Mol. Phys. **11**, 515 (1966).
21. Frenkel, J.: Kinetic theory of liquids, p. 192. New York: Dover 1955.
22. Hodgman, C. D. (Ed. in chief): Handbook of Chemistry and Physics, 41 st. ed. Cleveland: Chemical Rubber Publishing Co. 1959.
23. Nozaki, T., Shimoji, M.: Trans. Faraday Soc. **65**, 1489 (1969).
24. McCall, D. W., Douglass, D. C.: J. Phys. Chem. **69**, 2001 (1965).
25. Ascarelli, P., Paskin, A.: Phys. Rev. **165**, 222 (1967).
26. Vadovic, C. J., Colver, C. P.: Phys. Rev. B **1**, 4850 (1970).
27. Vadovic, C. J., Colver, C. P.: Phil. Mag. **21**, 971 (1970).
28. Vadovic, C. J., Colver, C. P.: Phil. Mag. **24**, 509 (1971).
29. Carnahan, N., Starling, K.: J. Chem. Phys. **51**, 635 (1965).
30. Ashcroft, N. W., Lekner, J.: Phys. Rev. **145**, 83 (1966).
31. Lo, R. E.: Anorg. Allgem. Chem. **344**, 240 (1966).
32. Ashcroft, N. W., Langreth, D. C.: Phys. Rev. **159**, 500 (1967).
33. Bowen, D. E., Thompson, J. C., Millett, W. E.: Phys. Rev. **168**, 114 (1968).
34. Cohen, M. H., Turnbull, C.: J. Chem. Phys. **31**, 1164 (1959).
35. Ott, A., Lodding, A.: Z. Naturforsch. **20a**, 1578 (1965).
36. Saxton, H. J., Sherby, O. D.: Am. Soc. Mech. Eng. **55**, 826 (1962).

Discussion

M. J. SIENKO It is important that the diffusion coefficient also be determined by measurements on ^{14}N since the proton results may be questioned on the basis that proton hop plus NH_3 rotation, may be just as effective a relaxation path as simple NH_3 translation. It is bothersome that the sodium results also indicate a 4:1 complex. One would think that the Na–NH_3 complex would be much less tightly bound than the Li–NH$_3$ complex. If the Na could easily slip out of its cage, it should show a high diffusion coefficient.

A. GARROWAY   The strong inequality between the longitudinal and transverse relaxation rates of the ammonia protons in these solutions indicates that the proton lifetime against interammonia exchange is greater than 10^{-4} sec. In this time a proton will diffuse much further on the ammonia molecule than it would by a proton hop plus NH$_3$ rotation.

If the sodium ion were less tightly bound to its cage of ammonia molecules, then a larger sodium self-diffusion coefficient would be observed. Of course, if the solvation interaction were much weaker, then a sodium D appropriate to an undressed sodium ion in liquid ammonia would be seen, in contradiction to the diffusion data. It is somewhat perverse to imagine that the sodium is solvated by 6 ammonia molecules in just such a fashion to mirror the diffusion results of the Li(NH$_3$) system, which indicate a solvation number of 4.

I should again stress that the diffusion results give a quantitative prediction of the solvation number only in the concentrated region, i.e. when a significant fraction of the ammonia molecules are bound. At lower concentrations (below about 8 MPM) the diffusion results indicate that the metal ion is solvated, but the exact solvation number cannot be determined. In particular the solvation number could be 6 near the critical concentration.

Cell e.m.f.'s in Metal–Ammonia Solutions*

K. ICHIKAWA** and J. C. THOMPSON

Abstract

We measured the e.m.f.'s of cells of Na–NH$_3$ solutions vs Na–Hg amalgams to determine the activity (chemical potential) as a function of the concentration of metal in liquid ammonia and of temperature. We paid particular attention to the region of phase separation. From these data we obtained values of the critical exponent δ. Using the known value of $\beta (= 0.50)$ for these solutions and scaling, we obtained an estimate for α and γ as well. The results are $\alpha = 0.05$, $\gamma = 1.0$, $\delta = 2.9$. Clearly, liquid conductors behave more like classical fluids than do nonconductors.

Introduction

The striking phase transition, so familiar to all who study metal–ammonia solutions, has received renewed interest since Sienko and Chieux [1] reported their detailed study of the shape of the coexistence curve at Colloque Weyl II. The parabolic shape [2] observed for $(T_c - T)/T_c \equiv \varepsilon \geq 10^{-2}$, and the transition to non-classical behavior for smaller ε have reinforced the impression that M–NH$_3$ solutions always yield unexpected and interesting data.

Critical phenomena in liquid-vapor transitions [3—5] involving many one-component fluids have been extensively studied. For a number of nonconducting fluids critical exponents quite close to the predictions of lattice-gas theory [6] are obtained. The behavior of metallic fluid alloys near critical points is still an open question both theoretically and from an experimental viewpoint. According to Egelstaff and Ring [4] the differences obtained between critical exponents for nonconducting fluids and liquid metals may be related to the range of the interatomic forces: the long-ranged Coulomb force dictating classical (or mean-field) behavior [3] for the metals. However, it should be noted that screening drastically reduces the range of interaction.

This paper describes results from e.m.f. measurements on Na–NH$_3$ solutions which led to determination of several critical exponents.

Experimental

Determination of the activity in Na–NH$_3$ solutions by e.m.f. measurements has been performed using beta–alumina, nominally described by the formula: Na$_2$O · 11 Al$_2$O$_3$ [7]. By using this material as a solid electrolyte [8, 9] the problem of finding an ionically conducting medium for Na$^+$ which is neither attacked by nor mixed with either Na or Na–NH$_3$ solutions, was successfully

* Supported in part by the U.S. NSF and by the R.A. Welch Foundation of Texas.
** On leave from Department of Chemistry, Faculty of Science, Hokkaido University, Sapporo, Japan and currently R.A. Welch Foundation Postdoctoral Fellow.

removed. Three types of concentration cells without transference, different from Russell and Sienko's [10], were employed:

cell A: Na(metal)|beta–alumina crystal|Na–Hg amalgam,
cell B: Na–NH$_3$|beta–alumina crystal|Na–Hg amalgam,
cell C: Na–NH$_3$(consolute)|beta–alumina crystal|Na–NH$_3$.

Thermodynamics provides relations between the measured potentials E_A, E_B, and E_C and the activity of Na, a_{Na}:

$$E_A + E_B = -(R^T/F) \ln a_{Na} , \tag{1}$$

and

$$E_C = -(RT/F) \ln[a_{Na}(c)/a_{Na}(x)] , \tag{2}$$

where R is the gas constant, F the Faraday, $a_{Na}(c)$ denotes the activity of the consolute (critical) composition, and $a_{Na}(x)$ the activity at concentration x. We will also use the notation a_c and a_x when there is no possibility for confusion over the species of interest.

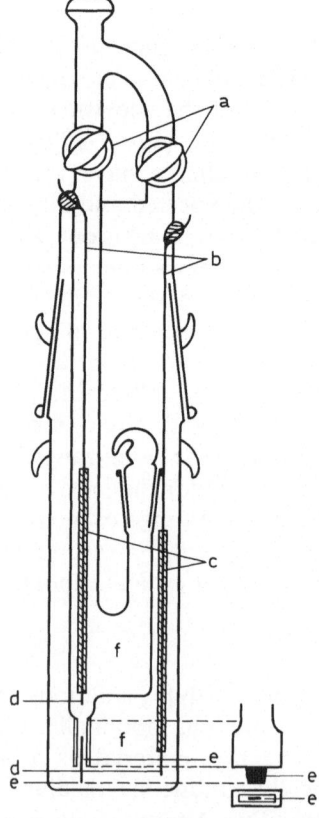

Fig. 1. Pyrex e.m.f. cell; a stopcock; b tungsten lead; c sealed by Pyrex; d platinum electrode; e Sodium Beta–Alumina (Na$_2$O · 11 Al$_2$O$_3$ single crystal) sealed by silicon rubber and Epoxy; f solutions

The preparations of Na–NH$_3$ solutions were similar to those previously used in our laboratory [11]. Sodium amalgams (1.5—2.0 at % Na) having less than the eutectic concentration 2.8 at % Na were prepared from triple distilled mercury and reagent-grade sodium metal. Contacts were made to the solutions with platinum electrodes and potentials were measured with a Data Precision Model 2400 digital voltmeter having a 10^3 MΩ input impedance and a Keithley Model 160 dvm with a 10 MΩ input impedance.

Figure 1 shows the cell design. The beta–alumina crystal was mounted at e using silicone rubber with current flow perpendicular to the c-axis. The two solutions are at f and the electrodes at c.

Results

In order to test the assumptions that the conductivity of beta–alumina is ionic in nature and that the sodium ions common to the samples of interest are mobile, we duplicated Russell and Sienko's [10] data on the Na–Hg amalgam (Fig. 2).

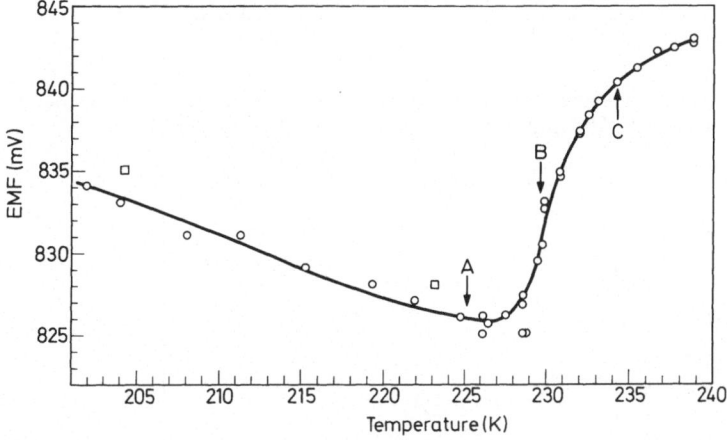

Fig. 2. Temperature dependence of e.m.f. of the cell A; ○ due to this work, □ due to Russell and Sienko. A denotes the eutectic temperature in the Hg rich region, B the liquidus temperature, and C the melting temperature of Hg

The agreement is within ±0.1 %. The two discontinuities are near the eutectic (A) and liquidus (B) temperatures, respectively; the melting point of pure Hg is at C [12].

Figures 3 and 4 show the temperature dependence of the potentials produced by cells B and C, respectively. The activity coefficients found from Eqs. (1) and (2) are self-consistent.

Discussion

The essentials of critical point behavior are best displayed in the language of critical exponents [3—5]. Chieux and Sienko [2] and Teoh, Antoniewicz, and

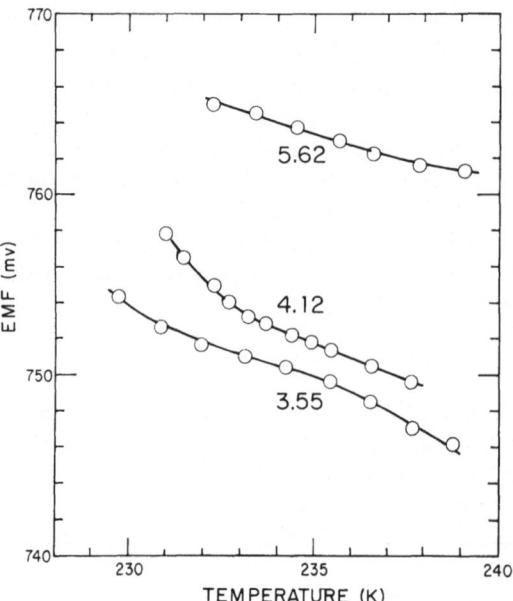

Fig. 3. Temperature dependence of e.m.f. in cell B for concentrations on the metal-rich side of the phase boundary. The numbers give the concentration in MPM

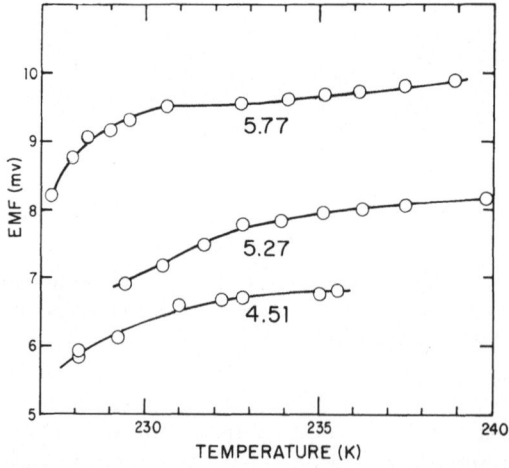

Fig. 4. Temperature dependence of e.m.f. in cell C for concentrations on the metal-rich side of the phase boundary

Thompson [13] found β from the relation:

$$|x - x_c| \sim \varepsilon^\beta .\tag{3}$$

We are able to report estimates for δ, γ, and α which are defined as follows: In a liquid–gas system δ is defined by [14]

$$P - P_c \sim |\varrho - \varrho_c|^\delta ,\tag{4a}$$

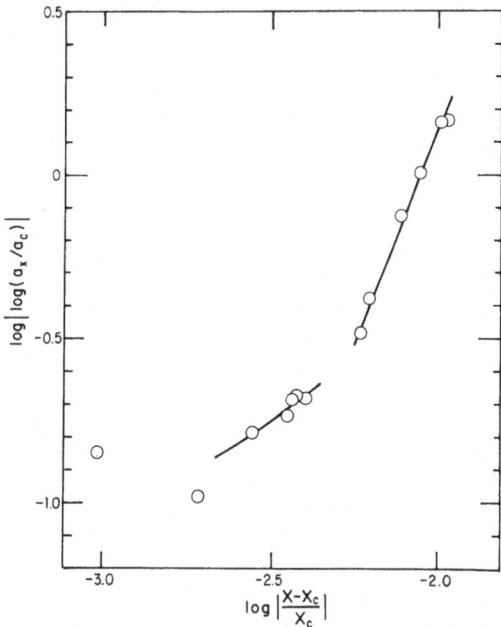

Fig. 5. $\log|\log(a_x/a_c)|$ vs. $\log|x/x_c - 1|$ for Na–NH$_3$ solutions at the critical temperature 231.5 °K

where ϱ is the density. The analogs of P and ϱ for a binary solution are chemical potential μ and concentration x [4]:

$$|\mu - \mu_c| \sim |x - x_c|^\delta . \tag{4b}$$

The chemical potential is, of course, related to the activity by $\mu = \mu_0 + RT \ln a$. In Fig. 5 the chemical potentials are plotted against the concentration and a value of 2.9 is obtained for δ. The steep slope does not appear to persist as the critical composition is approached. Similar results reported for Xe [15] and for Ga–Bi alloys [16] are shown in Fig. 6 along with the Na–NH$_3$ data. Note the shifts in ordinate and abscissa to bring the curves closer together. It is well known in one- or two-component systems that gravitational effects [17—19] can produce such deviations near the critical point, and such problems appear to exist in our work, as well as in the other data shown in the figure.

For partially miscible mixtures we can derive the following equation at the critical concentration

$$|\mu(T) - \mu(T_c)| \sim \varepsilon^{\beta\delta} . \tag{5}$$

We see in Fig. 7 that $\beta\delta$ changes as ε decreases below 10^{-2}. Recall that β transforms from 0.50 to 0.34 as ε decreases below 10^{-2} according to Chieux and Sienko [2].

From Griffiths-Rushbrooke relations [20] among the exponents, α and γ (which, respectively, describe anomalies in C_v and C_p in one-component systems) are found to be 0.05 and 1.0 for $\varepsilon < 10^{-2}$. A somewhat larger value is indicated for $\varepsilon > 10^{-2}$ but our errors are sufficiently large to prevent accurate determination.

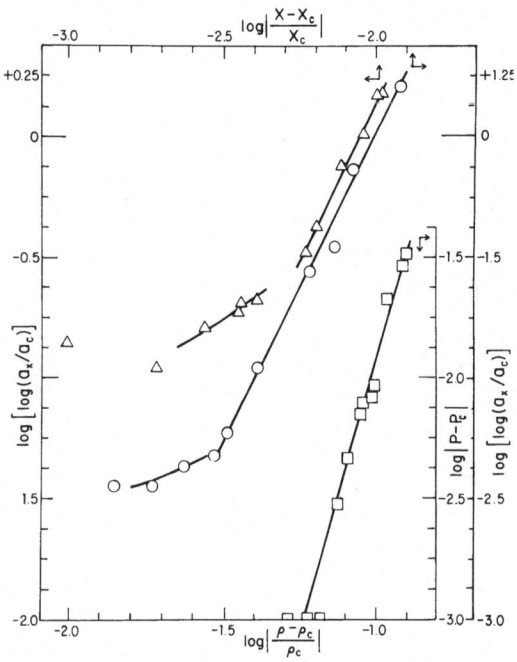

Fig. 6. $\log|\log(a_x/a_c)|$ vs. $\log|x/x_c - 1|$ for liquid Ga–Bi mixtures $\bigcirc$ and Na–NH$_3$ liquids $\triangle$, and $\log|P - P_c|$ vs. $\log|\varrho/\varrho_c - 1|$ for Xe $\square$, at or near the critical temperature

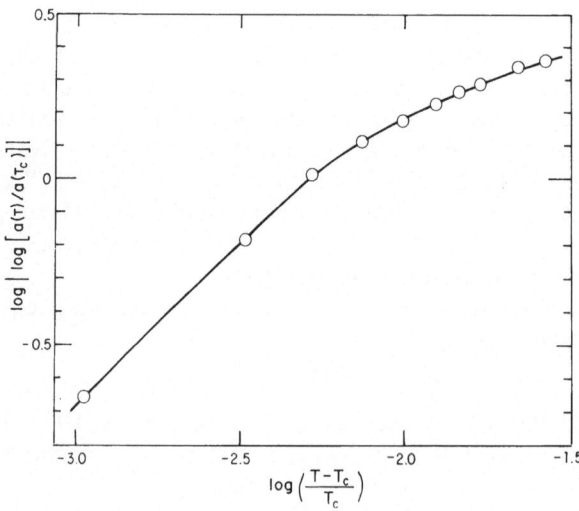

Fig. 7. $\log|\log[a(T)/a(T_c)]|$ vs. $\log|T/T_c - 1|$ for the critical concentration or 4.12 mole % Na in liquid NH$_3$

Table I. Comparison of the critical exponents among the various systems, including the calculated values

	$\varepsilon > 10^{-2}$			$\varepsilon < 10^{-2}$			
	α	β	γ	α	β	γ	δ
Na–NH$_3$	0.05[b]	0.50[a]	1.0[b]	0.7[b]	0.34[a]	0.1[b]	2.9
Xe	0.3[b]	0.33[c]		0.3[b]	0.33[c]	1.2[d]	4.2[d]
CCl$_4$–C$_7$F$_{14}$		0.33[c]			0.33[c]		
Na		0.42[c]					
Ga–Hg		0.59[e]			0.33[e]		3.1[e,f]
Classical theory	0	0.5	1	0	0.5	1	3

[a] Ref. [2].   [b] Calculated.   [c] Ref. [4].   [d] Ref. [15].   [e] Ref. [16].   [f] For Ga–Bi.

We do, nevertheless, confirm the crossover seen by Chieux and Sienko [2] near $\varepsilon = 10^{-2}$ as may be seen in Fig. 7.

The critical indices obtained here, together with other values, are collected in Table I. In the region where $\varepsilon > 10^{-2}$ each of the exponents is similar to those of other conducting liquids and to the results of mean-field theory. As the critical point is more closely approached, a crossover to non-classical values is observed. The existence of a crossover when $|x - x_c|$ becomes small is apparently obscured by gravitational effects on the phase separation, though it should be noted that the flattening is less abrupt in the data for the conducting fluids in Fig. 6 than in the Xe data.

The observation of mean-field behavior in conducting fluids at large ε has been attributed to the presence of the long-range Coulomb force [4, 21]. Screening, however, drastically reduces the range of the force in any aggregate of charged particles such as the system at hand. It is more likely that it is the long-ranged electron wave function that carries the effect of the ion–ion interaction to great distances and thereby restores the conditions of mean-field theory [22]. No simple, all inclusive explanation is immediately available. There are, after all, three quite different sorts of conducting systems which show phase separation: 1. those such as the present in which a M–NM transition is simultaneously observed [23]; 2. those such as Ga–Bi in which both phases are metallic; and 3. electrolytic solutions wherein ionic drift (rather than electronic motion) is the conduction process in both phases. It has already been shown that the anomaly in $d\sigma/dT$ is different in systems (2) and (3) [24]. One must expect, it would seem, differences in other critical point behavior.

Acknowledgements

We wish to thank Mr. H. R. Weiler of the Carborundum Company for providing beta–alumina crystals.

References

1. Sienko, M. J., Chieux, P.: In: J. J. Lagowski, M. J. Sienko (Eds.): Metal–ammonia solutions, p. 339. London: Butterworths 1970.
2. Chieux, P., Sienko, M. J.: J. Chem. Phys. **53**, 566 (1970).

3. Egelstaff, P.A.: An introduction to liquid state, p. 199. London, New York: Academic Press 1967.
4. Egelstaff, P.A., Ring, J.W.: In: H. N. V. Temperley, J. S. Rowlinson, G. S. Rushbrooke (Eds.): Physics of simple liquids, p. 253. Amsterdam: North-Holland 1968.
5. Stanley, H.E.: Introduction to phase transitions and critical phenomena. Oxford: Oxford University Press 1971.
6. Fisher, M.E.: J. Math. Phys. **5**, 944 (1964).
7. Whittingham, M.S., Helliwell, R.W., Huggins, R.A.: Final Technical Report, Office of Naval Research. Stanford University 1969.
8. Whittingham, M.S., Huggins, R.A.: J. Chem. Phys. **54**, 414 (1971).
9. Hsueh, L., Bennion, D.N.: J. Electrochem. Soc. **118**, 1128 (1971).
10. Russell, J.B., Sienko, M.J.: J. Am. Chem. Soc. **79**, 4051 (1957).
11. Kyser, D.S., Thompson, J.C.: J. Chem. Phys. **42**, 3910 (1965).
12. Hansen, M., Anderko, K.: Constitution of binary alloys. New York: McGraw-Hill 1958.
13. Teoh, H., Antoniewicz, R.R., Thompson, J.C.: J. Phys. Chem. **75**, 399 (1971).
14. Fisher, M.E.: In: M. S. Green, J. V. Sengers (Eds.): Critical phenomena, p. 21. Washington: National Bureau of Standards 1966.
15. Habgood, H.W., Schneider, W.G.: Can. J. Chem. **32**, 98, 164 (1954).
16. Predel, B.: Z. Physik. Chem. **24**, 206 (1960).
17. Ulykin, S.A., Malyshenko, S.P.: In: S. Gratch (Ed.): Adv. Thermophys. Properties at Extreme Temperatures and Pressures, p. 68. New York: Am. Soc. Mech. Eng. 1965.
18. Schmidt, E.H.W.: In: M. S. Green, J. V. Sengers (Eds.): Critical phenomena, p. 13. Washington: National Bureau of Standards 1966.
19. Lorentzen, H.L., Hansen, B.B.: In: M. S. Green, J. V. Sengers (Eds.): Critical phenomena, p. 213. Washington: National Bureau of Standards 1966.
20. Griffiths, R.B.: J. Chem. Phys. **43**, 1958 (1965).
21. Young, D.A., Alder, B.J.: Phys. Rev. A **3**, 364 (1971).
22. Antoniewicz, P.R.: Private communication.
23. Thompson, J.C.: Rev. Mod. Phys. **40**, 704 (1968).
24. Schurmann, H.K., Parks, R.D.: Phys. Rev. Letters **26**, 1790 (1971). — Stein, A., Allen, G.F.: Bull. Am. Phys. Soc. **17**, 278 (1972). — D'Abramo, G., Ricci, F. P., Menzinger, F.: Phys. Rev. Letters **28**, 22 (1972).

Discussion

M. J. SIENKO   Fluctuations just above the critical temperature may limit the electron motion to short, finite lengths along a string of suitably oriented cavity sites. Maybe there is no anion species! What we need is some suggestion as to the appropriate kind of experiment to do to obtain evidence for these fluctuation coherence lengths. Perhaps one could take directly the results that have come from percolation theory.

Impedence Behavior
of Metal–Ammonia Solid Metal Interface*

PAUL D. SCHETTLER, CRAIG L. VAN ANTWERP, JAMES A. HAMILTON,
JOAN E. THILLY, and JAMES D. SPEAR

Abstract

This paper is concerned with the electrical behavior of the metal–ammonia metal electrode interface with particular attention placed on results obtained from a three-electrode cell.

Introduction

In 1914 Kraus [1] pointed out that the electrolysis of a metal–ammonia solution was unusual in that the negative carrier in the solution passed into the anode without producing material effects and thus Faraday's laws could not be applied in the normal sense. In 1948 Laitenen and Nyman [2] studied the metal electrode–ammonia solution interface by means of polarography and the utilization of solid platinum electrodes. Utilizing voltage versus current curves of sodium in ammonia in the vicinity of the potential of zero current, they were able to obtain an electrode potential for the electron of between − 1.86 and − 1.92 Volt. They noticed no irreversibility or overvoltage effects, but more recently Schettler and Patterson [3] have reported a field dependence of metal–ammonia solutions that could only be interpreted in terms of time and voltage-dependent electrode-solution interface effect. Thompson and co-workers [4,5] likewise observed electrical behavior that is best understood in terms of phenomena associated with the solution-electrode interface.

More recently Schettler and Patterson [6] presented detailed results for the platinum-metal solution interface using for the most part cells designed for standard conductance work. Thus, although the results reported indicated the existence of both a time- and voltage-dependence of electrode-solution interface impedence, they were the algebraic sum of anodic and cathodic effects. It is of interest to separate these effects, and the present authors have considered several possibilities for appropriate differentiation. Among them is the use of electrodes which are significantly different in size or material, or alternatively, the utilization of a three-electrode cell in which one electrode is used as a noncurrent-carrying reference and placed close to the electrode of interest.

Experimental Procedures

Two different electrical measuring techniques were used. Standard bridge measurements were used for two-electrode cells as described elsewhere [6], the object being to find a combination of resistances and capacitances that would be

* The authors would like to thank the Research Corporation for a Frederick Gardiner Cottrell Project Grant and the National Science Foundation for URP grants which made this research possible.

identical to the response of the cell for the entire length of a pulse. For three-electrode cells, measurements were taken point by point as a function of time, utilizing a type 547 Tektronix oscilloscope with Type W plug-in for accurate time and voltage measurements. Alternatively, on occasion a photograph was made of the oscilloscope trace. For much of the work reported, a Philbrook Nexus Type 1016 high speed-high power operational amplifier was used in the voltage follower mode in conjunction with a General Radio Type 1340 Pulse Generator. In spite of widely varying cell impedence, the device delivered pulses of controllable constant current with only a few per cent deviation over the range of several tenths of a microsecond to at least 1 sec. Typically data were taken over a time span of six decades, from 0.05 μsec after the start of a pulse to about 0.1 sec. At times greater than this, voltage measurements at the reference electrode became irreproductible, presumeably because of convection effects.

The solutions were unstirred during the pulse. Appropriate precautions (either stirring or waiting between pulses) were taken to insure that each pulse was unaffected by previous pulses.

Solutions were prepared by standard techniques reported elsewhere [7]. When a supporting electrolyte was used, reagent grade sodium iodide was ground, vacuum dried and inserted into a special side arm of the cell and baked at 400° C and 10^{-6} torr overnight. No evidence of solution decomposition was observed in any run.

Temperature control was likewise accomplished by thermostatic techniques reported elsewhere [7], except that an Aminco bimetallic thermoregulator was used. Temperature was maintained to $\pm 0.1°$ measured by a calibrated copper constanthan thermocouple with a triple point cell reference. Cells were constructed of Pyrex utilizing 0.3 cm tungsten rod to nonex glass to uranium glass to Pyrex in the construction of the electrode to glass seals. The tungsten was completely encapsulated with nonex and then the glass ground away to expose a sheathed electrode surface. Reference and working electrodes were ground to a point before encapsulation; in this way extremely small electrode surfaces could be produced. In cases where electrode metals other than tungsten were desired the appropriate metal was either electroplated or vacuum evaporated into the electrode after construction. The electrodes were sealed at the end of a deep well so that only minor glassblowing was involved in inspecting, polishing or replating the electrodes and refitting them to the cell. The cell was constructed so that the reference electrode was placed in a position close to the working electrode, with the stationary electrode placed about 1 cm away.

Results

Figure 1 shows the set of chronopotentiometric curves for anodic and cathodic pulses utilizing a tungsten electrode in a solution ~0.03 molar in sodium and 1.3 molar in sodium iodide. In addition, Fig. 2 shows an oscillographic curve of the same solution against a larger electrode in order to obtain a curve of lower current density. Under the conditions of the experiment, the sodium iodide should serve as a supporting electrolyte, and electrostatic attraction should be virtually eliminated as a transport mechanism of the solvated electron as the

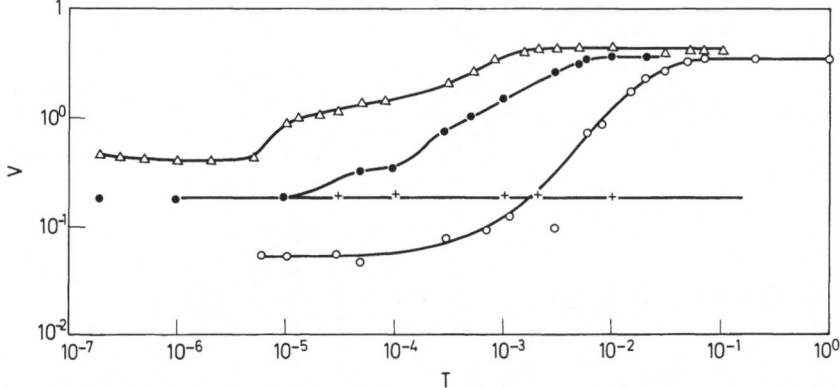

Fig. 1. Logarithm of voltage plotted against logarithm of time for four anodic current densities. —○— 0.129 amp/cm² anodic. —●— 0.466 amp/cm² anodic. —△— 1.28 amp/cm² anodic. —+— 0.475 amp/cm² cathodic. The working electrode material was tungsten utilizing gold-plated reference and stationary electrodes. Temperature was −56.5° C and the solution composition was ∼0.03 molar in sodium metal and 1.3 molar in sodium iodide as a supporting electrolyte ($X_{Na} = 6 \times 10^{-4}$ and $X_{NaI} = 3 \times 10^{-2}$)

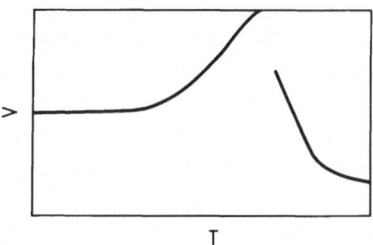

Fig. 2. An oscillographic trace of the voltage of the stationary gold electrode potential as a function of time. The current density was 0.014 amp/cm² with other conditions the same as in Fig. 1. The curve begins at $t = 0$ and the break corresponds to the end of the pulse at 0.72 sec

electro active species. If the transport mechanism is entirely diffusion-controlled, the following relationship holds [8]

$$i\tau^{1/2} = \frac{\pi^{1/2} F D^{1/2} C}{2} \tag{1}$$

where i is the current density, τ is the time at transition between plateaus, F is the Faraday, D is the diffusion constant and C is the concentration. If there is an equilibrium between two or more species, only one of which is electroactive, $i\tau^{1/2}$ can be expected to decrease with increasing current density, and the rate of decrease reflects the rate of transformation between species [9].

Figure 3 shows that apparently there is a decrease of $i\tau^{1/2}$ with increasing i for the system in question and as Delahay [10] pointed out, given a mechanism, one can then proceed to estimate rate constants. If one extrapolates the curve of Fig. 3 back to zero current, one obtains a diffusion constant for the electron

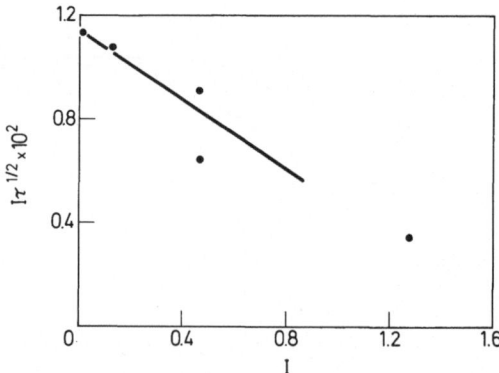

Fig. 3. Transition times obtained from inflection points of Figs. 2 and 3 were used to obtain $i\tau^{1/2}$ (amp sec$^{1/2}$/cm^2) as a function of current density. The two points at 0.46 amp/cm^2 reflect the two possibilities for the transition time of the middle curve of Fig. 2

of $6.0\,(\pm 2.4) \times 10^{-5}$ cm^2/sec which compares with 9.6×10^{-5} cm^2/sec calculated from conductance data.

A second feature of Fig. 1 is the contrast between the forms of the anodic and cathodic curves. This feature is fairly general; a series of runs is shown in Fig. 4 in which the differences between the cathodic curves are shown to have no time dependent effects over a range of temperatures and current densities. Runs were made with sodium iodide in ammonia before the addition of sodium. In contrast to these results in which in some cases the specific electrode resistance rose to $3.0 \times 10^2\,\Omega$ cm^2 or higher, the addition of a 0.03 molar sodium metal caused the

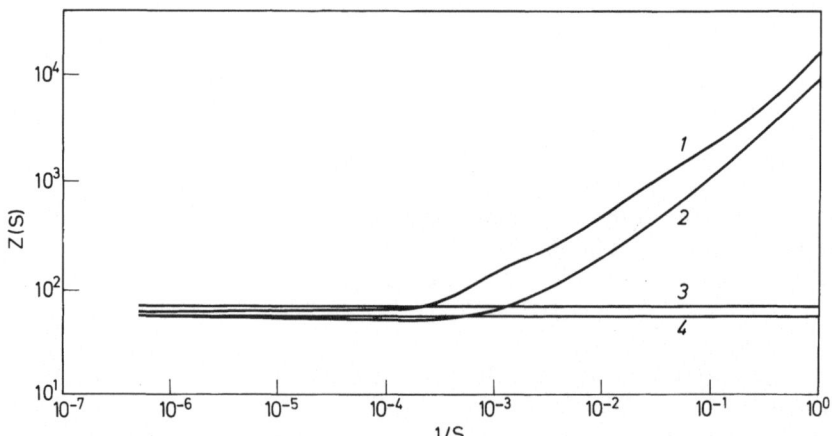

Fig. 4. A representative sample of 14 runs. $Z(s)$ is plotted against $1/s$ where $Z(s)$ is the ratio of the Laplace transform of the voltage divided by the Laplace transform of the current. The runs shown represent the extreme limits of runs between $+0.5$ V and -0.5 V and $-30°$ and $-40°$ at each voltage. (Some runs were made at $-56.5°$). *1* 0.5 volt anodic $-30°$ C. *2* 0.25 volt anodic $-40°$ C. *3* 0.25 volt cathodic $-40°$ C. *4* 0.50 volt cathodic $-30°$ C. The solution was 0.035 molar in Na and 1.14 molar in NaI ($X_{\text{Na}} = 9.1 \times 10^{-4}$; $X_{\text{NaI}} = 0.0302$). The area of the working electrode for the entire work was 2.46×10^{-3} cm^2 except as noted

resistance to drop to about $0.4\,\Omega\,cm^2$ for similar pulses, with virtually no change with time. These results are compatible with the polarographic work in liquid ammonia of Laitenen and Nyman [2] who showed that the cathodic production of solvated electrons in sodium iodide and other salts was inhibited for reasons which are only partially understood. His polarogram for sodium metal is steeply linear through the potential of zero current in contrast to his results for salts which have very low slopes at the potential of zero current.

Although it would appear plausible from these results that the solvated electron somehow catalyzes the cathodic reaction, such an explanation may be premature. All of the measured voltages are relative to the electrode potential of zero current which would be expected to have different values at different electron concentrations. At low overvoltages, the fundamental equation of electrochemical kinetics can be rewritten as

$$i = \frac{-k_r^0 F^2 \eta}{RT} \tag{2}$$

where i is the current density, F is the Faraday, η is the overvoltage and k_r^0 is the rate of the cathodic reaction at the potential of zero current. At this point of equilibrium, the requirement of zero net current dictates that the anodic and cathodic reactions be equal, i.e.,

$$k_r^0 = k_0^0 [e^-] \tag{3}$$

where $k_0^0 [e^-]$ is the rate of the anodic reaction. Thus for our conditions the effective electrode resistance would be inversely proportional to the concentration of the solvated electron, which would account for the dramatic decrease in cathodic impedence upon addition of solvated electrons.

Results of both Laitenen and myself suggest that by itself, this explanation may be too facile. Results from our laboratory suggest that under some circumstances avalanche behavior may be exhibited. Some of the two-electrode cell measurements utilizing a germanium electrode and without a supporting electrolyte exhibited an effect wherein the resistance of a cell was observed to decrease over the length of a square wave rather than to increase with time, which is normal. There is also some fragmentary evidence of a negative resistance region in voltage vs. current curves for this system. Along the same lines Laitenen's cathodic currents were smaller for solutions without solvated electrons than with, at appropriately comparable voltages. If the electron catalyzes the cathodic reaction, either by solvent affects or by forming an appropriate diamagnetic activated complex, such affects can be explained.

With this cell containing one germanium and one platinum electrode, significant rectification results were noted that were attributable to the solution-germanium interface. The cell was balanced against a circuit containing two resistances R_1 and R_2 in series with a capacitor C in parallel with R_2. Observed capacitances were about $30\ mfd/cm^2$ of electrode area, R_2 was as high as $220\ \Omega\ cm^2$ for anodic pulses and as low as zero for cathodic pulses. The effects observed were considerably greater than those observed previously with platinum [6].

It seems plausible from these results that the solvated electron in some sense catalyzes the cathodic reaction either by affecting the solvent structure in the vicinity of the electrode or by direct reaction of an electrode electron with a paramagnetic solvated electron to form some diamagnetic intermediate which may be identical to the diamagnetic species which accounts for the decrease of paramagnetism with increasing concentration.

A final feature of the curves in Fig. 4 is that they do not all extrapolate to the same high frequency value of resistance for different current densities. Specific experimental attention accorded to this phenomena resulted in the curve shown in Fig. 5. Two points are in order. What is plotted is solution resistance between

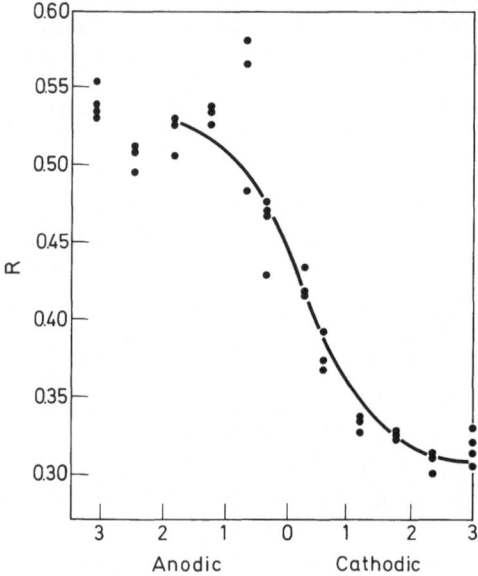

Fig. 5. Variation of specific electrode resistance (Ω cm^2) with current density (amp/cm^2). The measurements were made 10 microseconds after the beginning of a constant current pulse and are essentially identical to the resistance extrapolated back to zero time. The data were taken in an order such as to eliminate any possible time dependence. Other conditions were identical to those of Fig. 1

the reference electrode and the working electrode plus the resistance of the working electrode. It is significant that for tungsten at least 0.12 Ω cm^2 specific resistance must be associated with the electrode itself at low current densities. Insofar as the curve of Fig. 5 is approximately an odd function about zero current density, electrode resistance effects would be difficult to detect with a cell with two symmetrical electrodes. The overall effect as a function of applied voltage would be the sum of two terms, one greater than and one less than the value at zero field by the same amount, resulting in an overall effect that was independent of applied voltage. The extent to which similar effects occur with other metals is unknown. A second point is that these results are compatible with the earlier suggestion that the anodic reaction has some rate-limiting step associated with it in contrast to the cathodic reaction.

It is interesting to note that for the cathodic curves of Fig. 4 at $-30°$ and $-40°$, where the data are most reliable, very low temperature coefficients are observed in the range of 0.6 to 0.3%/° C. At short time intervals, the anodic temperature coefficients seem to be in the range of 2—4%.

Conclusions

The principles of microscopic reversibility would require that, under the same conditions, the anodic and cathodic reactions have the same reactive intermediate. On theoretical grounds, this requirement eliminates a cathodic reaction in which the electron tunnels into the solution and then is solvated in conjunction with a anodic reaction in which the electron tunnels from the trap to the electrode leaving a dissipating trap behind. (Incidently, such a mechanism would require an overvoltage equal to the solvation energy, about 1.4 eV, for conduction to occur.) One thus encounters difficulty in finding an activated complex that meets the above requirement and in addition has the ~ 0.03 eV ($\sim kT$) that is compatible with the low temperature coefficient of electrode impedance.

According to Kestner and Copeland [11] about 38% of the electronic wave function density is outside the central sphere. This wave function would not be expected to be symmetrically distributed but rather to respond to and intensify local fluctuations of the ammonia surrounding the central sphere. Thus, not only the central cavity but also surrounding "proto cavities" would be expected to contain nonstoichiometric quantities of electron wave function density. This conception can be extended to electrons leaving and entering the electrode surface with low activation energy. It is also conceivable that cathodic electrons can penetrate the solvent more readily if the solvent already contains solvated electrons due to the existence of "proto cavities". Such a model could also account for the apparent kinetic step in the anodic curves in that single electrons would have greater difficulty in leaving the solution than electrons acting in concert.

References

1. Kraus, C. A.: J. Am. Chem. Soc. **36**, 864 (1914).
2. Laitenen, H., Nyman, C. J.: J. Am. Chem. Soc. **70**, 3002 (1948).
3. Schettler, P. A., Patterson, A.: Am. Chem. Soc. **87**, 392 (1965).
4. Nasby, R. D., Thompson, J. C.: J. Chem. Phys. **49**, 969 (1968).
5. Kyser, D. S., Thompson, J. C.: J. Chem. Phys. **42**, 3910 (1965).
6. Schettler, P. D., Patterson, A.: In: Lagowski, J. J., Sienko, M. J. (Eds.): Metal–Ammonia Solutions, Colloque Weyl II. London: Butterworths 1969.
7. Schettler, P. D., Patterson, A.: J. Phys. Chem. **68**, 2870 (1969).
8. Sand, H. J. S.: Phil. Mag. **1**, 45 (1901).
9. Gierst, L.: Thesis, University of Brussels, 1952.
10. Delahay, P., Berzins, T.: J. Am. Chem. Soc. **75**, 2486 (1953).
11. Kestner, N. R., Copeland, D. A.: In: Lagowski, J. J., Sienko, M. J. (Eds.): Metal–Ammonia Solutions, Colloque Weyl II. London: Butterworth 1969.

Discussion

P. DELAHAY   Some eight or ten years ago we investigated (with Timmer, unpublished work) electron transfer at metal-solvated electron solution (liquid ammonia) interfaces. We used faradaic rectification in the MHz range but found only diffusion control. I should think that you are not detecting slow electron transfer in your experiments but rather some artifacts.

P. D. SCHETTLER   I disagree with your implication that the electrode kinetics are diffusion-migration controlled for all systems containing a metal electrode and the solvated electron. Considerable evidence from a variety of authors and utilizing a variety of apparatus would suggest otherwise. Your work suggesting that at least one electrode-solvated electron combination is diffusion controlled is very interesting in that it suggests that bulk phase kinetics are not involved and that the effects that myself and others have observed are purely surface effects. I would like to learn more of this work.

Absorption of Ultrasound in Metal–Ammonia Solutions*

D. E. BOWEN

Abstract

A new system for measuring the ultrasonic attenuation in metal–ammonia solutions is described. Preliminary measurements on pure ammonia and on lithium–ammonia solutions are reported. The ratio of the measured attenuation to the classical attenuation in pure ammonia at $-60°$ C is 13.02 indicating that ammonia is a Kneser liquid. The attenuation in the lithium–ammonia solutions rises to a maximum at the concentration of the critical point indicating critical sound absorption.

I. Introduction

Measurements of ultrasonic absorption in liquid systems are very useful in obtaining information as to the molecular processes in the liquid [1, 2]. Classically this absorption arises due to mechanisms associated with viscosity and thermal conduction and is given by the equation [1]:

$$\alpha_c/f^2 = (8\,\pi^2/3\,\varrho\,c^3)\,[\eta_S + 3(\gamma - 1)\,K/4\,C_p]\,, \tag{1}$$

where α_c is the classical absorption coefficient, ϱ is the density, c the velocity of sound, η_S the coefficient of shear viscosity, γ the ratio of specific heats, K the thermal conductivity, C_p the specific heat at constant pressure, and f the frequency of the ultrasonic wave. Experimental measurements of α/f^2 show that there are two types of deviations from the "classical absorption" given by Eq. (1) [2]:
1. α/f^2 is greater than that given by Eq. (1) and is not a function of frequency;
2. α/f^2 shows a dependence on frequency.
Tisza [3] pointed out that the deviations from classical behavior of type (1) can be accounted for by a volume viscosity, η_v, from which the absorption coefficient can be calculated by:

$$\alpha_v/f^2 = (2\,\pi^2/\varrho\,c^3)\eta_v\,. \tag{2}$$

The measured absorption is thus $\alpha_c/f^2 + \alpha_v/f^2$. It should be pointed out that this does not account for deviations of type (2) and at present ultrasonic means are the only way of measuring η_v.
Explanations of the type (2) deviations from the classical absorption usually make use of relaxation models assuming frequency dependent transport coefficients [4]. It is not possible, however, to account for all of the reported deviations in a simple manner.

* Research assisted by the Robert A. Welch Foundation, the Research Corporation and the University Research Institute of the University of Texas at El Paso.

Measurements of the absorption coefficient in liquid mixtures have shown that anomalous increases occur near critical points. Moderate success in explaining these increases has been obtained by Fixman [5] who assumed that critical fluctuations enter into the sound absorption through the strong functional dependence of the Ornstein-Zernike long-range correlation length upon the temperature change produced by the sound wave.

Measurements of the absorption of ultrasound in metal–ammonia solutions appear to be of interest from two viewpoints; both as a measure of the properties of the solutions themselves and as a test of the theories of ultrasound absorption. Metal–ammonia solutions exhibit a whole host of properties which should have considerable effect on the absorption of ultrasound [6, 7], with perhaps the most important being the existence of the miscibility gap [8]. While extensive sound velocity measurements [9] have been made, there have been no measurements of ultrasonic absorption reported to date.

We would thus like to describe a new ultrasonic system for the measurement of the absorption of ultrasound in metal–ammonia solutions and to report preliminary data on the absorption in pure NH_3 and in lithium–ammonia solutions.

II. Experimental

The experimental apparatus consisted of four major parts:
1. sample cell,
2. electronic equipment,
3. ammonia preparation apparatus, and
4. temperature regulation and measurement apparatus.

The ammonia preparation apparatus is a standard one and will not be described here. The temperature bath was a Lauda K-70 DW capable of temperature regulation to within $\pm 0.3°$ C over a $+40$ to $-77°$ C range. The temperature was measured by a Relco Model 1065 digital platinum resistance thermometer.

The sample cell is shown in Fig. 1. Two fused quartz rods served as the delay line and reflector. The upper rod is connected to a stainless steel bellows which allows the rod to be moved both up and down for measurements and angularly for parallelism adjustments. The lower portion of the bellows is connected to the sample cell with an "0" ring such that a high vacuum may be established in the cell and maintained as the upper rod is moved. The upper rod is 17 cm long with a spiral grove cut in it and has a coaxially plated 10 MHz X-cut quartz transducer sealed to it with Nonaq stopcock grease. The top of the delay line is attached to a mechanism (not shown in the Figure) which allows the rod to be moved vertically over a distance of up to 3 cm.

A block diagram of the electronics used is shown in Fig. 2. A standard single-ended pulse echo technique was used in which the RF pulses were generated by a Matec Model 6000 RF Pulse Generator and Receiver. This instrument is capable of producing pulses from 10 to 320 MHz at a pulse width of 1—20 μsec at a repetition rate of up to 500 Hz. The received pulses are sent to the Matec Model 2470 Automatic Attenuation Recorder [10]. This instrument provides the

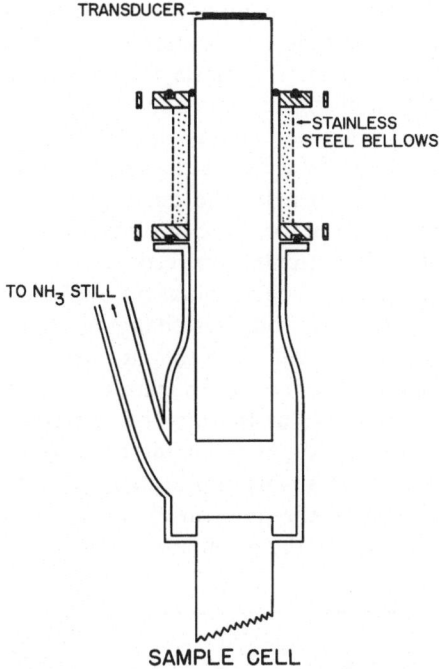

Fig. 1. Sample cell. The upper rod is 17 cm long and 3 cm in diameter. There is a spiral grove cut along the length of this rod to prevent spurious reflections

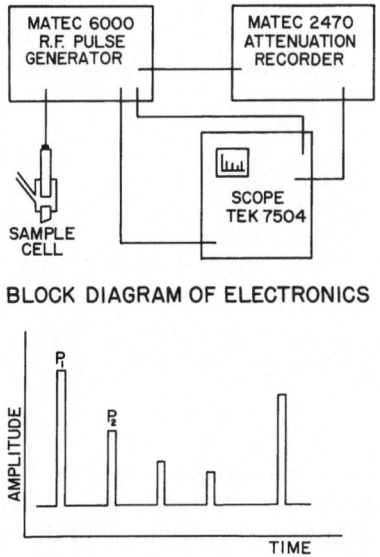

Fig. 2. Block diagram of electronics. Lower portion of Figure is oscilloscope pattern

necessary gates and delays to allow any two pulses from the pulse train to be selected. These two pulses are then peak detected with the first pulse's amplitude being maintained constant by an automatic gain control circuit. The two pulses are then fed into a differential logarithmic amplifier. The resultant voltage is the ratio of the two pulse amplitudes and is displayed on a meter calibrated in decibels. An example of the pulse-echo train is shown at the bottom of the figure. The first pulse (P_1) is the echo from the delay line-solution interface and the second pulse (P_2) is the reflection from the bottom rod-solution interface. If these two pulses are selected by the attenuation recorder the ratio of their amplitudes will be displayed on the meter. The attenuation coefficient for a given solution is obtained by moving the upper rod thus changing the distance through which the solution pulse must travel. The slope of the meter reading versus distance curve is then the attenuation coefficient. The parallelism of the surfaces could be adjusted by three leveling screws at the upper rod mount. The parallelism was adjusted for minimum attenuation at the highest frequencies used (usually 50 or 70 MHz). Measurements of the absorption in water at room temperature yielded values in good agreement with the literature [1]. Figure 3 is an example of the results obtained; in this manner the attenuation coefficient could be determined to within 2%.

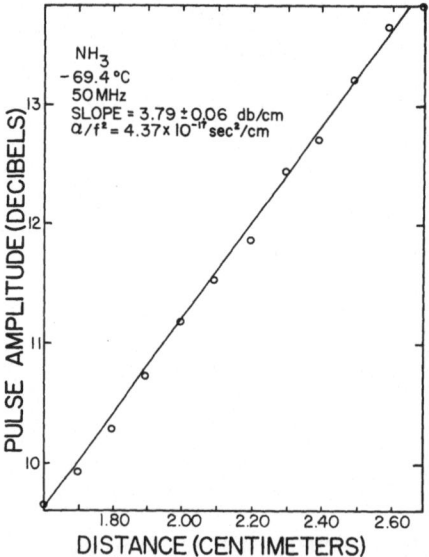

Fig. 3. Pulse amplitude *versus* distance for pure NH_3 at $-69.4°$ C. The pulse amplitude here is really the ratio of the amplitude of pulse P_2 to that of P_1; the amplitude of P_1 remains constant as the distance is varied

For the metal–ammonia solution a piece of lithium metal was prepared in an inert atmosphere box and then transferred to the sample cell with the aide of a glove bag. The solution was then prepared by distilling into the sample cell a known amount of dried ammonia. The solution was then diluted in successive steps.

III. Results and Discussion

Ammonia. The attenuation coefficient for pure liquid ammonia was measured as a function of temperature at a frequency of 50 MHz. The results of these measurements are shown in Fig. 4. Using the thermodynamic quantities given in Table 1 the classical absorption coefficient is calculated to be:

$$\alpha_c/f^2 = 0.303 \times 10^{-17} \text{ np sec}^2/\text{cm},$$

where at this temperature the measured value is

$$\alpha/f^2 = 3.95 \times 10^{-17} \text{ np sec}^2/\text{cm},$$

with the ratio

$$\alpha/\alpha_c = 13.02.$$

This ratio indicates that NH_3 should be classified as a "Kneser liquid" [1] in so far as its ultrasonic absorption is concerned. This implies that the reason for the excess absorption is thermal relaxation or a slow exchange between external and internal energy.

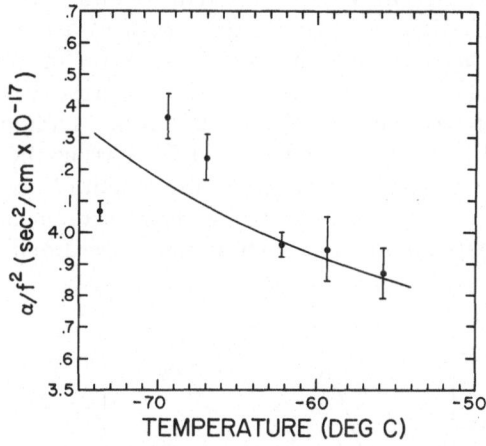

Fig. 4. Attenuation *versus* temperature in pure NH_3 at 50 MHz

Table I. Numerical values of quantities used in Eq. (1) for NH_3 at 213.15 °K (-60 °C)

Quantity	Value	Ref.
ϱ	0.7136 gm/cm^3	[11]
c	1890 m/sec	[12]
η_S	0.387 centipoise	[13]
γ	1.52	[12]
K	5.37×10^{-3} watt/cm Deg.	[14]
C_p	17.79 cal/mole Deg.	[15]

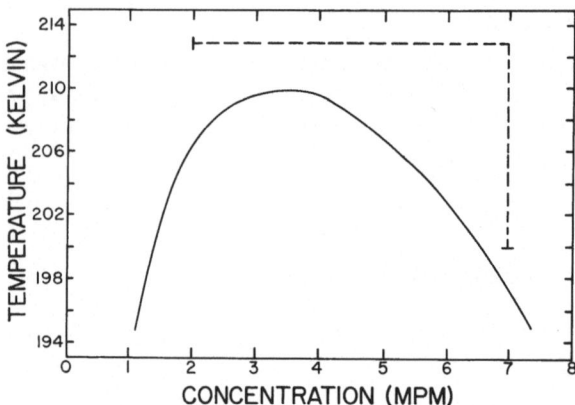

Fig. 5. Mixed phase region for lithium–ammonia solutions; the dashed line represents the regions covered in these experiments

Lithium–ammonia solutions. Figure 5 shows a portion of the phase diagram for the lithium–ammonia solutions [16]; the dashed line on this figure gives the range both in concentration and in temperature covered in these experiments. Figure 6 shows the results as a function of metal concentration at two frequencies at a temperature of $-60°$ C. Here we see that α/f^2 rises to a maximum at a concentration corresponding to the critical concentration for the lithium–ammonia solutions (4 MPM). The change with concentration is most striking at the lower frequencies; however, it was not possible to obtain measurements at the lower frequencies for the higher concentrations (6 and 7 MPM) as the attenuation was too small to provide a measurable change in pulse amplitude as the distance was varied. Similarly it was difficult to obtain good measurements at the higher frequencies when the attenuation was high as the pulse amplitudes were quite

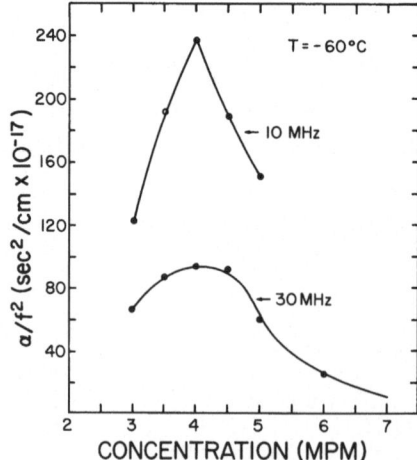

Fig. 6. Attenuation *versus* concentration for lithium–ammonia solutions at $-60°$ C

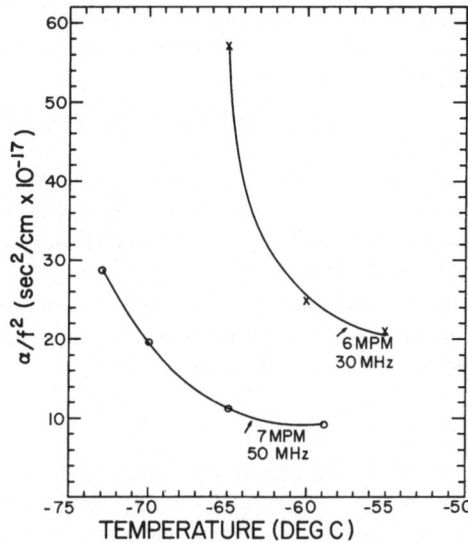

Fig. 7. Attenuation *versus* temperature for the lithium–ammonia solutions

small and the voltage ratio was noisy. The type of behavior shown in Fig. 6 is typical of the behavior seen in other critical systems near the critical point [17].

Figure 7 shows the temperature dependence of α/f^2 for two different concentrations and frequencies. We see that α/f^2 is a strong function of the temperature as the solution approaches a phase boundary. It is clear that the onset of the phase transition is evidenced at temperatures well above the usually reported phase boundary. It is felt that more information about this phenomenon can be obtained by a more thorough study of the temperature dependence of α/f^2 at temperatures and concentrations well away from phase boundaries.

Figure 8 shows the frequency dependence of α/f^2 for a 5 MPM solution at $-63.5°$ C. We see that α/f^2 is a definite function of the frequency, decreasing as the frequency is increased.

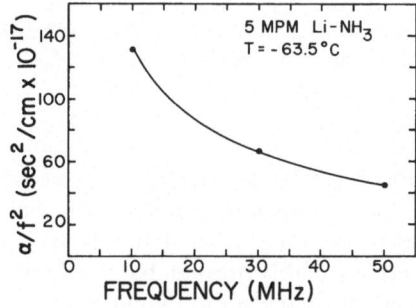

Fig. 8. Attenuation *versus* frequency for the lithium–ammonia solutions

Due to a lack of thermodynamic properties of the lithium–ammonia solutions it is difficult to evaluate the classical absorption as given by Eq.(1). However by making reasonable estimates we find that

$$\alpha_c/f^2 = 0.39 \times 10^{-17} \text{ sec}^2/\text{cm}$$

for a 5 MPM solution at −63.0° C. This should be compared with the values shown in Fig. 8. We see that the measured value is from about 2 to 3 orders of magnitude higher. Clearly we are looking at a combination of thermal relaxation processes (as for pure ammonia) and critical phenomena.

In conclusion we feel that these preliminary data indicate that considerable information as to the nature of metal–ammonia solutions can result from the further measurements which are presently in progress.

Acknowledgments

The author would like to acknowledge the assistance of Karl Marquardt whose expertise in making the many small parts made the performance of these experiments possible. Further assistance was provided by M. A. Priesand, S. Stanulonis, and L. B. Franceware.

References

1. Beyer, R. T., Letcher, S. V.: Physical ultrasonics. New York: Academic Press 1969.
2. Herzfeld, K. F., Litovitz, T. A.: Absorption and dispersion of ultrasonic waves. New York: Academic Press 1959.
3. Tisza, L.: Phys. Rev. **61**, 531 (1942).
4. Zwanzig, R.: Ann. Rev. Phys. Chem. **16**, 67 (1965).
5. Fixman, M.: J. Chem. Phys. **36**, 1961 (1962).
6. Cohen, M. H., Thompson, J. C.: Advan. Phys. **17**, 857 (1968).
7. Lepoutre, G., Sienko, M. J. (Eds.): Metal–ammonia solutions. New York: Benjamin 1964. Lagowski, J. J., Sienko, M. J. (Eds.): Metal–ammonia solutions. London: Butterworth 1970.
8. Chieu, P., Sienko, M. J.: J. Chem. Phys. **53**, 566 (1970).
9. Bowen, D. E.: In: Lagowski, J. J., Sienko, M. J. (Eds.): Metal–ammonia solutions, p. 355. London: Butterworth 1970.
10. This instrument is described in R. Truell, C. Elbaun, and B. B. Chick: Ultrasonic methods in solid state physics. New York: Academic Press 1969.
11. Cragoe, C. S., Harper, D. R.: Natl. Bur. Std. Sci. Papers **17**, 287 (1922).
12. Bowen, D. E., Thompson, J. C.: J. Chem. Eng. Data **13**, 207 (1968).
13. Hutchison, C. A., Jr., O'Reilley, D. E.: J. Chem. Phys. **52**, 4400 (1970).
14. Varlashkin, P. G., Thompson, J. C.: J. Chem. Eng. Data **8**, 526 (1963).
15. Overstreet, R., Giaque, W. F.: J. Am. Chem. Soc. **59**, 259 (1937).
16. Nasky, R. D.: Ph. D. Dissertation. The University of Texas at Austin, 1968.
17. Puls, M. P., Kirkaldy, J. S.: J. Chem. Phys. **54**, 4468 (1971).

Discussion

K. G. BREITSCHWERDT   Liquid ammonia, being an associated liquid, is definitely not a "Kneser liquid", i.e., the excess ultrasonic absorption above the classical value is not due to thermal relaxation, rather it should be ascribed to structural relaxation effects. You were probably misled by the high value of the ratio between experimental and classical absorption you obtained. Our measurements of this ratio gave a value of about three, which is close to what one finds for water. The frequency dependence of your absorption data indicates that there may be some kind of a low-frequency chemical relaxation.

D. E. BOWEN   We were certainly led to conclude that liquid ammonia was a Kneser liquid by the high value of the ratio. At this time we have no reason to doubt our value for the ratio although the identification of liquid ammonia as a Kneser liquid is probably not correct.

U. SCHINDEWOLF   I would like to stress what Breitschwerdt has said. The increase in sound absorption with decreasing frequency in the megacycle region is evidence for "chemical reactions" in the solutions, as Tamm and Eigen showed for ion pairing in aqueous solution. For future discussions it seems worthwhile for Bowen to extend his experiments to smaller concentrations, where ion pairing, "quadripole formation" and so on are assumed to occur. Such data might not only give independent experimental evidence for the different species discussed here but might also give the kinetics of their formation.

D. E. BOWEN   Such measurements are presently underway and we hope to report on them as soon as possible.

Note Added in Proof: A check of our calculations indicated an error in determining α_c/f^2. The correct value should be 2.27×10^{-17} np $\sec^2$/cm. The ratio α/α_c is then 1.74 instead of 13.02. Our identification of ammonia as a "Kneser" liquid is therefore incorrect.

The Electronic Structures of Disordered Materials

MORREL H. COHEN

Abstract

A brief review is given of current understanding of the electronic structures of disordered materials, with emphasis on work by the author and his associates. Formal arguments are omitted; simple physical arguments are used instead to build up a coherent conceptual structure. Disordered materials are briefly introduced followed by the notion of universal features in the electronic structures of classes of materials. These are summarized for crystals, and an elementary semiclassical argument is used to elicit them for disordered materials. The basic band model for disordered materials is introduced and used to explain the basic experimental facts about amorphous semiconductors. The history of localization theory, which gives the formal justification of the basic band model, is then reviewed and its present status assessed. Eggarter's theory of electron mobility of gaseous He is reviewed. Some recent results on localization in one dimension are presented and discussed. Localized states found by Eggarter and Kirkpatrick for lattices with vacancies are also discussed. Both the one-dimensional and the vacancy results shed light on the Cohen-Sak theory of the metal–semiconductor transition in disordered materials. Mott hopping between localized states in disordered materials is next reviewed, with application to the properties of homogeneously stacked organic or metallo–organic salts. Finally, remarks are made on the relevance of these considerations for the theory of metal-ammonia solutions.

I. Introduction

I have been asked to review the theory of electrons in disordered materials as my contribution to Colloque Weyl III, in which the original focus on metal–NH_3 solutions has been broadened to include electrons in fluids, generally. I do so here, omitting formal theoretical arguments and using simple physical arguments instead to build up a coherent conceptual structure. I have already published a number of such reviews [1—4], and it is impossible for me to do so again without repeating myself. In mitigation, the audience reached by the Proceedings of the Colloque Weyl is substantially different from previous ones, certain basic points have not yet been widely grasped even by experts in the field, and illustrations have been taken from as yet unreviewed and even unpublished work of my colleagues and myself.

II. Disordered Materials

In a perfect elemental crystal, all atoms are identical and arranged periodically in space. This periodic arrangement constitutes perfect, long-range translational order. In a perfect crystalline compound, each element has its own set of periodically arranged sites. There is long-range compositional, as well as translational, order. Finally, in magnetic materials some or all of the constituents have spins, and these are periodically oriented in perfect magnetic crystals. In summary, perfect crystals possess long-range order of translational, compositional and/or magnetic character.

Disorder can be defined as the absence of long-range order. It does not preclude the existence of short-range or local order. There are thus three kinds of disordered materials. Those with translational disorder only are, e.g., elemental gases, liquids and glasses, the latter two being two aspects of a condensed amorphous phase. Those with compositional disorder only are substitutional crystalline alloys. Those with magnetic disorder only are, e.g., ferromagnetic crystals above their Curie temperature. Two or even all three kinds of disorder can be present simultaneously as in liquid alloys. It should also be recognized that the introduction of compositional or magnetic disorder simply involves the breaking of translational order without the randomizing of atomic sites.

The well-known Bloch theorem holds for perfect crystals, which permits one to infer the existence of universal features in their electronic structures. The Bloch theorem follows directly from the existence of complete translational order, a universal structural feature of all perfect crystals. Are there also universal structural features of disordered materials which lead to universal features in their electronic structures? The answer to this question by argument and example is the theme of this review, as it has been of my earlier ones.

III. The Universal Features of Perfect Crystals

We consider the motion of electrons in perfect crystals in the single-particle picture; the explicit inclusion of electron-electron interactions does not change our conclusions in any essential way. The electronic motion is then governed by a single particle Hamiltonian

$$\mathcal{H} = \frac{p^2}{2m} + V(r) \tag{3.1}$$

where $V(r)$, the crystal potential, has the periodicity of the atomic arrangement. That feature of V alone permits one to prove the Bloch theorem

$$\mathcal{H} \, \Psi_{kn}(r) = E_n(k) \, \Psi_{kn}(r), \tag{3.2a}$$

$$\Psi_{kn}(r) = e^{ik \cdot r} \, U_{kn}(r). \tag{3.2b}$$

The wavefunction $\Psi_{kn}(r)$ has two factors, a phase factor $e^{ik \cdot r}$ and an amplitude factor $U_{kn}(r)$. The latter has the periodicity of the crystal. Thus there is long-range order both in the phase and amplitude of $\Psi_{kn}(r)$. The periodicity of $U_{kn}(r)$ implies that $\Psi_{kn}(r)$ is extended, i.e. extends everywhere within the material. The form (3.2b) together with the periodicity of $U_{kn}(r)$ is universal among crystals, as are its implications of long-range order in phase and amplitude and the existence of extended states only.

The one-electron energies $E_n(k)$ are functions of wave-vector k with values falling into allowed bands designated by the index n and separated by forbidden gaps. There are sharp band edges at which the density-of-states shows square root singularities in its dependence on energy. Within the bands there are also singularities in the density-of-states corresponding to various kinds of critical points in the dependence of E on $\boldsymbol{k}$. These are all universal features, also following directly from the translational symmetry of the crystal.

This translational symmetry is lost in disordered materials. Accordingly, we can expect all those features of crystals dependent specifically upon it to disappear as well. The long-range order of the phase of the wavefunction will presumably be replaced by short-range order; that is, there will be a finite phase-coherence length, essentially the mean free path. Similarly, the long-range order in the amplitude can disappear. One can expect both extended and localized states. Finally, the square root singularities in the density-of-states will disappear.

The breakdown of translational symmetry implies the existence of local variations in the configuration of the material. That is, fluctuations, microscopic or submacroscopic, are universal features of the structures of disordered materials. We must anticipate that these will play an essential role in establishing the universal features of the electronic structures of disordered materials.

IV. Semiclassical Motion of an Electron in a Random Potential

If disordered materials have universal features, these may be inferred from the study of very simple model systems. All disordered materials contain fluctuating atomic configurations. The potential $V(\boldsymbol{r})$ in $\mathscr{H}$ for a disordered material therefore contains at least a fluctuating part. One can deal with $V(\boldsymbol{r})$ itself, or an optical potential, pseudopotential, or some other model potential constructed so as to smooth out uninteresting short-range variations of $V(\boldsymbol{r})$. Now we shall be interested in such questions as whether there are localized states or extended states for an electron of given energy and whether there are energies of transition between localized and extended states. For such questions, the long-wave length fluctuations of the model potential play the essential role. Therefore, we can expect semiclassical model systems to exhibit the same features as fully quantum-mechanical ones. Accordingly we study the semiclassical motion of a particle in a random potential in the present section, a problem first studied by Ziman [5] in the present context.

Consider a potential V

$$V = V(\boldsymbol{r}) > 0 \tag{4.1}$$

which is everywhere positive. It is a random potential in the sense that its correlation function

$$C(\boldsymbol{R}) = \frac{1}{\Omega} \int d\boldsymbol{r} \, V(\boldsymbol{r}) \, V(\boldsymbol{r}+\boldsymbol{R}) - \frac{1}{\Omega^2} \int d\boldsymbol{r}' d\boldsymbol{r} \, V(\boldsymbol{r}) \, V(\boldsymbol{r}'+\boldsymbol{R}) \tag{4.2}$$

is a short-ranged function of the separation R so that values of the potential at sufficiently well separated points are uncorrelated. For simplicity, we shall set the range of C equal to zero. No essential change in the conclusions occurs when the range is sufficiently extended for the semiclassical approximation to be valid. The probability distribution of values of V throughout the entire volume Ω is $P(V)$. We shall suppose there to be no values of V below V_1 or above V_2,

$$P(V) = 0, \quad V \langle V_1, V \rangle V_2. \tag{4.3}$$

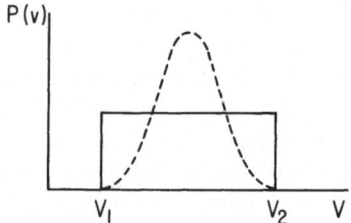

Fig. 1. Sketch of probability distribution $P(V)$ of values of potential V in a random semi-classical system for two cases: — rectangular, --- bell shaped

Otherwise $P(V)$ can have any reasonable dependence on V within V_1 and V_2. Two possibilities are sketched in Fig. 1.

The density-of-states is easy to calculate. The contribution to the density-of-states per unit volume for a region of potential V is

$$\propto (E - V)^{1/2}, \quad E > V \atop 0 \qquad , \quad E < V \Big\}. \tag{4.4}$$

Now $P(V)\,dV$ is just the volume fraction (i.e. fraction of Ω) having V between V and $V + dV$. Thus summing contributions like (4.4) from the entire material gives for the density-of-states

$$n(E) = \alpha \int_{V_1}^{E} P(V)\,(E - V)^{1/2}\,dV. \tag{4.5}$$

For a $P(V)$ like those in Fig. 1, it is easy to extract the salient features of $n(E)$ from (4.5). These are shown in Fig. 2. One notes that there is a band edge at V where if $P(V) \propto (V - V_1)^S$ for V V_1, then $n(E) \propto (E + V_1)^{S + 3/2}$, for $E \gtrsim V_1$. Above V_1 there is a tail and above that the main body of the density-of-states appears as a shifted free-particle density-of-states. For $E \gg V_2$, the shift is simply

$$\bar{V} = \int_{V_1}^{V_2} P(V)\,V\,dV \Bigg\} \tag{4.6a}$$

$$n(E) = \propto (E - \bar{V})^{1/2} \Bigg\}. \tag{4.6b}$$

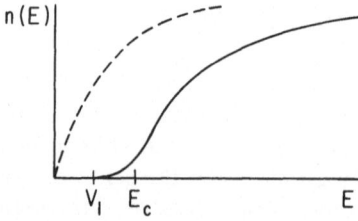

Fig. 2. Sketch of density-of-states $n(E)$ following from Eq. (4.5) for a $P(V)$ like that of Fig. 1 (solid curve) together with free-particle density-of-states (dashed curve) for comparison. E_c is the critical energy for percolation

Of general significance in the above are (1) the band edge at V_1 without a square root singularity in $n(E)$, (2) a tail on the density of states above V_1, and (3) a shift of all energies by $\bar{V}$ for $E \gg V_2$.

What are the characteristic states of motion in these various regions of the density-of-states? To answer this question, we must first recognize with Ziman [5] that we are dealing with a percolation problem. A point $\boldsymbol{r}$ is prohibited (P) to an electron of energy E if the potential there exceeds E; $\boldsymbol{r}$ is allowed (A) if $V(\boldsymbol{r}) < E$:

$$\left.\begin{array}{llllll} \boldsymbol{r} & P & \text{at} & E & \text{if} & V(\boldsymbol{r}) > E \\ \boldsymbol{r} & A & \text{at} & E & \text{if} & V(\boldsymbol{r}) < E \end{array}\right\}. \tag{4.7}$$

Now define $C(E)$ as the volume fraction of the material which is allowed to an electron of energy E. It follows from (4.7) and the definition of $P(V)$ that

$$C(E) = \int_{V_1}^{E} P(V) \, dV. \tag{4.8}$$

The corresponding dependence of $C(E)$ on E is sketched in Fig. 3. The essential features are that $C(E)$ remains zero until V_1 when it increases monotonically to unity at V_2 and remains at unity above V_2.

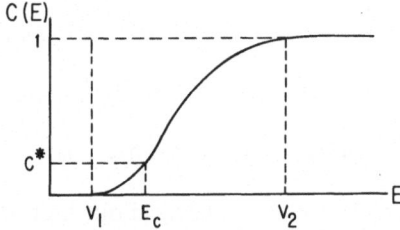

Fig. 3. Sketch of the allowed volume fraction $C(E)$ as a function of energy E for a $P(V)$ like those of Fig. 1. E_c is the critical energy for percolation when C^* reaches a value of $\simeq 0.2$

We have sketched out in Fig. 4 typical prohibited and allowed regions of space in two dimensions for various ranges of E, making use of the results of percolation theory and various numerical studies of related problems. For $E < V_1$ (Fig. 4a) all of the material is prohibited. The density-of-states vanishes. As E increases above V_1 (Fig. 4b), allowed regions appear localized about the deepest minima in $V(\boldsymbol{r})$. As E increases further, distinct allowed regions merge at saddle points in $V(\boldsymbol{r})$ until, at a critical energy E_c, a multiconnected channel emerges which, for higher energies (Fig. 4c), spans the entire material. The energy E_c is determined by the requirement that the volume fraction allowed reach the percolation threshold,

$$C(E_c) = C^*. \tag{4.9}$$

This particular percolation problem has not been worked out numerically. My best guess based on extrapolation from studies of percolation on lattices [6] is that (in three dimensions)

$$C^* \simeq 0.2. \tag{4.10}$$

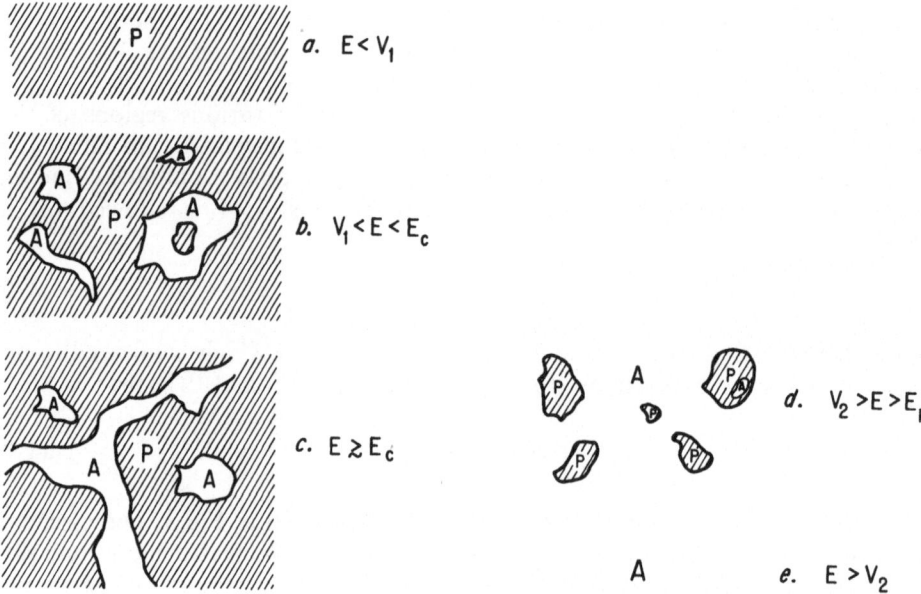

Fig. 4a–e. Sketch of allowed and prohibited regions for various ranges of energy E. See text for discussion

As the energy increases further it reaches a value E_p complementary to E_c,

$$C(E_p) = 1 - C^* \simeq 0.8 , \qquad (4.11)$$

when all prohibited regions become localized about high maxima $V(r)$ (Fig. 4d). Finally when E exceeds V_2, all of the material is allowed (Fig. 4e).

Table I summarizes the characteristics of the states of electronic motion associated with these various fluctuations in the potential for the corresponding energy ranges. Certain points are worth emphasizing. The extended states which appear at E_c are quite different from those we are familiar with in the Bloch theory of perfect or nearly perfect crystals. They do not extend with equal

Table I. Energy dependence of state characteristics for electrons in a random potential

Energy	Characteristic	
	Semiclassical	Quantum
$E < V_1$	No states	No states
$V_1 < E < E_c$	All states localized	All states localized
$E = E_c$	Extended (channel) states appear	Extended (channel) states appear
$E \gtrsim E_c$	Channel states + localized states	Channel states + channel resonances
$E = E_p$	Prohibited regions become localized	Extended states become simple
$E > V_2$	No prohibited regions	

amplitude everywhere in the material, but are confined to the percolation channels. We can define a modified percolation probability $C_P(E)$ as that part of the allowed volume fraction which lies on a percolation channel:

$$C(E) = C_L(E) + C_P(E), \qquad (4.12)$$

where $C_L(E)$ is that part of the allowed volume fraction comprised of localized regions. Percolation theory permits us to sketch $C_P(E)$ and $C_L(E)$ as in Fig. 5. Near E_c, $C_P(E)$ behaves as

$$C_P(E) \propto (E - E_c)^s, \qquad (4.13)$$

where $s < 1$ and s is approximately 0.6 for three dimensions. It is only when E reaches E_P that the last vestiges of channel-like behavior disappear and the states become entirely like the familiar extended states.

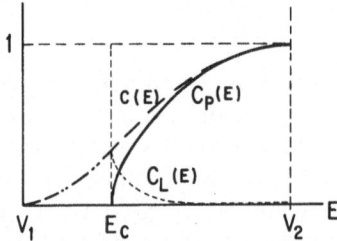

Fig. 5. Sketch of energy dependence of volume fraction of localized states $C_L(E)$, volume fraction of extended states $C_p(E)$, and allowed volume fraction $C(E) = C_L(E) + C_p(E)$

We are of course not interested in strictly semiclassical systems and must know how quantum effects modify the results of Table I. There are three quantum-mechanical effects to be taken into account:
1. zero point kinetic energy in the allowed regions;
2. spreading of the wavefunctions into the prohibited regions;
3. tunneling from allowed region to allowed region.
Effect 1. shifts the entire density-of-states and $C(E)$ curves upwards in energy. Effect 2. on the other hand tends to shift $C(E)$, in particular, downwards. These two effects tend to cancel, and only give quantitative corrections to the semi-classical results in any event. What is important is effect 3. because tunneling from one allowed region to another could destroy localization. The tunneling can be between two localized allowed regions, one localized and a portion of an extended region, or two portions of an extended region, as in Fig. 6a, b, and c, respectively. In the tunneling Hamiltonian formulation, tunneling between the two localized allowed regions of Fig. 6a is expressed as a transition between those quantum states of one region obtaining in the absence of the second region and the corresponding states for the second region in the absence of the first. Since the size, shape, and internal potential of the two regions are different, so are the two energy spectra. The tunneling is therefore nonresonant, with associated loss of amplitude. For energies below E_c, therefore, all states remain localized primarily in one allowed region (or in a small number of neighboring allowed

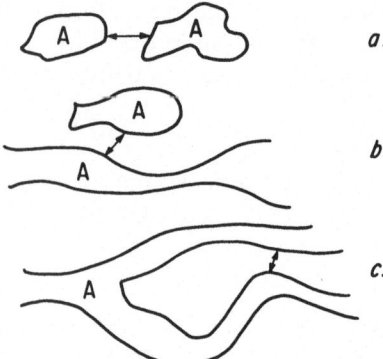

Fig. 6. Tunneling between semiclassical allowed regions: a Localized to localized; b localized to extended; c extended to extended

regions) and decay exponentially in amplitude from allowed region to allowed region outside the primary region or regions of localization. The situation is quite different for Fig. 6b. The extended states in the percolation channel are densely distributed in energy. Resonant tunneling can occur without loss of amplitude from the localized allowed region to the channel. As a result the localized states cease to exist as such above E_c, and become channel resonances instead, states with substantial amplitude in a localized region or regions and resonant tails in the percolation channel. The fraction of the density-of-states consisting of channel resonances is given very approximately by $C_L(E)$ and that consisting of the semiclassical channel states by $C_P(E)$. Both states are of course extended. Thus there is a sharp transition from localized states only at E_c. The third case, Fig. 6c, is uninteresting.

We summarize our most important results for the quantum-mechanical case also in Table I. We note that no states appear below V_1 also in the quantum case; V_1 is the Lifschitz limit of the spectrum [13].

It is important to remember that while the classical arguments given here are exact, the quantum arguments are not. Although we believe them to be correct and to be supported both by experiment insofar as we understand the experimental facts and by the formal theory, there is always the possibility of error in physical arguments of the kind used here. Accordingly, our proposals and interpretations in the sections to come must be viewed with final judgment reserved.

V. Basic Band Model

If we assume universality, the conclusions of the previous section lead us to a model in which fluctuations in disordered materials always lead to a tail of localized states at a band edge separated from extended states in the interior of the band by a critical energy E_c, as in Fig. 7. Inside E_c the extended states are channel-like or are local resonances with channel-like states. Further inside, the channel-like character disappears as to the resonances.

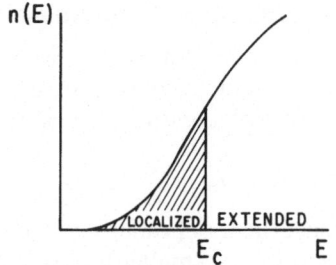

Fig. 7. Density-of-states near the band edge in the band model, showing a tail of localized states and a transition to extended states at E_c

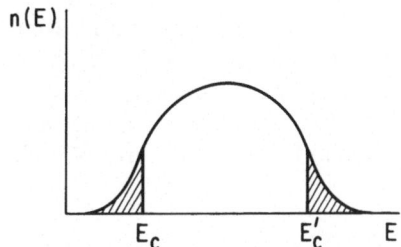

Fig. 8. Density-of-states for the basic band model (case of a single isolated tight-binding band) of a disordered material. Regions of localized states are shaded

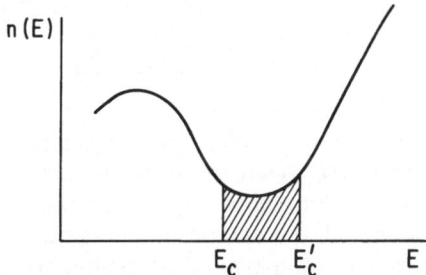

Fig. 9. Density-of-states showing localized states (shaded) entirely within a band. Mott's pseudogap lies between E_c and $E_{c'}$.

Thus, an isolated tight-binding band would have the form shown in Fig. 8 in the basic band model of a disordered material. More generally a region of localized states can occur entirely within a band, giving rise to Mott's pseudogap [9], as in Fig. 9.

In summary, disorder produces tails of localized states, as has been well known for a quarter of a century. Localization of all states in the band occurs if the disorder is strong enough, as shown by Anderson in 1958 [7, 8]. Otherwise, there is a transition from localized to extended states at a critical energy E_c, as realized by Mott in 1967 [9]. The energy E_c is in fact a mobility edge, as recognized by Cohen, Fritzsche, and Ovshinsky [10] in 1969 and discussed further below.

VI. Amorphous Semiconductors: Basic Experimental Facts

The first application of the concept developed thus far will be to the explanation of the following basic experimental facts concerning amorphous and also liquid semiconductors [11]. The electrical conductivity of these materials is of the intrinsic form,

$$\sigma = \sigma_0\, e^{-\Delta E/kT}, \tag{6.1}$$

even for alloys containing constituents of differing valences over wide ranges of composition. The pre-exponential σ_0 is smaller for amorphous materials than for related crystalline ones. The activation energies, on the other hand, are comparable.

The existence of an activation energy suggests the presence of an energy gap, and this is confirmed qualitatively by the optical absorption.

The thermoelectric power is normal in magnitude and usually positive. The Hall constant is small in magnitude and usually negative. The picture which is emerging from studies of the transport properties is that these materials behave as though they were low-mobility intrinsic semiconductors. There is, in addition, evidence of hopping conduction at low temperatures of high frequencies for some of the materials.

VII. Amorphous Semiconductors: Application of Basic Band Model

The simplest model of an amorphous or disordered semiconductor is the Wilson two-band model using bands of the form proposed in V, as shown in Fig. 10.

The Kubo-Peierls-Greenwood [12] formula becomes, when specialized to semiconductors

$$\sigma = \sum_{b=c,\,v} \int dE\, n_b(E)\, e\, \mu_b(E)\, f_b(E). \tag{7.1}$$

Here b is a band index, v for valence or c for conduction, $n_b(E)$ is the density-of-states of either band, $\mu_b(E)$ is the mean mobility of carriers of energy E and $f_b(E)$ is the occupation number for electrons in the conduction band or holes in the valence band. For a crystal, the activation energy arises upon integration of (7.1) directly from the existence of sharp band edges in $n_b(E)$. For the amorphous materials, sharp band edges do not exist in the sense that square root singularities in $n(E)$ occur. There can be well-defined spectral limits, the Lifschitz limits [13],

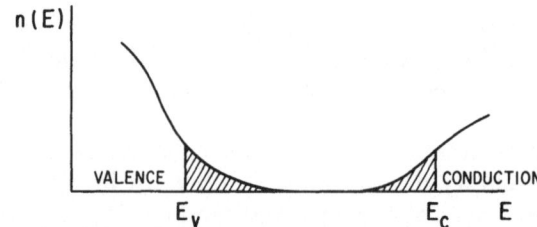

Fig. 10. Simplest model of a disordered semiconductor: a filled valence band with a tail of localized states (shaded) above E_v and an empty conduction band with a tail of localized states below E_c

but the density-of-states is small near there and in general goes to zero more rapidly than the square root of the energy difference. Under such conditions, the energy regions most important in (7.1) are well above the Lifschitz limits, and our considerations thus far are insufficient to explain how a well-defined activation energy can appear in σ in amorphous semiconductors.

The missing element was supplied by Cohen, Fritzsche, and Ovhinsky [10]. Anderson's theory [7] implies that $\mu_b(E)$ is identically zero at $T=0$ for localized states. At finite temperatures, conduction involving localized states can occur via phonon-assisted hopping, but $\mu_b(E)$ remains small for $E_v < E < E_c$, as in Fig. 11.

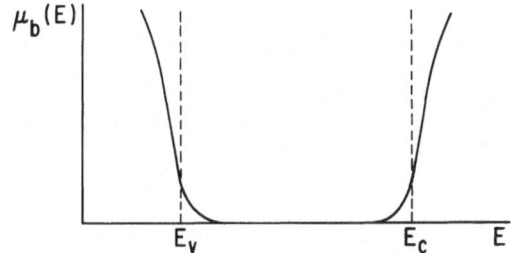

Fig. 11. Energy dependence of the mobility $\mu_B(E)$ in the Mott-CFO model of a disordered semiconductor

They suggested that E_v and E_c act as mobility edges, and the energy region E_v to E_c is a mobility gap. Integration of (7.1) would then yield an activation energy for σ which is either $E_c - E_F$ or $E_F - E_v$, depending on circumstances.

The carriers important in (7.1) are within a few kT of the $T=0$ mobility edges, depending on how rapidly μ increases with energy. Thus the important carriers can be expected to occupy channel states unless (1) these occur within a range of energy smaller than kT or (2) the mobility is so much greater at higher energies that the maximum in the integrand of (7.1) occurs outside the energy range of the channel states. For channel states, the mobility will have the general form

$$\mu_b(E) = g_b(E)\, h_b(E)\,, \quad E > E_c \quad \text{or} \quad E < E_v \tag{7.2}$$

where $g(E)$ is a slowly varying factor describing the motion of the carrier within the channel, and $h(E)$ is a geometric factor relating to the overall conduction of electricity through the channel. The latter can be roughly expressed in terms of the effective electrical length L_{eff} and cross-sectional area A_{eff} of the channel or channels accessible to carriers of energy E:

$$h(E) = \frac{A_{\mathrm{eff}}}{A}\, \frac{L}{L_{\mathrm{eff}}} \tag{7.3}$$

where A is the cross section of the specimen perpendicular to the deviation of mean current flow and L is its length in the direction of mean current flow (Fig. 12). Because cul de sacs do not contribute effectively to current flow (Fig. 12),

$$\frac{A_{\mathrm{eff}}}{A}\, \frac{L_{\mathrm{eff}}}{L} < C_p(E)\,; \tag{7.4}$$

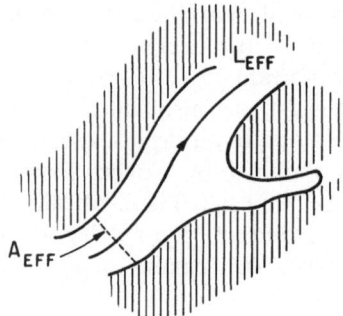

Fig. 12. Sketch indicating current flow in a channel. The shaded area is prohibited. Current flow is indicated by an arrow. The effective path length L_{eff} would be measured along the direction of current flow for a singly connected channel. The effective cross-sectional area A_{eff} is measured perpendicular to the direction of current flow and is an appropriate average. The cul de sac is not included in L_{eff} or A_{eff}

that is, the fraction of the volume effective in conduction is less than the percolation probability. Moreover, the percolation channel is tortuous near E_c,

$$L_{eff} > L. \tag{7.5}$$

Inserting (7.4) and (7.5) into (7.3), one arrives at the result that

$$h(E) < C_p(E). \tag{7.6}$$

This is consistent with numerical calculations by Kirkpatrick of the resistance of a simple cubic network of resistors with a fraction x missing [equivalent to $C(E)$ in our case]. Taking over his results, we propose the curves shown for $h(E)$ and $C_p(E)$ near E_c at $T = 0$ in Fig. 13. The concavity of $h(E)$ suggests that the activation energy ΔE of σ relates to an energy E_M^c or E_M^v at which the integrand of (7.1) has its maximum, $\Delta E = E_M^c - E_F$ or $E_F - E_M^v$, which may lie considerably inside E_c or E_v.

Depending on whether electronic motion within the channel is by propagation or diffusion, g is given by

$$g(E) \simeq \frac{4}{3} \frac{e\,\tau(E)}{m^*} \quad \text{or} \quad \frac{eD}{kT} \tag{7.7}$$

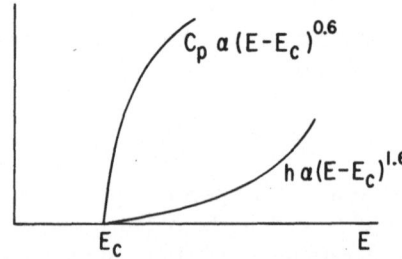

Fig. 13. Proposals for the energy dependence of the percolation probability, $C_p(E)$, and the geometric factor in the mobility, $h(E)$, based on Kirkpatrick's resistance network calculations

respectively, with

$$D \simeq \frac{1}{6} \frac{B}{h} a^2, \tag{7.8}$$

where B is the band width of the equivalent crystal (same short-range order) and a is an interatomic separation. A more accurate treatment of the diffusion regime has been given by Friedman [14].

This idea of a gradual transition in $\mu_b(E)$ to zero at E_b for $T = 0$ is disputed by Mott [15], who proposes instead that μ_b, in effect, is given by g_b alone and jumps to zero at E_b from eD_b/kT with D_b given by (7.8). This divergence in views of the mobility arises from a corresponding divergence in views on the nature of localization [16—18]. If we define $\mathscr{L}(E)$ as a mean range of localization of states of energy E, then $\mathscr{L}(E)$ goes to ∞ as $|E - E_b|^{-0.6}$ as E approaches an E_b from outside. Mott, along with others, takes this to mean that the wavefunction uniformly expands to cover the entire material as E_b is approached. Percolation considerations are ignored, and h is set equal to one. The semiclassical arguments of Section IV prove quite rigorously that this cannot be true in general. Thus, while $h_b(E)$ need not be given accurately by considerations of classical current flow through an inhomogeneous medium such as we have used to argue for the proposal of Fig. 13, it cannot be ignored. Recent work by Jortner and myself [19] on expanded liquid Hg and on liquid Te, moreover, indicates that the inhomogeneous medium arguments may in fact be quantitatively accurate for some systems.

A more detailed band model is now available for the tetrahedrally-bonded elemental amorphous semiconductors. The model, developed by Weaire and collaborators [20], contains topological disorder only and not quantitative disorder in bond distances and angles. Its simplified tight-binding form gives a poor representation of the crystal, but has interesting features in the amorphous case. An exact analysis shows that it has well-defined band edges with square root singularities in the density-of-states. The well-defined band edges coincident with those of the crystal are nothing more than a further example of Lifschitz limits already familiar in the study of binary alloys and other systems. The square root singularities are surprising and, at present, unique to this model among disordered systems. They would probably be eliminated in any more realistic band model, even with only topological disorder. However, what those findings do suggest and what appears to be confirmed experimentally is that well-annealed films of amorphous Si and Ge should have mobility edges close to the energies where the density-of-states drops precipitously.

VIII. Localization Theory

That considerations of the data and physical argumentation lead Mott and myself to quite different conclusions about the nature of the wavefunctions and of the mobility indicates the need for an underlying body of rigorous formal theory of the universal features of the electronic structures of disordered materials.

The concept of tails of localized states in disordered materials is well established. Refs. [9, 10] and [21—25] are representative either of the earliest or the most significant work in that regard.

The concept of bands of extended states with a finite range of phase coherence (mean free path) has been understood since Bloch.

Several years ago, the two principal problems were the demonstration of a sharp transition from localized to extended states at E_c and the demonstration that E_c is a mobility edge. The resolution of both rests on a fundamental paper (1958) of Anderson [7]. In it, he proved, but not with complete rigor, that all states in a simple tight-binding band are localized and that no diffusion occurs [$\mu(E)=0$ all E] when the randomness is sufficiently great. Mott in 1967 recognized that Anderson's work implied sharp energies of transition E_c for lesser degrees of randomness, and he synthesized the tails plus extended states into the basic band model [9, 10].

Anderson's 1958 paper [7] was difficult and a bit messy. Ziman in 1969 gave it a clear presentation [8c] and Thouless in 1970 gave it greater mathematical precision [8a]. Economou and Cohen [8b] improved the accuracy of the arguments and addressed the question of the mobility edges directly. While a completely rigorous location of the positions of the mobility edges is still lacking, there is little doubt left of their existence.

There is now grown up a subfield called localization theory within which studies of localization functions, positions of mobility edges, localization length, etc. are carried out. Unfortunately, accurate calculations of transport properties near mobility edges, or in energy regions which would become mobility gaps with greater randomness, have not yet been attained. Such calculations or their equivalent are required to settle the questions raised in the previous sections concerning percolation and the mobility.

IX. Mobility of an Electron in Gaseous He

One of the most interesting experiments on electronic motion in disordered materials is the measurement of the mobility of an excess electron in He^4 vapor as a function of He atom density and temperature by Sanders and Levine [26a]. Similar measurements were done at higher temperatures by Harrison [26b]. The results all have the form shown in Fig. 14. The electron mobility μ drops by about

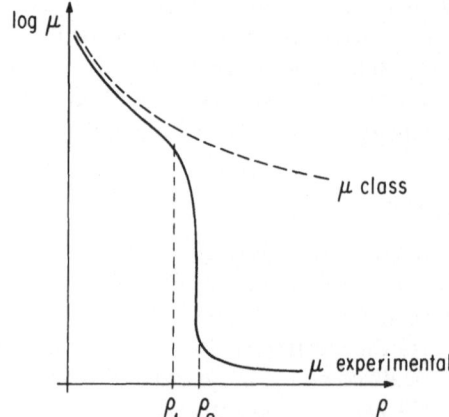

Fig. 14. Typical constant-temperature mobility curve for an excess electron in gaseous He. The dotted line is the prediction of classical kinetic theory (after Eggarter) [28]

six orders of magnitude below the semiclassical value in the density range 10^{20}—10^{21} He atoms/cc for $T \simeq 4\,°K$.

Eggarter [27, 28] has given a simple theory of electron states and transport for this density range which agrees within about 30% with the data except at the lower densities, where the discrepancy can rise to a factor of 2 (Fig. 15). The electron-He atom scattering length a is 0.62 Å so that $\varrho\, a^3 \ll 1$ throughout the transition range. Nevertheless, the electron energies of interest are sufficiently low that the electron interacts simultaneously with many atoms. As a consequence, it may be regarded as moving semiclassically in a continuous effective potential approximately given by

$$\tilde{V}(\boldsymbol{r}) = V_{\mathrm{WS}}(\tilde{\varrho}(\boldsymbol{r})) \tag{9.1}$$

where V_{WS} is the Wigner-Seitz potential and $\tilde{\varrho}(\boldsymbol{r})$ is a locally averaged density. Eggarter used the simple prescription of averaging ϱ over a cube of edge

$$L = c\, 2\pi (6\,m/\hbar^2\, E)^{1/2} \tag{9.2}$$

for an electron of energy E. The scaling parameter c was expected to be of order unity and was found to range from 1.02 to 1.4 depending upon temperature after

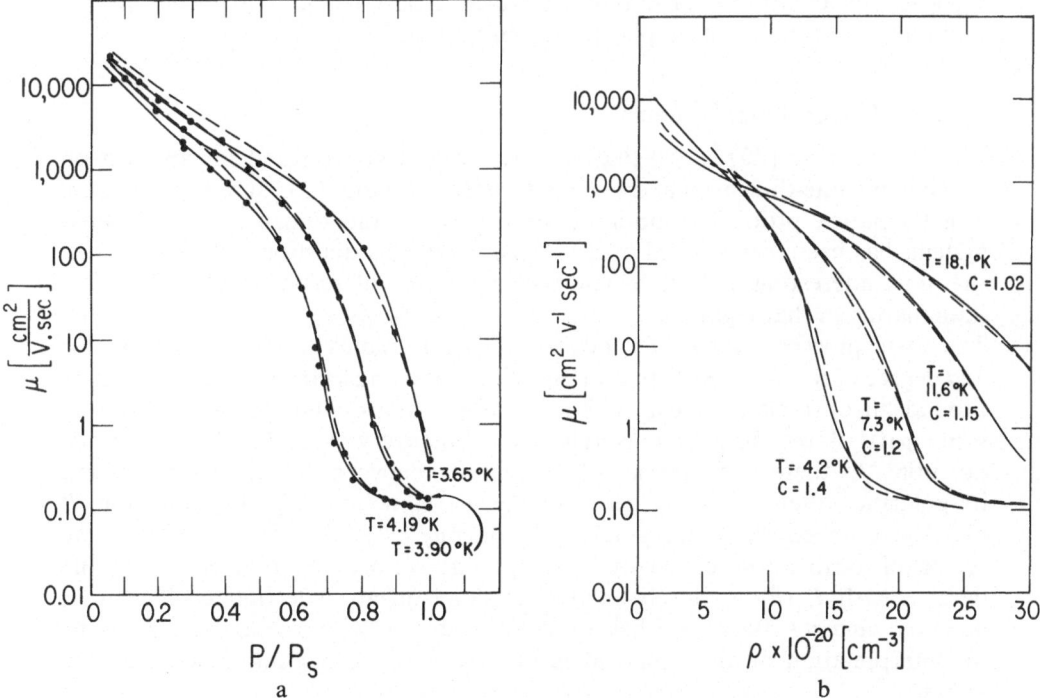

Fig. 15. Mobility μ of excess electrons on gaseous He; comparison between experiment and Eggarter's theory (after Eggarter [28]). a The dots are the measurements of Levine and Sanders; the dotted lines are Eggarter's calculations. The full lines have no theoretical significance. P is the experimental pressure and P_s the vapor pressure. b The full lines are Harrison's measurements; the dotted lines are Hernandez' calculations. Eggarter's scaling parameter c for the sampling length had to be changed slightly with temperature to maintain the fit. ϱ is the He number density

fitting the experimental mobilities to Eggarter's theory, Fig. 15. Since we are dealing with a gas, $\tilde{\varrho}(r)$ fluctuates randomly and non-negligibly from point to point. $\tilde{V}(r)$ is thus a rondom potential like that discussed in IV.

There is a tail of localized states associated with regions of low He atom density below a mobility edge E_c. The density of states is given by (4.5), and the mobility by (7.1). As the density increases, so does the potential V and the width of the tail. The function $n(E) e^{-\beta E}$ has one maximum, where it is rather sharply peaked. At low density, $\varrho < \varrho_1$ on Fig. (14), this maximum occurs well above E_c. The electron spends most of its time in extended states for which $C_p(E) \cong C(E)$, and the classical Langevin mobility is approached. For $\varrho_1 < \varrho < \varrho_2$, the peak in $n(E) e^{-\beta E}$ moves from the right to the left of E_c. The mobility decreases drastically as the electron spends more and more of its time trapped in localized states. For $\varrho > \varrho_2$, the electron spends practically all of its time localized in regions of lower than average density. Because the material is a gas, however, the trapped electrons are free to move together with the density fluctuations, and the mobility saturated at a value characteristic of these pseudobubbles.

Kirkpatrick [29] has criticized the detailed behavior of $\mu(E)$ used by Eggarter near E_c, $\mu(E) \propto C_p(E)$ instead of $(E - E_c)^{1.6}$ as described in Section IV. In the present case, however, the region near E_v does not play a significant role. Since Eggarter's calculation is prototypic, however, it is important to recognize this limitation in extending its application as in Section XII.

X. One-Dimensional Systems

Mott and Twose [30] stated that all states would be localized for an electron moving in a one-dimensional disordered system. Borland [31] subsequently gave a mathematical proof. The motion of an electron in one-dimensional disordered systems has been well studied over the years, but this question of localization is the most interesting and we review here a number of recent results which are illuminating in that regard.

The basic physical reason for localization of all states in one dimension for any degree of randomness is that in one dimension an electron cannot propagate around an obstacle as it can in 2 and 3 dimensions. Thus in one dimension, some part of the amplitude is reflected from every irregular change in the potential. As there is no special phase relation between reflections at different irregularities, there is steady decay of the wavefunction. The wavefunction Ψ must have an envelope decaying asymptotically as $e^{-|x|/L_d(E)}$ outside of some region of localization, of length $L_e(E)$. The above argument suggests that the decay length $L_d(E)$ is closely related to some mean reflection coefficient at potential changes. Bush [32] has applied this notion with substantial success to the interpretation of his numerical results for $L_d(E)$ of one-dimensional binary alloys.

Economou and Cohen [8b] have shown that the localization theory introduced by Anderson provides a convenient formal basis for making general and numerical arguments. Economou [32], Economou and Papatriantafillou [33, 34], and have proved that $L_d(E)$ is sharply distributed whereas $L_e(E)$ is not. They have obtained integral equations for these or related quantities, and Papatriantafillou, in particular, has found several results of considerable interest. Some results of Economou and Papatriantafillou [33] for a one-dimensional tight-binding disordered lattice are shown in Fig. 16. In addition to peaks in the density of states above the main body of the band which are due to states localized on B clusters, there are distinct correlated anomalies in $n(E)$ and $L_d(E)$ at several points within the main body of the band, notably at an energy $|V|$ below the middle of the pure A band, where V is the n-matrix element of $\mathscr{H}$, Fig. 16a. One notes that increases in $L_d(E)$ correspond to increases in $n(E)$. One can understand this readily in terms of the semiclassical considerations of IV. The larger $L_d(E)$, the larger the length of the system accessible to an electron of energy E. The density of states, however, is directly proportional to the fraction of the total length that is available to the electron semiclassically. Thus an increase in $L_d(E)$ implies an increase of $n(E)$. This is a result which will assume further importance in later sections as we build up a case for its generality and for

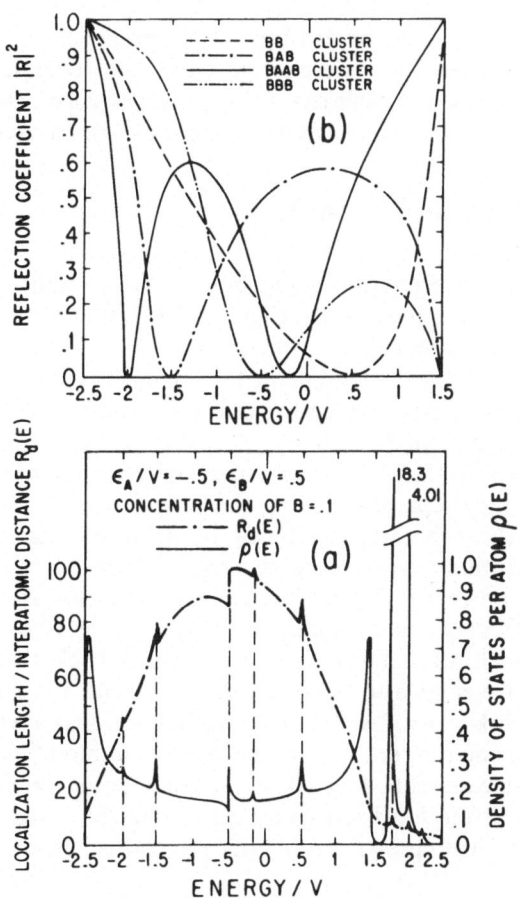

Fig. 16. a Average density-of-states and localization length for a tight-binding one-dimensional random alloy $A_{1-x}B_x$. b Reflection coefficients from different clusters embedded in an environment of A atoms

its importance in metal-semiconductor transitions. The behavior of $L_d(E)$ can in turn be readily understood in terms of the energy dependence of the reflection coefficient for various classes of clusters.

Papatriantafillou [34] introduced ordering of the ABAB type into an $A_{55}B_{45}$ alloy. With increasing order the jump in density of states at $E = -|V|$ turns into a dip and becomes the band edge of the perfect AB crystal with the excess A atoms giving rise to states in the gap. Thus the anomaly at $E = -|V|$ is the residuum in the disordered alloys of the band edge in the 50—50 ordered alloy and is associated with special clusters displaying that order. This result is also very important for the theory of metal-semiconductor transitions in disordered materials.

Finally, we mention Papatriantafillou's [34] correlation between $L_d(E)$ and $L_e(E)$ for binary alloys. He finds that $\langle L_e \rangle$ (an ensemble average must be taken because L_e is not sharply distributed) is given by

$$\langle L_e(E) \rangle \cong 2.3 \, L_e(E) \tag{10.1}$$

until L_d becomes too small. For example, L_e approaches the mean separation between impurities when the difference in single site energies becomes much larger than the band width, whereas L_d approaches zero. This is a result we shall need in our discussion of transport in one-dimensional materials.

XI. Randomly Distributed Vacancies

Consider an elemental crystal in which a fraction x of the atoms are missing. Suppose that we are dealing with a single, nondegenerate tight-binding band with nearest neighbor matrix elements only. Eggarter [35] showed that isolated clusters of vacancies could give rise to localized states in the middle of the band. Kirkpatrick and Eggarter [35] then went on to a detailed study of these states for finite x, elucidating their physical nature and their effect on the density-of-states and the neighboring extended states.

Take for illustration a simple square lattice with an isolated cluster of four vacancies. The vacancy separation along one diagonal is the diagonal itself. The separation along the other diagonal can be arbitrary, enclosing a string of $1, 2, \dots n, \dots$ sites entirely within the rectangle. States of amplitude $1/n$ and alternating in sign on that string but zero everywhere else are eigenfunctions of the Hamiltonian of energy zero (middle of the band) (Fig. 17).

In the perfect crystal, all states with wave-vector k such that $k_x \pm k_y = \pm \pi/a$, where a is the square edge, also have energy zero. With the cluster present, the extended states with $k = \pm \pi/2a \, (1,1)$ are modified into states which vanish along the string of sites inside the rectangle and which are unmodified elsewhere except for effects describable as scattering off the vacancy pairs terminating the string. The primary effect of the localized states on the neighboring extended states is exclusion through orthogonalization of the latter from the interior of rectangle.

In the Kirkpatrick and Eggarter study [35] for finite x, all possible clusters with the property illustrated above are present in finite concentration and interact. The net result, shown in Fig. 18, is a δ-function peak in the density-of-states at $E = 0$

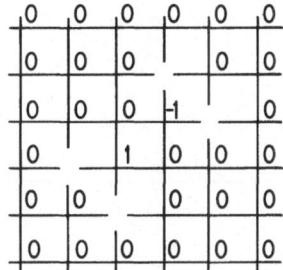

Fig. 17. Simple square lattice. Tight-binding, nearest-neighbors only. Cluster of four vacancies with localized state within them. Values of the wave-function amplitude (unnormalized) are indicated at each site

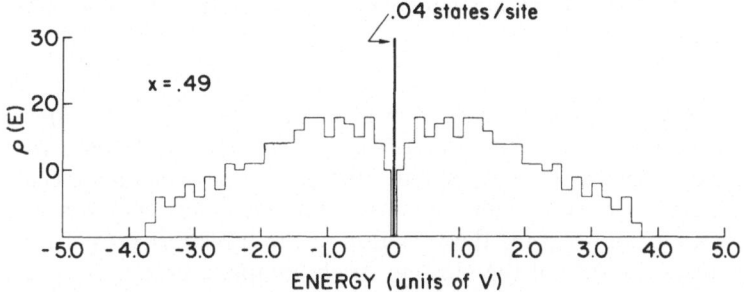

Fig. 18. Density-of-states histograms for $3D$ samples of $8 \times 8 \times 20$ atoms at a vacancy concentration of 0.49. There is a narrowing of the dip in the center of the band for concentrations such as this which are greater than the percolation concentration (0.31). After Kirkpatrick and Eggarter: Phys. Rev. B **6**, 3598 (1972)

surrounded by a dip down to zero which grows in width as x increases. We shall defer the interpretation of the dip until the next section. Here we simply note that the localized states and the associated dip in the density-of-states arise from effects of particular vacancy clusters upon the electronic motion at and near a particular energy, a circumstance already encountered for one-dimensional binary alloys in x.

A similar circumstance occurs in the model of disordered tetrahedral semiconductors of Weaire and collaborators [20]. The topology of tetrahedral networks is such as to permit similar strings to be set up, and there are δ-functions found in the density-of-states for that model and its derivatives. In that case, the δ-functions are artifacts of the band model and disappear for more accurate models. For more accurate models, in the present case, however, the bound states simply become resonances and the peak in $n(E)$ at $E = 0$ simply broadens.

XII. The Metal Semiconductor Transition

A number of liquid binary alloys, $A_{1-x}B_x$, of metallic constituents (valences n and m) show striking anomalies in their transport properties as functions of x at the virtual compound composition $A_m B_n$. The resistivity and Hall coefficients have

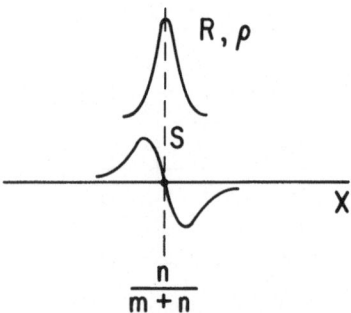

Fig. 19. Transport coefficients as functions of atomic fraction x of B in $A_{1-x}B_x$. The resistivity ϱ and Hall coefficient R shows maxima at the compound composition A_m, B_m, $x = n/(m+n)$, where the thermoelectric power vanishes in such AB systems as liquid TlTe

maxima at, and the thermoelectric power is odd at about $y = n/(m+n)$, vanishing between a maximum and a minimum (Fig. 19). Enderby [37] pointed out the importance of the formation of $A_m B_n$ clusters for this phenomenon. His argument that bound states associated with the clusters removed electrons from the Fermi sea, in this way accounting for the variation of the transport properties, was an overstatement of the role of the clusters. Such complete bound state formation is unnecessary to account for the facts. Mott proposed that a pseudogap [9] was formed near the compound composition but left the mechanism by which it was formed unspecified [37]. Cohen and Sak [38] proposed a mechanism for the formation of the pseudogap, or mobility gap, and went on to construct a theory of the transport properties.

Their argument runs as follows. Suppose for the sake of simplicity that all the B atoms present in the alloy are tied up in $A_m B_n$ clusters for $x < \dfrac{n}{m+n}$ and similarly for the B atoms for $x > n/(m+n)$. Thus the system consists of a liquid metal either of A or of B containing $A_m B_n$ clusters. Suppose the short-range order within the clusters to be good and that the corresponding $A_m B_n$ crystal is a semiconductor. A very large cluster would then behave like a macroscopic semiconducting inclusion within a metal. An electron incident upon the cluster from the metal would be Bragg-reflected if its energy lies within the energy gap (more accurately a pseudogap). The scattering cross section of a metallic electron by a cluster of realistic size would therefore show the energy dependence of Fig. 20, a peak coinciding with the pseudocrystalline band gap E_G centered at the Fermi energy of the compound composition, E_F^c. Thus the metallic electrons are excluded from the clusters when their energy lies within E_G. The region of exclusion increases as the compound composition is approached both because the volume fraction occupied by the clusters increases and because the Fermi energy approaches E_F^c. If the most probable cluster size is several Fermi wavelengths, the semiclassical arguments of IV can be applied. The density-of-states is given by

$$n(E) = C(E)\, n_0(E) \tag{12.1}$$

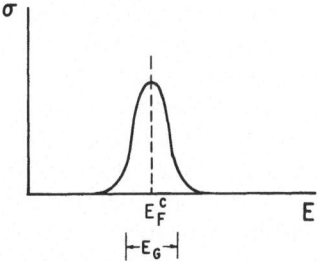

Fig. 20. Scattering cross section σ of an Am Bn cluster for a metallic electron as a function of energy E. The peak occurs midway within what would be the energy gap E_G of a large cluster with excellent local order at an energy equal to the Fermi energy E_F^c at the compound composition

where $C(E)$ is the allowed volume fraction, readily computed by the above arguments, and $n_0(E)$ is the metallic density-of-states. Percolation theory tells us that a pseudogap opens up near E_F^c when

$$\frac{n(E)}{n_0(E)} = C(E) = C^* \simeq 0.2, \qquad (12.2)$$

the critical value for percolation. As $C_{max} = C(E_F^c)$ decreases below C^*, the pseudogap runs between the mobility edges E_c and $E_{c'}$ where $C(E) = C^*$. The localized states lie in the metallic region, and exist because $C(E)$ is too low between E_c and $E_{c'}$ for percolation. All this is summarized in Fig. 21.

It is now a simple matter of explaining the transport properties. As x approaches $n/(m+n)$, the density-of-states at the Fermi energy continuously decreases. The Fermi energy enters the mobility gap, and when $E_{cmin} - E_F$ or $E_F - E_c$ exceeds $2kT$, full semiconducting behavior is established. Using Eggarter's theory of the transport properties [28], Cohen and Sak [38] were able to calculate x dependences of ϱ, R and S resembling those observed. It is probably necessary to improve the details of Eggarter's transport theory, however (cf. IV and IX).

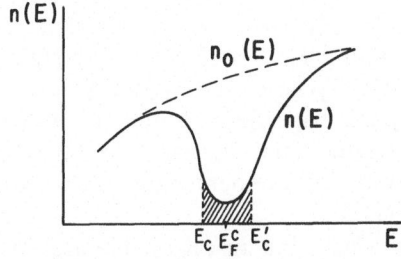

Fig. 21. Density-of-states $n(E)$, full line, as a function of energy in an energy range encompassing that in which the metallic electrons are excluded from the semiconducting clusters by Bragg reflection. $n_0(E)$ is the unperturbed metallic density of states. Their ratio is the allowed volume fraction $C(E)$. In the mobility gap between E_c and $E_{c'}$ [when $C(E) = C^*$, the critical value for percolation], the metallic states become localized, provided $C(E_F^c) > C^*$. Here E_F^c is the Fermi energy for the compound composition

The assumption of complete clustering used above is unnecessary. All of the above conclusions follow in detail for a more general model in which composition and configuration is allowed to fluctuate randomly. All that is necessary is that there is a range of compositions and configurations around the compound composition which are semiconducting in character. Then, provided the range of short-range order is somewhat longer than the Fermi wavelength or phase coherence length, whichever is shorter, semiclassical theory and the notion of the allowed volume fraction can be introduced. Moreover, the theory applies with minor modification to any system in which microscopic inhomogeneities in atomic composition or configuration give rise to a partition into semiconducting and metallic regions. Cohen and Jortner [39] have used these ideas in giving a general account of the five transport regimes encountered in a transition from metallic to semiconducting behavior in disordered materials. In doing so they generalized the effective medium theory [40, 41, 29] of conduction in random inhomogeneous materials to include the Hall effect and showed [42] that the Hall mobility could in some cases depend weakly on the volume fraction in the inhomogeneous metallic regime,

$$\sigma = f\,\sigma_0\,.$$

$$\left.\begin{aligned} f &= a + \left(a^2 + \frac{x}{2}\right)^{1/2} \\ a &= \frac{1}{2}\left|\left(\frac{3}{2}\,C - \frac{1}{2}\right)(1-x) + \frac{x}{2}\right| \\ x &= \sigma/\sigma_0\,, \end{aligned}\right\} \tag{12.3a}$$

$$R = h R_0\,, \tag{12.3b}$$

$$\mu = g\,\mu_0\,, \tag{12.3c}$$

$$h = g/f\,, \tag{12.3d}$$

$$g = f^{-1}\left|1 - \frac{(2f+1)(1-C)(1-xy)}{(2f+1)^2(1-C)+(2f+x)^2 C}\right|, \tag{12.3e}$$

$$y = \mu_1/\mu_0\,, \tag{12.3f}$$

where σ_0 is the metallic conductivity in the absence of inhomogeneities, and R_0 and μ_0 the corresponding Hall coefficient and Hall mobility. Using these ideas, Cohen and Jortner [19] were able to give a good account of the density dependence of the transport properties of liquid Hg at densities lower than those for which the diffusion regime held. Similarly, by taking $C(E_F)$ at various temperatures from Knight shift measurements in liquid Te, they were able to obtain a good fit of σ and R to (12.3).

What appears to be the weak link in all of this argumentation is the explicit quantitative use of semiclassical theory. In Eggarter's calculations, it was justified by the consideration that there are many scatterers within a deBroglie wavelength so that one can use a locally averaged density to compute a local optical or Wigner-Seitz potential which varies smoothly. For the metal-semiconductor transition the deBroglie (Fermi) wavelength λ_F is equal to the interatomic separation a within about 30% ($\lambda_F = 1.3a$ for valence 2, 1.2 for valence 3, and 1.0 for valence 4). As long as the range of short-range order is $\gtrsim 2a$, the semiclassical theory should work. More precisely, it is the phase coherence length, which becomes less than λ_F in the diffusion regime, which should be compared with the range of order.

XIII. Mott Hopping

In a disordered semiconductor, the Fermi energy E_F normally lies within the mobility gap. In a disordered one-dimensional system, if $n(E_F)$ is finite, the states there are localized within a band. At $T = 0\,°K$, the conductivity must then vanish. The proof of localization and of vanishing diffusion or mobility, however, ignores atomic motion. In particular, the argument that resonant tunneling, tunneling without loss of amplitude, is not possible, breaks down. Absorption or emission of phonons compensates for energy differences between neighboring localized states; resonant tunneling occurs; the mixed electron-phonon states become extended in their electronic components; and transport occurs. In other words, phonon-assisted hopping conduction takes place between localized states near E_F at finite T.

Mott was the first to give the theory of hopping conduction [43] and obtained all the essential results. Ambegaokar, Halperin, and Langer [44] based a more detailed analysis on the fundamental work of Miller and Abrahams [45]. The transition rate W_{ij} from state i to state j is given by

$$W_{ij} = v_0 e^{-\alpha R_{ij}} f_i (1 - f_j) \tag{13.1}$$
$$\cdot [n(|E_i - E_j|) u(E_j - E_i) + (1 + n(|E_i - E_j|)) (1 - u(E_j - E_i))]$$

where v_0 is a constant, α relates inversely to the decay length of the localized states, R_{ij} is their separation, f_i and f_j are Fermi factors, $n(x')$ is the Bose factor, E_i and E_j are the energies of states i and j and u is the Heaviside unit function. Thus the transition probabilities are functions of the random variables E_i, E_j, and R_{ij}. The Miller and Abrahams formulation of impurity conduction states that the conductance is that of a random network of conductances directly given by the W_{ij}, and it remains true in the present instance. Ambegaokar, Halperin and Langen [44, 46] argued that the process of current flow through the present network is much like a percolation process [29], and used percolation theory to obtain the results of Mott somewhat more precisely. Shante [47] has made the percolation arguments more precise and obtains

$$\sigma = \sigma_0 e^{-(T_0/T)^{(d+1)^{-1}}} \tag{13.2}$$

where d is the dimensionality of the system. Thus the exponent of $\frac{1}{T}$ in the exponential is $\frac{1}{2}$ in 1 dimension, $\frac{1}{3}$ in 2 dimensions, and $\frac{1}{4}$ in 3 dimensions (Mott's original results).

Mott's argument makes the origin of this temperature-dependent activation energy clear. When T is large, the energy dependent factors are less important, and jumps with smaller R_{ij} are favored. As T decreases, the energy dependent factors play an increasingly important role in restricting the jumps. As a consequence, jumps with larger and larger R_{ij} play a role because their energy dependent factors are more favorable. The effective activation energy therefore decreases with decreasing T, and the theory gives $k(T^d T_0)^{1/d+1}$ for the effective activation energy.

Mott hopping has been observed in a variety of amorphous semiconductors or insulators, but perhaps the most interesting case is discussed in the next section.

XIV. Almost One-Dimensional Conductors

There are now about 30 organic, metallo-organic, or inorganic compounds for which structural and chemical considerations strongly suggest that conduction or valence electron motion should be nearly one-dimensional [48]. Figure 22 is a highly schematized representation of a typical member of this group.

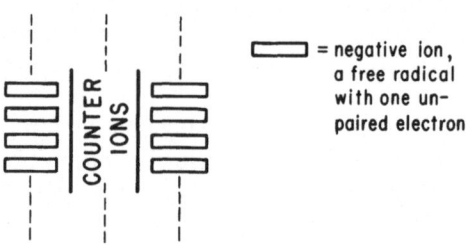

Fig. 22. Schematic rendering of the structure of a nearly one-dimensional conductor

One has a set of parallel one-dimensional stacks of planar negative molecule ions which are free radicals with one unpaired electron per ion. These are well separated by counter ions, either metallic or similarly stacked positive molecule ions. $(NMP)^+ (TCNQ)^-$ is an example of the latter case. The states of the excess electrons on the negative ions have appreciable nearest-neighbor overlap within the stack. Ignoring interaction between stacks, the levels occupied by the extra electrons broaden out into a half-filled, one-dimensional band, Fig. 23a. However, small matrix elements coupling stacks would reduce the height of the band-edge peaks to a finite value, and broaden them somewhat into peaks containing two saddle points each and bounded by square root singularities.

Such materials offer possibilities for the observation of phenomena uniquely characteristic of electron motion in one dimension, and they are currently under intense investigation.

Heeger and his colleagues [49] have called attention to the linear low temperature specific heat, the transport properties, the strong T-dependence of the susceptibility, and other magnetic properties of $NMP^+ TCNQ^-$. They propose that all of these can be understood as consequences of a Mott metal-insulator transition in a one-dimensional material in which a low temperature antiferromagnetic insulator transforms to a paramagnetic insulator and then to a paramagnetic metal as T increases. The transitions are gradual, of course, in one dimension. They claim the quantitative aspects are consistent with a simple Hubbard model.

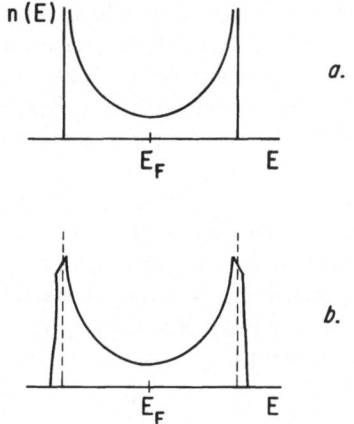

Fig. 23. a One-dimensional band expected for substance of Fig. 22 upon ignoring interstack interaction. b The additional structure introduced by interstack interaction

On the other hand, Bloch, Weissman, and Varma [50] have considered the transport properties, particularly the conductivity, of a number of these materials including $NMP^+ TCNQ^-$. They point out that there is reason to expect these materials to be disordered except for those which behave as simple insulators or semiconductors and show a Peierl's instability (bond alternation). Moreover, such materials are difficult to purify. They claim, therefore, that σ vs. T is quantitatively explicable as one-dimensional Mott hopping for which the exponent of $1/T$ in the exponent is $\frac{1}{2}$, Eq. (13.2). Indeed, if one carries out a least squares fit of σ to $\sigma_0 e^{-(T_0/T)^{1/n}}$ for various n using the available data, the best fit over the widest range of temperature is for $n = 2$. The thermoelectric power is consistent with this picture of Mott hopping. Bloch, Weissman, and Varma [50] conclude that one does not need to invoke the Mott metal-insulator transition to explain transport in $NMP^+ TCNQ^-$.

The resolution of his difficulty lies in combining the two points of view. Bloch, Weissman, and Varma are able to infer values of L_e/L_d of the order 4 to 10 from the T-dependence of σ. Papatriantafillou's calculations of L_e/L_d show that this range of values occurs primarily when there is a low concentration of strongly perturbing impurities. Moreover, Papatriantafillou and Cohen [51a] and Bush

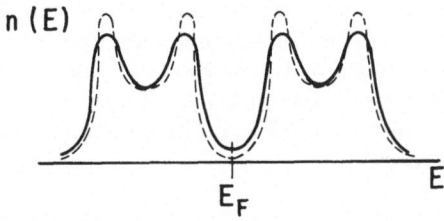

Fig. 24. Density-of-states versus energy for a magnetically split band broadened by impurities and other sources of disorder. The dotted line corresponds to a purer material than does the solid line

have carried out of dielectric screening which studies shows that the intrinsic disorder in $NMP^+ TCNQ^-$ associated with random orientation of the NMP^+ dipoles is not likely to be important for the transport properties except for very pure materials. Accepting the Heeger group arguments about antiferromagnetic ordering at low temperatures, what one is clearly dealing with is a magnetically split band, broadened by disorder probably originating from impurities, as in Fig. 24. Purifying the material would then reduce the broadening, decreasing the density-of-states at E_F, and decreasing the conductivity. This is precisely what has been observed to happen by Heeger et al. [49].

It is clear that what is needed is a theoretical treatment of the one-dimensional Hubbard model with disorder included to provide a more quantitative basis for the interpretation of experiment.

XV. Remarks on Metal–Ammonia Solutions

It is appropriate for me to conclude this article with remarks on metal–ammonia solutions [52]. Studies of the very concentrated solutions suggest that the electronic motion is that of the propagation regime down to concentrations perhaps as low as 8—9 MPM. Studies of the dilute solutions demonstrate clearly the existence of clusters. At lower temperatures, the solutions saturate, and there is a phase separation below a consolute temperature. Taking all this into account makes it very likely that the solutions are microscopically inhomogeneous, probably in the range 1—9 MPM, with volume fraction f of the material occupied by metallic clusters of a mean concentration of 9 MPM or higher, up to the saturation concentration, the remainder being liquid NH_3 containing a low concentration (under 1 MPM) of smaller complexes.

That being the case, one immediately understand the relation between the metal-nonmetal transition and the phase separation in M–NH_3 solutions from percolation considerations. At the metal-nonmetal transition f should be close to 0.2, the value required for percolation. Now f should be about 0.5 above the consolute temperature, so that the metal-nonmetal transition should occur above and slightly to the low concentration side of the consolute point on the $T-C$ phase diagram, as it in fact does.

With the possible concentrations inside the metallic clusters in the range 8—20 MPM, it is a simple matter to estimate the corresponding range of mean concentrations at which f might become 0.2. We obtain the range 2—3 MPM for the metal-nonmetal transition in this way, which is in excellent agreement with the observations. Moreover, the conductivity σ and Hall constant show roughly the dependence on concentration predicted by Eq. (12.3), when scattering off cluster boundaries is included by the method of Eggarter [30].

Our conclusion is that the metal–ammonia solutions undergo a metal-nonmetal transition similar to that discussed in § XII. The nonmetal is an electrolyte instead of a semiconductor, however.

Acknowledgment

In almost every aspect, this work has benefited greatly from discussions with many colleagues at Chicago and elsewhere, among whom are A. Bloch, R. Bush,

E. N. Economou, T. P. Eggarter, H. Fritsche, J. Jortner, E. S. Kirkpatrick, C. Papatriantafillou, P. Sen, V. K. Shante, C. Varma, and B. Weissman.
It is a pleasure to acknowledge the hospitality of the Cavendish Laboratory where the manuscript of this paper was prepared from the notes of my lecture at Colloque Weyl III.

References

1. Cohen, M. H.: In: Amorphous and liquid semiconductors, p. 391. N. F. Mott (Ed.): Amsterdam: North-Holland Publishing Company (1970).
2. Cohen, M. H.: J. Non-Crystalline Solids **2**, 432 (1970).
3. Cohen, M. H.: Proceedings of the Tenth International Conference on the Physics of Semiconductors. Cambridge, Mass. S. K. Keller, J. C. Hensel, F. Stern (Eds.), p. 645. Springfield: USAEC Division of Technical Information 1970.
4. Cohen, M. H.: Physics Today **24**, 26 (1971).
5. Ziman, J. M.: J. Phys. C **1**, 1532 (1968).
6. Shante, V. K. S., Kirkpatrick, S.: Advan. Phys. **20**, 325 (1971).
7. Anderson, P. W.: Phys. Rev. **109**, 1492 (1958).
8. a) Thouless, D. J.: J. Non-Crystalline Solids **8–10**, 461 (1972).
   b) Economou, E. N., Cohen, M. H.: Phys. Rev. **5**, 2931 (1972).
   c) Ziman, J. M.: J. Phys. C **2**, 1230 (1969).
9. Mott, N. F.: Advan. Phys. **16**, 49 (1967).
10. Cohen, M. H., Fritsche, H., Ovshinsky, S. R.: Phys. Rev. Letters **22**, 1065 (1969).
11. See for example Stuke, J.: In: Mott, N. F. (Ed.): Amorphous and liquid semiconductors. Amsterdam: North-Holland Publishing Company 1970.
12. Kubo, R.: J. Phys. Soc. Japan **12**, 470 (1957).
13. Lifschitz, I. M.: Advan. Phys. **13**, 483 (1964).
14. Friedman, L.: J. Non-Crystalline Solids **6**, 329 (1971).
15. Mott, N. F.: Phil. Mag. **26**, 1015 (1972).
16. Freed, K. F.: Phys. Rev. B **5**, 4802 (1972).
17. Lukes, T.: J. Non-Crystalline Solids **8–10**, 470 (1972).
18. Anderson, P. W.: Proc. Natl. Acad. Sci. US. **69**, 1097 (1972).
19. Cohen, M. H., Jortner, J.: Phys. Rev. Lett. **30**, 698 (1973).
20. Weaire, D., Thorpe, M. F., Heine, V.: J. Non-Crystalline Solids **8–10**, 128 (1972).
21. Mott, N. F.: Phil. Mag. **13**, 989 (1966).
22. Mott, N. F.: J. Non-Crystalline Solids **1**, 1 (1968).
23. Mott, N. F.: Phil. Mag. **17**, 1259 (1968).
24. Mott, N. F.: Phil. Mag. **19**, 835 (1969).
25. Mott, N. F.: Rev. Mod. Phys. **40**, 677 (1968).
26. a) Levine, J. L., Sanders, T. M.: Phys. Rev. **154**, 138 (1967).
   b) Harrison, H. R.: Ph. D. Thesis, University of Michigan, Ann Arbor (unpublished).
27. Eggarter, T. P., Cohen, M. H.: Phys. Rev. Letters **27**, 129 (1971).
28. Eggarter, T. P.: Phys. Rev. A **5**, 2496 (1972).
29. Kirkpatrick, S.: Phys. Rev. Letters **27**, 1722 (1971).
30. Mott, N. F., Twose, W. D.: Advan. Phys. **96**, 1208 (1954).
31. Borland, R.: Proc. Roy. Soc. (London) A **274**, 529 (1963).
32. Bush, R. L.: Phys. Rev. **56**, 1182 (1972).
33. Economou, E. N., Papatriantafillou, C.: Solid State Comm. **11**, 191 (1972).
34. Papatriantafillou, C.: Phys. Rev. B (in press).
35. Eggarter, T. P.: Unpublished.
36. Kirkpatrick, S., Eggarter, T. P.: Phys. Rev. B **15** (in press); Computational Methods For Large Molecules And Localized States in Solids, ed. Herman, F., McLean, A. O., Nesbet, R. K., p. 327, Plenum Press, 1972.
37. a) Enderby, J., Simmons, C. J.: Phil. Mag. **20**, 125 (1969) and Enderby, J.: Non-Crystalline Solids (in press).
37. b) Mott, N. F.: Phil. Mag. **22**, 1 (1970); **24**, 1 (1971).

38. Cohen, M. H., Sak, J.: Non-Crystalline Solids **8–10**, 696 (1972).
39. Cohen, M. H., Jortner, J.: To be published.
40. Bruggeman, D. A. G.: Ann. Phys. (Leipzig) **24**, 836 (1935).
41. Landauer, R.: J. Appl. Phys. **23**, 779 (1952).
42. Cohen, M. H., Jortner, J.: Phys. Rev. Lett. **30**, 696 (1973).
43. Mott, N. F.: Phil. Mag. **19**, 835 (1969); — Festkörperprobleme **9**, 22 (1969).
44. Ambegaokar, V., Halperin, B. I., Langer, J.: Phys. Rev. B **4**, 2612 (1967).
45. Miller, A., Abrahams, E.: Phys. Rev. **120**, 745 (1960).
46. Ambegaokar, V., Halperin, B. I., Langer, J.: J. Non-Crystalline Solids **8–10**, 492 (1972).
47. Shante, V. K. S.: To be published.
48. Leblanc, O. H.: In: Physics and chemistry of the organic solid state, Vol. III, Chapter 3.
    New York: Interscience 1967.
49. Epstein, A. J., Etemad, S., Garito, A. F., Heeger, A. J.: Phys. Rev. B **5**, 952 (1972);
    Etemad, S., Garito, A. F., Heeger, A. J.: Phys. Letters **40** A, 45 (1972);
    Chaiken, P. M., Garito, A. F., Heeger, A. J.: Phys. Rev. B **5**, 4966 (1972);
    Colemán, L. B., Cohen, J. A., Garito, A. F., Heeger, A. J. (in press);
    Ehrenfreund, E., Etemad, S., Coleman, L. B., Rybaczewski, E. F., Garito, A. F., Heeger, A. J.:
    Phys. Rev. Letters **29**, 269 (1972);
    Coleman, L. B., Cohen, J. A., Garito, A. F., Heeger, A. J.: Phys. Rev. B (in press);
    Ehrenfreund, E., Rybaczewski, E. F., Garito, A. F., Heeger, A. J.: Phys. Rev. Letters **28**, 873
    (1972);
    Ehrenfreund, E., Rybaczewski, E. F., Garito, A. F., Heeger, A. J., Pincus, P.: Phys. Rev. B
    (Jan. 1973).
50. Bloch, A., Weissman, B., Verna, C.: Phys. Rev. Letters (1971).
51. a) Papatriantafillou, C.: To be published.
    b) Bush, R. L.: To be published.
52. Cohen, M. H., Thompson, J. C.: Advan. Phys. **17**, 857 (1968).

Discussion

J. JORTNER   I would like to raise the issue as to whether the electronic states and the transport properties of a disordered system should be described in terms of a homogenous or inhomogenous model. The physical properties predicted by these two models should be vastly different. For the inhomogenous model Cohen has applied percolation theory, so that the conductivity, σ, is given by the mean field theory, (being linear in the allowed volume for high values of this volume $\sigma \propto C$) and is expected (in principle) to drop to zero at the critical percolation concentration when $C \approx 0.16 - 0.25$. In the case of the homogenous model in the strong scattering regime (where the mean free path is comparable to the interatomic spacing) Mott has introduced the pseudogap depth parameter, g, which monitors the ratio of the density of states at the Fermi energy to the free electron density of states. In this case $\sigma \propto g^2$, $R \propto g^{-1}$ and the metallic region is terminated according to Mott at $g < 1/3$. The experimental data on metallic vapors (a one component system) to be reported by Hensel and by Even can be adequately interpreted by the homogenous medium picture down to density of 9.3 grams/cm^3 for mercury. On the other hand percolation is crucial in the case of metal ammonia solutions where aggregates consisting of solvated electrons and metal ions are presumably formed.

G. LEPOUTRE   For metal–ammonia solutions, with two kinds of electronic carriers having conductivities, σ_1 and σ_2, the conductivity σ cannot be expressed as the sum of σ_1 and σ_2 if σ_1 and σ_2 differ by several orders of magnitude. Such a difference is indeed obtained if the values of σ_1 and σ_2 are chosen in the dilute and the concentrated range, kept constant in the transition range. But, in this transition range, could not σ_1 be much larger than in the dilute range and σ_2 be much smaller than in the concentrated range, with an homogeneity sufficient to yield $\sigma \approx \sigma_1 + \sigma_2$?

M. H. Cohen   Unless $\sigma_1 \approx \sigma_2$, $\sigma \neq \sigma_1 + \sigma_2$ in a medium as unhomogeneous as I believe the metal–ammonia solutions to be in the transition region.

K. Bar-Eli   Is it possible to explain the instability of dilute metal solutions, compared to that of the concentrated ones, by the fact that there are resonating transitions between allowed states in the latter, without loss of energy, while in the dilute solutions the transitions involve loss of energy, until the electron finds itself on the lowest possible state, namely, hydrogen?

M. H. Cohen   Yes, this is a very good suggestion. The extra stabilization of the solvated electrons within the metallic regions by resonant propogation from cavity to cavity, essentially a metallic contribution to the binding of the cluster, would reduce their chemical activity. The stabilization energy would be of the order 1/4 of the tight binding band width, which we have not been able to estimate yet.

Concentrated M–NH$_3$ Solutions: A Review*

J. C. THOMPSON

Abstract

A review is given of the properties of concentrated (> 10 MPM) M–NH$_3$ solutions, with particular attention paid to data obtained since Colloque Weyl II. Differences between solutes seem to result primarily from solvation effects. Comparison of transport and magnetic effects suggests that the temperature dependence results from density-of-states effects as in a pseudo-gap. Interpretation of these phenomena is given in terms of a model which supposes the solutions to be composed of solvated ions, nonsolvating NH$_3$ molecules and free electrons.

In reviewing the properties of concentrated ($x < 10$ MPM) metal-ammonia solutions for this talk I was struck by the wealth of data presented by Lepoutre [1] at Cornell three years ago. It seemed as if nothing was left for me. Indeed the list of accomplishments since that conference is short. There are no data on new effects; the diffraction data we so sorely needed then are still lacking. But the data base has been broadened so that the effects of ionic species are now revealed and one need no longer take, for example, conductivity data on Na – NH$_3$ for comparison with Hall coefficients on Li–NH$_3$. There have also been technical improvements on old experiments and therefore better data.

The foremost theoretical accomplishment has been the theory of conductivity developed by Ashcroft and Russakoff [2]. There has also been a broad survey of all electrically conducting fluids by Allgaier [3] which provides a better framework than we had previously against which to calibrate our results.

Before looking at data let me say a few words about phase diagrams. Figure 1 shows the phase diagrams for Li– and Cs–NH$_3$ solutions [4, 5]. It seems now well established that Cs and NH$_3$ are miscible in all proportions [6]. It seems highly probable that each of the metals will prove miscible if the temperature is raised to the normal melting point of the metal. Even Kraus [7] noticed that NH$_3$ was soluble in molten Na–K alloy. It would be interesting to see more of the high temperature phase diagrams. The Li–NH$_3$ solubility limits are those quoted by Bridges *et al.* [8] and are in disagreement with the curved limits reported by Lo [9]. I have long been struck by the apparent vertical rise of that limiting curve from the eutectic [10] and had deluded myself that it was unique. Recent reading has revealed to me that the eutectics of the alkali metal–alkali metal salt mixtures lie similarly close (10^{-7} molar) to the pure metal end [11], as you may see in Fig. 2. I don't know that any insight is to be derived from this similarity but I am somehow reassured by finding materials similar to M–NH$_3$ solutions.

Let me turn now to the new data on transport properties. Figure 3 shows the concentration dependence of the conductivity of all of the alkali metals [12]

---

* Research supported in part by the U.S. NSF and by the R. A. Welch Foundation of Texas.

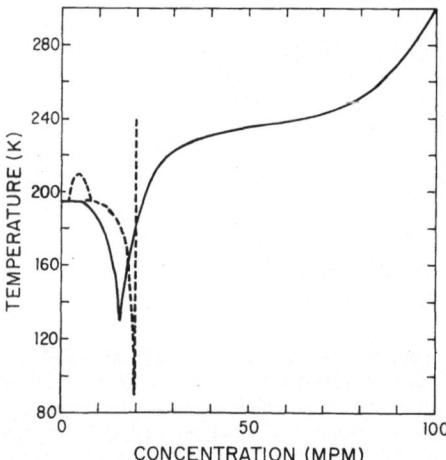

Fig. 1. Phase diagram for Li–NH₃ (dashed curve) and Cs–NH₃ (solid curve) solutions. Note the rapid rise in the solubility limit curve for $Li-NH_3$ at 20 MPM, i.e., at $Li(NH_3)_4$. The eutectic is quite close to the compound. Note also the complete miscibility of Cs and NH_3. Details of the Cs–NH₃ phase diagram have been given recently by Refs. [21] and [31]

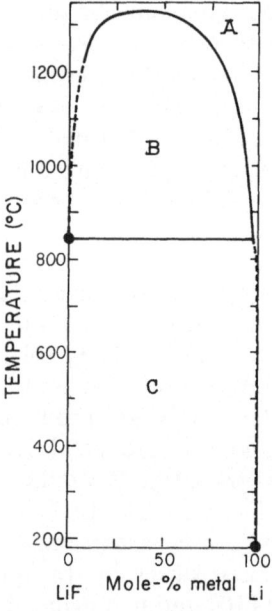

Fig. 2. Phase diagram for Li–LiF. Note the deep eutectic very near to the pure metal and the similarity to the eutectic at $Li(NH_3)_4$. The data are from Ref. [11]

in NH_3 at 240 °K (except for the Rb and Cs data, taken at 195 °K). If one considers only the concentration range common to all of the solutions, the order of conductivity is Li–Na–K–Cs–Rb. The temperature dependence is not expected to be strong enough to push the Rb–NH₃ conductivity above that for

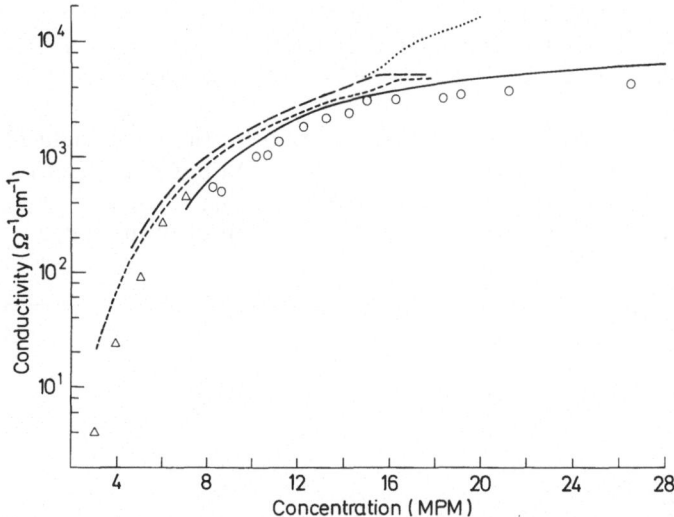

Fig. 3. Conductivity *vs* concentration for all alkali metal–ammonia solutions: ○ Rb–NH₃ at T = 195 °K using an ac electrodeless method; △ Rb–NH₃ extrapolated to 195 °K using an ac electrode technique; ——— Cs–NH₃ at T = 195 °K; ---- K–NH₃ at T = 240 °K; ——— Na–NH₃ at T = 240 °K; ··· Li–NH₃ at T = 240 °K. The temperature dependence is not expected to be strong enough to push the Rb–NH₃ conductivity above that for Cs–NH₃ solutions at 240 °K. Taken from Ref. [12]

Cs–NH₃ at 240 °K [5]. The similarities among the results are greater than the differences.

Clearly, atomic number (or atomic weight or ionic radius) is not critical for transport processes in these materials.

The temperature coefficients $\gamma = \sigma^{-1} \, d\sigma/dT$ are also remarkably similar [4, 5], as may be seen in Fig. 4. It is important, however, to note that as T in-

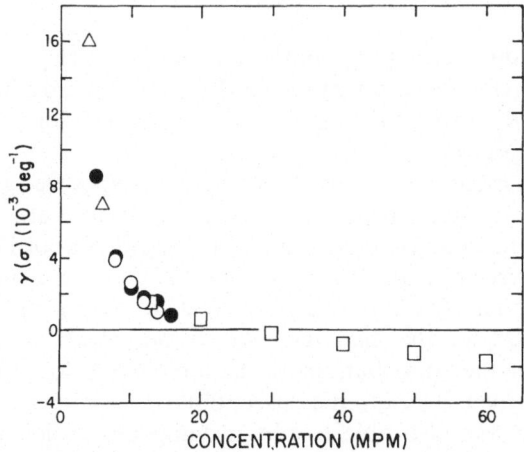

Fig. 4. The temperature coefficient of conductivity $\gamma(\sigma) = \sigma^{-1} \, d\sigma/dT$, △ Li–NH₃; ○ Na–NH₃; ● K–NH₃ and □ Cs–NH₃

creases above the NBP of NH_3 the sign of γ changes to negative at the higher concentrations [4, 5]. The solutions thus resemble pure *divalent* metals which exhibit a positive γ near the melting point [13, 14]. The solutions are not divalent, of course.

Lepoutre's group has now added more thermoelectric power data [15] which is compared to the older data [16, 17] in Fig. 5. Notice that when the data are corrected to absolute thermopower there is a sign change to *p*-type above 15 MPM; *n*-type behavior obtains, however, over most of the range.

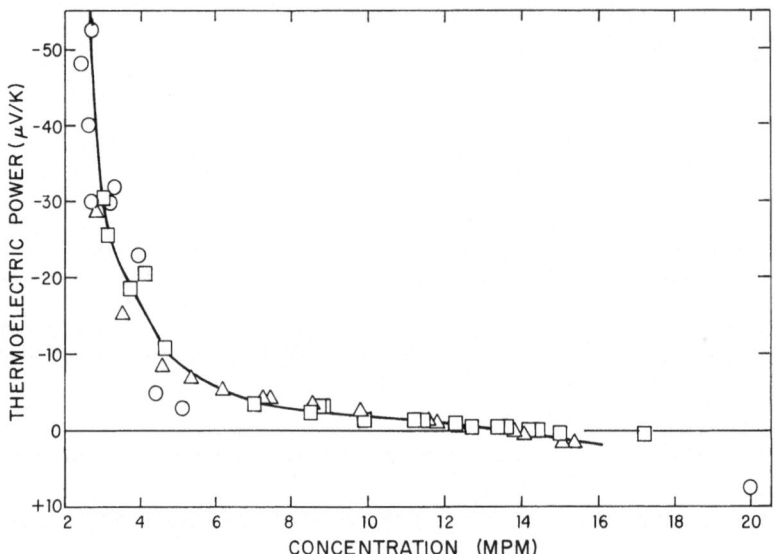

Fig. 5. The thermoelectric power according to Lepoutre and several co-workers.
○ Li–NH_3; △ Na–NH_3 and □ K–NH_3

Hall coefficient measurements [18] continue to be unexciting: all valence electrons are free. A careful re-examination of the data [19, 20] indicates a temperature coefficient $R_H^{-1} dR_H/dT$ near 0.1—0.2% per °K. This is quite close to the rate of change of density with temperature.

There are new magnetic data, again from the Lepoutre group at Lille. Lelieur [21] has measured the static susceptibility in both Na–NH_3 and Cs–NH_3 solutions. His data are consistent with the old data of Huster [22] and also with the more recent work of Suchannek [23], but are more detailed in the concentrated region. The analysis of such data always presents problems since one must correct for the diamagnetism of the solvent using Wiedeman's rule [24] and also for the contributions of the ion cores before the effects of the electron gas are revealed. The atomic susceptibilities calculated by Lelieur are shown in Fig. 6. He was also able to obtain temperature dependences, which are less influenced by the corrections, as shown in Fig. 7. These curves are strikingly similar to the conductivity temperature coefficients shown in Fig. 4.

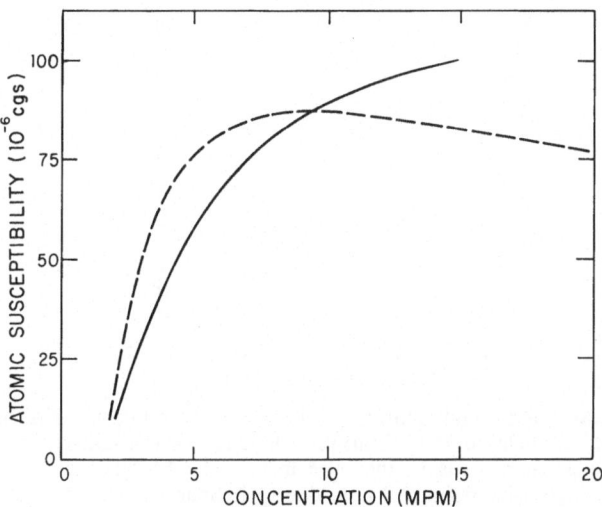

Fig. 6. The atomic susceptibility of Na–NH₃ (solid curve) and Cs–NH₃ (dashed curve) from Lelieur's thesis

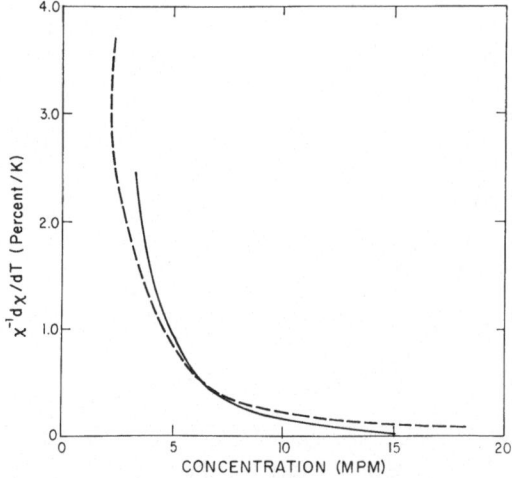

Fig. 7. The temperature coefficient of susceptibility according to Lelieur. The notation is as in Fig. 6

In a separate experiment Lelieur [21] has measured Knight shifts at the metal nucleus [25, 26] in Na–, Cs–, and one Li–NH₃ solutions. His results are shown in Fig. 8, where each shift is normalized to the value of the shift in the pure metal. Temperature derivatives are shown in Fig. 9 and again there is remarkable resemblance to Fig. 4.

The optical data are not so extensive, but now cover several solutes [27—29]. NFE behavior over the 0.5—5.0 eV range. No significant trends appear when the solute is varied.

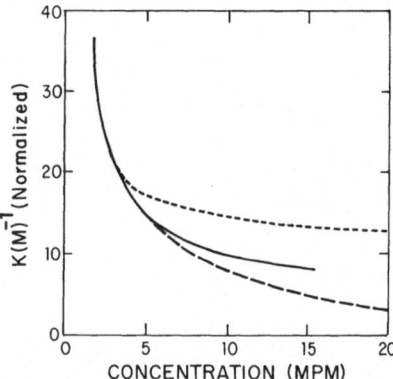

Fig. 8. The reciprocal of the Knight shift at the Li (short dashed curve), Na (solid curve) and Cs (long dashed curve) nuclei in NH_3 solutions. The data are normalized by dividing by the Knight shift at the given nucleus in the pure metal. The Li–NH_3 curve is this author's extrapolation of a single data point. All data due to Lelieur

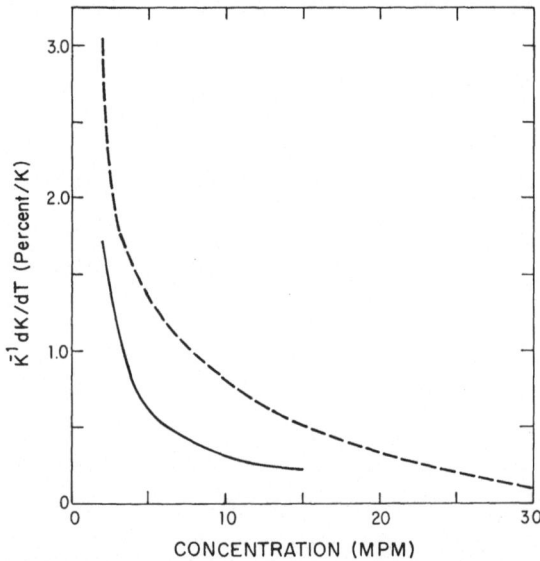

Fig. 9. The temperature coefficient of the Knight shift according to Lelieur. The notation is as in Fig. 6

Nonelectronic phenomena have been even less studied than the electronic over the past few years. New sound speed data [29—31] are summarized in Fig. 10. The larger (heavier?) solutes depress the sound speed further than the smaller (lighter). Bowen [29] has found the minimum only in Li and Ca solutions despite careful searches in the other systems. The sound speed C is always observed to be a linearly decreasing function of temperature and may be described by an equation of the form $C = C_0 - C_1 T$. The values of the coefficients show no noticeable trend with solute.

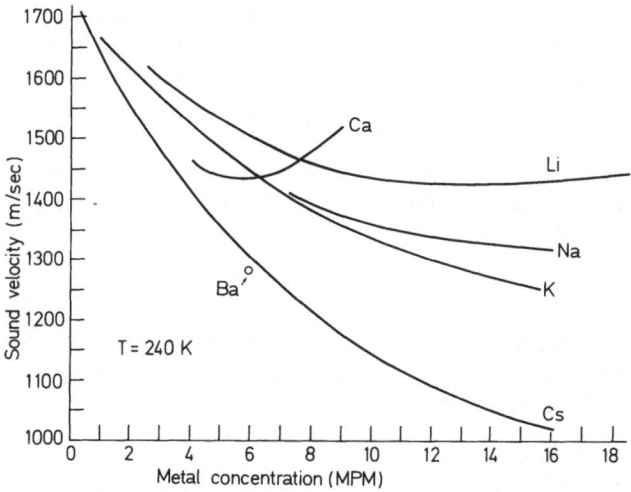

Fig. 10. The speed of sound in several M–NH₃ solutions at 240 °K according to Bailey and Bowen, Ref. [31]

New density measurements are underway in Professor Lepoutre's laboratory but are as yet unpublished.

What now can be made from this data, new and old? First, it is informative to put these results into context using the classification scheme for liquid, electronically conducting materials proposed by Allgaier [3]. He describes three basic types. Class A fluids are good metals having σ in excess of $5000\,\Omega^{-1}\,cm^{-1}$, are exemplified by Na, and may be described by NFE theory [14]. The mean free path Λ is long and the electron-ion interaction weak. Most transport parameters have values and signs typical of solid metals. Class B liquids have σ between 100 and $5000\,\Omega^{-1}\,cm^{-1}$, and in most cases $d\sigma/dT$ is positive (as in semiconductors), though there is no other indication of thermal activation of carriers. The thermopower S and Hall coefficient often have opposite signs. In class C liquids, nonmetallic behavior occurs. The boundary, $\sigma = 100\,\Omega^{-1}\,cm^{-1}$, between classes B and C is an estimate of the lowest conductivity values for which it is still possible to describe charge transport by conventional extended electronic states [32–34]. The concept of propagation with occasional scattering breaks down in class C and models are based on hopping or tunneling of electrons between localized states [34].

Most concentrated M–NH₃ solutions fall into class B, only highly concentrated Li–NH₃ and Cs–NH₃ solutions attain class A. The M–NH₃ solutions do not appear to satisfy the conditions for class C because electrolytic effects take over upon dilution.

Concentrated metal–ammonia solutions, then, exemplify a transitional regime, where the application of either the nearly free electron theory or percolation models is likely to be troublesome.

Some points towards a model of the solutions can be made without any specific details:

1. The temperature-independent Hall coefficient together with the linear temperature dependence of the conductivity points towards a mobility or mean free path which is a linearly increasing function of temperature (below ~ 20 MPM).

2. The Hall mobility [20] defined as $\mu_H = R_H \sigma$ is sufficiently large ($5 \leqq \mu_H \leqq 50$) that hopping processes are unlikely [33—35].

3. The optical data on the plasma resonance lead to estimates of the effective mass [27] not significantly different from the free electron mass.

The Knight shift $K(M)$ is defined as

$$K(M) = (8\pi/3) \chi_p (N_0)^{-1} \langle |\Psi(M)|^2 \rangle \tag{1}$$

where N_0 is Avogadro's number, $\Psi(M)$ is the electron wavefunction at nucleus M, $\langle \rangle$ indicates an average over the Fermi surface, and χ_p, the atomic, paramagnetic susceptibility given by

$$\chi_p = \mu_B^2 N(E_F), \tag{2}$$

where μ_B is the Bohr magneton. If χ_p is known, then $\langle |\chi(M)|^2 \rangle$ can be determined. This Lelieur has done [21] upon the assumption that $\chi_p = 3/2 \chi_{total}$ (i.e., that the Landau diamagnetism is given by 1/3 the Pauli paramagnetism. The assumption is consistent with direct measurements [21]). The results are shown in Fig. 11 normalized by the pure metal. As might be expected the effect of the change in ion is not small. There is also a temperature dependence as displayed in Fig. 12.

Comparison of the magnetic and transport data permits one to add the following conclusions:

4. None of the properties is *strongly* dependent upon the solute ion. The conductivity, particularly, varies little among the alkali metal solutes.

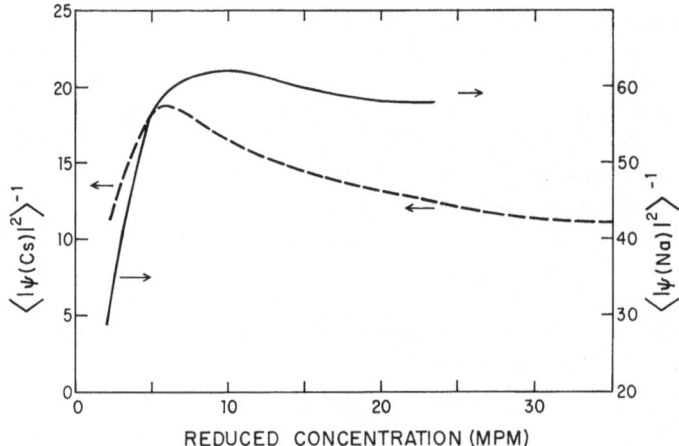

Fig. 11. The reciprocal of the square of the wavefunction at the metal nucleus, according to Lelieur. The data are normalized by dividing by the same quantity in the pure metal. Notation is as in Fig. 6; note that the two curves use different ordinates as marked by the arrows. The abscissa is given by $x/(1-zx)$ where x is the mole fraction and z is the co-ordination number of the ion: for Na, $z = 4$; for Cs, $z = 8$

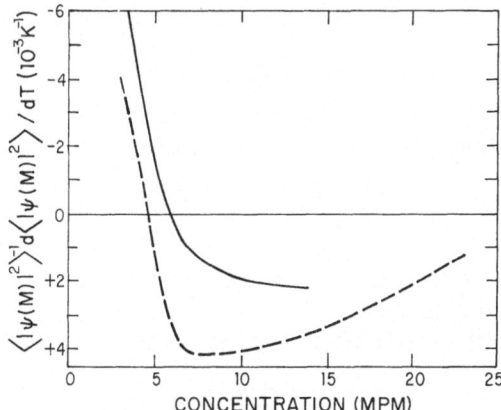

Fig. 12. The temperature coefficient of the square of the wavefunction according to Lelieur. The notation is as in Fig. 6

5. The susceptibility is sufficiently close to the free electron result to reinforce the presumption of an effective mass ratio near unity.

6. The close similarity of the temperature dependences of the susceptibility and of the conductivity strongly suggests that they have the same origin. The only common contributor to each is the density of states.

7. There is, nevertheless, a slight dependence of $\Psi(M)$ on temperature, and a significant solute effect.

These and other observations find qualitative and even quantitative explanation in terms of a simple model which is a natural outgrowth of the standard treatment of ionic solutions [36]. The solution is presumed to be a mixture of solvated metal ions, of ammonia molecules and of free electrons. A solvated ion is a (positive) ion surrounded by several oriented NH_3 molecules; complexes such as $Li(NH_3)_4^+$ will be introduced. The binding forces are at least partially electrostatic (charge-dipole). Those NH_3 molecules which are not involved in the sheath about the positive ion at any given instant are considered "free".

The absence of a strong solute effect on the conductivity may be attributed to the concealment of the specific ionic pseudopotential (see below) by the solvation sheath. Scattering by the free NH_3 molecules must also be considered.

The change in $\Psi(M)$ from Na to Cs is attributable not only to the change in the character of the ion core but also to differences in solvation. Cs^+ would be expected to have a less well defined solvation layer because of its larger size. The increased effect of T on $\Psi(M)$ when Cs is substituted for Na is a consequence of the weaker binding of the NH_3 to the Cs.

Considerable success has been had in deriving the concentration dependence of conductivity and compressibility using a simple, straightforward application of the Ziman model [14] of liquid metals [2, 37]. The resistivity is written as an integral over the product of a structure factor $a(k)$ and a form factor $u(k)$, where k is the momentum transfer [35].

$$\frac{1}{\sigma} = (3\pi/e^2 h V_F^2)\,\Omega\langle F(k)\rangle, \tag{3}$$

where V_F is the Fermi velocity, Ω is the atomic volume, and $\langle F(k) \rangle$ denotes

$$(4k_F^4)^{-1} \int_0^{2k_F} F(k)\, k^3\, dk\,,$$

where k_F is the Fermi momentum. The thermopower is given by

$$S = \frac{\pi^2 k_B^2 T}{3\gamma E_F} \left[3 - 2\, \frac{F(2k_F)}{\langle F(k)\rangle}\right], \tag{4}$$

where k_B is Boltzmann's constant, E_F the Fermi energy, and the quantity in the square bracket is usually denoted ξ.

For a one-component fluid $F(k)$ has the simple form

$$F(k) = a(k)\, |u(k)|^2\,. \tag{5}$$

A two-component system, such as the M–NH_3 solutions, requires a more complex formula:

$$\begin{aligned}
F(k) =\ & (1-x)\, |u_1(k)|^2\, a_{11}(k)\,, \\
& + 2[x(1-x)]^{\frac{1}{2}}\, u_1(k)\, u_2(k)\, a_{12}(k)\,, \\
& + x\, |u_2(k)|^2\, a_{22}(k)\,,
\end{aligned} \tag{6}$$

where the a_{ij} are partial structure factors*.

Schroeder and Thompson [37] and also Ashcroft and Russakoff [2] have estimated the a_{ij} using the Percus-Yevick treatment of a hard-sphere fluid [38]. This requires estimates of the packing fraction η, the ratio of the radii of the two constituents (solvated ion and free NH_3), and of course, knowledge of the concentration.

A calculation of the pseudopotential form factor is beyond the present capabilities of the theory. One must therefore search for plausible and workable approximations for the potentials of both the solvated ion and the NH_3 molecule.

Russakoff and Ashcroft [2] represented the solvated ion by a screened Coulomb potential, and included the *solvating* NH_3 molecules as point dipoles with appropriate positions and orientations. The free, non-solvating NH_3 molecules were treated as point dipoles. They argue that the effacacy of electron screening is such as to reduce the tendency of dipoles to align so that angular orientations are independent of spatial positions. Russakoff and Ashcroft then take advantage of the cancellation which occurs when random dipole potential terms are treated in first Born approximation [2, 39]. The sums which lead to dipole-dipole structure factors for spherically symmetric potentials then lead to independent scattering and a k-independent contribution to $F(k)$. Cross terms involving the ion-dipole partial structure factor average to zero on the same basis. Only the ion-ion structure factor remains and is calculated from Percus-Yevick theory.

---

* The structure factors $a_{ij}(k)$ are related to the Fourier transform of the radial distribution functions $g_{ij}(r)$. The radial distribution function $g_{11}(r)$ is defined such that $n_1 g_{11}(r)\, 4\pi r^2\, dr$ gives the number of type 1 particles in a spherical shell of radius r centered about a type 1 particle; n_1 is the average density of type 1 particles. g_{22} refers to type 2 particles with respect to type 2, and g_{12} (which is equal to g_{21}) refers to type 2 particles with respect to type 1. Naturally for a one-component system there is only one structure factor.

Russakoff and Ashcroft adjust the moment of the bound NH$_3$ dipole to 125% of the free dipole moment to account for the polarization produced by the field of the ion. The dipole contributions to the field of the solvated ion tend to cancel the ionic and thus account for the long mean free path in saturated solutions.

Schroeder and Thompson [37] used a repulsive δ-function approximating to the pseudopotential, ignored the dipole field and obtained results equivalent to Russakoff and Ashcroft. Schroeder and Thompson argue that the high dielectric constant of the NH$_3$ together with the free electron screening make any long ranged potential unlikely. The dipole potential, in any event, will be much weaker than the Coulomb and it seems more likely that the repulsion from filled orbitals will dominate. Schroeder and Thompson therefore chose the simplest form factors for the ammonia and solvated ion potentials: constant form factors, which are equivalent to δ-function potentials. The constants were fitted empirically.

The agreement in either calculation must be regarded as exceedingly good, as may be seen in Fig. 13 especially when the wide range of conductivities is considered. The immediate implication is that the large change of σ with x within the metallic range is primarily a consequence of the change in local order [i.e., $a(k)$] with x and of the decrease in the free NH$_3$ fraction. The change in electron concentration, or k_F, is not so important. The "dressing" of each ion by a sheath of NH$_3$ molecules is doubtless responsible for the slight dependence of the conductivity, among other parameters, on the particular solute considered.

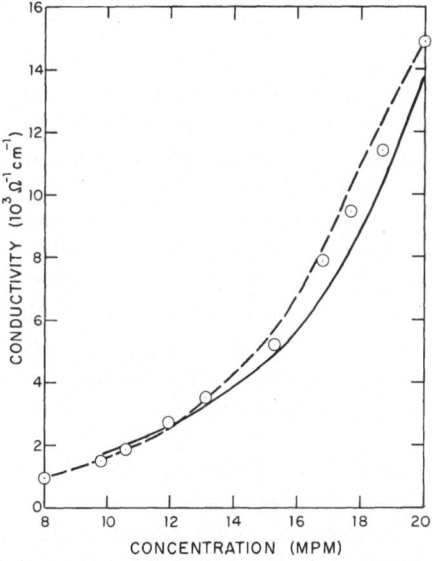

Fig. 13. A comparison of calculated and experimental conductivities for Li–NH$_3$ solutions at 240 °K. The solid curve is due to Ashcroft and Russakoff who adjusted the NH$_3$ dipole moment to improve their fit while the dashed curve is due to Schroeder and Thompson who forced a fit at 20 MPM by adjusting the strength of the solvated ion potential and at a single, lower concentration. Other details are given in the text. The data are denoted by ○

The thermopower agrees with free electron calculations as to order of magnitude [1, 20] but the sign changes near 15 MPM present a problem. Presumably Eq. (4) should permit a more precise calculation and, indeed, calculations based on Eq. (4) are able to produce both positive and negative values for S in pure liquid metals [14]. Such calculations often require that the pseudopotential form factor depend upon energy, however. Attempts to apply Eq. (4) to $M-NH_3$ solutions [37] have not been too successful thus far, though fairly close agreement has been obtained [37] when an empty-core potential is used [37]. The model potentials used thus far are just too simple to be expected to yield close agreement in so sensitive a parameter as S; the qualitative interpretation seems firm, nevertheless.

One of the most persistent problems plaguing attempts to fit $M-NH_3$ solutions into the framework of the NFE or Ziman theory [14] of liquid metals has been the positive values of the temperature coefficient of the conductivity. If γ is defined by

$$\gamma = \frac{1}{\sigma} \, d\sigma/dT,$$

then $\gamma < 0$ for $0 > x > 20$ MPM for all solutions [4, 40] at 240 K. While negative values of γ have been reported for nearly saturated $Li-NH_3$ solutions and for $Cs-NH_3$ solutions more concentrated than 30 MPM, the general trend [4, 5] is for $\gamma > 0$. In the early days this sign led to the suggestion that concentrated $M-NH_3$ solutions were semiconductors. Hall effect measurements have eliminated the latter possibility and it is also important to note that σ is a linear [20] (not exponential) function of T. Vexing though this result may be in terms of NFE theory it is the rule rather than the exception among liquid conductors of fairly low conductivity. Figure 14 shows some data compiled from the references quoted by Allgaier [3]. Though the solutions do not precisely mimic the other data, the sense is clearly the same. We must thus look to general explanations rather than to specific ones. That is, the phenomenon is sufficiently widespread that it seems reasonable to find an explanation of broad applicability rather than one tied to highly detailed models.

Parenthetically, let me point out that ad hoc explanations can be constructed within the NFE theory for models such as that of Ashcroft and Russakoff [2]. Indeed, even the simple model developed by Schroeder [37] can be used to illustrate what must be done; an ad hoc assumption of a temperature dependent potential is required. The net change is near -0.15% per degree. While temperature (or energy) dependent pseudopotentials are a common part of many models of liquid metals [38], such an approach would seem too detailed to be realistic.

One may then ask what more general mechanisms are available.

Catterall and Mott [41] have listed several:

a) Thermal fluctuations in density or polarization of the free NH_3.

b) Scattering of the Conwell-Weisskopf type from the charged cation as modified by the decrease in the dielectric constant as temperature increases.

c) A drop in $a(k)$ for ion-ion correlation due, again, to the drop in dielectric constant. Only the small number of NH_3 molecules not involved in solvation would be expected to contribute to this screening.

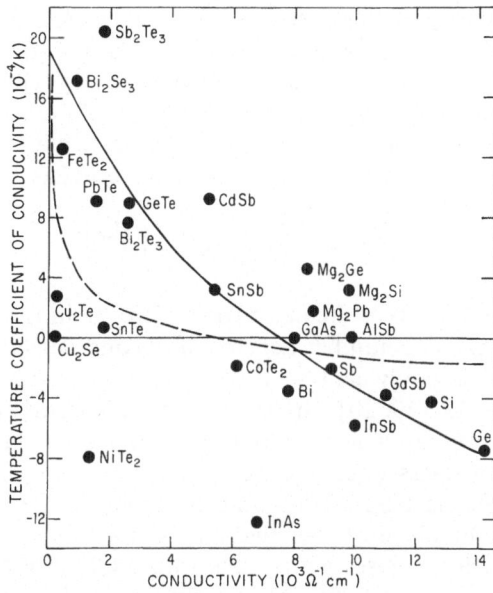

Fig. 14. The temperature coefficient of conductivity as a function of conductivity for several materials as compiled from references given in Ref. [3]. The solid line simply indicates the trend while the dashed curve gives similar data for Cs–NH₃ solutions from Ref. [5]

d) Scattering by the density fluctuations which exist *between* solvated ions [1]. The NH₃ molecules oriented by positive ions tend to produce a polarization well between the cations. These defects would be less well defined as T increases and thus would scatter less.

None of their explanations appears to have the requisite generality, expecially when the trend toward $\gamma > 0$ when $\sigma < 5 \times 10^3 \, \Omega^{-1} \, cm^{-1}$ reported by Allgaier [3] is considered.

It must also be noted that a positive γ among "good" liquid metals (e.g., Hg and Zn) is a structural effect possible only in divalent metals [14] where $2k_F$ lies close to the first peak in $a(k)$. But structural effects of that sort seem to be ruled out in complex binary mixtures such as this one.

It seems that a consistent and plausible answer can be obtained by building upon the similarities in the temperature coefficients of conductivity and suscep- tibility [21]. The one factor common to both properties is the *density-of-states*. The susceptibility χ contains $N(E)$ explicitly, and it is the only part of χ which can provide a temperature dependence. The NFE formulae for σ presume the density-of-states to be of the form $N(E) \sim E^{\frac{1}{2}}$ and temperature *independent*. The data require, instead, that $N(E)$ have a temperature dependence. While such an assumption provides a ready explanation of the effect of T upon σ and χ it is less obvious that no effect on R_H [20] or the optical $\varepsilon(\omega)$ [27] should be observed.

If one follows Mott [35] who has introduced the density-of-states explicitly into the transport phenomena via the factor

$$g = N(E_F)/N(E_F)_{f.e.}, \qquad (7)$$

where $N(E)$ is the density-of-states and $N(E)_{\text{f.e.}}$ is the free electron density-of-states then g must enter not only σ (as g^2) and χ (as g) but also R_H and $\varepsilon(\omega)$. However, the effect of g on $\varepsilon(\omega)$ is expected to be negligible [42] so long as g exceeds 0.5 and thus any temperature dependence of $\varepsilon(\omega)$ might easily have been missed. One expects [43, 44] g to enter R_H as g^{-1}, but R_H also has a temperature dependence due to the density alone. Thus one has

$$\frac{1}{R_H}\frac{dR_H}{dT} = \frac{1}{n}\left|\frac{dn}{dT}\right| - \frac{1}{g}\left|\frac{dg}{dT}\right|, \tag{8}$$

where n is the number density of free electrons. Here, as in the other properties, there is a mixture of mass or number density effects and density-of-states effects. The observed dR/dT is equivalent to dn/dT both as to sign and magnitude within experimental error [19, 20]; the effect of dg/dT seems small. There is no trend with concentration equivalent to that seen in χ or σ. This point, then, is unresolved. Density-of-states effects could also provide an explanation of the deviations between free electron models and the observed thermopower.

It nevertheless seems highly probable that the temperature effects may be explained by $N(E)$, which increases with T due to the filling of a pseudogap.

In Figs. 11 and 12 we saw the $\langle|\Psi|^2\rangle$ extracted from the Knight shift at Na and Cs by Lelieur [21]. As the system tends more and more towards pure "solvated ionium", more and more of the wavefunction piles up on the solvated ion. The differences there, apart from those obviously attributable to ion size, may be assigned to solvation effects. In particular, the concentration dependence is sufficiently close when the reduced concentration is used, that one must conclude the wavefunctions to be quite similar and determined by the medium (the free NH_3) separating the solvated ions. Perhaps it is the potential wells between the ions [1] which are washed out at high temperatures and concentrations.

Other solvation effects appear in the compressibility [37, 45]. The Percus-Yevick, hard sphere structure factor $a(k)$ yields as $k \to 0$ an estimate of the compressibility. Best fits of the hard sphere calculation to the measured compressibilities yield the coordination numbers implicit in the following formulae for the solvated cations:

$$Li(NH_3)_4^+, \quad Na(NH_3)_4^+, \quad K(NH_3)_6^+, \quad \text{and} \quad Cs(NH_3)_8^+.$$

These results are generally consistent with geometrical expectations and chemical experience [15, 36].

Solutions of divalent metals show no essential difference from those of alkali metals [4]. The conductivity is a somewhat weaker function of concentration. The probable reason is the fact that the free ammonia fraction [2, 45] changes less rapidly at the lower metal concentrations characteristic of the metallic phase of divalent solutions. Bowen [8] ascribes the minimum in the sound speed to compound formation. One expects models similar to those developed for the monovalent solutions to be equally successful here.

We finally arrive at a model which considers the solutions to be composed of solvated metal ions, free NH_3 molecules (i.e., NH_3 molecules not involved in any solvation layer at a given instant), and free electrons. Though this mixture is not so good a metal as Na (at least below 20 MPM) and must therefore be placed in Allgaier's [3] class B, the qualitative deviations from NFE behavior are few. There appears to be a pseudogap in the density of states which has a noticeable effect on the temperature dependence of the conductivity and susceptibility. The origin of this pseudogap is uncertain though it may possibly arise from the weak attraction of an electron towards the space between two solvated cations where the positive ends of the oriented NH₃ molecules create a well somewhat similar to that which traps a solvated electron [1].

Diffraction studies would aid in the resolution of this as well as many other outstanding problems. It is also time to study the divalent solutions in more detail and to explore the saturation curves above room temperature.

References

1. Lepoutre,G., Lelieur,J.P.: In: Lagowski,J.J., Sienko,M.J. (Eds.): Proc. Collloque Weyl II, p. 247. London: Butterworths 1970.
2. Ashcroft,N.W., Russakoff,G.: Phys. Rev. A **1**, 39 (1970).
3. Allgaier,R.S.: Phys. Rev. **185**, 227 (1969); — Phys. Rev. B **2**, 2257 (1970). — Thompson, J.C., Allgaier,R.S.: Phys. Rev. A **2**, 1103 (1970).
4. Schroeder,R.L., Thompson,J.C., Oertel,P.L.: Phys. Rev. **178**, 298 (1969).
5. Castel,J., Lelieur,J.P., Lepoutre,G.: J. Phys. **32**, 211 (1971).
6. Arias-Limonta,J.A., Varlashkin,P.G.: J. Chem. Phys. **52**, 581 (1970).
7. Kraus,C.A.: J. Am. Chem. Soc. **43**, 749 (1921).
8. Bridges,R., Ingle,A.J., Bowen,D.E.: J. Chem. Phys. **52**, 5106 (1970).
9. Lo,R.E.: Z. Anorg. Allgem. Chem. **344**, 230 (1966).
10. Morgan,J.A., Schroeder,R.L., Thompson,J.C.: J. Chem. Phys. **43**, 4494 (1965).
11. Bredig,M.A.: Mixtures of metals with molten salts. In: Blander,M. (Ed.): Molten salt chemistry, p. 367. New York: Wiley 1964.
12. Sharp,A.C., Davis,R.L., Vanderhoff,J.A., LeMaster,E.W., Thompson,J.C.: Phys. Rev. A **4**, 414 (1971).
13. Cusack,N.E.: Rep. Prog. Phys. **26**, 361 (1963).
14. Ziman,J.M.: Phil. Mag. **16**, 551 (1967).
15. Damay,P.: Thesis (Lille), unpublished (1972).
16. Dewald,J.T., Lepoutre,G.: J. Am. Chem. Soc. **76**, 3369 (1954).
17. Damay,P., Depoorter,M., Lepoutre,G.: In: Lagowski,J.J., Sienko,M.J. (Eds.): Proc. Colloque Weyl II, p. 233. London: Butterworths 1970.
18. Vanderhoff,J.A., Thompson,J.C.: J. Chem. Phys. **55**, 105 (1971).
19. Kyser,D.S., Thompson,J.C.: J. Chem. Phys. **42**, 3910 (1965).
20. Nasby,R.D., Thompson,J.C.: J. Chem. Phys. **53**, 109 (1970).
21. Lelieur,J.P.: Thesis (Lille), unpublished (1972) and this conference.
22. Huster,E.: Ann. Physik **33**, 477 (1938).
23. Suchannek,R.G., Naiditch,S., Klejnot,O.J.: J. Appl. Phys. **38**, 690 (1967).
24. Myers,W.R.: Rev. Mod. Phys. **24**, 15 (1952).
25. O'Reilly,D.E.: J. Chem. Phys. **41**, 3729 (1964).
26. Acrivos,J.V., Mott,N.F.: Phil. Mag. **24**, 19 (1971).
27. Somoano,R.B., Thompson,J.C.: Phys. Rev. A **1**, 376 (1970).
28. McKnight,W.H., Thompson,J.C.: Bull. Am. Phys. Soc. **17**, 368 (1972). — Mueller,W.E., Thompson,J.C.: In: Lagowski,J.J., Sienko,M.J. (Eds.): Proc. Colloque Weyl II, p. 293. London: Butterworths 1970.
29. Bowen,D.E.: J. Chem. Phys. **51**, 1115 (1969).
30. Thompson,J.C., Oré-Oré,C.R.: J. Chem. Phys. **54**, 2279 (1971).
31. Bailey,K.E., Bowen,D.E.: J. Chem. Phys. **56**, 4809 (1972).
32. Kaplan,J., Kittel,C.: J. Chem. Phys. **21**, 1429 (1953).
33. Mott,N.F.: Phil. Mag. **19**, 835 (1969).

34. Cohen, M. H.: Lectures at NATO Summer School on Amorphous Semiconductors, Gent, Belgium, 1969 (unpublished). — J. Non-Crystalline Solids **4**, 391 (1970).
35. Mott, N. F., Davis, E. A.: Electronic processes in non-crystalline materials. Oxford: Clarendon Press 1971.
36. Conway, B. E., Barradas, R. G.: Chemical physics of ionic solutions. New York: Wiley 1966. — Hinton, J. F., Amis, E. S.: Chem. Rev. **71**, 627 (1971).
37. Schroeder, R. L., Thompson, J. C.: Bull. Am. Phys. Soc. **13**, 397 (1968); — Phys. Rev. **179**, 124 (1969).
38. Ashcroft, N. W., Lekner, J.: Phys. Rev. **145**, 83 (1966). — Ashcroft, N. W., Langreth, D. C.: Phys. Rev. **156**, 685 (1967); — Phys. Rev. **159**, 500 (1967).
39. Davis, H. T., Schmidt, L. D., Minday, R. M.: Phys. Rev. A **3**, 1027 (1971).
40. Cohen, M. H., Thompson, J. C.: Advan. Phys. **17**, 857 (1968).
41. Catterall, R., Mott, N. F.: Advan. Phys. **18**, 665 (1969).
42. Faber, T. E.: Advan. Phys. **15**, 547 (1966); — Intro. Theory of Liquid Metals. Cambridge: Cambridge University Press 1972.
43. Even, U., Jortner, J.: Phys. Rev. Letters **28**, 31 (1972); — Phil. Mag. **25**, 715 (1972).
44. Friedman, L., Mott, N. F.: J. Non-Crystalline Solids **7**, 103 (1972).
45. Thompson, J. C.: Phys. Rev. A **4**, 802 (1971).

Discussion

M. J. SIENKO   I should like to support your suggestion that the positive temperature coefficient of susceptibility is mainly due to a change in the density-of-states. At Cornell we have measured the susceptibility of concentrated lithium–ammonia solutions and we find that the slight increase (4%) from 94—194 °K can be accounted for by assuming a variation of band width due to volume expansion of the solution with temperature [J. Chem. Phys. **56**, 4756 (1972)].

J. C. THOMPSON   I welcome your support; however, the required effects in the concentrated solutions considerably exceed those attributable to volume expansion. Some other source must be found.

M. H. COHEN   I should like to offer for consideration an additional mechanism for the temperature and concentration dependence of electronic properties in the concentrated range. At low temperatures and concentrations below saturation, the structure should not be greatly different from that of (e.g.) the compound $Li(NH_3)_4$. There are well-developed electron cavities (after all, the excess volume associated with the electron remains throughout this range), and these are well ordered with respect to the metal ions to minimize the Madelung energy (as reduced to screening by the NH_3). This gives a tight binding band associated with motion of electrons from cavity to cavity. At higher temperatures, this short-range order is washed out, allowing the possibility that the cavities come closer together. The band width would go up, density-of-states go down, conductivity go up, etc. This effect of the washing out of short-range order with increasing temperature with consequent increase of conductivity via a variety of mechanisms is quite common and may well be one of the explanations underlying the Allgaier correlation.

While I am prepared to accept the nearly free electron model well above the "compound" or "complex" composition, I find that the above arguments for a tight binding model are more convincing in the range near and below that composition. Since your calculations were fitted at two points and Ashcroft's at one, it might be better to view them more as interpolation procedures. Then, in my view, the deficiency in the calculation lies in ignoring Coulomb correlation and its temperature dependence on a_{12}.

Finally, I might add that I see no way to get a pseudogap at the Fermi energy within the structural models we have been discussing.

J. JORTNER   I would like to comment on the conjecture that the electrical conductivity, σ, of concentrated metallic solutions is determined by the pseudo-gap depth. When the mean free path considerably exceeds the spacing between the scattering centres, the cancellation theory for the conductivity (Faber and Ziman) applies, and σ is independent of the pseudogap depth. Furthermore, as pointed out by Even, this interpretation is in variance with the Hall effect data.

Concerning the comments made by Cohen, I think that the pressure dependence of σ reported by Schindewolf at Colloquim Weyl II supports the tight binding model for the formation of the conduction band in concentrated solutions.

J.C. THOMPSON   I am well aware that Mott's g-factor cancels out the con-ductivity when the mean free path is long. It is well known (Refs. [19] and [20]) that the mean free path ranges from 10—70 Å over the 10—20 MPM range, in Li–NH₃ solutions. What I suggest is not too far from what Cohen is talking about, that is, the regions between solvated ions in a concentrated solution (Ref. [1]) are really similar to the cavities which trap electrons in dilute solutions. These shallow, fluctuating traps will wash out as T increases, pushing more of the electrons onto the solvated ion and into the solvated ion band.

Nuclear Magnetic Resonance Studies of Cesium–Ammonia Solutions

J. P. Lelieur

Abstract

The ^{133}Cs and ^{14}N Knight shifts have been measured in cesium–ammonia solutions, versus metal concentration and temperature. A by-product of the experimental results is a liquid–solid phase diagram at high concentrations; evidence is found that the solid is pure cesium.

I. Introduction

Cesium is known to be soluble in liquid ammonia at any concentration, while for other alkali metals saturation occurs at about $15-20$ MPM (mole per cent of metal). Physical properties can then be followed continuously versus cesium concentration in cesium–ammonia solutions. Among different methods, NMR provides a powerful way to study the environment of nuclei in condensed matter. We performed NMR experiments for Cs–NH$_3$ solutions, measuring Knight shifts of ^{133}Cs and ^{14}N nuclei. O'Reilly [1, 2] previously measured ^{133}Cs Knight shifts, but for concentrations below 2 MPM, and at 300 °K. The experiments were relative to more concentrated solutions; they were performed at variable temperatures to obtain various temperature coefficients. The phase diagram of Cs–NH$_3$ solutions has been determined from electrical conductivity studies [3]. Our work by NMR was also undertaken to clarify the phase diagram for concentrations greater than the eutectic.

II. Experimental

Nuclear resonances were observed with a VARIAN Associates D P 60 spectrometer, by continuous radio-frequency irradiation of the sample. The experiments were done at fixed frequency (about 7 MHz for ^{133}Cs resonance and at 3 MHz for ^{14}N resonance). The frequency of the radio-frequency unit was stabilized by a quartz oscillator. The magnetic field was produced by a 12-inch VARIAN electromagnet. The magnetic field measurements were taken with a F 8 A VARIAN gaussmeter which uses the nuclear resonance of proton or deuton. The Knight shift (K) is defined by:

$$K = \frac{H_r - H_S}{H_r},\tag{1}$$

where H_r is the magnetic field for which resonance occurs in the reference solution and H_S, the magnetic field at which resonance occurs in the Cs–NH$_3$ solutions at fixed frequency. For ^{133}Cs Knight shift, the reference is a saturated ClCs aqueous solution. For ^{14}N Knight shift, the reference is a liquid ammonia sample.

Measurements were taken at various temperatures. The sample was cooled with a temperature-regulated, gaseous nitrogen stream. The temperature of the sample was measured by thermocouple situated as close to the sample as possible.

The solutions were prepared by the usual methods: cesium was displaced from its chloride with calcium and then triply distilled; ammonia was dried on an alkali metal and distilled on cesium; the composition of the solutions was determined by chemical analysis [4]. It was observed that Knight shift measurements are much less sensitive to decomposition than magnetic susceptibility measurements. It was found however than cesium-ammonia solutions were much more stable when all the glassware was cleaned with ammonium fluoride.

For concentrations greater than about 4 MPM asymmetric line shapes were observed. This asymmetry is due to the skin effect and is characteristic of metals [4]. To escape the skin effect, almost all resonance studies in metals were performed on small particles embedded in an oil. In metal–ammonia solutions, it was impossible to use such techniques [4]. We used cylindrical samples 6 mm in diameter. Their diameter was greater than the skin depth above about 4 MPM. The skin effect causes the center of the resonance line to be slightly displaced. This effect has been computed by Chapman et al. [5]. In Cs–NH_3 solutions, this effect is small (~ 10 ppm) because the line width is about 200 mG; however, it was taken approximately into account [4].

III. Experimental Results

A. Variations of Cesium Knift Shift [K(Cs)] Versus Metal Concentration

The experimental Knight shifts of the cesium nucleus are plotted in Fig. 1 versus temperature for the samples studied. Although Fig. 1 implicitly contains the variations of the K(Cs) with concentration and temperature, it essentially provides evidence of the liquid-solid equilibrium. In effect, for any sample, for temperatures below a given temperature T_L, the values of the Knight shift depend only upon temperature. The temperature T_L, variable with concentration, defines a point of the liquidus. All of the measurements for temperatures below the liquidus are situated on the AC curve. For each point of this curve, there is another resonance signal situated on the AB line. This means that for temperatures below the liquidus, two resonance signals are observed. These observations are characteristic of a liquid-solid equilibrium. We shall see later that the second resonance signal (on line AB in Fig. 1) corresponds to pure cesium.

From the experimental points in Fig. 1, the variations of K(Cs) versus metal concentration at $-40°$ C were plotted in Fig. 2. This curve deals only with Knight shifts in the liquid phase. For concentrations above 50 MPM, $-40°$ C is already under the liquidus; in this case, the results in the liquid phase were extrapolated down to $-40°$ C. It can be seen in Fig. 1 that the K(Cs) increases from some ppm for weakly concentrated solutions to about 15000 ppm for pure cesium. The values of K(Cs) increase rapidly up to about 25 MPM, and show an inflexion between 30 and 35 MPM in the same concentration range in which the electrical conductivity also shows an inflexion point.

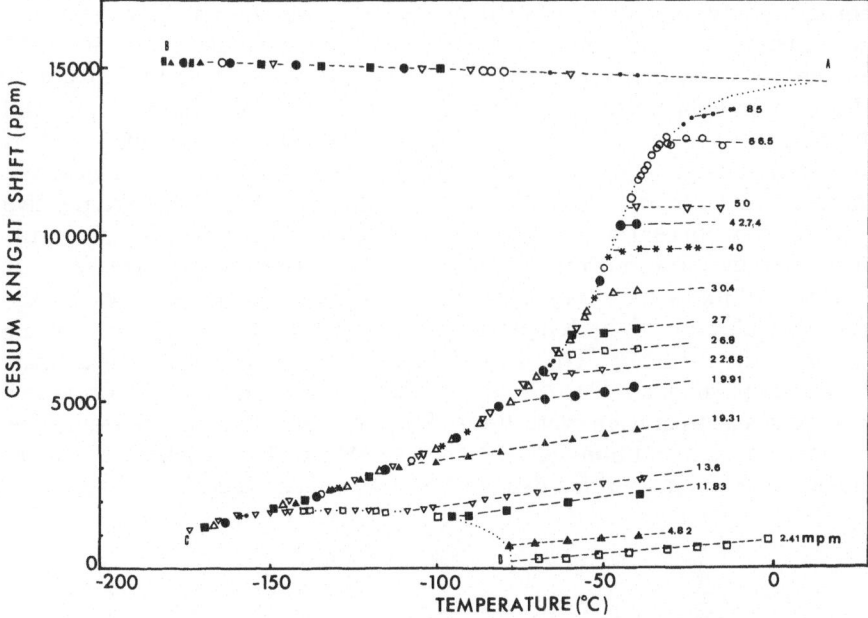

Fig. 1. Cesium Knight shift versus temperature for the samples studied

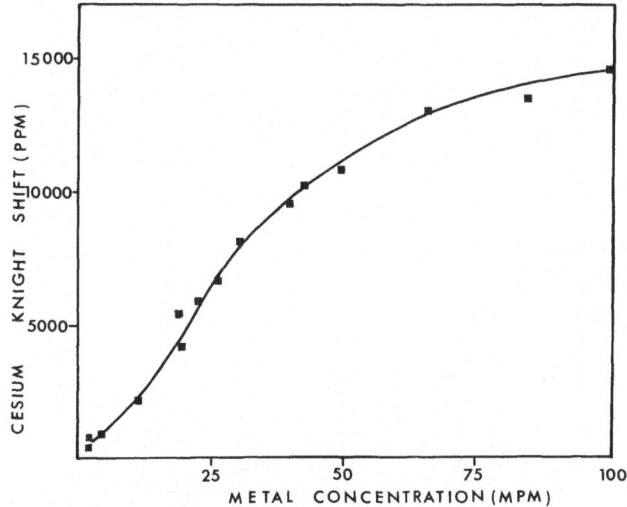

Fig. 2. Cesium Knight shift versus metal concentration at $-40°$ C

B. Variations of the Cesium Knight Shift Versus Temperature

The variations of $K(Cs)$ with temperature are also implicitly contained in Fig. 1. We define the temperature coefficient of the $K(Cs)$ at $-60°$ C by:

$$\gamma[K(Cs)]_{-60°C} = \left\{ \frac{1}{K(Cs)} \frac{dK(Cs)}{dT} \right\}_{-60°C}. \qquad (2)$$

The corresponding values are plotted in Fig. 3. It was found than the $K(Cs)$ always increases with temperature, for concentrations below about 50 MPM and for temperatures below about $-30°$ C. We cannot exclude the possibility that above room temperature for instance, $K(Cs)$ goes through a maximum and that its temperature coefficient becomes negative, but cesium–ammonia solutions are not stable enough, even when prepared very carefully, to permit accurate enough measurements. It can be seen in Fig. 3 that $\gamma[K(Cs)]$ has large positive values for weakly concentrated solutions. This quantity decreases rapidly in the intermediate concentration range when metal concentration increases; for concentrated solutions the decrease is smooth. Recall that for pure cesium the temperature coefficient of the Knight shift is negative ($-1.9 \cdot 10^{-4}$ deg^{-1} in the solid phase, $-3 \cdot 10^{-4}$ deg^{-1} in the liquid phase). In the temperature range usually studied (from $-80°$ C to about $-20°$ C), it was found that the variations of $K(Cs)$ versus temperature were linear for very concentrated solutions. For the less concentrated solutions (2.41, 2.55, 4.82 MPM) this variation was more rapid than linear; in fact, in this case, $\ln K(Cs)$ was linear versus $1/_T$ [4].

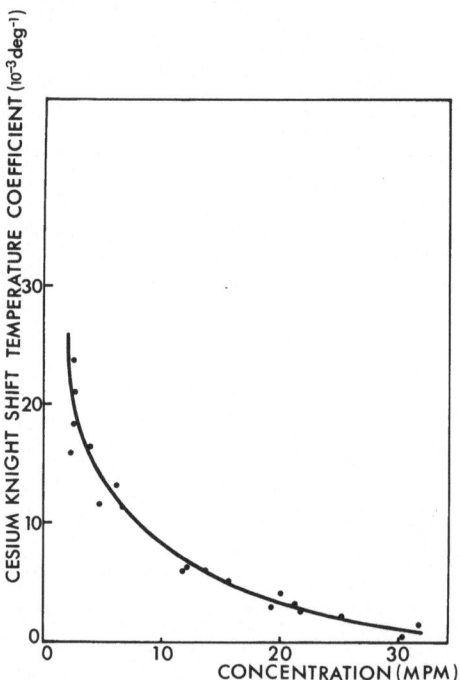

Fig. 3. Cesium Knight shift temperature coefficient versus metal concentration at $-60°$ C

C. Variations of Nitrogen Knight Shift [$K(N)$] Versus Metal Concentration

Measurements of the $K(N)$ versus metal concentration and temperature have been taken for Cs–NH$_3$ solutions and for 4 Na–NH$_3$ solutions. For very concentrated Cs–NH$_3$ solutions, measurements were difficult because the nitrogen nuclear resonance signal is very weak; results have been obtained for concentrations below 40 MPM. Variations of $K(N)$ versus metal concentration at $-60°$ C are shown in Fig. 4 (these values are relative to the liquid phase). In Fig. 4, $K(N)$ is seen to increase rapidly versus metal concentration up to about 20 MPM, and then $K(N)$ becomes approximately constant and equal to 910 ppm. Although the experimental signals were weak, we think the constancy of $K(N)$ from about 20 to 40 MPM is a real effect. Figure 4 also shows that the nitrogen Knight shift does not depend on the cation in solution. This confirms previous observations of O'Reilly [1, 2].

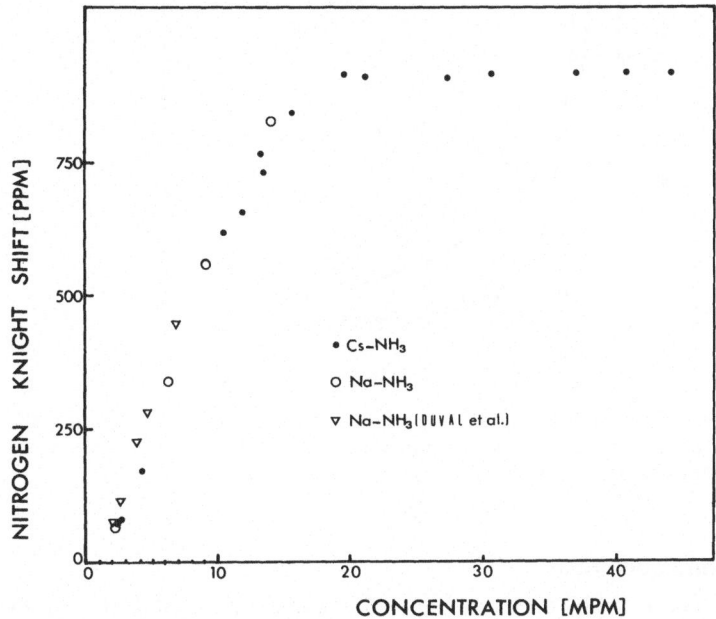

Fig. 4. Nitrogen Knight shift versus metal concentration at $-60°$ C in Cs–NH$_3$ solutions

D. Variations of the Nitrogen Knight Shift Versus Temperature

The temperature coefficient of the $K(N)$ has been found to be always positive, as that of the $K(Cs)$. Similarly we define the temperature coefficient of the $K(N)$ by:

$$\gamma[K(N)]_{-60°C} = \left\{ \frac{1}{K(N)} \frac{dK(N)}{dT} \right\}_{-60°C}. \tag{3}$$

Variations of this quantity versus metal concentration are plotted in Fig. 5, with the temperature coefficients of the $K(Cs)$ and of the paramagnetic susceptibility of cesium in liquid ammonia. It is seen that $\gamma[K(N)]$ has large values for weakly concentrated solutions, decreases rapidly up to about 4 MPM and then decreases smoothly. It must also be emphasized that the variations of $\gamma[K(N)]$ are approximately identical to those of the temperature coefficient of the paramagnetic susceptibility.

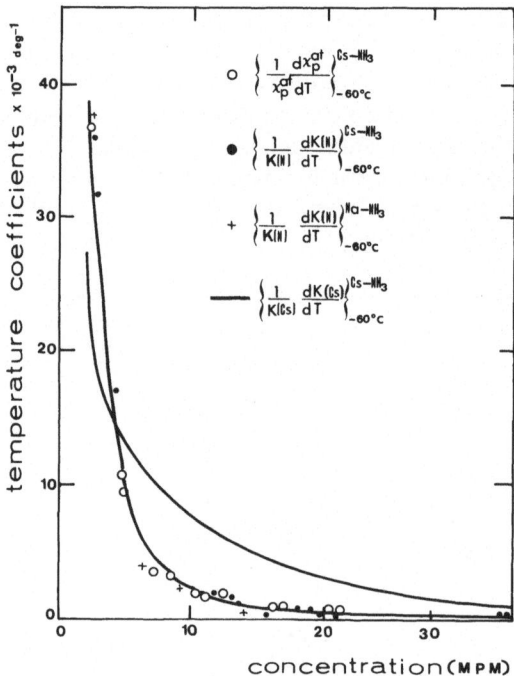

Fig. 5. Nitrogen Knight shift temperature coefficient versus metal concentration at $-60°$ C, compared to the cesium Knight shift temperature coefficient and to the paramagnetic susceptibility temperature coefficient

E. Liquid-Solid Equilibrium

Above (A), it has been shown from Fig. 1 that the experimental measurements of the cesium Knight shift show the existence of a liquid-solid equilibrium. In Fig. 6 are plotted most of the experimental points corresponding to line AB of Fig. 1, but at a different scale. Knight shifts of pure cesium versus temperature are also plotted in Fig. 6. These values are about 40 ppm greater than the Knight shifts of line AB assigned to the "solid" resonance signal. We see that this difference of 40 ppm is slightly greater than the dispersion of experimental points shown in Fig. 6. Therefore we can conclude that the "solid" is pure cesium. It is possible that a few ammonia molecules are still in solution inside. But it is impossible that this "solid" should be a compound of the type $Cs–(NH_3)n$ where n should be an integer.

The eutectic point is situated at a concentration of 15.7 ± 0.5 MPM. The determination of the temperature of the eutectic is more difficult, because the variations of the cesium Knight shift show some peculiarities at that point, as plotted in Fig. 7. For concentrations above the eutectic concentration, $K(Cs)$ goes along the ABEC curve when the temperature decreases. For concentrations below the eutectic concentration, $K(Cs)$ follows the A'BDEC curve. For all of the "liquid" resonance signal disappears at $-175°$ C. One therefore thinks that $-175°$ C is the eutectic temperature. The trend of the curves plotted in Fig. 7,

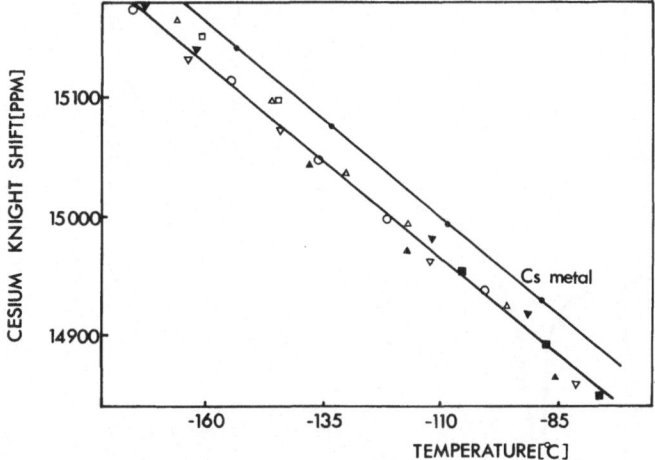

Fig. 6. Cesium Knight shift for pure cesium (.) and for the "solid" in Cs–NH$_3$ (different symbols represent different samples) versus temperature

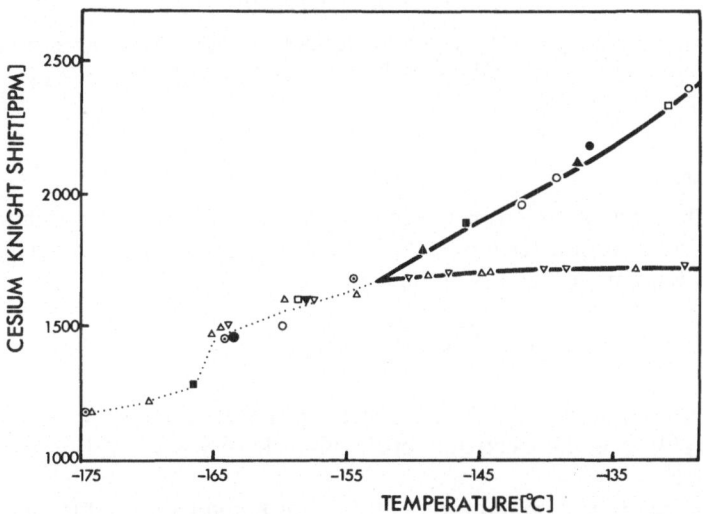

Fig. 7. Cesium Knight shift in the neighborhood of the eutectic versus temperature (different symbols represent different samples)

however, is surprising. In any case, the previously reported eutectic temperature of $-143°$ C [3] is not correct.

In Fig. 1, it is seen that the variations of the $K(Cs)$ versus temperature for a given sample change their slope at the liquidus, above the eutectic concentration where the "solid" is pure cesium, and below the eutectic concentration where the "solid" is NH$_3$. The phase diagram of Cs–NH$_3$ solutions can therefore be plotted in Fig. 8. The values of the cesium Knight shift at the liquidus and Fig. 2 have been used to determine more accurately the concentrations of Fig. 8.

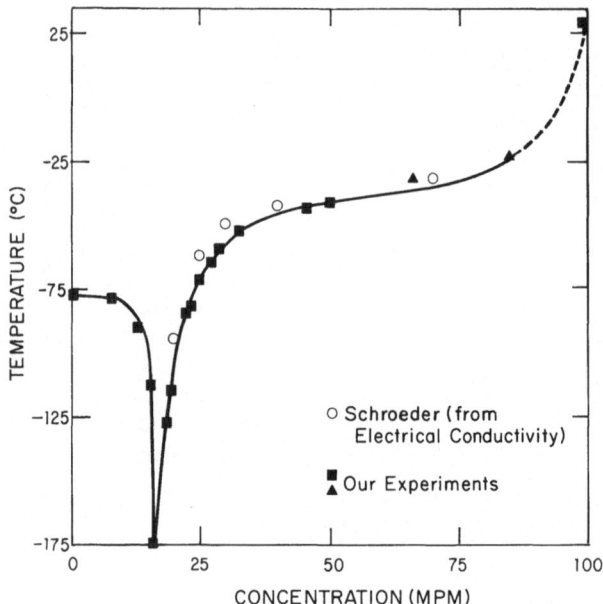

Fig. 8. Phase diagram of Cs–NH$_3$ solutions (temperature versus metal concentration). ■, ▲ = our results and ○ = Schroeder's results from electrical conductivity

IV. Discussion

The Knight shift originates from the contact term of hyperfine interaction [8]. This contact term is nonzero only for the s character of the wavefunction. Generally the Knight shift can be expressed by:

$$K = \frac{8\pi}{3} \frac{\chi_p^{At}}{N_{AV}} \langle |\Psi_F(0)|^2 \rangle . \qquad (4)$$

N_{AV} is Avogadro's number; χ_p^{At}/N_{AV} is therefore the paramagnetic susceptibility per atom. $\langle |\Psi_F(0)|^2 \rangle$ is the electronic probability density averaged on the Fermi electronic states.

To appreciate the order of magnitude of the metal Knight shifts which have been measured in liquid ammonia solutions, it is interesting to plot the values of the ratio:

$$\frac{K(M)}{K(M-NH_3)} \qquad (5)$$

where $K(M)$ is the Knight shift of pure metal M, and $K(M-NH_3)$ is the Knight shift of metal M in NH$_3$. With previously published Knight shifts [6, 7], the values of the above ratio are reported in Fig. 9. The ratios are identical for weakly concentrated solutions, while deviations appear for concentrated solutions. At first sight, it appears that, for weakly concentrated solutions, the electron wavefunction is weakly dependent on the cation, but for very concentrated solutions, the wavefunction is more dependent on the cation in solution. In fact,

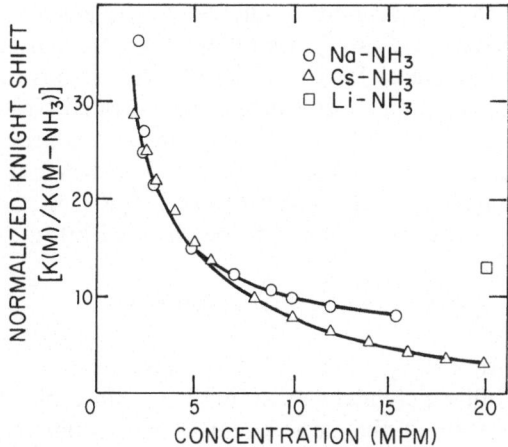

Fig. 9. $K(M)/K(M-NH_3)$ for Li–NH$_3$ [6], Na–NH$_3$ [7], Cs–NH$_3$ (this work) versus metal concentration

as seen from Eq. (4), the electron density at the nucleus and the paramagnetic susceptibility intervene simultaneously in the Knight shift. Using our magnetic susceptibility results [4, 9], it was possible to compare the electron density at the Na and Cs nuclei. The quantities $\langle|\Psi_F(0)|^2\rangle$, relative to Na and Cs in Na–NH$_3$ and Cs–NH$_3$ solutions, are reported in Fig. 10. Theses quantities have a different order of magnitude, due in part to the nuclei themselves. The trends are approximately the same below 4 MPM: $\langle|\Psi_F(0)|^2\rangle$ decreases when metal concentration increases; this trend can probably be related to the electron

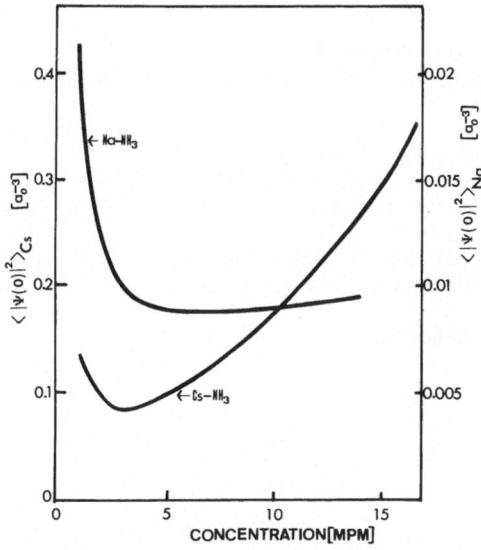

Fig. 10. $\langle|\Psi_F(0)|^2\rangle$ for Cs in Cs–NH$_3$ at $-40°$ C (left-hand scale) and $\langle|\Psi_F(0)|^2\rangle$ for Na in Na–NH$_3$ at $-30°$ C (right-hand scale) versus metal concentration

delocalization and to the corresponding change of the electron wavefunction. Above 4 MPM the trends of $\langle|\Psi_F(0)|^2\rangle$ are different for Na and Cs: a very slight increase for Na, a strong increase for Cs. This difference means that the electron wavefunction has about the same nature in the concentrated Na–NH$_3$ solutions, while this nature changes in Cs–NH$_3$ solutions. This fact is probably related to differences in the solvation of respective cations. However no quantitative analysis has as yet been done for the variations of $\langle|\Psi_F(0)|^2\rangle$.

For the nitrogen Knight shift, Eq. (4) has to be changed into:

$$K(N) = \frac{8\pi}{3} \frac{\chi_p^{At}}{N_{AV}R} \langle|\Psi_F(0)|^2\rangle_N \tag{6}$$

where R is the mole ratio [NH$_3$]/[Na] [4]. One must divide by R to obtain the paramagnetic susceptibility per nitrogen atom. The values of $\langle|\Psi_F(0)|^2\rangle_N$ have been calculated to have about the same order of magnitude, equal to $0.4\,a_0^{-3}$ from 5—15 MPM, in Na–NH$_3$ and Cs–NH$_3$ [4]. For Cs–NH$_3$ solutions, the nitrogen Knight shift has been found approximately constant between 20 and 40 MPM. From Eq. (6) this means that $\chi_p^{At}\langle|\Psi_F(0)|^2\rangle_N$ varies as R does. In fact χ_p^{At} has been found to vary very slowly in this concentration range. Therefore $\langle|\Psi_F(0)|^2\rangle_N$ varies approximately as R from about 20 to 40 MPM, i.e. the electron density to the nitrogen nucleus varies linearly with the ammonia concentration.

Analysis of the temperature coefficients of nitrogen and cesium Knight shifts [4] indicates that they are much larger than corresponding free electron values. The temperature coefficients of $\langle|\Psi_F(0)|^2\rangle$ are found to have the same trends and values for Na and Cs; these values are much larger than in the corresponding pure metals. A quantitative description of the electronic wavefunctions is needed to interpret these experimental results.

References

1. O'Reilly, D. E.: J. Chem. Phys. **41**, 3729 (1964).
2. O'Reilly, D. E.: In: Lepoutre, G., Sienko, M. J. (Eds.): Metal–ammonia solutions. New York: Benjamin 1964.
3. Schroeder, R. L., Thompson, J. C., Oertel, P. L.: Phys. Rev. **178**, 298 (1969).
4. Lelieur, J. P.: Thèse doctorat, Orsay, 1972.
5. Chapman, A. C., Rhodes, P., Seymour, E. F. W.: Proc. Phys. Soc. **70**, 4B, 345 (1957).
6. Haynes, R., Evers, E. C.: In: Lagowski, J. J., Sienko, M. J. (Eds.): Metal–ammonia solutions. London: Butterworths 1970.
7. Duval, E., Rigny, P., Lepoutre, G.: Chem. Phys. Letters **2**, 237 (1968).
8. Abragam, A.: Les Principes du magnétisme nucléaire, Presses Universitaires de France, 1961.
9. Lelieur, J. P.: In this colloquium.

Conductivity of Concentrated Metal–Ammonia Solutions in the Frequency Range 0—70 GHz

K. G. BREITSCHWERDT and H. RADSCHEIT

Abstract

The conductivity of sodium–ammonia and lithium–ammonia solutions has been measured in the frequency range 0—70 GHz for concentrations between 0.01 and 4 MPM. It may be expressed in terms of a real conductivity and a complex permittivity. The deviation of·the low-frequency conductivity from an Arrhenius plot indicates that the conduction mechanism is not a simple activated process in the concentration range 0.1—3 MPM. The increasing dielectric constant in the same concentration range may be interpreted on the basis of a second relaxation process with longer relaxation time in addition to the dielectric relaxation of the ammonia molecules. Possible explanations for the experimental results are given.

Introduction

The high-frequency properties of metal–ammonia solutions can be described using in Maxwell's equation either a real conductivity σ and a complex dielectric constant $\varepsilon = \varepsilon' - i\varepsilon''$ or a complex conductivity $\sigma^*(\omega)$ as a function of frequency:

1.
$$\operatorname{rot} \boldsymbol{H} = \varepsilon\varepsilon_0 \frac{\partial \boldsymbol{E}}{\partial t} + \sigma \boldsymbol{E} , \tag{1}$$

where for a harmonic wave

$$\boldsymbol{E} = \boldsymbol{E}_0 e^{i\omega t}$$

one obtains

$$\operatorname{rot} \boldsymbol{H} = \boldsymbol{E}[i\omega\varepsilon\varepsilon_0 + \sigma] , \tag{2}$$

or

2.
$$\operatorname{rot} \boldsymbol{H} = \varepsilon_0 \frac{\partial \boldsymbol{E}}{\partial t} + \sigma^* \boldsymbol{E} , \tag{3}$$

where for a harmonic wave one obtains

$$\operatorname{rot} \boldsymbol{H} = \boldsymbol{E}[i\omega\varepsilon_0 + \sigma^*] . \tag{4}$$

Comparison between Eqs. (2) and (4) yields

$$\sigma^*(\omega) = (\varepsilon - 1)\,\varepsilon_0 i\omega + \sigma , \tag{5}$$

indicating that, in this picture, the imaginary part of the dielectric constant now contributes to the frequency-dependent real part of $\sigma^*(\omega)$, while the real part of the dielectric constant causes a phase shift. In the following we interpret our results using notation 1.

Experimental

The low-frequency conductivity was measured with the usual bridge techniques. The complex high-frequency permittivity may be determined from the complex propagation constant $\gamma = \alpha + i\beta$ of an electromagnetic wave in a wave guide filled with the metal–ammonia solution, where α is the attenuation constant and β is the phase constant of the microwave. Real and imaginary parts of the dielectric constant can be expressed in the following way:

$$\varepsilon' = \left(\frac{\lambda_0}{2\pi}\right)^2 (\beta^2 - \alpha^2) + \left(\frac{\lambda_0}{\lambda_c}\right)^2, \tag{6}$$

$$\varepsilon'' = \left(\frac{\lambda_0}{2\pi}\right)^2 2\alpha\beta - \frac{\sigma}{\varepsilon_0 \omega}, \tag{7}$$

where λ_0 is the free-space wavelength and λ_c the cut-off frequency of the empty wave guide.

An alternative method is to insert a capillary filled with the metal–ammonia solution into microwave resonators with various resonance frequencies and to measure the shift of their resonance frequencies and the decrease of their Q factors. From these data again the real and imaginary parts of the complex dielectric constant can be determined in the following way:

$$\varepsilon' = 1 + 2J_1^2(y_0) \frac{R^2(\omega_0 - \omega_0')}{\varrho^2 \omega_0}, \tag{8}$$

$$\varepsilon'' = J_1^2(y_0) \frac{R^2(\Delta' - \Delta)}{\varrho^2 \omega_0} - \frac{\sigma}{\varepsilon_0 \omega_0}. \tag{9}$$

where ω_0 and ω_0' are the resonance frequencies of the resonator, Δ and Δ' the half widths of the resonance curves without and with the sample, respectively, y_0 the zero of the Bessel function of order zero, R the radius of the cylindrical resonator and ϱ the radius of the cylindrical sample.

Equations (8) and (9) are an approximate solution to the problem, giving results with an error limit of about 2% as long as the condition

$$2.4 \sqrt{\varepsilon'} \frac{\varrho}{R} < 1 \tag{10}$$

is fulfilled. Otherwise a numerical method has to be employed.

Results

The d.c. conductivity [1—4] versus $1/T$ for different sodium concentrations is shown in Fig. 1. The maximum of the activation energy is clearly noticeable. In the intermediate concentration range a deviation of the experimental points from a straight Arrhenius plot exists with a tendency of decreasing activation energy with decreasing temperature. Because of the phase separation only a small temperature range could be covered in the Na–NH$_3$ solutions at higher concentrations and a deviation is hard to detect. In Li–NH$_3$ solutions, where a wider temperature range is available, again a systematic deviation is found (Fig. 2). The deviation is stronger and occurs at higher temperatures for the

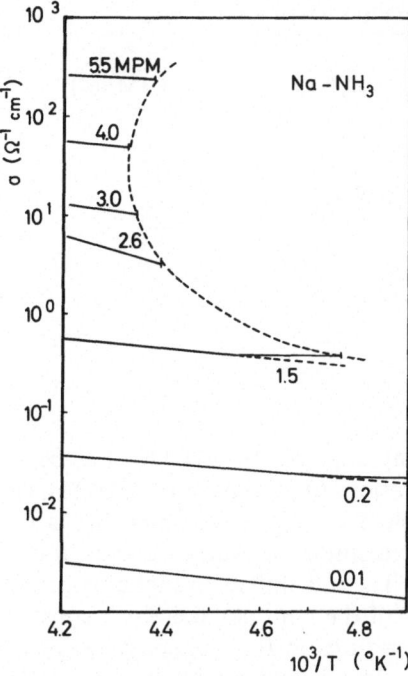

Fig. 1. Conductivity versus temperature for Na–NH$_3$ solutions

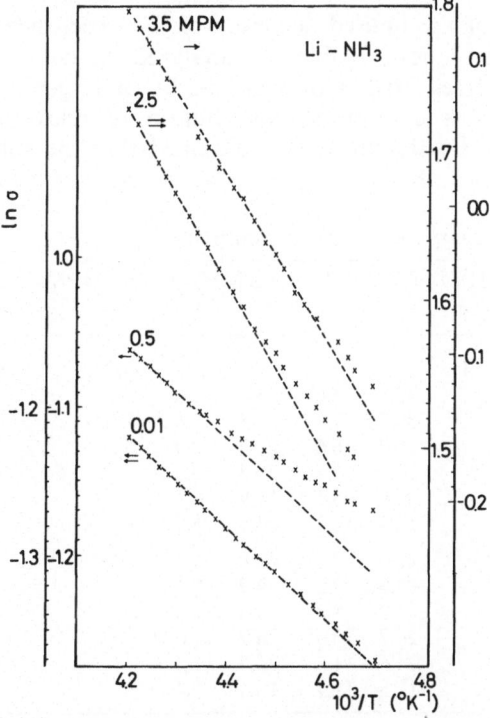

Fig. 2. Conductivity versus temperature for Li–NH$_3$ solutions

Table I. Activation energies

Na–NH$_3$		Li–NH$_3$	
C (MPM)	E_a (eV)	C (MPM)	E_a (eV)
0.01	0.101	0.01	0.063
0.2	0.064	0.5	0.059
0.9	0.056	1.5	0.073
1.5	0.120	2.0	0.081
2.0	0.137	2.5	0.106
2.6	0.136	3.5	0.120
3.1	0.130	4.0	0.218
4.0	0.117		
5.5	0.042		

medium concentrations investigated (0.5 MPM). It is barely noticeable for the lower and higher concentrations (0.01 and 3.5 MPM). The values of the activation energies obtained from these plots are presented in Table I for Na–NH$_3$ and Li–NH$_3$ solutions. The maximum activation energy for Li–NH$_3$ seems to be higher compared to Na–NH$_3$ and shifted to higher concentrations.

Real and imaginary parts of the complex dielectric constant for different metal concentrations and temperatures in the frequency range covered are presented in Table II for Na–NH$_3$. The frequency dependence of these values indicates that dielectric relaxations [5—7] exist in the concentration range 0.2—1.3 MPM, at least one of which is due to the orientational polarization of the "free" NH$_3$ molecules. Because of the limited accuracy of the experimental data a unique analysis of the results is not possible. We analyzed the data assuming that only two relaxation processes exist. However, neither a larger number of discrete relaxation times nor a continuous distribution of relaxation times can be completely excluded. Furthermore, it is assumed that the solvent relaxation is

Table II. Real and imaginary parts of the permittivity

T (°C)	f (GHz)	0.2 MPM		0.7 MPM		1.2 MPM		1.5 MPM		2.0 MPM		3.1 MPM	
		ε'	ε''	ε'	ε''	ε'	ε''	ε'	ε''	ε'	ε''	ε'	ε''
−40	1.2	23.0	1					42.0		36.0		20.0	
	2.5			32.3	3.6	33.6	3.6						
	6	21.0	1	30.7	7.3	30.8	6.8	410.0		35.0		−120.0	
	10			26.0	6.6	26.7	8.9						
	23			21.8	7.1	20.4	8.0						
	35			20.4	8.5	16.9	4.7						
	70			16.4	10.2	14.5	5.1						
−75	2.5			41.0	2.8	42.8	2.5						
	6			34.9	7.5	39.9	8.5						
	10			31.8	8.8	34.3	11.4						
	23			26.7	10.6	26.9	12.0						
	35			24.6	11.3	22.5	11.2						
	70			17.5	11.7	18.4	8.4						

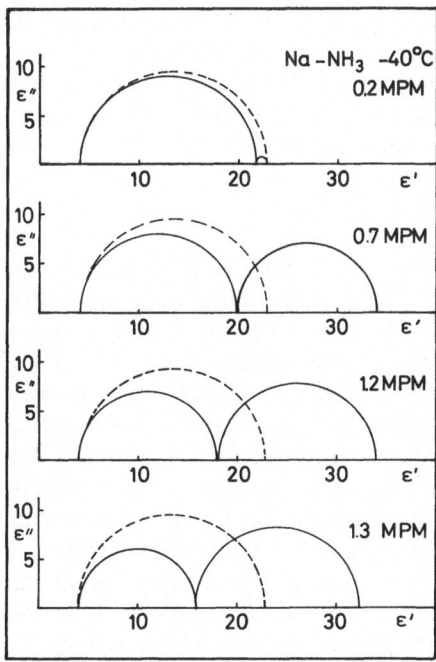

Fig. 3. Cole-Cole diagrams for Na–NH₃ solutions; ——— semicircle for pure ammonia

not influenced by the presence of solvated ions and electrons. This latter assumption is, however, known to be valid only in a first approximation [8]. In the so-called Cole-Cole diagram a simple Debye relaxation process is represented by a semicircle in the complex plane of ε. If several simple relaxation processes exist they may simply be added to one another.

At a concentration of about 0.2 MPM a relaxation in addition to the orientational relaxation of the ammonia molecules becomes noticeable (Fig. 3). The strength of this additional relaxation increases with increasing concentration while the relaxational strength of the "free" ammonia molecules decreases. The relaxational

Table III. Relaxation strengths and relaxation times

T °C	C (MPM)	$(\Delta\varepsilon)_{NH_3}$	$(\Delta\varepsilon)_D$	$10^{12} \times \tau_{NH_3}$ (sec)	$10^{11} \times \tau_D$ (sec)
−40	0.0	19	—	1.2	—
	0.2	18	1.3	1.2	—
	0.7	16	14	1.2	1.7
	1.2	14	16	1.2	1.4
−75	0.0	24	—	2.0	—
	0.2	24	1	2.0	—
	0.7	23	13	2.0	2.1
	1.2	22	18	2.0	1.6

strengths of both the NH_3 and the additional relaxation are smaller at higher temperatures, the relaxation time of the additional relaxation decreases with concentration and temperature (Table III). It can be seen from Table II that for concentrations above 1.5 MPM the real part of the dielectric constant increases very strongly then decreases again and finally reaches negative values [9].

Discussion

The deviation of the $\ln \sigma - 1/T$ plot from a straight line in the intermediate concentration range seems to indicate that no simple activated process is present at lower temperatures. The thermally activated hopping process for the electron transport in disordered systems suggested by several authors [10—16] shows such a behavior. In this case a so-called four-dimensional hopping process exists where the activation energy decreases with decreasing temperature due to larger hopping distances combined with accidentally lower activation energies. Although localization because of disorder is not expected to be large in systems with states of s-like wavefunctions such as metal–ammonia solutions, there may be localization as a consequence of density fluctuations. If such a model applies, a plot $\ln \sigma$ versus $1/T^{1/4}$ should give a straight line in the intermediate concentration range. While the individual curves of Fig. 2 are straightened out in such a plot, the experimental points are still not on a single line.

At higher concentrations, around 4 MPM, where the transition to the metallic state is assumed to be completed, the experimental points are again on a straight Arrhenius line. Also, at concentrations below 0.01 MPM, where one expects a simple diffusion mechanism for the charge transport, the experimental points are on an Arrhenius line.

The increase of the dielectric constant with concentration at about 0.1 MPM indicates the formation of aggregates in the metal–ammonia solutions which, according to Fig. 3 and Table III, give rise to a dielectric relaxation with relatively long relaxation time. If one adopts the cluster model for this relaxation, the hopping probability or the mobility of the electrons in the high-concentration clusters could not — because of the relatively long relaxation time — be too high and certainly not metal-like. At any rate, the clusters in this concentration range would be subject to density fluctuations and would have to be looked at from a dynamic point of view. Judging from the structural relaxation times in liquid ammonia the clusters can form and disappear during the process of the dielectric relaxation [17]. The increasing size of the clusters with increasing concentration and their decreasing stability with temperature could explain the behavior of the dielectric constant. The relatively small difference of the relaxation times at -40 and $-75°$ C leads to a small activation energy for the relaxational process. This activation energy is certainly smaller than the one for the d.c. conductivity at the same concentration, which would be mainly determined by the conduction mechanism between the clusters. The lower relaxation time at higher concentrations could be explained with a higher hopping probability for these concentrations.

At a concentration of about 1.5 MPM the real part of the dielectric constant becomes very large, possibly because the clusters are not subject to fluctuations any more, then ε' decreases again around 2 MPM and is negative at 3.1 MPM indicating the nonmetal-metal transition in this concentration range. The frequency dependence of ε' at 1.5 MPM cannot be explained with the simple electron gas model, it is more resonance-like.

The concentration of dipoles involved in a particular dielectric relaxation process may be calculated from the relaxational strength of the individual process. With the assumption that the static dielectric constant is much larger than 1, which is certainly fulfilled for the metal–ammonia solutions investigated, the following expression can be derived from the Onsager-Böttcher relation [18]:

$$(\Delta \varepsilon)_i = 4\pi \frac{N_i m_i^2}{(1 - A_i)\, 3kT}\,, \tag{11}$$

where m_i is the dipole moment and N_i the concentration of species i. The constant A_i depends on the special microscopic shape of the dipoles. From the decreasing $(\Delta \varepsilon)_i$ for the solvent molecules with concentration, one obtains that about 15 NH_3 molecules per dissolved metal atom cannot take part in the dielectric relaxation process, i.e., these molecules are irrotationally bound in the strong electric fields of metal ions and electrons.

References

1. Lagowski, J. J., Sienko, M. J. (Eds.): Metal–ammonia solutions, Proc. Colloque Weyl II. London: Butterworths 1970.
2. Catterall, R., Mott, N. F.: Advan. Phys. **18**, 665 (1969).
3. Cohen, M. H., Thompson, J. C.: Advan. Phys. **17**, 857 (1968).
4. Schindewolf, U., Bödekker, K. W., Vogelsgesang, R.: Ber. Bunsenges. Phys. Chem. **70**, 1161 (1966).
5. Breitschwerdt, K. G., Radscheit, H.: Phys. Letters **29** A, 381 (1969).
6. Breitschwerdt, K. G., Radscheit, H.: Ber. Bunsenges. Phys. Chem. **75**, 644 (1971).
7. Breitschwerdt, K. G., Radscheit, H.: Z. Angew. Phys. **32**, 276 (1971).
8. Breitschwerdt, K. G., Schmidt, W.: Z. Naturforsch. **25** A, 1467 (1970).
9. Mahaffey, D. W., Jerde, D. A.: Rev. Mod. Phys. **40**, 710 (1968).
10. Anderson, P. W.: Phys. Rev. **109**, 1492 (1958).
11. Mott, N. F., Twose, W. D.: Advan. Phys. **10**, 107 (1961).
12. Mott, N. F.: Advan. Phys. **16**, 49 (1967).
13. Mott, N. F.: Phil. Mag. **17**, 1259 (1968).
14. Mott, N. F.: J. Non-Crystalline Solids **1**, 1 (1968).
15. Mott, N. F.: Phil. Mag. **19**, 835 (1969).
16. Brenig, W., Döhler, G., Wölfle, P.: Z. Physik **246**, 1 (1971).
17. Breitschwerdt, K. G.: Z. Naturforsch. (in press).
18. Quoted in: Brown, W. F.: Handbuch der Physik, Vol. 17. Flügge, S. (Ed.). Berlin-Göttingen-Heidelberg: Springer 1956.

Discussion

M. H. COHEN   If metallic clusters are clearly present in the solutions by 1 MPM and have grown to the percolation limit by ~ 5 MPM, one can expect that by 2 or 3 MPM substantial numbers of clusters have begun the process of elaborating a percolation channel. The resonance you observe could then be a Mie resonance or a plasma resonance brought down to your frequency by the low depolarizing factor of an incipient channel. It would be interesting to estimate the depolarizing factor required.

K. G. BREITSCHWERDT   We have measured only a portion of the total dispersion curve so far, therefore the type of resonance and the resonance frequency are not certain as yet. As soon as we know the dispersion characteristics in detail we will calculate the depolarizing factor required.

Strange Magnetic Behavior and Phase Relations of Metal–Ammonia Compounds*

Thérèse David, William Glaunsinger, Saul Zolotov, and M. J. Sienko

Abstract

The magnetic susceptibility for the ammonia compounds of lithium, calcium, strontium, and barium has been measured down to $1.5\,°K$. In the liquid and fcc solid states, $Li(NH_3)_4$ behaves nearly as a free-electron metal; in the hexagonal solid, it behaves as a temperature-dependent paramagnet with suggestion of a Néel point at $10\,°K$. When supplemented with detailed ESR line-shape studies, the results indicate a localized moment of the order of that of a single electron, which orders antiferromagnetically at low temperatures. In their low-temperature behavior, the susceptibilities of $Ca(NH_3)_6$, $Sr(NH_3)_6$, and $Ba(NH_3)_6$ correspond roughly to that of $Li(NH_3)_4$ and can be interpreted in terms of low carrier densities arising from almost complete zone-filling; at higher temperatures, an anomalously large increase in susceptibility is observed, possibly due to electron unpairing. There is clear magnetic evidence for a phase transition in $Ba(NH_3)_6$ at $76\,°K$. Further phase and structure studies as well as ESR studies under pressure are indicated. DTA-pressure studies are reported for the $Li–NH_3$ system.

Introduction

It is a remarkable fact that the alkali metal lithium, the alkaline earth metals calcium, strontium, and barium, and the rare earth elements europium and ytterbium form near-stoichiometric compounds of the type $Li(NH_3)_4$ and $Ca(NH_3)_6$. The compounds, being of the "expanded-metal" type, wherein the metal atoms can be regarded as spatially separated by the dielectric ammonia, are of great interest to solid state physics since they appear to represent materials just barely on the metallic side of the nonmetal–metal transition. Unfortunately, experimental work with these materials is extraordinarily difficult: the component metals are reactive and hard to purify, the solutions from which the compounds form are thermodynamically unstable, the phase relations are ambiguous and difficult to establish because of hysteresis effects on thermal cycling. Still an impressive number of reproducible data are accumulating so that the beginnings of a self-consistent picture are starting to emerge. Important points of structure, phase-relation and property behavior remain to be settled, but the early anticipation that we are dealing here with a system of major theoretical interest appears to be borne out.

This paper is composed of four parts: Part I reviews the lithium–ammonia problem with particular reference to recent magnetic studies on the solid and liquid states of $Li(NH_3)_4$; Part II describes some recent ESR findings on $Li(NH_3)_4$ which suggest at least a partial explanation of the findings in Part I; Part III summarizes some new DTA-pressure studies of $Li(NH_3)_4$ which settle some of

* This research was sponsored by the National Science Foundation through grant GP–6246 and was supported in part by the Air Force Office of Scientific Research and the Materials Science Center at Cornell University.

the phase questions but raise others; finally, Part IV gives preliminary findings on the magnetic susceptibility of the alkaline earth–ammonia compounds, their tentative interpretation, and their possible application to the problem of $Eu(NH_3)_6$.

I. Magnetic Susceptibility of $Li(NH_3)_4$ in the Range 1.5—194 °K

The phase diagram of the lithium–ammonia system shows a deep eutectic at 88.8 °K and 20 MPM of lithium [1]. Unlike the other alkali metals, which form conventional eutectic mixtures of metal and ammonia, the lithium eutectic appears to be associated with compound formation, probably as $Li(NH_3)_4$. The evidence is largely indirect: large negative heat of solution for Li in NH_3 [2], low vapor pressure of the saturated solution [3], persistence of bronze color at the liquid-to-solid transition, high conductivity of the solid [4]. More direct evidence comes from breaks in the conductivity-temperature curve [5], 50 % greater heat capacity than sum of components, and X-ray diffraction patterns that are neither Li nor NH_3 [6]. Mammano and Sienko [6] have made an X-ray study on solid $Li(NH_3)_4$ at 77 °K. They found a cubic phase stable above 82 °K with $a_0 = 9.55$ Å and a hexagonal phase stable below 82 °K with $a = 7.0$ Å and $c = 11.1$ Å. Kleinman *et al.* [7] also performed X-ray measurements at 77 °K and found $a = 7.12$ Å and $c = 11.29$ Å. Both investigators found 1.585 for the c/a ratio of the hexagonal phase.

Electrical conductivity measurements of Morgan *et al.* [5] indicate the melting point of $Li(NH_3)_4$ is 89.6 °K with solid–solid phase transitions at 82 °K and 69 °K. Mammano and Coulter [1] measured heat capacities of concentrated lithium–ammonia solutions from 60—110 °K and observed first order transitions at 82.2 °K and 88.8 °K but no 69 °K transition.

Magnetic susceptibilities of five different lithium–ammonia samples, ranging in composition from 18.7 to 37.1 MPM of Li, have recently been reported by Glaunsinger *et al.* [8]. They measured the susceptibilities by the Faraday method over a field range up to 13 kG and a temperature range down to 1.5 °K. Particular precautions were taken to assure high purity of the materials and precise establishment of the thermal cycling. Magnetic susceptibilities were also determined for pure lithium and for pure ammonia over the whole range of temperature.

Figure 1 shows typical results, given here as a plot of magnetic susceptibility per mole of $Li(NH_3)_4$ after correction for any excess Li or NH_3 in the sample, the diamagnetism of the Li^+ core, and the diamagnetism of the bound ammonia. As can be seen the susceptibility is small and positive; it is essentially independent of temperature in the liquid state (above 89 °K) and in the cubic state (between 82 °K and 89 °K), but it becomes a strong function of temperature in the hexagonal solid (below 82 °K). At 194 °K the magnetic susceptibility per mole of $Li(NH_3)_4$ ranged from 68.5×10^{-6} to 97.9×10^{-6} for all the samples studied. A 5 % drop in susceptibility occurs at the liquid-to-solid transition; a further 25 % drop occurs at the cubic-to-hexagonal transition. In all cases, hysteresis was observed at both the 82° and 89 °K transitions. On heating, the solid–solid transition occurred at 82.2—84.0 °K and the solid–liquid at 88.9—91.0 °K; on cooling, 74.2—76.4 °K

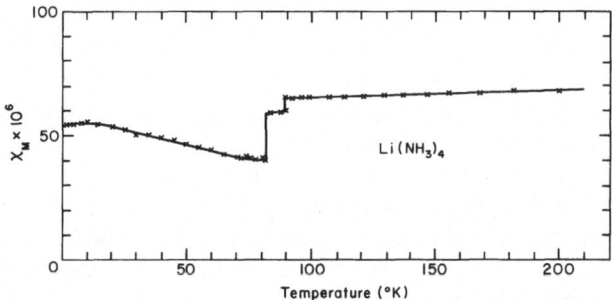

Fig. 1. Magnetic susceptibility per mole of Li(NH$_3$)$_4$ versus temperature

and 85.9—84.0 °K, respectively. Contrary to what has been reported in heat capacity studies, the solid–solid transition required only about 15 min for completion. Also, in contrast to low-temperature resistivity studies, the susceptibility at 4.2 °K was found to be reproducible to 1 or 2% independent of the rate of cooling through the solid–solid transition.

The susceptibility behavior can be broken down into three regions: liquid state, from 194 °K to 89 °K; cubic phase of the solid, from 89 °K to 82 °K; hexagonal phase, from 82 °K to 1.5 °K. In the liquid state, the susceptibility can be accounted for quite satisfactorily by application of the Pauli-Peierls equation for the susceptibility of a quasi-free electron gas for which the effective mass approximation is valid:

$$\varkappa = \frac{4\,m^*\,\mu_0^2}{h^2}\,(3\,\pi^2\,n)^{1/3}\left(1 - \frac{m_0^2}{3\,m^{*2}}\right).$$

Here $\varkappa$ is the susceptibility per unit volume, m^* is the effective mass, μ_0 is the Bohr magneton, m_0 is the electron rest mass, and n is the carrier density. The observed susceptibility for Li(NH$_3$)$_4$ can be fitted quantitatively by choosing the effective mass m^* to be equal to 1.83 m_0, which is very near to the value $m^* = 1.66\,m_0$ appropriate for pure Li [9]. No temperature dependence is expected for a Pauli-Peierls gas. A very slight decrease observed in our values may be due to small changes of density with temperature.

The decrease observed in susceptibility at the freezing point may be fully accounted for by the effect of changing density on the application of the Pauli-Peierls equation. If the density of the cubic phase is taken to be 0.572 g/cm^3, as calculated from the X-ray measurements [6], and that of the melt is taken to be 0.525 g/cm^3, as obtained by extrapolating liquid density data [10], the per cent decrease in susceptibility is predicted to be 5.6% compared to an observed change of 5.7%.

For the cubic phase of Li(NH$_3$)$_4$ the average value of m^* calculated from the Pauli-Peierls equation is 1.83 m_0. The fact that the magnetic susceptibility remains constant is in accord with the quasi-free electron model in the absence of bandwidth effects.

The large decrease observed in susceptibility at the cubic-to-hexagonal phase transition is more difficult to explain. Using 0.53 g/cm^3 for the density of the

hexagonal phase, as calculated from the X-ray parameters, one predicts a 4.9%
· *increase* in susceptibility at the cubic-to-hexagonal transition. The observed
change is opposite in direction and greater by a factor of five. Since the X-ray
data were of poor quality, they are suspect. In fact, on thermodynamic grounds,
the observed DTA-pressure studies described in Part III indicate that the density
of the hexagonal phase must be greater than that of the cubic phase. Using
0.572 g/cm^3 for the cubic phase, one predicts 0.61 g/cm^3 for the hexagonal phase.
If the density of the hexagonal phase is indeed 0.61 g/cm^3, the Pauli-Peierls treatment would predict a 4.4% *decrease* in susceptibility. The change would be in
the right direction but still far short in magnitude. If the decrease in susceptibility
were attributed to a change in effective mass, m^* would have to become about
$1.1 \, m_0$ for the hexagonal phase.

The most intriguing aspect of Li(NH$_3$)$_4$ is the temperature dependence of χ
that appears in the hexagonal phase. There are two unexpected features: one is
the Curie-Weiss dependence between approximately 20 and 60 °K, the other is
the flattening out below 10 °K. Fitting the observed data to a law of the form
$\chi = C/(T - \Theta)$ and calculating $\mu_{\text{eff}} = 2.828 \, C^{1/2}$ leads to a μ_{eff} value of 0.294 Bohr
magneton per Li(NH$_3$)$_4$. However, the straightforward application of the Curie-
Weiss formalism may not be significant here.

Resistivity and magnetoresistance measurements [4] on hexagonal Li(NH$_3$)$_4$
have been interpreted in terms of a completely compensated metal with relatively
low electron density [$n < 1.47 \times 10^{20}$ as compared to 4.2×10^{21} if one assumes
one free electron per Li(NH$_3$)$_4$]. In such case, Li(NH$_3$)$_4$ should be treated as a
nearly-degenerate electron gas, for which the temperature dependence of the
Fermi-Dirac statistics would need to be considered in calculating the magnetic
susceptibility. The ratio of the volume susceptibility at zero temperature, $\varkappa(0)$,
to the volume susceptibility at finite temperature, $\varkappa(T)$, is given by

$$\frac{\varkappa(0)}{\varkappa(T)} = \frac{3}{2} \frac{T}{T_F} \frac{F_{1/2}(\alpha)}{F'_{1/2}(\alpha)},$$

where T_F is the Fermi temperature and $F_{1/2}$ is the Fermi integral expressed in
terms of α, the ratio of chemical potential to kT. Figure 2 shows the degree of fit
that can be obtained between observed volume susceptibilities and those calculated for a dilute electron gas characterized by Fermi temperature 128.1 °K. This
corresponds to a carrier density of 6.3×10^{19} per cm^3 and $m^* = 5.2 \, m_0$ on a 1-
band model or to a carrier density of 3.2×10^{19} per cm^3 and $m^* = 3.3 \, m_0$ on a
2-band model. The value of n for the two-band model is lower than the
1.47×10^{20} per cm^3 upper limit set by McDonald and Thompson on the basis
of their resistivity and magnetoresistivity studies. Carrier mobilities come out to
be about 60 and 0.8 m^2/Volt sec but one cannot say whether the electron or
hole mobility is the larger.

The above treatment considers the hexagonal phase of Li(NH$_3$)$_4$ as a simple
metal in which the average inter-electron distance $r_s = (1/a_0) (3/4\pi n)^{1/3}$ is so
great ($r_s = 7.34$ if $n = 4 \times 10^{21}$ per cm^3) that Coulomb correlations compensate
for exchange. As r_s is increased, the gain in energy associated with correlations
between electrons of parallel spin nearly equals that due to correlations between

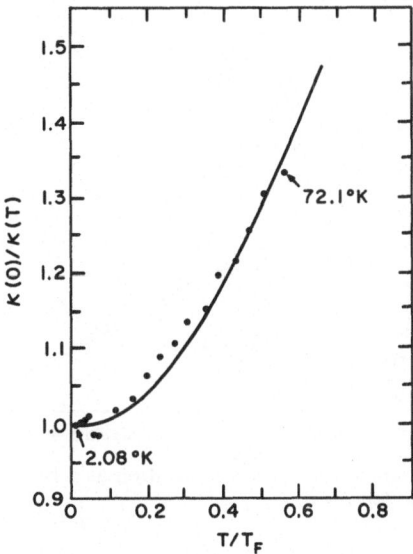

Fig. 2. Reciprocal of volume susceptibility $\varkappa(T)$ vs. temperature for $Li(NH_3)_4$. Points represent experimental data: solid line is calculated assuming Fermi temperature T_F is 128 °K

electrons of antiparallel spin. Hubbard's Hamiltonian for introducing the effects of correlation is of the form [11]:

$$\mathscr{H} = \sum_{k,\sigma} E(k)\, n_{k\sigma} + U \sum_i n_{i\uparrow} n_{i\downarrow}$$

where $E(k)$ are the one-electron energies, $n_{k\sigma}$ are the operators which give the number of electrons of spin σ in the one-electron state k, $n_{i\sigma}$ are the operators indicating the number of valence electrons of spin σ on the ion core at R_i, and U is the intra-atomic Coulombic repulsion. With complete neglect of magnetic ordering, Hubbard finds for the case of a single half-filled band that a continuous insulator-to-metal transition occurs as the ratio of U to bandwidth W passes through a critical value of 0.87.

Cyrot [12] has studied the insulator-to-metal transition taking into account spin density fluctuations which build up magnetic moments on each site. The resulting phase diagram is shown in Fig. 3. The cross-hatched region is the one believed relevant to $Li(NH_3)_4$. To the left, where U/W is small, the system would be Pauli metallic; to the right, the magnetic behavior would depend on U/W. Although the Li–Li distance in hexagonal $Li(NH_3)_4$ is about 6.9 Å compared to 3.0 Å in Li metal, an unusually narrow bandwidth is not expected. The $Li(NH_3)_4$ molecule can be visualized as a $Li(NH_3)_4^+$ ion inside a sheath of expanded 3s-electron charge density. The radius of $Li(NH_3)_4^+$ is estimated at about 2.5 Å, and dielectric expansion of the 3s orbital places the density maximum at about 4 Å, sufficient to give large overlap between adjacent molecules. Hence W is probably not much different from its value in metallic lithium. To estimate U,

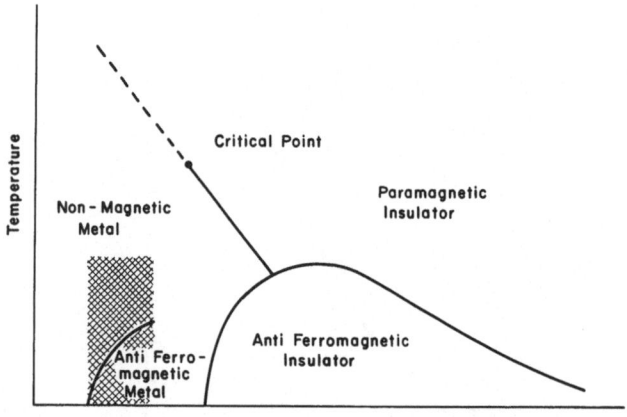

Fig. 3. Schematic phase diagram showing phase variation as a function of temperature and Coulomb repulsion relative to bandwidth

one can use the collective-electron model of Pines [13] to get a screening length, $\lambda_c = 0.797\, r_s^{1/2}$, of 2.1 Å from the bulk density of $Li(NH_3)_4$ or 4.9 Å on a two-band model for nearly degenerate electron gas with equal effective masses for electrons and holes. Since $\lambda_c = 1.43$ Å for pure lithium, the appreciably larger value in $Li(NH_3)_4$ indicates enhanced short range electron-electron interactions and hence a larger value of U than would have been expected with an ordinary metal. The observation of Cate and Thompson [14] of a strong T^2 component dominating a T^5 component in the resistivity from 1.5—4.2 °K lends support to the above arguments. In any case, it is not unreasonable that U/W for $Li(NH_3)_4$ may be large enough to enter the cross-hatched region.

In the cross-hatched region, Cyrot has shown there are two possibilities for appearance of magnetism: 1. the nonmagnetic phase undergoes first order transition to an antiferromagnetic one below a temperature T_N; 2. localized moments appear in the nonmagnetic phase and interact through a Ruderman-Kittel mechanism to produce antiferromagnetic ordering below T_N'. The localized moments are stable only below a temperature $T_L(T_L > T_N)$, but no phase transition occurs at T_L. If the first possibility applies to $Li(NH_3)_4$, it suggests that, whereas the high-temperature magnetic behavior of the hexagonal phase is like that of a nearly degenerate electron gas, the low-temperature behavior may be associated with antiferromagnetism. If the second possibility applies, partial localization of electrons could occur at the cubic-to-hexagonal transition and give rise to small effective localized moments that could couple antiferromagnetically below 10 °K.

Which of the above models is most appropriate for $Li(NH_3)_4$ remains to be decided. Several key parameters are still missing, e.g., the band structure cannot be computed until the space group ambiguity has been removed. Recent neutron diffraction studies [15] at 77 °K and 4.2 °K failed to disclose evidence for localized moments, but subtle line intensity changes that were observed may eventually yield some information. In the meantime, other experimental data leading to a full computation of the Fermi surface would be desirable. Unfortunately, the prospect of getting singly crystals of $Li(NH_3)_4$ for the necessary pertinent studies seems quite remote.

II. ESR Lineshape Studies on $Li(NH_3)_4$

The very high electrical conductivity of $Li(NH_3)_4$ makes difficult the study of ESR signals but the direct connection of paramagnetic interactions between conduction electrons, skin-depth penetration, lineshapes, and relaxation times has made possible deduction of some useful results on the $Li(NH_3)_4$ problem. Details of this investigation will be published elsewhere; only a brief summary is given here.

First and second derivative detection of electron spin resonance absorption has been used to study the conduction electrons in $Li(NH_3)_4$. An X-band reflection spectrometer operating in a balanced mixer configuration with a rectangular cavity in the TE_{102} mode was modified to allow sample cooling from 190 °K to about 15 °K. Cylindrical samples were used with dimensions large (0.3 cm) compared to skin depth δ and spin depth δ_e (of the order of 10^{-4} cm). Modulation frequencies of 100 kHz and 1 kHz were employed. In metals, the only electrons that contribute to the resonance must lie within the skin depth and have energies within kT of the Fermi energy; high sensitivity is thus required, but this was not a problem here.

The central parameter in the lineshape function is $R = (T_D/T_2)^{1/2}$ where T_D is the time required for an average electron to diffuse across δ and T_2 is the electron spin relaxation time. The phase shift of the microwave magnetic field within the metal combined with diffusion of the magnetization via the conduction electrons produces asymmetric lineshapes. Analysis of the asymmetry parameters, linewidth at half-maximum, peak-to-peak linewidth, and resonant magnetic field displacement can be used to extract the pertinent parameters T_D, T_2, and H_O.

A typical second derivative resonance is shown in Fig. 4. The specimen was 25 MPM of Li in NH_3 in the liquid phase at 98.5 °K. The frequency was

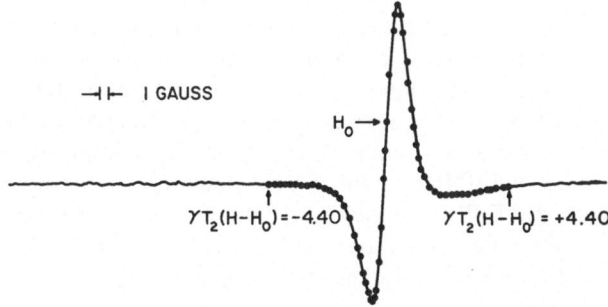

Fig. 4. Recorded trace of the second derivative of the $Li(NH_3)_4$ resonance. The frequency was 9.2230 GHz and the temperature was 98.5 °K. The dots are values computed taking $T_D/T_2 = 0.94$

9.2230 GHz. As can be seen, the recorder trace can be well reproduced by a computation in which T_D/T_2 is taken to be 0.94. Other relevant parameters deduced from the lineshape are as follows: $\Delta H = 3.01 \pm 0.07$ gauss, $T_2 = 1.51 \pm 0.04 \times 10^{-8}$ sec, $T_D = 1.42 \pm 0.04 \times 10^{-8}$ sec, $g = 2.00153 \pm 0.00003$, $\Delta g = 7.9 \pm 0.3 \times 10^{-4}$. The g-shift is small but much larger than that in pure lithium. The proximity of the g-shift to the theoretical value for sodium suggests that spin-orbit coupling in $Li(NH_3)_4$ approximates that in pure sodium.

The effective mass m^* of the conduction electrons may be calculated from the relation

$$\frac{m^*}{m} = \frac{2 T_D (v_F l)_{free}}{3 \delta^2},$$

where m, v_F and l are the free electron mass, Fermi velocity, and mean free path, respectively. Using the conductivity data of Morgan et al. [5], T_D as obtained above gives $m^*/m = 1.7 \pm 0.2$, which is in excellent agreement with the value found in Part I. For the cubic solid phase, agreement is fully as good.

Temperature studies which are only now being fully analyzed indicate some curious results. On cooling, the ESR signal decreases from 190—150 °K; at the freezing point, there is observed a 2-fold decrease, attributable to change in skin depth. On going from the cubic to the hexagonal phase there is another 60% drop, also attributable to skin depth. From 74 °K to 68 °K there is a 5-fold decrease, with another weak line showing up at 68 °K. This line, which disappears on heating to higher temperature, may be due to metallic lithium. From 68 °K to 50 °K there is a 10-fold decrease in the line intensity. It is still visible at 23 °K but disappears by 15 °K.

With an overall sample composition of 20 MPM of Li, only the cubic line is seen above 82 °K, the hexagonal line below 82 °K. When the composition is 15 MPM of Li, a mixture of cubic and hexagonal is observed below 82 °K but only cubic above 82 °K. When the lithium content is extremely low, about 0.05 MPM, only the cubic is seen over the whole range below 89 °K; in other words, there is no transition at 82 °K. Although fragmentary, these observations indicate that the solid-state portion of the $Li-NH_3$ phase diagram needs to be redrawn.

The paramagnetic susceptibility of the conduction electrons can be deduced directly from the ESR lineshape by standard procedures. When this is done for the 25 MPM of Li in NH_3 sample, a real minimum in the $1/\chi$ vs T curve appears at about 20 °K. This is very suggestive of a Neél temperature. Furthermore, the slope of the $1/\chi$ vs T curve leads to an effective moment of 0.03 Bohr magneton per mole of $Li(NH_3)_4$, but if we calculate n by comparing the ESR-observed χ_{hex} with the ESR-observed χ_{cubic} alone (where we know $n = 4 \times 10^{21}$), then we deduce that n lies between 1 and 2×10^{18} per cm^3, which would lead to a μ_{eff} between 1.2 and 1.9 Bohr magnetons per electron.

The implication is very strong, then, that the flattening in χ vs T observed at low temperature and mentioned in Part I is not attributable to nearly degenerate statistics but to exchange ordering of the antiferromagnetic type.

III. Pressure-Dependent Phase Changes in the Li–NH₃ System

The phase diagram proposed by Mammano and Sienko for the $Li-NH_3$ system shows as one possibility an equilibrium at 88.8 °K between liquid $Li-NH_3$ and three solid phases, viz., pure ammonia, cubic $Li(NH_3)_4$, and pure lithium. The intriguing prospect that 88.8 °K might represent a quadrupole point was the motivating influence for a pressure study of the system. Specifically, since a two-component system with four phases in equilibrium should be invariant, application of pressure would not be expected to shift the 88.8 °K transition. Furthermore, the pressure dependence of the 82.2 °K transition should be calculable from known thermodynamic data. Finally, it was hoped that a variable pressure study would cast light on the mysterious 69 °K transition, observed in the conductivity studies [5] but not in the thermal studies [1]. Because of inherent great sensitivity, the differential thermal analysis (DTA) technique was selected for defining the phase transition temperatures. A helium piston providing up to 30 atmospheres was the means for the pressure variation.

Details of the experimental set-up and manipulation will be described elsewhere. The essential features of the design were two opposed copper constantan thermocouples, one of which was in the sample and the other in a reference vessel filled with copper powder. Supercooling was a problem, so, except as noted below, peak positions were based on heating curves, although both cooling and heating curves were always determined.

Samples of pure lithium, on cooling, showed a phase transition peak at ∼69 °K. The peak was sharp and would shift as much as ±2 °K from run to run. On heating cycles, the transition was observed to occur at ∼128 °K; it was broad and difficult to detect. It is believed that this lithium peak is associated with the martensitic transition between *bcc* and *hcp* structures of Li metal [16].

With addition of ammonia to the lithium, the lithium peak became smaller and two other peaks, at 82.4 and 88.5 °K, appeared; these correspond to the solid-

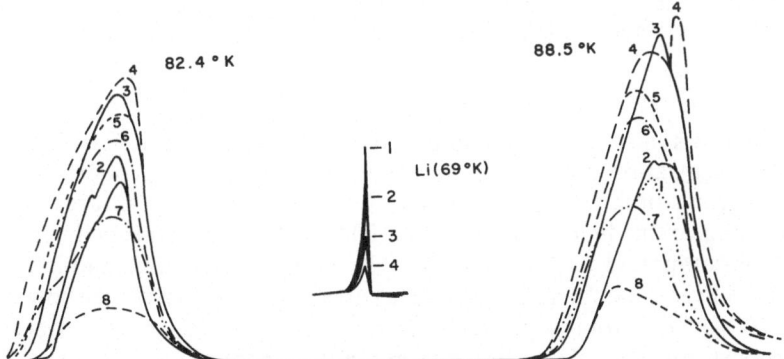

Fig. 5. DTA traces for lithium–ammonia mixtures. The ordinate represents the amplified output of the two opposed thermocouples, one in the sample and the other in the copper powder reference. The abscissa is time. The scan is to higher temperatures except for the inset, which shows, on cooling, the martensitic peak due to excess lithium. Numbers on the traces identify the overall concentrations as follows: 1, 75 MPM of Li; 2, 62%; 3, 45%; 4, 25%; 5, 16%; 6, 14%; 7, 10%; 8, 6%. In all cases, pressure is equilibrium vapor pressure of the corresponding mixture

solid transition and to the melting point of $Li(NH_3)_4$, respectively. As more ammonia was added, the two peaks became larger and the lithium peak, smaller. When dilution reduced the lithium concentration below 20 MPM, the lithium peak disappeared entirely and a new peak at 195.7 °K corresponding to excess ammonia appeared. Figure 5 shows the successive "zero-pressure" DTA traces on sequential addition of ammonia to lithium. As can be seen, for mixtures on the concentrated side of 20 MPM, the high-temperature peak seems to be composed of two subsidiary peaks. The subsidiary peak separation, which is greatest when the mixture is close to 20 MPM, reaches about 0.3 °K. With subsequent dilution of the sample, the peak splitting disappears. There is some faint indication of similar splitting in the peak of the 82 °K transition.

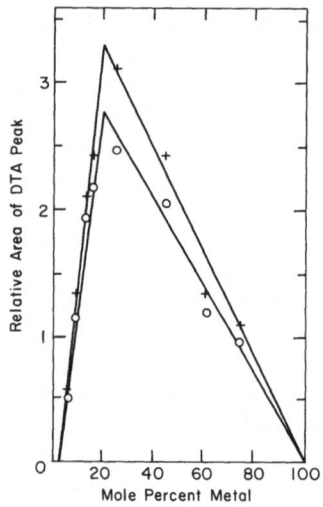

Fig. 6. Relative heat effect as measured by area under DTA peak as a function of total MPM of Li in various $Li–NH_3$ mixtures. Circles refer to 82 °K transition; crosses, to 89 °K transition

Figure 6 shows how the areas under the DTA peaks varied with the total lithium concentration of the samples. Since the peak area is proportional to the ΔH of the transition, the approximately linear relation of peak area versus Li concentration on the high lithium side (> 20 MPM) simply reflects the increasing amount of compound $Li(NH_3)_4$ formed by adding successive amounts of ammonia to a fixed amount of lithium. The observed linearity on the low lithium side (> 20 MPM) is totally unexpected. From the way the samples were made, by successive dilution of lithium, the amount of compound $Li(NH_3)_4$ formed should have become constant once enough NH_3 had been added to combine with all the lithium. Any excess NH_3 over that needed to form $Li(NH_3)_4$ would not contribute to the heat effect at either the 89 °K or 82 °K transition since the excess NH_3 would not have changed phase isothermally. Remarkable also is the fact that extrapolation of the two linear segments in Fig. 6 (< 20 MPM and > 20 MPM) gives an intersection at 20 MPM just corresponding to the compound $Li(NH_3)_4$. It is believed that the peak areas, which were measured both by weighing cut-out areas and by use of a planimeter, are known to about 5%. The ratio of the 89 °K to 82 °K transitions varied from a high of 1.26 to a low of 1.09 for the various compositions. According to the work of Mammano and Coulter, the ratio of the molar enthalpies is $553/520 = 1.06$. The fact that the DTA-determined ratio is higher may reflect the fact that in the DTA experiment the solid-solid transition may be incompletely achieved before melting occurs.

Figure 7 shows how the DTA-determined transition temperatures varied with pressure. Although the system was not designed to operate at such high temperature, the melting point of pure ammonia was measured as a function of pressure to check the accuracy of the system. As shown by the dashed curve in Fig. 7,

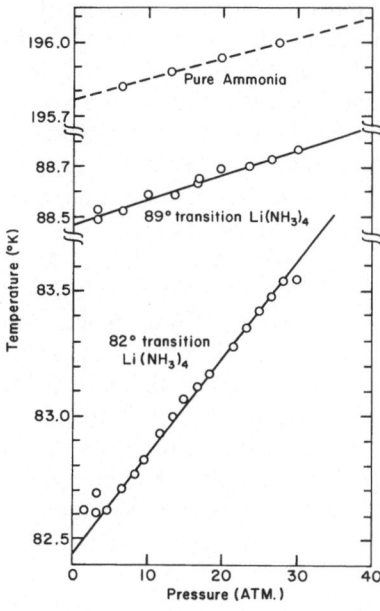

Fig. 7. Temperature of transition as a function of imposed helium pressure. Circles refer to the 89 °K and 82 °K transitions of $Li(NH_3)_4$, respectively, dashed curve refers to melting transition of pure ammonia

the observed $\Delta T/\Delta P$ of NH_3 was linear with imposed pressure and amounted to 0.0081 deg/atm. From the Clapeyron equation $\Delta T/\Delta P = T \Delta V/\Delta H$ and the known ΔV for pure ammonia [17], the observed value of $\Delta T/\Delta P$ leads to a ΔH value of 1350 cal/mole, in excellent agreement with the reported best value of 1376 cal/mole [18].

The other two lines in Fig. 7 show the pressure variations of the phase transitions of $Li(NH_3)_4$. Both are linear and correspond to 0.010 deg/atm and 0.038 deg/atm, respectively. The data shown are for a mixture corresponding to 18.5 MPM of Li. Four compositions, all on the dilute side of 20 MPM, were investigated; they all gave the same result. Extrapolation of the transition temperatures to zero pressure gave 88.5 and $82.4 \pm 0.1\,°K$ for the two transitions; these are to be compared to the Mammano and Coulter values [1] of 88.79 and 82.18 °K, respectively.

Once $\Delta T/\Delta P$ has been determined, it can be combined with the observed ΔH and the Clapeyron equation to give values of ΔV at the two transitions. Using the data quoted above, we obtain $\Delta V = 10.0\,cm^3/mole$ at the 82 °K transition and $\Delta V = 2.6\,cm^3/mole$ at the 89 °K transition. The fact that both these numbers are positive indicates that at each transition the density of the high-temperature phase is less than the density of the low-temperature phase. The X-ray density data, admittedly of poor quality, appear to be in error in predicting the cubic phase to be more dense than the hexagonal phase.

A further curious observation was that the 18.5 MPM lithium sample represented in Fig. 7 showed peak splitting of the 82 °K transition at elevated pressure. This splitting, which increased with increasing pressure, showed one of the peaks increasing and moving to higher temperature while the other decreased and shifted only slightly. The value of $\Delta T/\Delta P$ observed for the shifting peak correspond to that observed on the more dilute mixtures. The temperature split of the two peaks, which amounted to 1.4 °K at 30.6 atm, reversibly decreased on reduction of the pressure and vanished at one atm. Slight hysteresis was observed in the ratio of the area of the two peaks.

The conclusions of the above investigation are that, although the two transitions observed at 88.5 °K and 82.4 °K are consistent with the well-established transitions found for $Li(NH_3)_4$, there was no DTA evidence for a $Li(NH_3)_4$ transition at 69 °K as reported by Morgan et al. [5]. A transition at 69 °K was found only in mixtures that have excess lithium. The fact that the 69 °K transition disappeared on dilution supported the idea of compound formation. Plots of integrated peaks as function of concentration further supported this idea and indicated that the compound that is formed around 88.5 °K undergoes another transition at 82.4 °K. No evidence was found to indicate that there is a quadrupole point at 88.8 °K.

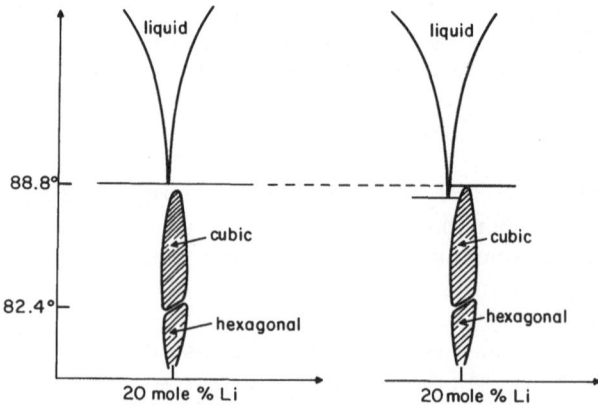

Fig. 8. Schematic possibilities for phase relations in the Li–NH₃ system near the 89 °K and 82 °K transitions

The exact significance of the "zero-pressure" splitting of the 89 °K peak and possibly also of the 82 °K peak remains to be established. Also, it is not clear how one should reconcile the composition of the compound $Li(NH_3)_4$ with the Mammano-Coulter observation that the heat effects peak at $Li(NH_3)_{4.15}$. In any event, the previously suggested phase diagram will have to be modified. Figure 8 shows schematically two versions of what the actual situation might be. In one case, the compound formation of $Li(NH_3)_4$ is shown completely submerged in the solid state, occurring at 88.5 °K, some 0.3 °K below a $Li-NH_3$ eutectic at $Li(NH_3)_{4.15}$. In the other case, the 0.3 °K splitting is shown to correspond to peritectic formation of $Li(NH_3)_4$ at 88.8 °K followed by eutectic formation [perhaps NH_3 plus $Li(NH_3)_4$ but in any event averaging $Li(NH_3)_{4.15}$] at 88.5 °K. To complicate matters, there is almost certainly a stoichiometry range for the compound with an offset in composition when the cubic-to-hexagonal conversion occurs. Figure 8 suggests these additional features, though, of course, the scales should not be taken seriously. Further studies of the $Li-NH_3$ system, particularly as a function of pressure, by DTA, ESR and heat capacity measurements are clearly desirable. The extraordinarily large pressure effects already observed at 30 atmospheres suggest that even modest pressure increments would prove very fruitful.

IV. Magnetic Susceptibilities of the $M(NH_3)_6$ Compounds

The phase rule studies of alkaline earth metal ammonia solutions by early workers [19] have confirmed the existence of solid compounds of the type $M(NH_3)_6$. There is little information in the literature on these solid compounds, which, unlike $Li(NH_3)_4$, have congruent melting points above room temperature. McDonald et al. [20] studied the electrical conductivity of $Ca(NH_3)_6$ and found a transition at 45 °K with a residual resistance of 3Ω cm at 4.2 °K. Powder X-ray investigation of Ca, Sr, and Ba hexaammines at 233 °K were reported by Cagle and Holland [21] and repeated at 77 °K by Mammano and Sienko [22]. The structure was found to be body-centered cubic and unchanged between 77 °K and 233 °K. The magnetic susceptibility of $Sr(NH_3)_6$ was measured by Oesterreicher et al. [23]. They found $\chi_M = 8.2 \times 10^{-3}$ at 4.2 °K, crossing over to diamagnetic above 22 °K.

We have measured by the Faraday method two specimens of Ca in NH_3 (10.27 and 14.97 MPM of Ca, respectively), one specimen of Sr in NH_3 (19.46 MPM of Sr), and one specimen of Ba in NH_3 (9.99 MPM of Ba). Samples were prepared from the central portion of a disc of the appropriate metal (99.5% pure), cut in a glove bag under inert gas. After transfer to a Spectrosil capsule, ammonia was distilled in, and the capsule sealed. Sample compositions were determined from weight of metal, weight of full capsule, and weight of empty capsule after clean break.

Figures 9, 10, and 11 show, respectively, the observed molar susceptibilities of $Ca(NH_3)_6$, $Sr(NH_3)_6$, and $Ba(NH_3)_6$, after corrections for excess components and diamagnetic contributions have been applied. At high temperatures, behavior is quite similar. Room temperature susceptibility decreases as temperature is lowered, becoming diamagnetic in the case of $Sr(NH_3)_6$ and $Ba(NH_3)_6$ but not

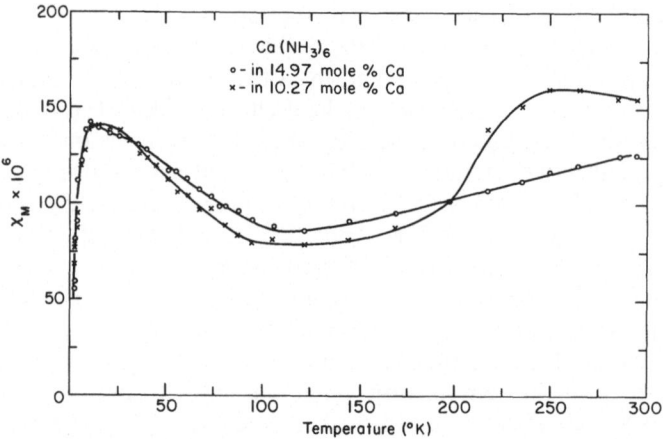

Fig. 9. Magnetic susceptibility per mole of $Ca(NH_3)_6$ versus temperature

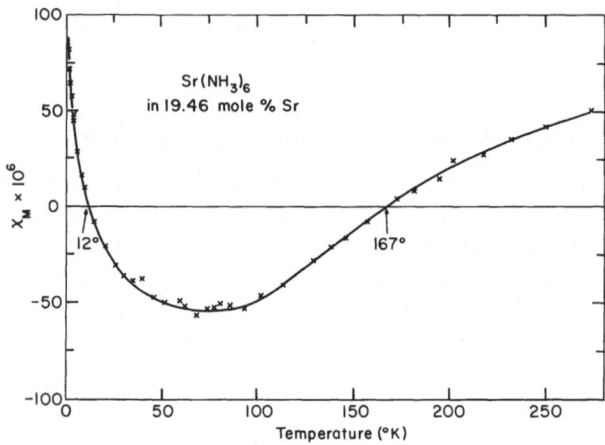

Fig. 10. Magnetic susceptibility per mole of $Sr(NH_3)_6$ versus temperature

in the case of $Ca(NH_3)_6$. At low temperature, the susceptibility first increases, then appears to flatten out, except in the case of $Ca(NH_3)_6$ where there is a steep drop in susceptibility below 10 °K. At 76 °K, $Ba(NH_3)_6$ appears to show a discontinuous change between positive and negative values of the susceptibility.

At low temperatures $Li(NH_3)_4$, $Ca(NH_3)_6$, $Sr(NH_3)_6$, and $Ba(NH_3)_6$ all resemble each other in showing a flat susceptibility that subsequently decreases at higher temperature. It is tempting to speculate that this feature reflects a similarity in the band structure of hex $Li(NH_3)_4$ with that of body-centered cubic $M(NH_3)_6$. The hexagonal $Li(NH_3)_4$ has two molecules per primitive cell; hence, the two electrons from the two $Li(NH_3)_4$ just fill the zone. The body-centered $M(NH_3)_6$ has one molecule per primitive cell, but there are two electrons, so the zone is again just filled. Slight spillover into the next higher zone would account for the slight paramagnetism, the temperature dependence

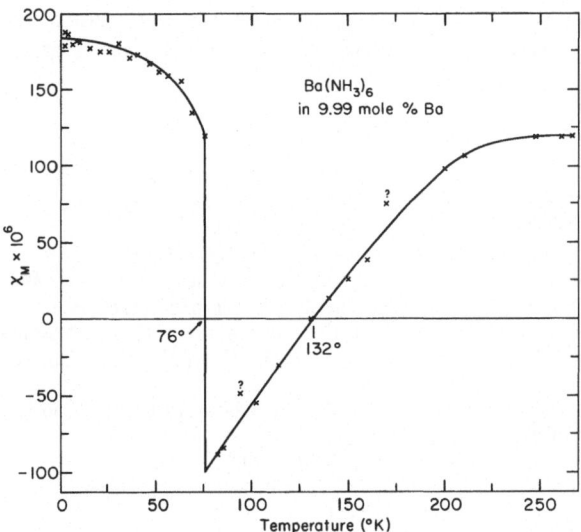

Fig. 11. Magnetic susceptibility per mole of $Ba(NH_3)_6$ versus temperature

presumably being reciprocal to the type shown in Fig. 2. At high temperature, except for $Li(NH_3)_4$, there is again an increase in susceptibility. The reason for this is not clear. If the low-temperature behavior is connected with a constant number of moments then the high-temperature behavior may reflect a thermally activated increase in the number of these moments. In any case, it almost looks as if the cubic-hexagonal and melting transitions in $Li(NH_3)_4$ interrupt what may have been similar behavior.

The very steep drop in the χ_M of $Ca(NH_3)_6$ below 10 °K and the very steep rise in the χ_M of $Sr(NH_3)_6$ below 12 °K are not understood. It may be that these features reflect flaws in the correcting procedure. The $Sr(NH_3)_6$ case is particularly uncertain because the sample contained excess strontium metal, for which the temperature dependence of χ has not been clearly established. Trace impurities could dramatically affect the low-temperature portions. Precision measurements on materials of scrupulous purity are now in progress.

The broad minimum in the χ_M vs T curve for $Ca(NH_3)_6$ at about 110 °K and for $Sr(NH_3)_6$ at about 70 °K does not show up for $Ba(NH_3)_6$, presumably because there is some kind of phase transition at 76 °K. It is interesting to note that Mammano and Sienko, when making their low-temperature X-ray studies of $M(NH_3)_6$, were actually looking for such a transition. They did not find it, but only went down to 77 °K!

The above findings, though still very preliminary, suggest an application to the problem of $Eu(NH_3)_6$. It was found there [23] that partial replacement of Eu by Yb did not appear to change the observed Curie temperature for ferromagnetic ordering, even though magnetic dilution should have depressed T_c. It may eventually turn out that the observed T_c was really due to a ferromagnetic contaminant such as $Eu(NH_2)_2$, but, if not, one will have to conclude that the ferromagnetic interaction is long range and almost certainly of the Baltensperger-de Graaf type as appropriate to Boltzmann statistics in contrast to Fermi-Dirac.

References

1. Mammano,N., Coulter,L.V.: J. Chem. Phys. **47**, 1564 (1967); **50**, 393 (1969).
2. Coulter,L.V., Monchik,L.: J. Am. Chem. Soc. **73**, 5687 (1951).
3. Marshall,P.R., Hunt,H.: J. Phys. Chem. **60**, 732 (1956).
4. McDonald,W.J., Thompson,J.C.: Phys. Rev. **150**, 602 (1966).
5. Morgan,J.A., Schroeder,R.L., Thompson,J.C.: J. Chem. Phys. **43**, 4494 (1965).
6. Mammano,N., Sienko,M.J.: J. Am. Chem. Soc. **90**, 6322 (1968).
7. Kleinman,L., Hyde,S.B., Thompson,C.M., Thompson,J.C.: In: Lagowski,J.J., Sienko, M.J. (Eds.): Metal–ammonia solutions, Coloque Weyl II, p. 229. London: Butterworths 1970.
8. Glaunsinger,W.S., Zolotov,S., Sienko,M.J.: J. Chem. Phys. **56**, 4756 (1972).
9. Ham,F.S.: Phys. Rev. **128**, 2524 (1962).
10. Lo,R.E.: Z. Anorg. Allg. Chemie **344**, 230 (1966).
11. Hubbard,J.: Proc. Roy. Soc. A **276**, 238 (1963); A **281**, 401 (1964).
12. Cyrot,M.: J. Phys. (Paris) **33**, 125 (1972).
13. Pines,D.: Solid State Phys. **1**, 367 (1955).
14. Cate,R.C., Thompson,J.C.: J. Phys. Chem. Solids **32**, 443 (1971).
15. Chieux,P., Sienko,M.J.: Unpublished data.
16. Barrett,C.S.: Acta Cryst. **9**, 671 (1956).
17. McKelvy,E.C., Taylor,C.S.: Bur. Stand. Sci. Papers Nr. 465, 655 (1923). — Durrant,A.A., Pearson,T.G., Robinson,P.L.: J. Chem. Soc. 730 (1934).
18. Jander,J.: Anorganische und allgemeine Chemie in flüssigem Ammoniak, p. 11. Braunschweig: Vieweg 1966.
19. See, for example the review by Catterall,R.: In: Lepoutre,G., Sienko,M.J. (Eds.): Metal–ammonia solutions, Colloque Weyl I, p. 53. New York: W. A. Benjamin 1964.
20. McDonald,W.J., Thompson,J.C., Bowen,D.E.: Bull. Am. Phys. Soc. **9**, 735 (1964).
21. Cagle,F.W., Holland,H.J.: Presented at the 145th National Meeting of the American Chemical Society, New York (September 1963).
22. Mammano,N., Sienko,M.J.: J. Solid State Chem. **1**, 534 (1970).
23. Oesterreicher,H., Mammano,N., Sienko,M.J.: J. Solid State Chem. **1**, 10 (1969).

Discussion

M. H. COHEN These metal–NH_3 compounds are extremely interesting and ought to play an important role in extending our understanding of the metallic state. The value of r_s in these materials is about the same as that of Cs. Moreover, in Cs the volume of space where the atomic potential (made up of several pseudopotentials) is not constant, is larger than for these materials. Thus they are very low density, nearly-free-electron metals and, as such, are very important. They are not completely free-electron-metals, otherwise the structured transformation would take place at lower temperatures, if at all, and there would be no great change in magnetic susceptibility at the transition, in contrast to what is observed. Thus the Fermi surface in the HCP phase may be substantially different from the NFE electron sphere, as reconstructed to fit into the first zone à la Harrison — and as found for Mg, Zn and Cd. However, ignoring that possibility, we can have quite a flat piece of Fermi surface, the 3rd zone lens. Now a flat piece of Fermi surface and a strong electron-electron interaction because of the large r_s give the preconditions for a spin density wave, as has been observed for Cr. Your peak in χ_{ESR} with the attendant presumption of an AFM transition may be evidence for the existence of a spin density wave state (SDW).

Overhauser has postulated an SDW for the nearly-free-electron-metal K and has lately changed his mind in favor of a charge density wave. The bulk of evidence, both theoretical and experimental, is against such a suggestion, in my opinion. So, you may be reporting here the first observation of an SDW in a free-electron-metal. However, the flattening of the 3 rd zone lens by the crystal potential and the large r_s may provide the essential physical basis rather than merely the free electron character.

M. J. SIENKO   The r_s for $Li(NH_3)_4$ is 7.25 a_0, considerably greater than the value of 5.6 a_0 characteristic of cesium. It is quite probable that the NH_3 structure affects the potential more than we have assumed. Also onset of NH_3 rotation about the C_{3v} axis may be responsible for some of the transitions we observe.

J. JORTNER   Whether these materials exhibit a free elctron behavior depends on the strength of the pseudopotential. In such expanded solid metal studied by Sienko the electron-positive ion pseudopotential will be considerably weaker than in the alkali metals. Thus these materials may exhibit the first example for the applicability of the theory of Cohen and Wiser concerning phase transformation in monovalent metals characterized by large r_s values.

On Superconducting Metal–Ammonia Complexes

JUANA V. ACRIVOS, S. F. MEYER, and T. H. GEBALLE

Abstract

The system $M^1 - NH_3$ (M^1 = metallic dichalcogenide) has been studied. Based on the theory of electron donor-acceptor complexes it is assumed that the intermolecular forces between the crystalline layer material and small intercalate molecules are due to dispersion (or van der Waals type interactions) and to charge transfer interactions. It is postulated that the factors which favor intercalation are low ionization energy of the highest filled spin orbital of the intercalate donor, high work function or electron affinity for the lowest empty spin orbital of the metallic acceptor, and that symmetry and directional and size effects are of utmost importance. The results are applied to interpret the superconducting properties of intercalated $2H$–TaS_2 and $2H$–$NbSe_2$.

I. Introduction

Lewis type bases (e.g. ammonia and pyridine) have been intercalated into metallic crystalline layer materials (e.g. $2H$–TaS_2 and $NbSe_2$) by Gamble et al. [1, 2]. The critical temperature for superconductivity T_c of these compounds has been correlated to the pK_a of the Lewis bases by DiSalvo [3]. M–NH_3 (M = alkali and alkaline earth metals) intercalation into layered type compounds by Acrivos et al. [4]. Somoano et al. [5] and Revelli et al. [6] can reversibly change semiconductors into metals.

In order to investigate a larger class of intercalated crystalline layer superconductors and to determine how intercalation affects T_c, it is important to determine the parameters which govern the formation of these compounds. The latter may be accomplished from a study of the nature of the molecular complexes formed between the intercalate and the layer material, using the concept of electron donor-acceptor interactions.

The theory of molecular complexes was originally developed by Mulliken [7, 8]. Recently, Krugler et al. [9] and McConnell et al. [10] treated the properties of collective electronic states in linear chain molecular crystals by an elegant extension of this theory using the methods of molecular field operators.

Appropriate to the Colloque Weyl, this work describes the properties of $M^1 - NH_3$ where M^1 is a layered dichalcogenide (Fig. 1). In the 2H hexagonal close-packed phase shown, the transition metal atoms form in planes sandwiched between two chalcogen layers. The unit cell consists of two such sandwiches separated by a distance of the order of twice the chalcogen van der Waals radius. The intercalate opens up the van der Waals gap and the complex is stabilized by charge transfer and long-range interactions. The theory of electron donor-acceptor complexes will be used to surmise and predict the effects of chemical intercalation on the superconducting properties measured for these compounds.

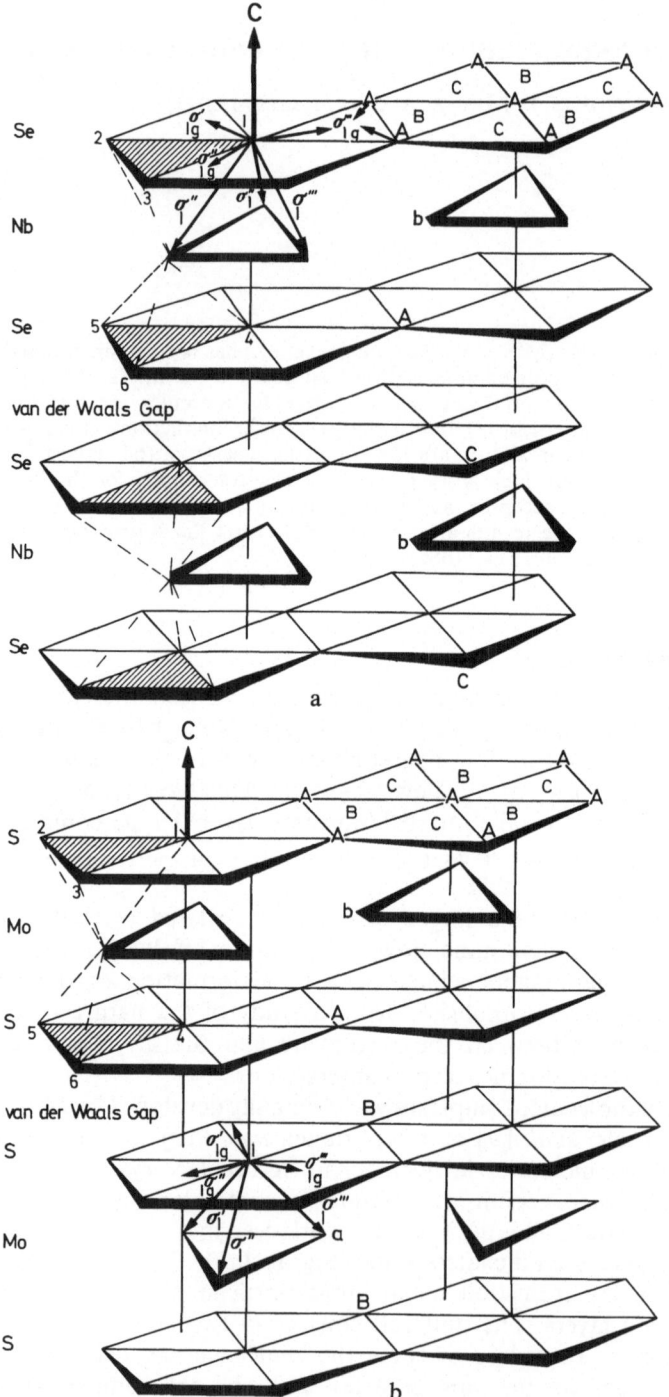

Fig. 1a and b. Structure of transition metal dichalcogenides in the 2H-hexagonal close packed phase. The σ_i are the directed hybrid orbitals centered at the chalcogen

II. Theory

A schematic energy level diagram for the process of intercalation is shown in Fig. 2. Here the intermolecular forces between the intercalate and the layer material are due to dispersion (or van der Waals interactions) and to charge transfer interactions. The initial state of the system, before the interaction is turned on is

$$|i) = \Psi_D^0 \cdot \Phi_A^0 \tag{1}$$

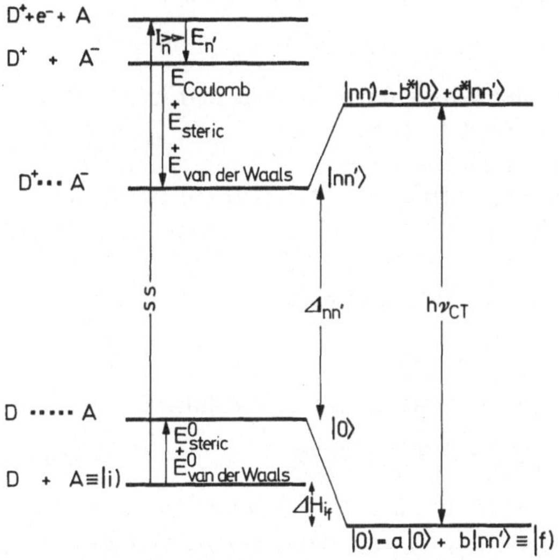

Fig. 2. Energy level diagram for the intercalation process

where Ψ_D^0 and Φ_A^0 are the free donor D and free acceptor A Slater-determinant wavefunctions. For the sake of simplicity and without loss of generality, the following discussion assumes that the donor is the intercalate with electrons occupying valence band spin orbitals while the dichalcogenide is the acceptor and has electrons occupying conduction band spin orbitals.

In the Mulliken theory the ground state of a nonionic complex is a no-bond wavefunction which takes into account the effects of the dispersion or van der Waals forces. This leads to an admixture into $|i)$ of intramolecular excited configurations, i.e.,

$$|0\rangle = |A - D\rangle = \Psi_D^0 \cdot \Phi_A^0 + \cdots \tag{2}$$

where the exact form of the excited configurations may be obtained using the methods of molecular field operators [9, 11].

The ground state energy differs from that of the initial state by the dispersion terms. In Fig. 2 $E_{\text{van der Waals}}^0$ is the energy necessary to open the van der Waals gap in the layer material (in order to accommodate the intercalate) and to break for example the hydrogen bonds, lowered by the intermolecular dipolar interactions. E_{steric}^0 is the energy term describing the direction-of-approach effects. These terms are assumed to be positive in Fig. 2. However, they may be negative when long-range dipolar interactions between the intercalate molecules are present. The change in entropy is expected to be negative because the intercalate molecules go into ordered states between the layers of the material [3].

Thus the intercalated system is stabilized mainly by the admixture of ionic states

$$|nn'\rangle = |D_n^+ - A_{n'}^-\rangle \tag{3}$$

resulting from the excitation of an electron in one of the highest filled valence band spin orbitals of D (say Ψ_n) into one of the lowest unoccupied conduction band spin orbitals of A (say Φ_n'). This state is assumed to lie above the ground state by

$$\Delta_{nn'} \cong I_n - E_{n'} + E_{nn'} \,(\text{Coulomb}) . \tag{4}$$

The steric and van der Waals terms shown in Fig. 2 in the ground and excited ionic states, are assumed to have canceled out. I_n is the ionization energy, $E_{n'}$ is the electron work function or electron affinity and $E_{nn'}$(Coulomb) is the ion-ion $(D_n^+ - A_{n'}^-)$ Coulomb attraction. All of these terms are known roughly from experimentation. $I_n = 10$ to $7\,\text{eV}$ for molecular donors [12, 13], $E_{n'} \cong 4.1\,\text{eV}$ for NbSe$_2$ [14].

Thus, given a charge separation of 4 to 5 Å, $E(\text{Coulomb}) \cong 2\,\text{eV}$ leads to

$$\Delta_{nn'} \cong 1 \text{ to } 4\,\text{eV} .$$

The largest off-diagonal elements are the Mulliken resonance integrals:

$$H_{nn'} = \langle D - A | \hat{H} | D_n^+ - A_{n'}^- \rangle = \langle 0 | \hat{H} | nn' \rangle \tag{5}$$

where H is the total Hamiltonian. $H_{nn'}$ depends on the overlap $\langle 0 | nn' \rangle = S_{nn'} \ll 1$ [15]. This means then that for $H_{nn'}$ sufficiently small $(H_{nn'} \ll 1\,\text{eV})$ a perturbation expansion in powers of $H_{nn'}/\Delta_{nn'}$ is valid [9]. Thus, a resonance stabilized final state is

$$|f\rangle = |0\rangle - \sum_{nn'} (H_{nn'}/\Delta_{nn'}) |nn'\rangle + \cdots \tag{6}$$

The energy is an average over all electrons, u.e.,

$$E_f \cong \left\langle H_{00} - \sum_{nn'} |H_{nn'} - S_{nn'} H_{00}|^2 / \Delta_{nn'} + \cdots \right\rangle_{\text{Av}} . \tag{7}$$

In an isolated D–A molecular complex $|H_{nn'}| \ll 1\,\text{eV}$, however, cooperative phenomena will enhance the charge transfer contribution. The change in enthalpy for the process shown in Fig. 2 is then

$$\Delta H_{if} = E_{\text{van der Waals}}^0 + E_{\text{steric}}^0 + \Delta(pV)_{if} - \left\langle \sum_{nn'} |H_{nn'} - S_{nn'} H_{00}|^2 \, \Delta_{nn'} \right\rangle_{\text{Av}} + \cdots \tag{8}$$

where $\Delta(pV)_{if}$ represents pressure volume changes.

The physical significance of Eq. (8) is that (since the change in entropy is expected to be less than or equal to zero) the free energy which drives the process arises mainly from the charge transfer resonance terms and lon-range dipolar interactions. Consequently, the criteria for intercalation are as follows:

1. Low ionization energy I_n and high electron affinity $E_{n'}$, favor intercalation.

2. In order that the matrix element of the interaction $H_{nn'} = \langle 0 | \hat{H}' | nn' \rangle$ does not vanish, $|0\rangle$ and $|nn'\rangle$ must transform in the same group theoretic and spin-orbit representation; and

3. Directional and size effects are of utmost importance because they determine the magnitude of the overlap $S_{nn'}$ and consequently the magnitude of $H_{nn'}$. Also, they determine long-range dipolar interactions.

The self-consistency of the theory may be tested using the charge transfer data assuming $|b|^2 \cong 0.2 \pm 0.1$, taken from the increase Δn in the number of conduction electrons per niobium atom, found in (pyridine)$_\frac{1}{2}$ · NbS$_2$ (Ehrenfreund et al. [16]) together with the enthalpies reported by DiSalvo [3], if, for negligible overlap:

$$\varepsilon^{(2)} = \left\langle \frac{|H_{nn'}|^2}{\Delta_{nn'}} \right\rangle \cong \Delta_{nn'} |b|^2 \tag{9}$$

where

$$|b|^2 = \left\langle \left| \frac{H_{nn'}}{\Delta_{nn'}} \right|^2 \right\rangle.$$

This follows from the second order approximation. Here (see Ruderman and Kittel [17] and Bloembergen and Rowland [18]):

$$\varepsilon_{nn'}^{(2)} \cong \sum_{D,A} \int \frac{dk}{(2\pi)^3} \int \frac{dk'}{(2\pi)^3} \left\{ \exp\left(i(k'-k)\cdot R_{DA}\right) \frac{H_{nkn'k'} H_{n'k'nk}}{E_{n'}(k') - E_n(k)} \right\} + \text{c.c.} \tag{9'}$$

Also:

$$|b|^2 \cong \sum_{D,A} \int \frac{dk}{(2\pi)^3} \int \frac{dk'}{(2\pi)^3} \left\{ \exp\left(i(k'-k)\cdot R_{DA}\right) \frac{H_{nkn'k'} H_{nkn'k'}}{[E_{n'}(k') - E_n(k)]^2} \right\} + \text{c.c.} \tag{10}$$

where R_{DA} is the donor-acceptor separation. The averaging must take into account the axial symmetry of the crystals, i.e., the matrix element

$$H_{nkn'k'} = \int d\tau \, \Psi_{nk}^* \hat{H} \, \Phi_{n'k'}$$

depends on the angle Θ_{kc} between the wave vector k and the crystallographic axis c shown in Fig. 1. The Ψ_{nk} are axially symmetric Bloch functions where

$$V^{-1} \int d\tau \, \Psi_{nk}^* \Psi_{nk} \to 1$$

and V is the sample volume. The excitation energy in the denominator is also orientation dependent [19]. In the intercalate layer $|k_z| < k_0$, where $k_0 r_0 = \pi$ [20] and r_0 is the thickness of the intercalate layer. In the metallic layer the excited electrons have energies above those in the intercalate layer and $|k'| > k_{\text{Fermi}}$. The complex integrals cannot be evaluated without the explicit knowledge of the terms involved. However, the assumption in Eq. (9) follows from (9') and (10) using the approximations:

$$|H_{nkn'k'}|^2 \cong |H_{nn'}|^2 \, \overline{v_0}^2 \quad \text{and} \quad \langle |E_{n'}(k') - E_n(k)|^{-2} \rangle \Delta_{nn'} \cong \langle |E_{n'}(k') - E_n(k)|^{-1} \rangle$$

where $H_{nn'}$ is the charge transfer matrix element for the isolated complex, $\overline{v_0}^2$ is the averaged square molecular volume of the complex, and only one transition (Ψ_n to $\Phi_{n'}$) is involved.

Given $|b|_{\text{exp}}^2 = 0.2 \pm 0.1$ for (pyridine)$_\frac{1}{2}$ · 2H–NbS$_2$ (Ehrenfreund et al. [16]), it then follows that $\varepsilon^{(2)} = 0.8 \pm 0.4$ eV or 18 ± 9 kcal/mol when $\Delta_{nn'} \cong 4$ eV. The experimental enthalpy change is $(\Delta H_{if})_{\text{exp}} = 10$ to 20 kcal per half mole of intercalate [3]. ΔH_{if} in Eq. (8) is the sum of $\varepsilon^{(2)}$ plus the dispersion terms. Here

$E^0_{\text{van der Waals}}$ is assumed to be less than $4\,\text{kcal/mol}$, the energy necessary to re-arrange the dichalcogenide layers [21]. Thus, the experimental enthalpies measured directly and obtained from the charge transfer data are equal within the large limits of the uncertainty.

The following experiments were carried out to test the effects of charge transfer on the superconducting critical properties of intercalated layer compounds. They are based on two basic premises: (1) that intercalates are electron donor-acceptor complexes and (2) that chemical interactions may be used to control the superconducting properties of a compound. The BCS theory [22, 23] relates the superconducting properties with the lattice vibrational modes and the density-of-states near the Fermi surface, which in some cases are measurable, and with the electron-phonon interaction and the screened Coulomb interaction, which are less accessible. The effects of static and dynamic charge transfer and of conjugation of the metallic free-electron system with that of organic molecules have been investigated by McConnell *et al.* [10, 24, 25] in vanadium and aluminum films deposited in a substrate together with organic donor and acceptor molecules. These molecules can be divided into two classes, enhancers and depressors of T_c, respectively [26].

Gamble *et al.* [2] have shown that there is a direct relationship between T_c and the number of donor-acceptor bonds by studying a series of intercalated aliphatic amines in $2H\text{-}TaS_2$.

III. Experimental Results

The superconducting transition data given in Tables I and II for intercalated $2H\text{-}TaS_2$ and $2H\text{-}NbSe_2$ respectively were taken using an oscillating detector [27] to measure both the real and imaginary parts of the complex susceptibility $\chi = \chi' - i\chi''$ at $\sim 75\,\text{kHz}$ and described elsewhere [28]. The $2H\text{-}TaS_2$ and $2H\text{-}NbSe_2$ powders and single crystals were obtained according to the procedure described by DiSalvo [3]. Intercalation of ammonia and amines was carried out using high vacuum techniques [29]. The temperature dependence of the frequency shift (χ') and the power losses (χ'') are shown in Fig. 3. Skin effects mix χ' and χ'' so that the observed signal χ_e is mixed.

The reversibility of the intercalation process is illustrated by samples T–6, T–6' and T–6'' in Table I and N–4 and N–5 in Table II.

In $NH_3\text{-}2H\text{-}TaS_2$ the rates of cooling influence the superconducting transition. This indicates that different thermodynamic states are formed. Figure 4 shows energetically plausible arrangements of the ammonia molecules in the plane of the van der Waals gap. Assuming that the Ta atoms remain directly over each other and that the expansion of the c-axis on intercalation is $2 \times 3.03\,\text{Å}$ the Ta–Ta separation in the c-direction is $9.07\,\text{Å}$. It then follows that if the isolated NH_3 molecular parameters remain unchanged in Fig. 4a, the nearest N–Ta distance is $4.3\,\text{Å}$ while the nearest H–S distance is $3.0\,\text{Å}$. This favors interaction between the molecular orbitals $3a_1$ of NH_3 (lone pair) and the atomic orbital $5d_{z^2}$ of Ta as well as chalcogen and proton orbital interaction. Figure 4b shows another possible arrangement which appears to have a higher energy barrier of formation

Table I. Data for 2H–TaS$_2$ intercalates

Sample number (Aggregate)	Intercalate (Symmetry group)	Intercalate I_n in eV from photoelectron spectra (b), (c) (Symmetry species)	Heat treatment: room temperatures to t(°C), P(time)	SC transitions T_c(K)		X-ray data[d-f] a-axis (Å)	c-axis (Å)	Stoichiometry (Intercal.)·(TaS$_2$)
J-5 (Powder)	Control		300°	0.8		3.31	12.10[d]	
T-6[a] (single crystals $c \parallel \boldsymbol{H}_{osc}$)	Ammonia: NH$_3$ (C$_{3v}$)	$(3a_1)$: 10.16[b] $(1e)$: 14.8	a) −190° (<5 min) b) −190° (12 hrs) c) NH$_3$ cryopumped at −190° d) NH$_3$ vac. pumped 230°, 10^{-1} torr (21 hrs)	3.68 3.8 3.8 <1.3	— 5.1 4.1	3.33	18.14[e]	1 : 1[e-g]
T-6'[a]	T-6d + ND$_3$		a') same as a above b') same as b above a'') same as a above	4.3 4.5 4.5	4.8 4.85 5.0			
T-6''[a]	T-6d + NH$_3$							
I-48 (Powder)	Methylamine	9.18[c]	a) −190° (<5 min) b) −190° (12 hrs)	1.3 1.3	4.8 5.4	3.33	18.74[f]	
T-100 (Powder)	Phosphine (C$_{3v}$)	$(5a_1)$: 9.9[b] $(2e)$: 13.0	a) −15° (2 months) b) Rm. Temp. (2 wks)	>1.0 1.3				

a The same crystals were used throughout. – b Taken from the data of Turner et al. [13]. – c Taken from the data of Gamble et al. [41]. – e Taken from the data of Franklin et al. [12]. – d Taken from the data of DiSalvo [3]. – f Taken from the data of Gamble et al. [2]. – g The state of aggregation is of great importance in the measurement of T_c for the intercalated materials. The following statistical sampling illustrates the NH$_3$, ND$_3$ isotope effect on T_c measured on pairs of samples prepared from identical starting compound.

Sample No.	Intercalate	T_c measured after heat treatment similar to T-6		
		a (after preparation) °K	a (months later) °K	b (months later) °K
NMR-6 (xtal)	NH$_3$ (saturated)	3.55	3.85	
NMR-7 (xtal)	ND$_3$ (saturated)	3.35	3.7	
M-22 (powder)	NH$_3$ (saturated)		3.6	3.7
M-21 (powder)	ND$_3$ (saturated)			3.6
M-17 (powder)	NH$_3$ (unsaturated)	3.6	3.6	
M-18 (powder)	ND$_3$ (unsaturated)	3.0	3.0	
Gamble et al. [2]	NH$_3$–TaS$_2$, T_c = 4.2			

Table II. Data for 2H–NbSe$_2$ intercalates

Sample number	Intercalate	Intercalate I_n in eV from photoelectron spectra[a]	Heat treatment: room temperature to t (°C), P (time)	SC transitions T_c (°K) (width)	X-ray data a-axis (Å)	c-axis (Å)	Stoichiometry (Intercalate) · NbSe$_2$
J–2	Control		300°	7.2	3.45	12.54[b]	
N–1	Ammonia	$(3a_1)$: 10.16[a] $(1e)$: 14.8	a) Rm. Temp. b) 100° c) X-ray d) Deintercalated by TGA	6.6 (broad) 0.6 7.0 (broad)		18.5 15.66 12.5	1 : 1
N–4	NH$_3$		a) Rm. Temp. (3 mo) b) Deintercalated at Rm. Temp., 1 atm (1 day)	<1.3 6.1			
N–5	ND$_3$		a) Rm. Temp. (3 mo) b) Deintercalated at Rm. Temp., 1 atm (1 day)	<1.3 6.7			
N–9	Cyclopropylamine	7[a] 8 9	a) 50° b) 650° (TGA)	1.0 <1	3.46	21.5	

[a] Taken from the data of Turner *et al.* [13].
[b] Taken from the data compiled by Wilson and Yoffe [41].

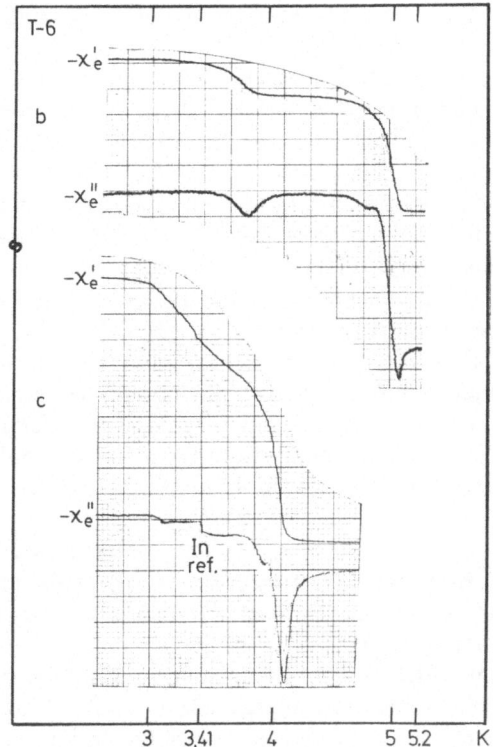

Fig. 3. Frequency shift and effective power losses (χ'_e) and (χ''_e) in arbitrary units (see Maxwell and Strongin [40]) vs T for NH_3–$2H$–TaS_2 after two different heat treatments. Skin effects mix χ' and χ'' so that χ'_e and χ''_e represent only signals in quadrature. T–6b shows the superconducting transition after slow cooling. T–6c shows the transition after cryopumping NH_3 to $-190°$C with the sample held at room temperature. An indium reference transition is used for calibration

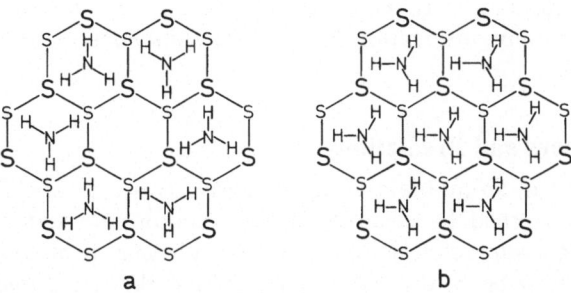

Fig. 4. Energetically possible arrangement of NH_3 molecules in the van der Waals gap of $2H$–TaS_2. Here the sulfur atoms above and below the gap (sites A and C respectively) are shown. The nitrogen atoms of ammonia occupy the B sites in the van der Waals gap, directly above and below are the transition metal in the neighboring layers. a shows neighboring NH_3 with antiparallel dipoles and b shows planar NH_3 molecules

because it requires that the NH_3 take on a planar configuration. The barrier to inversion is 0.23 eV [30].

However, tunneling is expected to reduce this value. Long-range interactions between the intercalate molecules are present in both models shown in Fig. 4. These models are not inconsistent with the structure data on solid and liquid ammonia, which has nearest neighbors at 3.56 Å [31—33].

The lower superconducting transitions in Table I were obtained by deintercalating the sample by condensing the ammonia to $-190°$ C. X-ray data on N–1c in Table II suggest that these are second stage intercalates in which every other van der Waals gap is filled with ammonia molecules, as has been previously observed in the case of intercalated pyridine [2].

These thermodynamic states with different T_c may be compared with the Mulliken concept of "inner and outer complexes" in, say, the pyridine-iodine complex [34]. In the outer complex the amount of charge transfer is less than for the "inner complex". Thus, the higher T_c state could be an analogue of the "inner complex" because the state with the highest charge transfer is expected to have a higher T_c.

The relatively high T_c obtained for 2H–TaS_2 intercalated with methylamine is a result in support of the low-ionization-energy-of-the-intercalate criterion. Here the amount of the charge transfer $|b|^2$ is assumed to be higher than for ammonia $\Delta_{nn'}$ is lower by 1 eV.

However, other terms which control the formation of the charge transfer complex obscure the low donor ionization energy effect. Thus, although phosphine has lower ionization energies than ammonia, its intercalation is very difficult. Here the barrier to inversion is 1.3 eV [35] and the P–H bond distance is 1.4 Å. These terms may give unusually high contributions to E^0_{steric} in Fig. 2. The low T_c reported in Table I are conjectured to be due to a high barrier for intercalation.

Intercalation of 2H–$NbSe_2$ lowers the superconducting critical temperature in all the cases studied in Table II. On the basis of the photoemission data of 2H–$NbSe_2$ [14], the charge transfer complexes formed should lower the density-of-states at the Fermi surface.

Isotope effects are present in the NH_3 and ND_3 intercalates T–6, T–6′ and T–6″, N–4 and 5 in Tables I and II. This is attributed not so much to an enhancement of the electron-phonon interaction as to a zero point energy effect of the chemical equilibrium.

IV. Consequences and Predictions

The intercalation of ammonia and amine derivatives into 2H–TaS_2 and $NbSe_2$ appears to be governed by electron donor-acceptor interactions. The physical significance is that superconduction and other physical properties of the above-layer materials may be controlled chemically and that experiments to test new theoretical developments may be prepared easily.

The energy level diagram shown in Fig. 2 makes possible the following predictions: (1) the effect of increasing pressure on nonionic ground state complexes will increase the amount of charge transfer $|b|^2$ by lowering the energy of the excited ionic state. Here T_c is expected to increase for the 2H–TaS_2 complexes. The opposite will be true for ionic ground state complexes. Thus, a study of the pressure dependence of T_c can give important information on the nature of the complex. (2) The same increase of charge transfer with pressure should lead to a decrease in the c-axis electrical resistivity. The recent discovery [36] of the minimum in the c-axis resistivity of carefully prepared pyridine-intercalated $2H-TaS_2$ near 20 °K, which has subsequently been confirmed by us, may be a direct consequence of the charge transfer. If so, Kondo-like scattering (for review see Kondo [37]) due to the fractional unpaired spin contribution from paramagnetic excited states should increase with pressure. The ESR studies of photo-excited charge transfer complexes [38] have verified the nature of the paramagnetic excited state of nonionic ground state molecular complexes. In more conventional three-dimensional systems Kondo scattering is a pair-breaking mechanism which rapidly destroys the superconductivity. In the almost-two-dimensional crystals being discussed here, only a very small fraction of the conduction band states which have velocity components along the c-axis are involved. Thus, the superconducting condensation energy should not be appreciably affected, although the coupling between layers, for example, Josephson tunneling, should be. Carefully controlled experiments will be necessary to investigate such possibilities. (3) Charge transfer CT optical transition should occur at:

$$hv_{CT} \cong \Delta_{nn'} + 2\varepsilon^{(2)}.\tag{11}$$

For the complex $(pyridine)_{\frac{1}{2}} \cdot 2H-NbS_2$, $hv_{CT} = 4.8$ to 5.6 eV from $|b|^2_{exp} = 0.2 \pm 0.1$ and $\Delta_{nn'} \cong 4$ eV. The feasibility of such measurements are favored by the strong CT transitions (extinction coefficient $= 5 \times 10^4$) observed for the pyridine-iodine complex at 235 nm by Reid and Mulliken [34]. The effects of charge transfer excitations on the superconducting properties of the above layer compounds may be significant.

Finally, (4) the nature of the chemical equilibrium:

$$D_v + A \rightleftarrows D_v - A \tag{12}$$

where D_v is the intercalate and A is the layer compound may be studied by isotopic substitution in the intercalate. The different values of T_c obtained in Tables I and II for NH_3 and ND_3 intercalation may be qualitatively explained from the ratio of the equilibrium constants K for Eq. (12). Thus:

$$r_{H/D} = \frac{K(NH_3-TaS_2)}{K(ND_3-TaS_2)} = \left(\frac{q^0(ND_3)}{q^0(NH_3)}\right)^v \frac{q^0(H\text{-complex})}{q^0(D\text{-complex})}\tag{13}$$

where the q^0 are the partition functions for the respective compounds in their standard states. The higher T_c observed for the $NH_3-2H\text{-}TaS_2$ complex would follow from an increase in the charge transfer due to a displacement of equilibrium (12) to the right for the latter relative to the $ND_3-2H-TaS_2$ complex. Using the Redlich-Teller product rule for isotope substitution and the NH_3 and ND_3 vibration frequency data [39] it follows that $q^0(ND)/q^0(NH_3) \gg 1$ at 5 °K. The lower frequency vibrations in the solid complex are not expected to be as strongly influenced by the isotope substitution in the intercalate and therefore $r_{H/D} \gg 1$. However, one cannot eliminate the possibility of isotope effects in the electron-phonon interaction.

Note Added in Proof:

Support for the models shown in Fig. 4 is obtained from the recent transmission electron microscopy measurements by Carter [42] and Carter and Williams [43] on ammonia intercalated $2H-NbSe_2$ and $2H-MoS_2$. These have led the authors to suggest: a) that a monolayer of ammonia is present in the van der Waals gap of the intercalated dichalcogenide; b) that the rings observed in the diffraction pattern below $-70°$ C (NH_3 m.p. $= -79°$ C) are due to turbostratically stacked ammonia with an a-axis of 4.2 to 4.4 Å; c) that the superlattice observed on intercalation (with scattering centers in a hexagonal array a(axis) $= 5.27$ Å) is due to rearranged impurities and d) that an additional superlattice is present in $NH_3-2H-NbSe_2$ but not in $NH_3-2H-MoS_2$.

Acknowledgements

The support of NSF GP 21293 and F. G. Cottrell Research Corporation grants in aid for research are gratefully acknowledged. Research at Stanford was supported by U.S. Airforce Office of Scientific Research, Office of Aerospace Research, under Grant No. AFOSR 68-1510C.

References

1. Gamble, F. R., DiSalvo, F. J., Klemm, R. A., Geballe, T. H.: Science **168**, 568 (1970).
2. Gamble, F. R., Osiecki, J. H., Cais, M., Pisharody, R., DiSalvo, F. J., Geballe, T. H.: Science **174**, 493 (1971).
3. DiSalvo, F. J.: Ph. D. Dissertation (1971), Stanford University.
4. Acrivos, J. V., Liang, Y., Wilson, J. A., Yoffe, A. D.: J. Phys. Chem. **3**, 118 (1971).
5. Somoano, R. B., Rembaum, A.: Phys. Rev. Letters **27**, 402 (1971).
6. Revelli, J., Phillips, A.: Bull. Am. Phys. Soc. **17**, 22 (1972).
7. Mulliken, R. S.: J. Am. Chem. Soc. **72**, 600 (1950).
8. Mulliken, R. S., Person, N. B.: Ann. Rev. Phys. Chem. **13**, 107 (1962).
9. Krugler, J. I., Montgomery, C. G., McConnell, H. M.: J. Chem. Phys. **41**, 2421 (1964).
10. McConnell, H. M., Hoffman, B. M., Thomas, D. D., Gamble, F. R.: Proc. Natl. Acad. Sci. (Wash.) **54**, 371 (1965).
11. Agranovich, V. M., Makhtiev, M. A.: Sov. Phys. Solid State **13**, 2284 (1972).
12. Franklin, J. L., Dillard, J. G., Rosenstock, H. M., Herron, J. T., Draxl, K., Field, F. H.: Ionization potentials, appearance potentials and heats of formation of gaseous positive ions. NSRDS-NBS 26, U.S. Govt. Printing Office, Wash. D.C. 1969.
13. Turner, D. W., Baker, C., Baker, A. D., Brundle, C. R.: Molecular photoelectron spectroscopy. London: Wiley-Interscience 1970.
14. McMenamin, J. C., Spicer, W. E.: To be published and private communication 1972.
15. Murrell, J. N., Randic, M., Williams, D. R.: Proc. Soc. A, **284**, 566 (1964).
16. Ehrenfreund, E., Gossard, A. C., Gamble, E. R.: Phys. Rev. B **5**, 1708 (1972).
17. Ruderman, J. A., Kittel, C.: Phys. Rev. **96**, 99 (1954).
18. Bloembergen, N., Rowland, T. J.: Phys. Rev. **97**, 1679 (1955).
19. Bromley, R. A., Murray, R. B., Yoffe, A. D.: J. Phys. Chem. **5**, 259 (1972).
20. Mott, N. F., Jones, H.: The theory of the properties of metals and alloys. Oxford: Clarendon Press 1936.
21. Huisman, R.: Ph. D. Dissertation, 1969, Groningen.
22. Bardeen, J., Cooper, L. N., Schrieffer, J.: Phys. Rev. **108**, 1175 (1957).
23. McMillan, W. L.: Phys. Rev. **167**, 331 (1967).
24. McConnell, H. M., Hoffman, B. M., Metzger, R. M.: Proc. Natl. Acad. Sci. (Wash.) **54**, 371 (1965).
25. McConnell, H. M., Gamble, F. R., Hoffman, B. M.: Proc. Natl. Acad. Sci. (Wash.) **57**, 1131 (1967).

26. Hoffman, B. M., Gamble, F. R., McConnell, H. M.: J. Am. Chem. Soc. **89**, 27 (1967).
27. Schwalow, A. L., Devlin, G. E.: Phys. Rev. **110**, 1011 (1958).
28. Meyer, S. F.: To be published 1972.
29. Acrivos, J. V., Azebu, J.: J. Mag. Res. **4**, 1 (1971).
30. Swalen, J. D., Ibers, J. A.: J. Chem. Phys. **36**, 1914 (1962).
31. Olovsson, J., Templeton, D. H.: Acta Cryst. **12**, 832 (1959).
32. Reed, J. W., Harris, P. M.: J. Chem. Phys. **35**, 1730 (1961).
33. Krugh, R. F., Petz, J. L.: J. Chem. Phys. **41**, 890 (1964).
34. Reid, C., Mulliken, R. S.: J. Am. Chem. Soc. **76**, 3869 (1954).
35. Weston, R. E.: J. Am. Chem. Soc. **76**, 2645 (1954).
36. Thompson, A. H.: Private communication (1972). Also reported by Gamble, F. R., APS March 1972.
37. Kondo, J.: Solid State Physics **23**, 184 (1969), Seitz, Turnbull, Ehrenreich, Eds. New York: Academic Press.
38. Ilten, D. B., Calvin, M.: J. Chem. Phys. **42**, 3760 (1965).
39. Herzberg, G.: Infrared and raman spectra of polyatomic molecules. New York: D. Van Nostrand 1954.
40. Maxwell, E., Strongin, M.: Phys. Rev. Letters **10**, 212 (1968).
41. Wilson, J. A., Yoffe, A. D.: Advan. Phys. **18**, 193 (1969).
42. Carter, C. B., Thesis, M. S.: Imperial College. London (1972).
43. Carter, C. B., Williams, P. M.: Phil. Mag. (1972). In press.

Discussion

W. H. Koehler   How does PH_3 intercalation into $2H-TaS_2$ effect its superconducting transition?

J. Acrivos   Although the ionization energy for the highest filled orbital in PH_3 is less than that for NH_3 (see Table I), the solid solution of the former shows a very small enhancement of T_c after several weeks of reaction at room temperature. This may be due to the high barrier to inversion in PH_3 of 1.3 eV, compared to that for NH_3 of 0.23 eV. This in addition to the increase in the P–H bond length to 1.4 Å may not allow the orientation of the PH_3 molecules in the van der Waals gap of the metal to form the same type of complex as for NH_3.

M. J. Sienko   What effect has the intercalation of the NH_3 on the conductivity in the normal (not superconducting) state? One would predict an increase due to electron motion across and in the gap via the "cavities" formed by the NH_3 molecules.

J. Acrivos   We have not measured it. However, a minimum in the conductivity in the direction of the c-axis at 20 °K discovered by A. H. Thompson in the complex (pyridine)$_{\frac{1}{4}}$ $2H-TaS_2$ may be a direct consequence of charge transfer.

Metallic Vapors

F. HENSEL

Abstract

A discussion of existing data of the electrical conductivity and the thermoelectric power of potassium, cesium and mercury as a function of the density at supercritical temperatures is given. It is proposed that a pseudogap is formed in the expanded supercritical metals and that the states in the pseudogap become localized at small metal densities. The relation between the thermodynamic phase transition from a dense liquid to a low density vapor and the metal–nonmetal transition will be discussed.

I. Introduction

Metal–nonmetal transitions have been observed in a number of different fluids including solutions of alkali metals in liquid ammonia, amines and polyethers [1, 2], and solutions of electropositive metals in their molten halides [3]. For these fluids a simple increase of concentration to sufficiently large values can cause the behavior of the system to change from nonmetallic to metallic. All of these examples, however, contain two or more components. It is therefore necessary to take into account the influence of the dominant liquid solvent on the minority component which exhibits the effect to be investigated.

On the other hand compressed metal vapors offer a unique possibility for studying this transition with a one-component system, if the vapor density can be varied continuously over a wide range. Such continuous variation of density is only possible, however, if the vapor is compressed above the critical temperature T_c which terminates the vapor pressure curve. This is very similar to the metal–ammonia and metal–salt solutions, in which the transition can only be observed at comparatively high temperatures, of course above the critical temperature T_c of the solubility gap. Below T_c the coexistence of two liquid phases is observed: one blue nonmetallic phase, and one bronze possibly metallic phase.

Figure 1 shows the equation of state [4—6] and the liquid-vapor phase behavior of the metal mercury. Again below the critical temperature $T_c = 1490°$ C a phase separation in a low density vapor phase with a relatively small conductivity and a highly conducting dense liquid phase is observed. Above the critical temperature T_c the homogeneous supercritical fluid can be compressed continuously to liquid-like densities without phase separation. It has recently been demonstrated for divalent mercury, monovalent cesium and potassium, by measurements of the electrical conductivity, thermoelectric power and in part of the Hall effect and the optical absorption, that a sufficiently large compression at temperature T larger than T_c can change the vapor system from nonmetallic to metallic behavior [7—11]. All properties of the vapor are affected by this transition to some extent and some, radically. However, the inevitable pressures are relatively high up

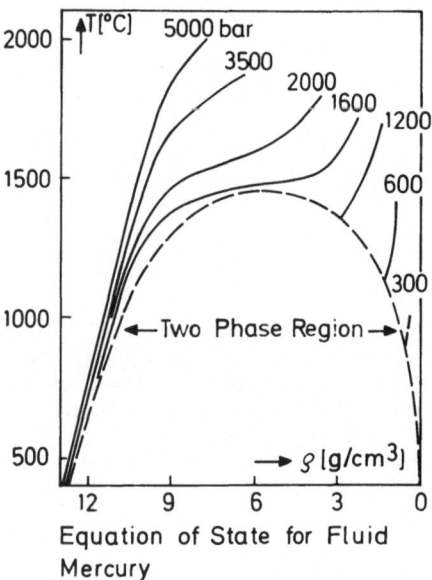

Fig. 1. Equation of state of fluid mercury

to more than three times the critical pressure. This is demonstrated by the isobars in Fig. 1.

Table I gives critical temperatures and pressures for a few selected typical metals. The critical temperatures are unusually high and only three (mercury, cesium, potassium) have been measured at present. The estimated values for heavy metals are probably beyond the possibilities of present static experimental methods.

The aim of this paper is to analyze and discuss the existing experimental results in terms of the recent theories of the metal–nonmetal transition in disordered and fluid systems [12, 13].

II. Divalent Metals

For divalent crystalline metals the Bloch-Wilson model of noninteracting electrons predicts a metal–nonmetal transition if their volumes are sufficiently expanded. With increasing distance between the divalent atoms the overlap between the

Table I

Metal	Critical temperature [°K]	Critical pressure [bar]
Hg	1760	1510
Cs	2020	110
K	2200	155
Pb	5400[a]	850[a]
Fe	10000[a]	—
W	23000[a]	10000[a]

[a] Estimated values.

first and second electronic energy bands decreases and eventually a gap appears between them, giving a transition to a nonmetallic state [14].

Measurements of the optical absorption edges in mercury at supercritical temperatures in terms of the density have been made in order to test this model for compressed divalent atoms in the disordered state. Transparency in the visible and infrared spectral regions tells us that there exists a gap between the valence and the conduction bands, i.e., for mercury between the 6s- and 6p-states. Figure 2a (see p. 493) shows the transparency of a 100 μ thick mercury film at different densities (the small numbers indicate the mercury densities in g/cm³, the temperature is constant 1800°). At the relatively small density of 0.7 g/cm³ the vapor is transparent in the visible spectral range of the tungsten lamp used as the light source in this experiment. With increasing density the expected red shift is observed. At the high temperatures applied in this experiment the radiation emitted by the dense mercury vapor is relatively large and can be observed (Fig. 2b, see p. 493). Again the expected shift is observed. With increasing density the color changes from violet to nearly white, indicating the gradual transition to a state at which the absorption edge of mercury is in the infrared spectral range. The experimental details were described elsewhere [8].

A quantitative analysis of these observations shows that the valence band (the 6s-band) and the conduction band (the 6p-band) cross at a density of about 5.3 g/cm³. At this density the d.c. conductivity (Fig. 3) is about $10^{-1}\,\Omega^{-1}\,cm^{-1}$ which is very small and certainly nonmetallic. This low nonmetallic conductivity can be explained by the concept of localized states suggested by Mott for noncrystalline materials. It was proposed by Mott that the gap separating the valence and conduction bands in an expanded solid must be replaced in a disordered system like a fluid metal by a "pseudogap" i.e., a minimum in the density-of-states $N(E)$ in the vicinity of the Fermi energy. Specifying the depth of the "pseudogap" in terms of the parameter g, defined as $g \equiv N(E_F)/N(E_F)_{free}$ which relates

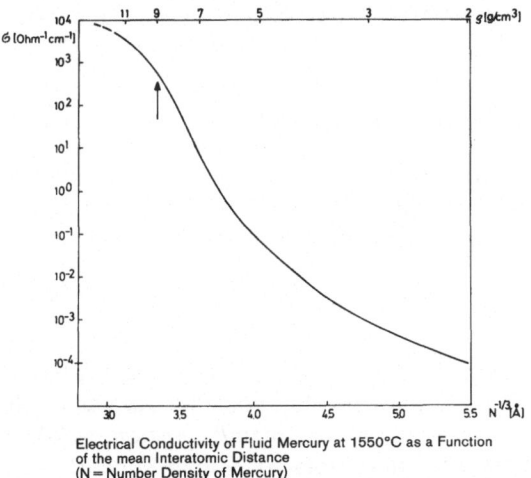

Electrical Conductivity of Fluid Mercury at 1550°C as a Function
of the mean Interatomic Distance
(N = Number Density of Mercury)

Fig. 3. Electrical conductivity σ versus the cube root of the reciprocal number density N. The upper scale gives the mass density in g/cm³. The temperature is constant 1550° C

the density-of-states at the Fermi energy E_F to the free electron density-of-states $N(E_F)_{free}$, Mott predicted that the states in the gap will be localized if the ratio g drops below about 1/3 resulting in a gradual transition from metallic to non-metallic behavior [15].

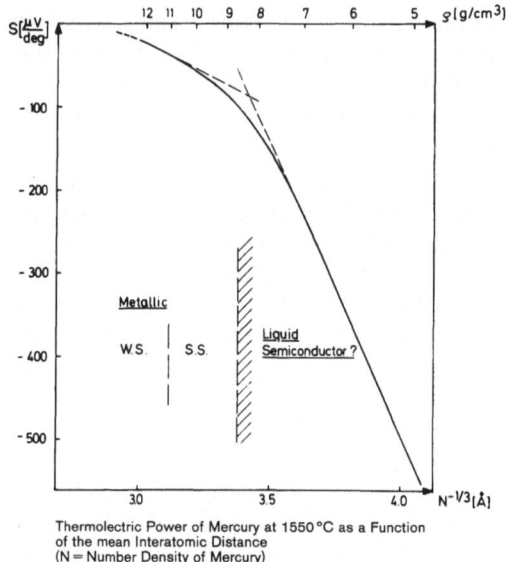

Thermoelectric Power of Mercury at 1550 °C as a Function
of the mean Interatomic Distance
(N = Number Density of Mercury)

Fig. 4. Absolute thermoelectric power S versus the cube root of the number density N. The upper scale gives the mass density in g/cm³. The temperature is constant 1550° C

The characteristic experimental conduction behavior of supercritical mercury is shown in Figs. 3 and 4, which show the conductivity and the thermoelectric power as a function of the cube root of the reciprocal number density (which is proportional to the mean interatomic distance). The upper scales give the mass density of mercury in g/cm³. The conductivity σ varies continuously from 10^{-4} to about $10^{+4}\,\Omega^{-1}\,cm^{-1}$ if the density is increased from 2 to 13.5 g/cm³, and the thermoelectric power S varies from $-550\,\dfrac{\mu V}{deg}$ to $-24\,\dfrac{\mu V}{deg}$ for an increase in the density from 5 g/cm³ to 12.0 g/cm³. Three different regions in Figs. 3 and 4 can be distinguished. At high densities between 13.5 g/cm³ and 11 g/cm³ the electron mean free path L calculated from the free electron formula for σ

$$\sigma = \frac{S_F \cdot e^2 \cdot L}{12 \cdot \pi^3 \cdot \hbar} \tag{1}$$

(S_F = free electron Fermi surface, e = electron charge) is larger than the interatomic distance $a \approx N^{-1/3}$, and the Hall constant R_H has the free electron value [11]. This is the weak scattering (w.s.) metallic region, in which σ and S should be explainable within the framework of Ziman's nearly-free-electron theory. Recent calculations of Evans [16] at normal densities of liquid mercury near 13.5 g/cm³ show that in this range of density one can predict the volume and the temperature dependence of σ if one uses an energy-dependent scattering cross section of the mercury ion.

However, with decreasing density the electron mean free path L drops down and at 11 g/cm^3 it is comparable with the mean interatomic distance $a \approx N^{-1/3}$. It is not likely that the Ziman model should hold here where the electron ion interaction is so strong. It was first stated by Mott [15] that for densities smaller than 11 g/cm^3, where L is about $a \approx N^{-1/3}$, in the strong scattering (s.s.) metallic region the conductivity σ and the termoelectric power S depend on the density-of-states at the Fermi energy. The simple formula (1) is then replaced by

$$\sigma = \frac{S_F e^2 a g^2}{12 \pi^3 \hbar} \tag{2}$$

($g = N(E_F)/N(E_F)_{\text{free}}$). For mercury in which the electrical current is determined by electrons with energies near E_F the thermoelectric power S [17] is given by

$$S = \frac{\pi^2}{3} \cdot \frac{k^2 T}{e \cdot E_F} \cdot X \tag{3}$$

with

$$X = \left(\frac{\partial \ln \sigma(E)}{\partial \ln E} \right)_{E = E_F}. \tag{4}$$

In the strong scattering range where L is about a and where σ is given by formula (2), the only terms which vary with E are S_F and g. Thus it is expected that in this region the form of the density-of-states determines S, i.e.

$$X = 2 \cdot \left[\frac{\partial \ln N(E)}{\partial \ln E} \right]_{E = E_F}. \tag{5}$$

The parameter g which measures the depth of the pseudogap can be determined with formula (2) from the experimental σ-values. The calculated g values are in excellent agreement with those obtained independently by Even and Jortner [18] from measurements of the Hall effect. Figure 5 gives a plot of g versus the mercury density; g drops continuously from 1 to 1/3 if the density varies from 11 to 9 g/cm^3. For g smaller than 1/3, according to Mott, the states at the Fermi energy should be localized. Then at the high temperatures applied in the present experiments conduction should be determined by electrons which are lifted from localized states near E_F to extended states across an energy E_c which separates the range in which the states are localized from that in which the states are extended. If hole conduction plays an unimportant role, then σ and S [19] should be given by

$$\sigma = \sigma_0 \cdot e^{-(E_c - E_F)/kT}, \tag{6}$$

$$S = k/e \cdot \left(\frac{E_c - E_F}{kT} + \text{const} \right) \tag{7}$$

(k = Boltzmann constant, e = electron charge). However, formula (7) is only valid if $E_c - E_F$ is larger than $4kT$, which is 0.62 eV at 1800 °K. Furthermore $E_c - E_F$ is temperature dependent and cannot be identified with a constant activation energy, even not for constant density. However, $E_c - E_F$ in formula (7)

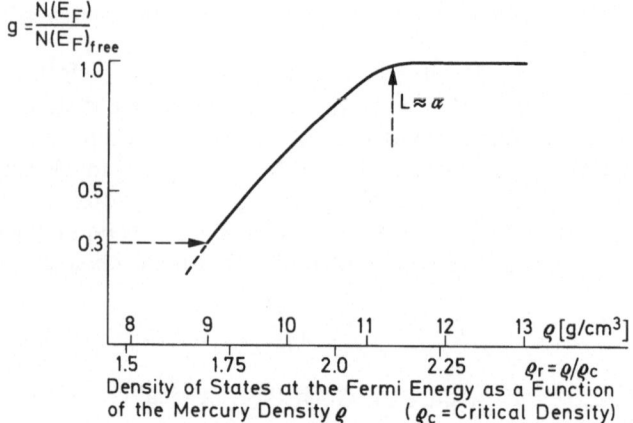

Fig. 5. Mott's *g*-value of mercury as a function of the density at the constant temperature 1550° C (ϱ_c is the critical density, $\varrho_c = 5.3 \text{ g/cm}^3$)

can be replaced by σ from Eq. (6) giving

$$\ln \sigma = \text{const} + e/k \cdot S . \tag{8}$$

Figure 6 shows a logarithmic plot of σ versus S for densities smaller than 8 g/cm³. A straight line is observed with a slope $e/k = 1.16 \cdot 10^{+4}$ deg/Volt. Thus for densities smaller than 8 g/cm³, σ and S are consistent with the Eqs. (6) and (7), i.e., with excitation of electrons from localized states at E_F to extended states above E_c. The main current is carried by excited electrons as in a liquid semiconductor. This, in fact, is consistent with the observed rapid increase of $\left(\dfrac{\partial \ln \sigma}{\partial T} \right)_V$ and

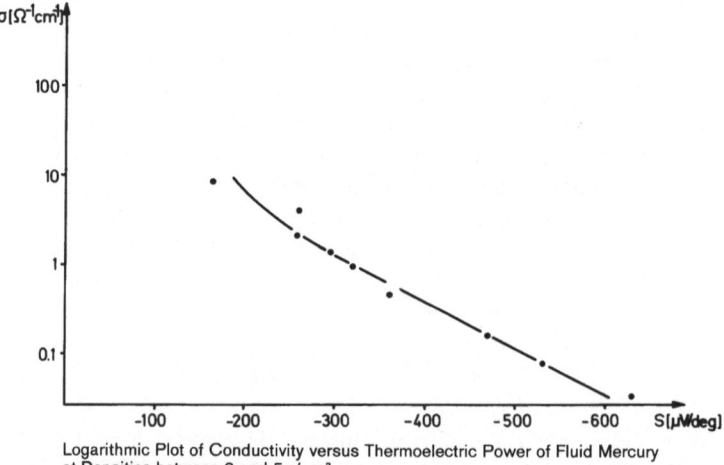

Fig. 6. Logarithmic plot of conductivity versus absolute thermoelectric power S for mercury densities between 8 and 5 g/cm³. (The slope of the curve is $e/k = 1.16 \cdot 10^{+4}$ deg/Volt)

$$\left(-\frac{\partial \ln \sigma}{\partial \ln V}\right)_T$$ ($V=$ volume) with decreasing density for densities smaller than 9 g/cm³ [9]. It seems that the gradual transition from nonactivated metallic to activated semiconducting conduction in divalent mercury occurs at a density 9 g/cm³, where g is about 1/3. Thus, on the basis of this argument, the metal–semiconductor transition in fluid mercury occurs relatively far away from the critical point at a reduced density $\varrho/\varrho_c = 1.7$ (Fig. 5) and no relationship between the metal–nonmetal transition and the liquid-vapor phase transition can be detected. The critical density ϱ_c is 5.3 g/cm³.

III. Monovalent Metals

In the case of a fluid monovalent metal the situation is more complicated. For an array of monovalent atoms, whether crystalline or not, a metal–nonmetal transition can only occur as a consequence of the interaction between the electrons [14, 20]. The formal treatment of this transition is due to Hubbard [20]. He finds that if the distances between the atoms exceed a critical value, a splitting of the conduction band into two bands occurs. The lower band is occupied and the upper band is empty and the material is a nonmetal with an activation energy for conduction. As already discussed in the case of divalent metals, however, for a noncrystalline array of atoms the two bands lead to a minimum in the density-of-states. Thus the metal–nonmetal transition takes place when the states at the Fermi energy become localized, which should occur when g falls below a value of about 1/3.

The characteristic conduction behavior of the fluid alkali metal potassium is shown in Fig. 7 as a function of the pressure at constant sub- and supercritical temperatures. Each of the points is an average of several independent determinations of pressure, temperature and conductivity. At slightly supercritical temperatures a continuous but steep decrease of conductance with decreasing pressure

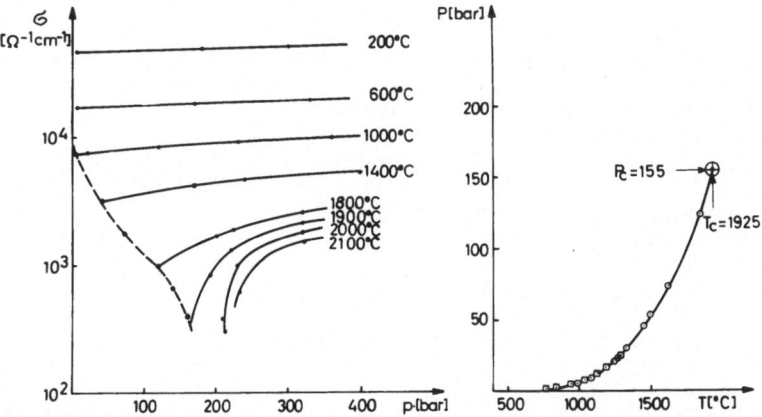

Fig. 7. The electrical conductivity of potassium at sub- and supercritical temperatures as a function of the pressure and the vapor pressure curve of liquid potassium

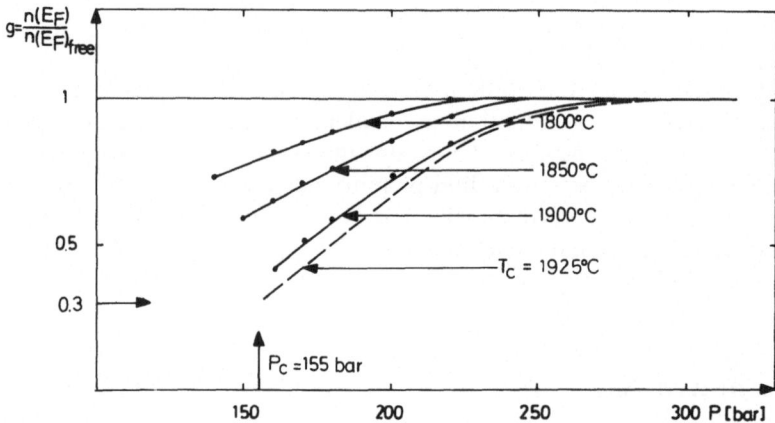

Fig. 8. Mott's g-value of potassium as a function of pressure at different constant temperatures

is observed, indicating the transition to a nonmetallic state. At the critical point, which terminates the vapor pressure curve, shown at the right side of Fig. 6 ($p_c = 155$ bar, $T_c = 1925°$ C), the conductivity σ is about $250\,\Omega^{-1}\,cm^{-1}$. This is a relatively large value compared with the conductivity of mercury at the critical point, which is about $10^{-1}\,\Omega^{-1}\,cm^{-1}$. From the measured values the parameter g has been calculated using formula (2). Figure 8 shows a plot of g versus the pressure at different constant temperatures. The critical value $g \approx 1/3$, for which localization should set in, is observed in the vicinity of the critical point. The same behavior is observed for the alkali metal cesium.

References

1. Lepoutre, G., Sienko, M. J.: Metal–ammonia solutions. New York: Benjamin 1964.
2. Thompson, J. C.: Rev. Mod. Phys. **40**, 704 (1968).
3. Waddington, T. C.: Non-aqueous solvent systems. New York: Academic Press Inc. 1965.
4. Hensel, F., Franck, E. U.: Ber. Bunsenges. Phys. Chem. **70**, 1154 (1966).
5. Kikoin, I. K., Sechenkov, A. R.: Phys. Metals Metallogr. **24**, 5 (1967).
6. Postill, D. R., Ross, R. G., Cusack, N. E.: Advan. Phys. **16**, 493 (1967).
7. Hensel, F., Franck, E. U.: Rev. Mod. Phys. **40**, 697 (1968).
8. Hensel, F.: Ber. Bunsenges. Phys. Chem. **75**, 847 (1971).
9. Schmutzler, R. W., Hensel, F.: Ber. Bunsenges. Phys. Chem. **76** (1972) in press.
10. Freyland, W. F., Hensel, F.: Ber. Bunsenges. Phys. Chem. **76**, 347 (1972).
11. Even, U., Jortner, J.: Physics Letters A (1972).
12. Mott, N. F.: Phil. Mag. **19**, 835 (1969).
13. Mott, N. F.: Phil. Mag. **24**, 1 (1971).
14. Mott, N. F.: Phil. Mag. **6**, 287 (1961).
15. Mott, N. F.: Phil. Mag. **13**, 989 (1966).
16. Evans, R.: J. Phys. C. Metal Phys. Suppl. **2**, 137 (1970).
17. Cutler, M.: Phil. Mag. **25**, 173 (1972).
18. Even, U., Jortner, J.: Phil. Mag. **25** (1972).
19. Mott, N. F., Davis, E. A.: Electronic processes in non-crystalline materials. Oxford: Clarendon Press 1971.
20. Hubbard, J.: Proc. Roy. Soc. (London) A **281**, 401 (1964).

Metal-Nonmetal Transition in Expanded Liquid Mercury

Uzi Even and Joshua Jortner

Abstract

We report the results of an experimental study of the electronic transport properties (Hall effect and electrical conductivity) of expanded liquid mercury in the density range 13.6—8.5 g cm^{-3} (temperature region 30—1500 °C and pressure region 1—1900 atm). From the correlation of the Hall effect and the conductivity data three distinct conduction regimes were identified in this one-component system: a) the weak scattering metallic region (13.6—11 g cm^{-3}); b) the strong scattering metallic region (11.0—9.2 g cm^{-3}); c) the localization regime (< 9.2 g cm^{-3}). Detailed quantitative information has been obtained for the formation of the pseudogap and the metal–nonmetal transition in this disordered system.

I. Introduction

Concentrated metal–ammonia solutions [1] provide a classical example for the metal–nonmetal transition in a disordered system, which is induced by concentration changes. In spite of the impressive amount of detailed experimental data concerning the static [2], the transport [2, 3] and the optical properties [1] of intermediate range and concentrated metal–ammonia solutions, a coherent physical description of the metal–nonmetal transition is not yet available. Old theoretical work has considered Mott's classical argument [4] concerning the effects of long range screening, subsequent work [5] focussed attention on the Hubbard band gap originating from correlation effects, while recent studies [6] attempted to apply Mott's ideas concerning the formation of a pseudogap in a disordered system. The following specific questions are pertinent:

a) The nonmetal–metal transition in metal–ammonia solutions occurs from a nonmetallic state involving spin-paired localized electrons [1]. It is interesting to inquire whether this transition can be described in terms of the classical Bloch-Wilson band overlap model.

b) In a disordered liquid system the transition between the metallic and the nonmetallic state may be fuzzy and not exhibited by sharp "breaks" in the transport (such as conductivity) properties. It is interesting to inquire how the transition density can be specified in terms of experimental observables.

c) Does the metal–nonmetal transition occur in a homogeneous system, or is it preceded by the formation of large aggregates?

It appears to us that in view of the complexity of metal–ammonia solutions, physical insight into the nature of metal–nonmetal transition in this two-component system can be obtained from the study of this transition in a simple one-component system. We report the study of the metal–nonmetal transition in expanded liquid mercury, leading to some general relations and correlations which pertain to the problem of metal–nonmetal transition in a disordered system originating from band overlap effects rather than from correlation effects. These general correlations obtained for the one-component system do not apply to metal–ammonia solutions, thus providing evidence that the metal–nonmetal transition in this system is preceded by aggregation.

A study of the metal–nonmetal transition induced by density changes in a one-component system can be conducted in expanded liquid metals and supercritical dense metal vapors at high temperatures and pressures [7]. In view of the relatively low thermodynamic critical point of mercury (Table I), this system was chosen for the study of the electrical transport properties of an expanded liquid metal. We shall demonstrate that the combination of the Hall effect and the electrical conductivity data provide direct information concerning the formation of a pseudogap and the metal–nonmetal transition in a disordered system.

Table I. Thermodynamic critical points of some metals

Metal	Hg	Cs	Rb	K	Na	Pb	Sn
Critical temperature (°K)	1763	2023	2057	2223	2573	5400	8720
Critical pressure (atm)	1510	110	157	160	350	850	2100
Ref.	a), b), c)	d)	e)	e)	e)	f)	f)

a) Franck, E., Hensel, F.: Phys. Rev. **147**, 109 (1966).
b) Kikoin, I. K., *et al.*: Soviet Phys. JETP **22**, 89 (1966).
c) Birch, F.: Phys. Rev. **41**, 641 (1932).
d) Rhenker, H., Hensel, F., Franck, E. U.: Phys. Letters **30**A, No. 9 (1969).
e) Dillon, I. G., Nelson, P. A., Swanson, H. S.: J. Chem. Phys. **44**, 4229 (1966).
f) Grosse, A. V., Ivory, J.: Nucl. Chem. **22**, 23 (1961).

II. Metal–Nonmetal Transition in a Divalent Liquid Metal

A. General Considerations

When the distance between divalent atoms in a crystalline metal is continuously increased, the overlap between the electronic bands decreases, resulting in a metal-nonmetal transition. In a disordered system the Van Hove singularities in the density-of-states function at the band edges are smeared out. Mott [8] and Cohen *et al.* [9], have suggested that in amorphous semiconductors the band edges exhibit tails in the energy region which is forbidden in the corresponding perfect crystalline solid. Thus the gap separating the valence band and the conduction band in the solid is replaced in the disordered system by a pseudogap, i.e. a minimum in the density-of-states near the Fermi energy. When a divalent liquid metal is gradually expanded a pseudogap is formed (Fig. 1). When the density-of-states in the pseudogap is relatively high (i.e. only somewhat lower than the free electron value), the states in the vicinity of the Fermi energy are delocalized, while when the density-of-states in that energy range is lowered, localized states will appear and the conductivity energy edge is located at higher energies. A coherent description of this physical situation was provided by Mott [10], who introduced a pseudogap depth parameter, g, relating the density-

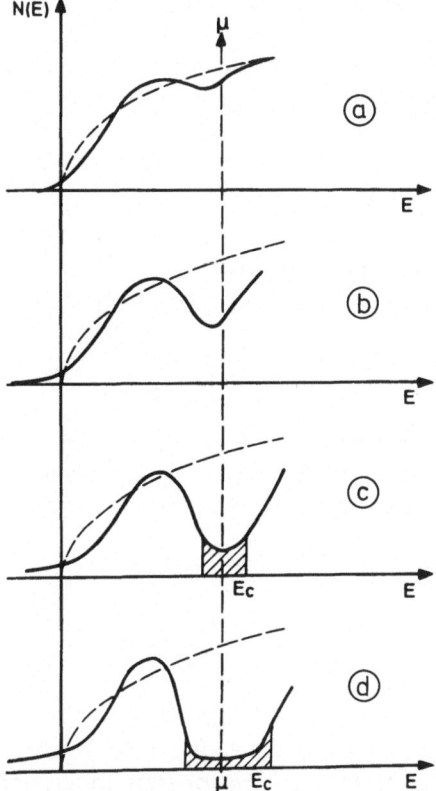

Fig. 1. A schematic representation of the density-of-states of mercury at different densities (solid curve) as compared with the free electron density-of-states (dashed curve). a Normal liquid at room temperature, which exhibits small deviations from the fe model. b The strong scattering regime where the pseudogap opens up, while the pseudogap depth parameter exceeds 0.3. c The localization regime where localized states appear near the Fermi energy. d Dilute fluid characterized by a large localization region. μ is the Fermi energy while E_c represents the mobility edge. The dashed areas represent localized states

of-states $N(E_F)$ at the Fermi energy to the free electron (fe) value

$$g = N(E_F)/N(E_F)_{fe}. \tag{1}$$

Utilizing Anderson's model [11] for cellular disorder Mott has predicted [10] that metallic conductivity occurs for

$$g \gtrsim 1/3 \tag{2}$$

while for lower g values localized states will appear in the pseudogap, so that the system will exhibit the features of a semiconductor. Focussing attention on the electrical transport properties of the system, three regions can now be distinguished:

a) The weak scattering metallic region where the mean free path exceeds the interatomic spacing, which prevails in liquid metals at normal temperatures and pressures. In this region we expect that g does not appreciably differ from unity.

b) The strong scattering metallic region where the mean free path is comparable to the interatomic spacing. In this range a pseudogap is formed in the disordered system and $0.3 \lesssim g \lesssim 1$.

c) The semiconductor region which is characterized by localized states in the pseudogap and where $g < 0.3$.

B. The Weak Scattering Metallic Region

The metallic conductivity, σ, can be in general expressed in the form

$$\sigma = \frac{e^2 S \Lambda}{12\pi^3 \hbar} \tag{3}$$

where Λ is the mean free path and S represents the area of the Fermi surface. When the density-of-states at the Fermi surface deviates from the free electron value (i.e. $S \neq S_{fe}$) one can introduce the pseudogap as a scaling parameter

$$S = S_{fe} g^2 \tag{4}$$

where upon

$$\sigma = \frac{e^2 S_{fe} \Lambda g^2}{12\pi^3 \hbar} . \tag{4a}$$

In the weak scattering limit $\Lambda > a$. In this case the free electron mean free path Λ, is conventionally calculated by the Ziman theory [12] which is based on a first order perturbation expansion (i.e. the Born approximation). As was demonstrated by Faber [13] a second order expansion leads to the relation

$$\Lambda = \Lambda_{fe}/g^2 \tag{5}$$

so that

$$\sigma = \frac{e^2 S_{fe} \Lambda_{fe}}{12\pi^3 \hbar} . \tag{6}$$

Thus, in the weak scattering region the electrical conductivity is independent of the pseudogap depth, g.

To the best of our knowledge no adequate theory has yet been provided for the Hall coefficient, R, and for the Hall mobility $\mu = |R| \sigma$ in the weak scattering region. This problem was handled by Ziman [12] and by Fukuyama et al. [14]. A plausibility argument can be provided by considering the Hall coefficient, R, for the free electron case

$$R = \frac{12\pi^3}{e m^* S V_F} \tag{7}$$

where m^* is the effective mass and V_F represents the Fermi velocity. For a spherical Fermi surface one gets the conventional expression

$$R = \frac{1}{ne} . \tag{8}$$

As pointed out by Ziman [12], Eq. (7) was derived utilizing the conventional expression for the Lorentz force $F = e V \times H$, where H is the magnetic field. It was argued [12] that when $g \neq 1$ one can apply Edward's [15] recipe replacing the velocity operator V_k by the current operator J_k so that (for any wave-vector k) $e V_k \rightarrow J_k$. This substitution does not affect the electrical conductivity [Eq. (6)], while the Hall coefficient is scaled by the reciprocal pseudogap depth, so that

$$R = \frac{1}{ne} g^{-1}. \tag{9}$$

On the other hand, when the velocity operator in the Lorentz force is replaced by the current operator, an additional g factor is introduced into the Hall coefficient whereupon

$$R = \frac{1}{ne} g^{-2}. \tag{10}$$

Equation (10) was obtained by Fukuyama et al. [14] who derived the Hall conductivity for a nearly-free-electron system utilizing many body perturbation techniques. In conclusion, it is important to point out that:

a) From the point of view of general methodology it is important to notice that in the weak scattering metallic region no direct correlation exists between the electrical conductivity and the Hall coefficient. When the density is lowered over a small region (so that S_{fe} is practically invariant) the change in the electrical conductivity reflects the decrease in the free electron mean free path, Λ_{fe}. On the other hand deviations of the Hall coefficient from its free electron value may be tentatively assigned to changes in g.

b) From the pratical point of view it is easy to set a lower limit for the electrical conductivity which characterized the weak scattering region. A cursory examination of Eq. (6) indicates that for a normal liquid metal ($n \approx 5 \times 10^{22}$ cm^{-3}) $\sigma \gtrsim 3000 \, (\Omega \, \text{cm})^{-1}$. On the other hand for metal ammonia solutions $\sigma \gtrsim 1000 \, (\Omega \, \text{cm})^{-1}$.

c) For the special case of liquid mercury at normal temperature and pressure $\sigma = 10^4 \, \Omega^{-1} \, \text{cm}^{-1}$ and condition (b) for the weak scattering region is satisfied. Greenfield's [16] experimental Hall effect data ($R = R_{fe}$ within 3%) indicate that in this system $g = 1$.

C. The Strong Scattering Metallic Region

It was argued by Mott [17] that the mean free path cannot be lower than the near lattice spacing, a, so that the strong scattering regime is specified by $\Lambda = a$. Making use of Eqs. (3) and (4) the conductivity takes the form

$$\sigma = \frac{e^2 S_{fe} a}{12 \pi^3 \hbar} g^2. \tag{11}$$

A tentative use of Eq. (9) for the strong scattering region as proposed by Straub et al. [18]

$$R = R_{fe} g^{-1} \tag{12}$$

yields a direct correlation between the Hall coefficient and the pseudogap depth, and hence leads to a direct relation between σ and R, i.e. $\sigma \propto \left(\dfrac{R_{fe}}{R}\right)^2$. A systematic derivation of this result, which rests on the random phase approximation was recently provided by Friedman [19]. This model originally proposed by Cohen [20] and by Mott [21] assumes that the strong scattering situation, in which scattering of the electron occurs on each lattice site, can be described in terms of states where the electronic wavefunctions do not exhibit any phase correlation between different sites. This model was utilized by Hindley [22] and by Friedman [19] for the calculation of the electrical conductivity and of the Hall coefficient in a degenerate electron gas and in amorphous semiconductors.

The basic physical idea underlying this approach involves the representation of the wavefunction $|K\rangle$ in the tight binding approximation

$$|K\rangle = \sum_n a_{Kn}|n\rangle \tag{13}$$

where $|n\rangle$ represent an atomic wavefunction. The expansion coefficients a_{Kn} are recast in the form

$$a_{Kn} = |a_{Kn}| \exp(i\phi_{Kn}) \tag{14}$$

where ϕ_{Kn} is a random phase. Thus the off-diagonal matrix elements of the velocity operator V_x (in any x direction) will vanish, i.e.

$$\langle K'|V_x|K\rangle = \sum_n \sum_{n'} a_{K'n'} a_{Kn} \langle n'|V_x|n\rangle = 0; \quad (K' \neq K). \tag{15}$$

Utilizing the Kubo-Greenwood formula [23], the electrical conductivity and the Hall coefficient can be expressed in terms of the components of the conductivity tensor

$$\sigma = \sigma_{xx} = \frac{2\pi}{3} \left(\frac{e^2}{\hbar a}\right) Z a^6 J^2 N(E_F)^2, \tag{16}$$

$$R = \frac{\sigma_{xy}}{\sigma_{xx}^2 H} = \frac{3\eta \bar{Z}}{Z^2 e J N(E_F)}, \tag{17}$$

where Z is the number of nearest neighbors, $\bar{Z}$ represents the number of triangular closed paths around each lattice site, η corresponds to a geometrical factor, J is the interatomic exchange integral between nearest neighbors (which is kept constant considering just cellular disorder) and finally a is the lattice spacing which for a diatomic liquid metal is $n = 2a^{-3}$.

Defining an auxiliary parameter

$$X = a^3 J N(E_F), \tag{18}$$

one gets for the electrical conductivity

$$\sigma = 2\pi \left(\frac{Z}{3}\right) \left(\frac{e^2}{\hbar a}\right) X^2, \tag{19}$$

the Hall coefficient is

$$R = \frac{6\eta}{Z} \left(\frac{\bar{Z}}{Z}\right) \frac{1}{ne} X^{-1} \tag{20}$$

while the Hall mobility, μ, is given by

$$\mu = R\sigma = 2\pi \left(\frac{\bar{Z}}{Z}\right) \left(\frac{3\eta}{Z}\right) \left(\frac{ea^2}{\hbar}\right) X. \tag{21}$$

From Eqs. (18)—(20) we can conclude that in the strong scattering limit one can express the electrical conductivity and the Hall mobility in terms of the deviation (R_{fe}/R) of the Hall coefficient from the free electron value
We can also assert that in the strong scattering region the Hall effect directly monitors the pseudogap depth. Equation (19) can be written in the form

$$\frac{R}{R_{\text{fe}}} = 4\eta F \left(\frac{\bar{Z}}{Z}\right) g^{-1} \tag{22}$$

where F is the band filling factor

$$F = \frac{E_{\text{F}}}{W}$$

and W is the band width. In a monovalent metal $F = 1/2$, while in divalent metal where the metallic properties originate from overlap between bands, a unique determination of F is difficult but we may get $0.5 < F < 1.0$, setting $\bar{Z} = Z$ and $\eta = 1/3$ we get from Eq. (21)

$$\frac{R_{\text{fe}}}{R} = \varkappa g \tag{23}$$

where

$$0.7 < \varkappa < 1.5. \tag{24}$$

Our experimental results for expanded liquid mercury indicate that $\varkappa \approx 1$ for that system.
From the foregoing discussion we conclude that the strong scattering metallic regime is characterized by the following features:
a) The Hall coefficient provides a direct measurement of the pseudogap depth.
b) The electrical conductivity is directly proportional to the square of the pseudogap depth, so that a direct correlation exists between the conductivity and the Hall coefficient.
c) The Hall mobility is directly proportional to the pseudogap depth.
d) For expanded divalent liquid metals ($a \sim 3$ Å) the strong scattering regime will be characterized by the conductivity $300 < \sigma < 3000 \, \Omega^{-1} \, \text{cm}^{-1}$.

D. Expanded Liquid Mercury in the Strong Scattering Region

It will be useful to provide numerical estimates for the electrical transport properties of an expanded divalent liquid metal in the strong scattering situation. We shall make use of values

$$\eta = 1/3; \quad \bar{Z} = Z.$$

The coordination number in normal liquid mercury at room temperature was determined by Rivlin et al. [24] who get

$$Z = 10,$$

$$a = 3 \,\text{Å}.$$

Invoking the assumption that these parameters are slowly varying with moderate density changes we get from Eqs. (18)—(20)

$$\sigma = 2100 \left(\frac{R_{\text{fe}}}{R}\right)^2 \Omega^{-1}\,\text{cm}^{-1}, \tag{25}$$

$$\mu = 0.18 \left(\frac{R_{\text{fe}}}{R}\right) \text{cm}^2/\text{Volt sec}. \tag{26}$$

In order to correct these results for density changes [expressed in terms of the parameter a in Eqs. (18)—(20)], we shall define the normalized conductivity. σ_n, and the normalized mobility, μ_n, in the form

$$\sigma_n = \sigma \left(\frac{d_0}{d}\right)^{1/3} = 2100 \left(\frac{R_{\text{fe}}}{R}\right)^2 \Omega^{-1}\,\text{cm}^{-1}, \tag{27}$$

$$\mu_n = \mu \left(\frac{d_0}{d}\right)^{-2/3} = 0.18 \left(\frac{R_{\text{fe}}}{R}\right) \text{cm}^2/\text{Volt sec}, \tag{28}$$

where d_0 and d represent the mercury density at room temperature and at higher temperature and pressure, respectively. It is expected that Eqs. (26) and (27) will be valid over the entire strong scattering region while, at densities below the metal–nonmetal transition the transport properties will exhibit deviations from these relations.

III. Experimental Procedures

We measured the Hall coefficient and the electrical conductivity of subcritical mercury over the temperature range 20—1500° C (in the pressure range 1—2000 atm). The Hall coefficient was measured by the double a.c. method. An a.c. current of ~ 5 amp at 20 Kc/sec and an a.c. magnetic field of ~ 100 gauss at 1 Kc/sec produced a Hall voltage of about $\sim 2 \times 10^{-9}$ volt at the sum and at the difference frequencies. The low intensity of the magnetic field, which was homogeneous over the sample volume to better than 0.1%, and the high frequencies employed, prevented parasitic signals due to hydrodynamic currents. Extensive (two stage) passive filtering was employed to prevent intermodulation distortion at the detection system. The total system noise figure was about 6 dB (producing an input equivalent noise of 10^{-10} volts after integration of 10 sec). The Hall current was generated by a 20-watt power amplifier designed to produce low noise at the difference frequency, characterized by a low FM index. The magnetic field was generated by a 3 kW power amplifier, specified by very low total harmonic distortion (0.1%) and FM index. The signals was filtered, amplified by a low noise amplifier, filtered again and detected by a lock-in amplifier. A calibration signal (of controlled amplitude and phase) was taken from the Hall current and the magnetic field sources, and injected in series with the Hall voltage sources, until a null was obtained at the output of the lock-in amplifier, thus avoiding frequent calibration checks and reducing the loading of the Hall voltage source. The general outline of the electronic system is presented in Fig. 2.

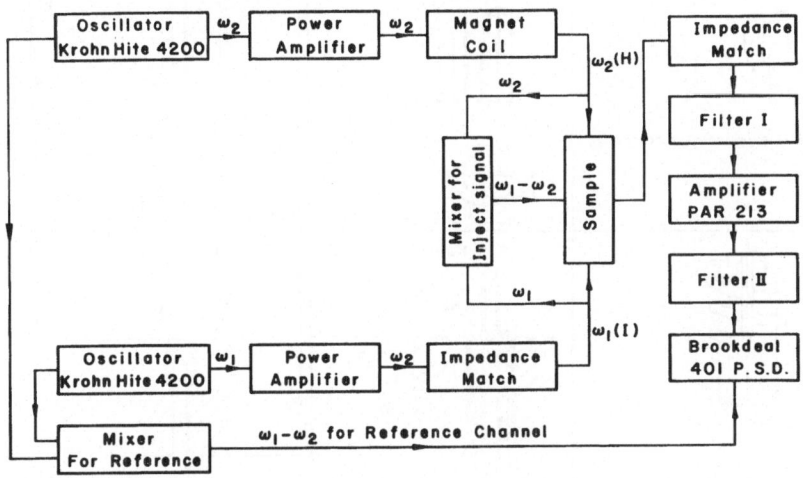

Fig. 2. A Block diagram of the electronic system for Hall effect measurements

The absolute value of $R(T = 30°\,C)$ at room temperature (accounting for the finite sample size) was within 6% of Greenfield's [16] result. Our results in the temperature region 30—300° C are consistent with previous data. Relative values of the Hall signal $[R/R(T = 30°\,C)]$ were obtained to better than 5%. Linearity of the Hall voltage with the magnetic field, the current and cell thickness were verified. The results were identical upon heating and upon cooling of the cell. Resistivity of the mercury was measured by the Four Probe method at a frequency of 10 c/sec, eliminating contact resistance errors. Values of the relative resistivity (to the room temperature value) are accurate to within 2%. The alumina cell was a 15×10 mm rectangle, relatively thick, 0.5 mm, sealed to a long (150 mm) alumina, open-end, six-bore tube, containing the molybdenum electrodes, which do not react with the hot mercury. The cell was sealed by a special high alumina glass frit at 1400° C under high vacuum (for details see Fig. 3). The helium leak-tested cell and the long mercury column (150 mm) slowed the dissolution of argon under high hydrostatic pressure into the mercury in the hot zone. The mercury cell was placed (see Fig. 4) in a temperature controlled furnace (stabilized within $\pm 0.2°$ C and measured within the absolute accuracy of $\pm 3°$ C in the region of 20—1500° C) surrounded by a water-cooled magnet coil, and the whole assembly was placed in an autoclave (Fig. 5). The mercury was thus under hydrostatic pressure of compressed, high-purity argon. Pressure up to 1900 atm was generated by a two-stage diaphragm compressor, and measured to ± 2 atm. Samples were taken from triple distilled mercury and the cell was filled under vacuum to eliminate trapped or dissolved gases.

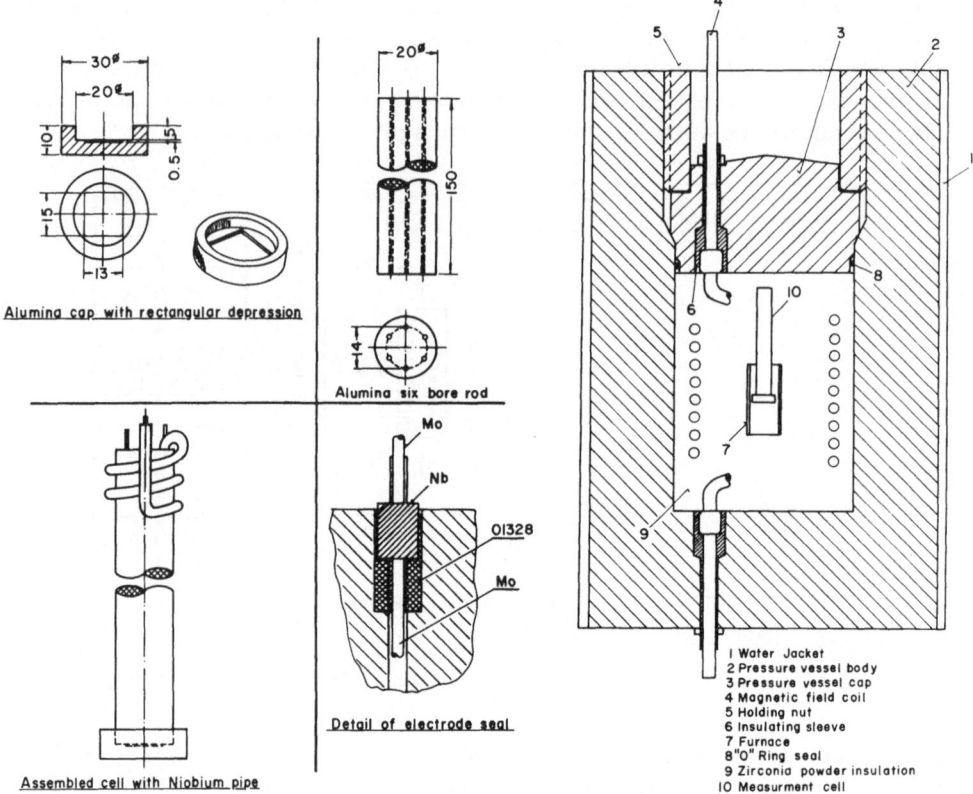

Fig. 3. The alumina cell for high temperature electrical measurements

Fig. 4. Schematic representation of furnace and cell

IV. Hall Effect and Electrical Conductivity in Expanded Liquid Mercury

We have measured the Hall effect and the electrical conductivity of subcritical mercury in the temperature region 20—1500° C and in the pressure range 1—1900 atm. This region corresponds to the density range 13.6—8.5 g cm^{-3}. The density data were obtained from a recent work of Hensel and Schmutzler [25], who measured simultaneously the density and the electrical conductivity. Near the critical point a small ($\sim 0.5\%$) uncertainty in the temperature measurement leads to a considerable error in the density. Therefore we have utilized our measured electrical conductivity data, which are determined quite accurately, as a common variable for comparison with Hensel's data. From this conductivity data we determined the density. Our experimental results for the density dependence of the electrical transport properties of expanded liquid mercury are presented in Figs. 6 and 7 where we have displayed the Hall voltage, V_H, normalized to the room temperature value ($V_{RT} = 3.1$ nV), the Hall mobility $\mu = R\sigma$, the electrical conductivity and the Hall coefficient R/R_{fe} normalized to the free electron value. The experimental results can be summarized as follows:

Fig. 5. Autoclave for high-temperature, high-pressure Hall effect measurements

a) The Hall voltage (Fig. 6) exhibits a linear increase with $1/d$ in the density region $d = 13.6$—$11.0 \, \mathrm{g \, cm^{-3}}$ and a much steeper increase with decreasing density in the range $8.5 < d < 11 \, \mathrm{g \, cm^{-3}}$.

b) The Hall coefficient normalized to the free electron value (Fig. 7) assumes the value $R/R_{\mathrm{fe}} = 1$ in the density range 13.6—$11.0 \, \mathrm{g \, cm^{-3}}$, whereupon in this density region liquid mercury exhibits free electron behavior. In the density region $9.2 < d < 11 \, \mathrm{g \, cm^{-3}}$ the Hall coefficient exhibits a considerable deviation from the free electron value and $(R/R_{\mathrm{fe}}) = 3$ at $d = 9.2 \, \mathrm{g \, cm^{-3}}$. For $d < 9.2 \, \mathrm{g \, cm^{-3}}$ (R/R_{fe}) exhibits a very fast increase with decreasing density.

c) The electrical conductivity (Fig. 7) exhibits a decrease (by a factor of ~ 3) in the density range 13.6—$11.0 \, \mathrm{g \, cm^{-3}}$, where $R/R_{\mathrm{fe}} = 1$. In the density range 11.0—$9.2 \, \mathrm{g \, cm^{-3}}$ the electrical conductivity decreases by one order of magnitude from $3000^{-1} \, \mathrm{cm^{-1}}$ to $300^{-1} \, \mathrm{cm^{-1}}$.

d) The Hall mobility exhibits a decrease (by a numerical factor of ~ 3) in the density range 13.6—$11.0 \, \mathrm{g \, cm^{-3}}$ which follows the conductivity change in that region. In the density range 11.0—$9.2 \, \mathrm{g \, cm^{-3}}$ the Hall mobility decreases by a numerical factor of 2.5. Finally, in the density range $d < 9.2\ \mathrm{g\ cm^{-3}}$ (where both σ and (R/R_{fe}) are rapidly varying functions of d), the Hall mobility is a slowly varying function of the density.

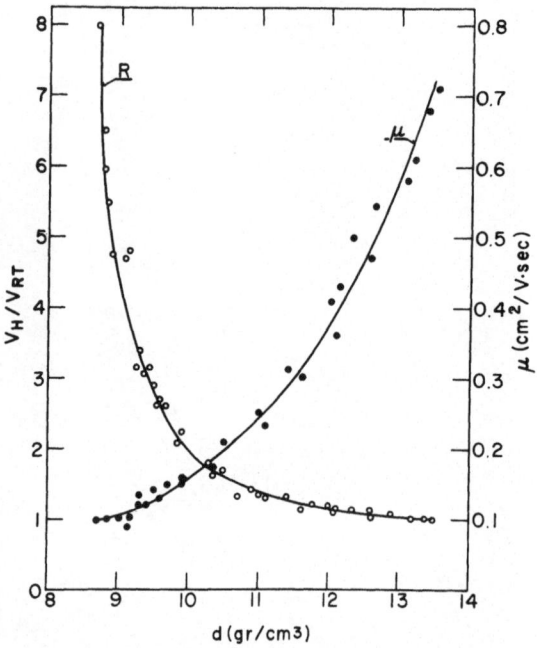

Fig. 6. The density dependence of the Hall voltage (normalized to the room temperature value) and of the Hall mobility in expanded liquid mercury

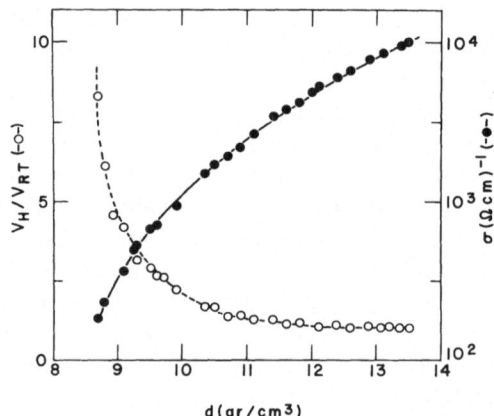

Fig. 7. The density dependence of the Hall coefficient (normalized to the free electron value) and of the electrical conductivity in expanded liquid mercury

V. Electronic Transport in Expanded Liquid Mercury

On the basis of our combined Hall effect and conductivity data we can distinguish three distinct conductions regimes in expanded liquid mercury:

A. *Density Range* $11.0 < d < 13.6$ g cm^{-3}. The electrical conductivity is $3000 < \sigma < 10^4 \Omega^{-1}$ cm^{-1}, the Hall mobility is $0.24 < \mu < 0.71$ cm^{-2} volt^{-1} sec^{-1}, while the Hall coefficient corresponds to the free electron value $R/R_{fe} = 1$.

B. *Density Range* $9.2 < d < 11.0$ g cm^{-3}. The electrical conductivity is $300 < \sigma < 3000 \Omega^{-1}$ cm^{-1}. The Hall mobility exhibits a moderate density dependence $0.1 < \mu < 0.24$ cm^2 volt^{-1} sec^{-1}. The Hall coefficient exhibits marked positive deviations from the free electron value $1 < R/R_{fe} < 3$.

C. *Density Range* $d < 9.2$ g cm^{-3}. The electrical conductivity is $\sigma < 300 \Omega^{-1}$ cm^{-1}, the Hall coefficient exhibits a sharp increase with a small density decrease, while the mobility is practically constant.

On the basis of the theoretical considerations presented in Section II these three conduction regimes can be classified as follows:

A. The Weak Scattering Regime ($11.0 < d < 13.6$ g cm^{-3})

In this region the transport properties of liquid mercury are amenable to theoretical description in terms of the free electron model. The high conductivity can be handled by Ziman's theory [12] [which can be recast in the form of Eq. (6)], where the mean free path (or the scattering time) can be expressed in terms of the Born approximation, utilizing Evan's [26] pseudopotential. The mean free path thus obtained is $\Lambda \approx 3a = 7\text{Å}$ and the self-consistency condition for the applicability of the weak scattering picture is satisfied.

No good theory is as yet available for the Hall coefficient in the weak scattering regime. If Ziman's Eq. (9) [or alternatively Eq. (10)] is to be believed, then our experimental results $R/R_{fe} = 1$ imply that $g = 1$ in this regime. The calculations of the density-of-states for normal liquid mercury by Evans [26] and by Chan and Ballantine [27] indicate that at $d = 13.6$ g cm^{-3} $g \approx 0.9$—1.0.

The decrease of the electrical conductivity over the weak scattering regime (accompanied by a parallel decrease in μ) reflects the decrease of the mean free path from $\Lambda = 7\text{Å} \approx 3a$ at $d = 13.6$ g cm^{-3} to $\Lambda = 2\text{Å} \approx a$ at $d = 11.0$ g cm^{-3}. At the lower density limit of the weak scattering regime $\Lambda = a$. Stated in an alternative manner, the condition for the applicability of the weak scattering model $k_F \Lambda \gg 1$ (where $k_F \propto \sqrt{E_F} = 1.16\text{Å}^{-1}$ the Fermi wave number) breaks down for $d = 11.0$ g cm^{-3}, where $\Lambda k_F = 2.3$. Thus the density $d = 11.0$ g cm^{-3} represents the onset of the strong scattering regime.

B. The Strong Scattering Regime ($9.2 < d < 11.0$ g cm^{-3})

The identification of this regime rests on the following qualitative arguments:

a) The change of the conductivity (300—$3000 \Omega^{-1}$ cm^{-1}) in this density region is consistent with Mott's general arguments [10, 21] for the magnitude of the conductivity in the strong scattering regime.

b) The positive deviations of R/R_{fe} from the free electron value are consistent with Eq. (12) or the more elaborate form of Eq. (22). In Fig. 8 we present the density

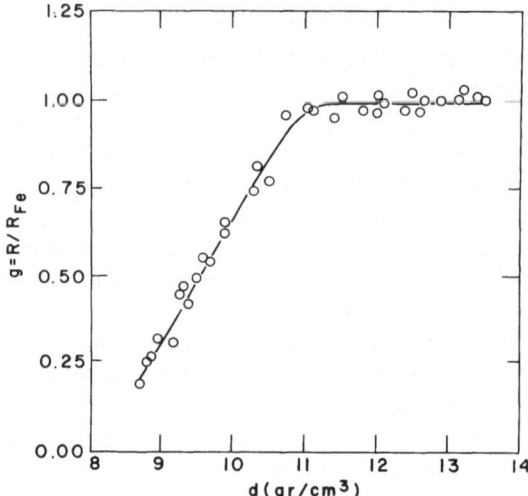

Fig. 8. The density dependence of the pseudogap depth parameter in expanded liquid mercury

dependence of g in the strong scattering region. It should be stressed that g is temperature independent at constant density, as expected.

c) A qualitative correlation of the electrical transport properties indicates that in this density range σ decreases by a factor of ~ 8, (R/R_{fe}) increases by a factor of ~ 3 while μ decreases by a factor of ~ 2.5. This behavior is consistent with the approximate functional dependence [see Eqs. (11) and (12)] $\sigma \propto g^2$, $(R/R_{fe}) \propto g^{-1}$ and $\mu \propto g$, where g varies in the range $g = 1.0$—0.3 (see Fig. 8).

Detailed quantitative physical information concerning the strong scattering regime is obtained by utilizing our results to test Friedman's random phase approximation [19]. Making use of Eqs. (18)—(20) we plot in Figs. 9 and 10 the normalized conductivity and the normalized mobility vs. (R_{fe}/R). In the density range 11.0—9.2 g cm^{-3} the following experimental correlations hold

$$\sigma_n = 2600\, g^2 \, \Omega^{-1} \, cm^{-1},$$

$$\mu_n = 0.24\, g \, cm^2 \, volt^{-1} \, sec^{-1}.$$

These experimental relations are in excellent agreement with the theoretical predictions given by Eqs. (26) and (27).

We conclude with the following comments:

a) Figs. 9 and 10 clearly exhibit the onset and the termination of the strong scattering regime. These data are obtained from the correlation of two transport properties (σ and R) and cannot be obtained from "breaks" in the conductivity alone.

b) The lower density limit for the strong scattering regime ($d = 9.2$ g cm^{-3}) corresponds to $g = 0.32$. The critical $g \lesssim 0.32$ value for the termination of the strong scattering regime is in excellent agreement with Mott's theoretical prediction for the onset of localized states at $g \lesssim 1/3$.

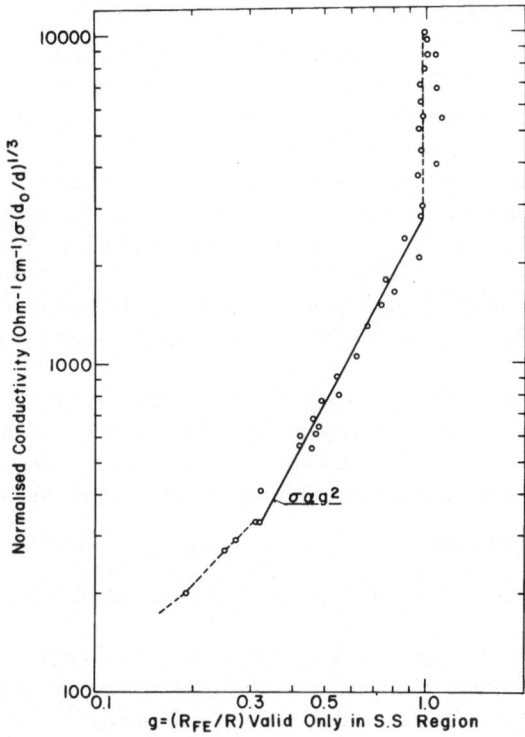

Fig. 9. The dependence of the normalized conductivity on the pseudogap depth parameter. In the strong scattering regime $\sigma_n \propto g^2$

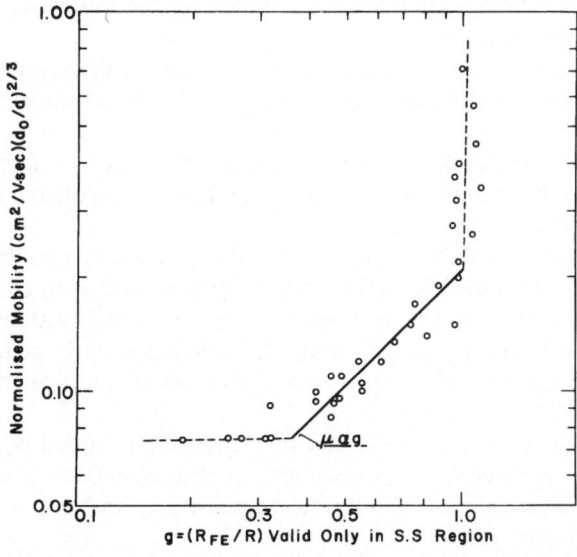

Fig. 10. The dependence of the normalized mobility on the pseudogap depth parameter. In the strong scattering regime $\mu_n \propto g$

C. The Localization Region ($d < 9.2\,\mathrm{g\,cm^{-3}}$)

The Friedman theory for the strong scattering regime [Eqs. (11) and (12) or alternatively Eqs. (18)—(20)] breaks down at $d < 9.2\,\mathrm{g\,cm^{-3}}$. In this region where $g < 1/3$ localized states are expected to appear in the pseudogap near the Fermi energy. The appearance of such localized states in this disordered system implies semiconducting behavior, whereupon electron transport will be governed by thermal excitation above the conduction edge (see Fig. 1). Supporting physical information for the semiconducting nature of expanded liquid mercury in the density region $d < 9.2\,\mathrm{g\,cm^{-3}}$ is obtained from the following experimental data:

a) The slow variation of the mobility, μ, over the density region 9.2—$8.5\,\mathrm{g\,cm^{-3}}$ is reminiscent of the behavior of liquid semiconductors, where μ is temperature independent.

b) The density dependence (at constant temperature) and the temperature coefficients of the electrical conductivity as recorded by Schmutzler [28] exhibit a sharp increase at $d < 9\,\mathrm{g\,cm^{-3}}$ indicating the onset of activated transport.

c) From the correlation between the thermopower S and the electrical conductivity for the semiconductor region $\ln(\sigma/\sigma_0) = \left(1 - \dfrac{eS}{K_\mathrm{B}}\right)$ Schmutzler [28] has demonstrated that the onset of the semiconducting behavior occurs for $d < 7.8\,\mathrm{g\,cm^{-3}}$.

Finally, we would like to point out a serious discrepancy between our value of $d = 9.2\,\mathrm{g\,cm^{-3}}$ for the onset of localized states and the extrapolation procedure employed by Hensel [29] for his optical data which yields $d = 5.5\,\mathrm{g\,cm^{-3}}$.

VI. Summary of Transport Properties of Expanded Liquid Mercury

We have demonstrated the applicability of combined Hall effect and conductivity data to derive direct physical information concerning the following features of a disordered one-component system.

a) Three distinct conductivity regimes involving the weak scattering metallic region, the strong scattering metallic region and the localized states region could be unambigiously identified.

b) The critical densities from the transition between the weak and strong scattering metallic regions and between the strong scattering metallic region and the localized state region were determined.

c) In a divalent liquid metal, the metal–nonmetal transition originates from interband overlap effects described in terms of the Wilson model. In the case of expanded liquid Hg, $g = 0.32$ at the transition from the strong scattering metallic regime to the localization region, and the corresponding critical density $d = 9.2\,\mathrm{g\,cm^{-3}}$ can be identified with the metal–nonmetal transition in this system.

d) The critical density for metal–nonmetal transition in expanded liquid mercury $(d_\mathrm{CM} = 9.2\,\mathrm{g\,cm^{-3}})$ considerably exceeds the thermodynamic critical density $(d_\mathrm{CT} = 4.5\,\mathrm{g\,cm^{-3}})$. There is no correlation between d_CM and d_CT in contrast to some theoretical conjectures.

e) The high value of $d_\mathrm{CM} = 9.2\,\mathrm{g\,cm^{-3}}$ in liquid mercury can be rationalized in terms of the large energy gap (6 eV) between the s and p atomic states of Hg.

Thus large exchange integrals (and consequently high densities) are required for interband overlap. On the other hand the critical density for metal–nonmetal transition for Cs vapor [7] is considerably lower, $d_{CM} = 0.45$ g cm^{-3}. If, in the Cs system, spin pairing, which results the formation of Cs_2 molecules, takes place, the metal–nonmetal transition in the latter system will occur via overlap of the Σg and Σu bands of the diatomic molecule. The level spacing of the molecular Σg and Σu states in $Cs_2 \approx 1$ eV so that the metal–nonmetal transition in this system will occur at a considerably lower density than in the case of Hg.

VII. Brief Comments on the Transport Properties of the Metal–Ammonia Solutions

It is important to notice that the general picture presented herein for the strong scattering and the weak scattering metallic regimes in the one-component system of expanded liquid Hg considers explicitly a homogenous physical system. In particular the identification of the strong scattering regime in any physical system will provide evidence for a) the validity of a homogenous model for the strong scattering metallic state; b) occurrence of the metal–nonmetal transition at the termination of the strong scattering regime.

Several attempts were reported by Acrivos and Mott [7] and by Acrivos [30] to derive the pseudogap depth parameter g for metal–ammonia solution from experimental transport and magnetic susceptibility data. In the work of Acrivos and Mott [7] every physical property yields different g values, so that no coherent picture of this system can be obtained. In a latter work Acrivos [30] attempted to apply the correlations expressed in Eqs. (11) and (12) to the Hall coefficient and conductivity. Unfortunately, the Hall effect data of Kyser and Thompson [3a] used by Acrivos were grossly modified in the later work of Thompson *et al.* [3b, 3c].

We have attempted to correlate the available experimental data for concentrated metal–ammonia solutions with the predictions of the strong scattering theory.

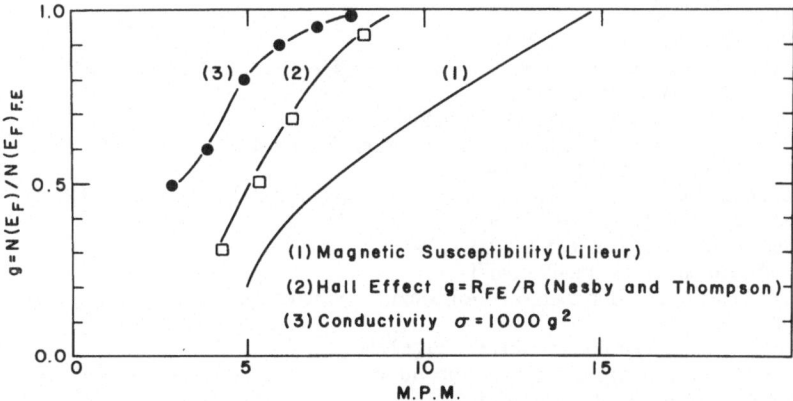

Fig. 11. Summary of attempts to derive the g parameter for concentrated metal ammonia solutions

In Fig. 11 we present our attempts to derive the pseudogap g depth parameter from the following experimental data:

a) the Hall mobility data of Nasby and Thompson [3] calculated according to Eq. (12);

b) from magnetic susceptibility data obtained in the interesting work of Lelieur [31].

c) We have also attempted to derive the g value from conductivity data for concentrated (4—9 MPM) metal–ammonia solutions. In Fig. 11 we analyzed the available conductivity data, fitting them to the relation $\sigma = 1000\,g^2\,\Omega^{-1}\,\text{cm}^{-1}$ in a desperate attempt to obtain g values which are close to those derived from the Hall effect.

The large discrepancy between those g values derived from different physical observables provides conclusive evidence that the homogeneous model is inapplicable for metal–ammonia solutions. In these systems aggregation effects are of crucial importance, affecting the transport and the nature of the metal–non-metal transition

Acknowledgements

We are indebted to J. Magen and M. Levine for their technical help and assistance.

References

1a. Metal–ammonia solutions Lepoutre, G., Sienko, M.J. (Eds.). New York: Benjamin 19 .

1b. Metal–ammonia solutions Lagowski, J.J., Sienko, M.J. (Eds.). London: Butterworth 1970.

2. Lepoutre, G., Lelieur, J.P.: Ref. [1b], p. 247.

3a. Kyser, D.S., Thompson, J.C.: J. Chem. Phys. **42**, 3910 (1965).

3b. Nasby, R.D., Thompson, J.C.: J. Chem. Phys. **53**, 109 (1970).

3c. Vanderhoff, J.A., Thompson, J.C.: J. Chem. Phys. **55**, 105 (1971).

4. Mott, N.F.: Phil. Mag. **6**, 287 (1961).

5. Cohen, M.H., Thompson, J.C.: Advan. Phys. **17**, 857 (1968).

6. Acrivos, J.V., Mott, N.F.: Phil. Mag. **24**, 19 (1971).

7. Hensel, F.: This issue.

8. Mott, N.F.: Phil. Mag. **17**, 1259 (1968).

9. Cohen, M.H., Fritzsche, H., Ovshinsky, S.R.: Phys. Rev. Letters **22**, 1065 (1969).

10. Mott, N.F.: Phil. Mag. **13**, 989 (1966).

11. Anderson, P.W.: Phys. Rev. **109**, 1492 (1958).

12. Ziman, J.M.: Properties of liquid metals. London: Taylor and Francis 1967.

13. Faber, T.E.: Advan. Phys. **15**, 547 (1966).

14. Fukuyama, H., Ebisawa, H., Wada, J.: Progr. Theoret. Phys. (Kyoto) **42**, 497 (1968).

15. Edwards, S.F.: Phil. Mag. **3**, 1020 (1958).

16. Greenfield, A.J.: Phys. Rev. **135**A, 1589 (1964).

17. Mott, N.F.: Advan. Phys. **16**, 49 (1967).

18. Straub, W.I., Roth, H., Bernard, W., Goldstein, S., Mulkern, J.E.: Phys. Rev. Letters **21**, 752 (1968).

19. Friedman, L.: Non-crystalline Solids **6**, 329 (1971).

20. Cohen, M.H.: Proc. Symposium on semiconductor effects in amorphous solids. Amsterdam: North Holland Publishing 1970.

21. Mott, N.F.: Advan. Phys. **16**, 49 (1967).

22. Hindley, N.K.: J. Non-crystalline Solids **5**, 17 (1970).

23. Kubo, R.: J. Phys. Soc. Japan **12**, No. 6 (1957).
24. Rivlin, V. G., Waghorne, R. M., Williams, G. I.: Phil. Mag. **13**, 1169 (1970).
25. Hensel, F., Schmutzler, R. W.: J. Non-crystalline Solids (in press).
26. Evans, R.: J. Phys. C. Metal Phys. Suppl. No. 2 S-137 (1970).
27. Chan, T., Ballantine, L. E.: Can. J. Phys. **50**, 813 (1972).
28. Schmutzler, R. W.: Doctoral Thesis. Karlsruhe University, Germany.
29. Hensel, F.: Phys. Letters **31**, No. 2, p. 88 (1970).
30. Acrivos, J. V.: Phil. Mag. **24** (1971).
31. Lelieur, P. J.: Ph. D. Thesis. Lille 1972.

General Discussion of Papers by Hensel and Even/Jortner

M. H. COHEN   The Mott criterion for the opening of a pseudogap that g should be 0.3 is a most significant contribution to this area. However, Mott presented no derivation, the critical value of 0.3 was an inspired guess, and the physical mechanism underlying the development of the pseudogap was not understood at that time. The discussion of the metal-semiconductor transition by Sak and myself, summarized in my own contribution to this conference, gives a very simple physical basis for the opening of the pseudogap in those materials. That argument is completely generalizable. In all disordered materials, it is the long-range components of fluctuation which govern the transition from localized to extended states. Therefore, a semiclassical theory can be used. All space can be partitioned into allowed and forbidden regions according to whether fluctuation in configuration raises the potential locally above the electron energy. The density in the semiclassical theory is just proportional to the available volume. Thus $g(E)$ is nothing other than the fraction of available volume $C(E)$

$$g(E) \equiv C(E). \tag{1}$$

The critical condition at which a pseudogap opens up occurs when $C(E)$ takes on its critical value for percolation, P_c

$$g^*(E_p) = P_c. \tag{2}$$

This may be regarded as a derivation of Mott's criterion. Now the value of P_c for this particular percolation problem is not precisely known; it is somewhere between 0.16 and 0.30. My best guess is 0.2 or

$$g^*(E_p) \simeq 0.2 \tag{3}$$

a value somewhat less than Mott's.

J. JORTNER   The correlation between the electrical conductivity and the Hall effect data provides a coherent picture for transport in the strong scattering regime of expanded liquid Hg in the range $1 > g > 0.3$. The correlations based on Friedman's random phase approximation break down for $g < 0.3$ (density lower than $9 \, \text{g cm}^{-3}$) indicating the possible onset of localized states according to Mott's ideas. We believe that the electronic states and electronic transport in expanded liquid Hg (at least down to $9 \, \text{g cm}^{-3}$) can be adequately described in terms of a homogenous medium rather than by percolation effects (i.e. inhomogeneous systems). We attempted to fit our experimental transport data to Kirkpatrick's percolation theory. In the range $0.4 < g < 1.0$ we expected $\sigma \propto (g - g_c)$ and $R \propto 1/g(g_c \simeq 0.33)$. This fitting was unsuccessful.

J. ACRIVOS   How does the percolation theory approach to M–NH$_3$ explain the rf and static susceptibility data of J. P. Lelieur at the onset of the metal–nonmetal transition? This also gives $g = N(E_F)/N(E_F)_{\text{free el}} \cong 1/3$. At the concentration where $g \sim 1/3$ ($\simeq 5 \, \text{MPM}$), electrical conductivity equals about $300 \, \Omega^{-1} \, \text{cm}^{-1}$ which is the estimated minimum conductivity for extended states. From analysis of transport properties the mobility μ_d of mobile electrons is about $1 \, \text{cm}^2/\text{V sec}$.

J. JORTNER   We have attempted to correlate the transport properties of concentrated metal–ammonia solutions in terms of the homogenous strong scattering model attempting to use the relations $\sigma \propto g^2$ and $R \propto g^{-1}$ for the conductivity and for the Hall effect data. These correlations do not hold for metal–ammonia solutions, providing strong evidence that in this case clustering results in an inhomogenous system which should be handled in terms of percolation theory.

M. H. COHEN   The Friedman strong scattering theory is a very important contribution to this field, second only to Mott's idea of the pseudogap. However, it suffers from one deficiency.

The deficiency is that the transfer integral J is taken as constant whereas it is strongly dependent on nearest-neighbor configuration in liquid Hg. Thus J has to be taken as a density-dependent average quantity. Different averages of J are involved in the different quantities considered. Thus the elimination of J from the final formulae given by Even and Jortner for σ_1, μ_n, and R is to some extent misleading.

The intrinsic limitation of the theory is that it assumes the amplitude of the wavefunction to be the same, on the average, for each atom. This is in fact correct for $g(E_F) = C(E_F)$ well above the percolation value of $P_c \simeq 0.2$. As P_c is approached, the amplitudes of the extended states are appreciable only within the percolation channels, the volume fraction of which is going rapidly to zero. The system becomes microscopically heterogeneous, and the Friedman theory no longer holds. Thus the value of g, 0.3, at which the strong scattering theory breaks down should give an upper limit to $g^*(E_F)$ itself, which, as I indicated above, may be as 0.2. I infer from the data that percolation without pseudogap formation is important for conduction in the range $P_c < g < 0.3$.

J. JORTNER   We agree that Friedman's theory handles a homogeneous system in terms of the tight binding approaching and then invoking the random phase approximation for the amplitudes of the wavefunctions on the lattice sites. Furthermore, Friedman considers only cellular disorder, while the interatomic exchange integrals are constant at any given density. Once this simplified model is accepted the transport properties (at each density) can be expressed in terms of a single auxiliary reduced quantity $X = a^3 J N(E_F)$.

M. H. COHEN   As the density is reduced and $g(E_F)$ goes below P_c, a gap is formed, the boundaries of which E_c and E'_c are determined by

$$g(E_c) = g(E'_c) = g^* = P_c .$$

Thus the density-of-states remains constant at the mobility edges E_c and E'_c as the density is reduced. The conduction takes place through percolation channels via carriers with energies within kT of E_c and E'_c. Thus the factors determining the mobilities of the charge carriers are the same, independent of density or of the width of the pseudogap. This is, I believe, the explanation of the approximate constancy of the Hall mobility once $g(E_F)$ drops below 0.3, because, even for $P_c < g(E_F) < 0.3$ the conduction process remains the same, independent of density.

J. JORTNER   I would like to comment on the conductivity equation in the semiconducting region which is given by $\sigma = \sigma_0 \exp[-(E_c - E_F)/kT]$ where according to Mott $\sigma_0 \approx 300 \, \Omega^{-1} \, cm^{-1}$. E_F is the Fermi energy while E_c corresponds to the energy of the mobility edge. Unfortunately, one cannot extract $(E_c - E_F)$ from the temperature dependence of σ, because as pointed out by H. Fritsche a few years ago, E_c is temperature dependent, whereupon $E_c = \alpha + \beta T$ and $\sigma = \sigma_0 \exp(-\beta/k) \exp(-\alpha/kT)$. A way to determine the temperature dependence of E_c is to utilize the temperature dependence of the thermoelectric power in the semiconducting region.

F. HENSEL This may be an interesting possibility.

J. JORTNER Hensel has carefully distinguished between divalent and monovalent metallic fluids. I would like to raise the question whether a monovalent metal vapor is indeed monovalent? This question has been privately discussed with Hensel and Even. If you take a dense vapor of Cs or K, diatomic molecules (Cs_2 or K_2) may be formed. The binding energy of such diatomics is ~ 0.5 eV so that they will be quite stable. This mechanism for spin pairing bears some analogy to spin pairing (via quadrupole ion pairs) in metal–ammonia solutions. If diatomic molecules are indeed formed in low density alkali metal vapor the metal–nonmetal transition will occur via overlap of bonds as in the case of mercury. The relatively low critical density for metal–nonmetal transition in Cs vapor (0.47 gm cm^{-3}) can then be rationalized in terms of the low spacing between the Σg and Σu bands in Cs_2 which is of the order of 1 eV. Magnetic susceptibility studies will establish whether this picture is valid.

Metal–Nonmetal Transition and Exciton Screening

AHARON GEDANKEN, BARUCH RAZ, UZI EVEN, and JOSHUA JORTNER

Abstract

We discuss a spectroscopic criterion for the observation of the insulator-metal transition in a two-component system which is based on the disappearance of Wannier exciton states in the metallic region. We report the effect of exciton screening in the vacuum ultraviolet spectra of mercury/xenon mixtures where the Xe Wannier states are abruptly washed out at $55\% \pm 5\%$ of mercury.

Metal–nonmetal (MNM) transitions in ordered and disordered systems [1] have been experimentally induced by the following physical effects:

a) structural modifications [2] (i.e. changes of temperature and pressure),

b) the application of external fields [3],

c) Concentration changes in two-component systems (i.e. doped semiconductors [4], metal–ammonia solutions [5], sodium-argon mixtures [6],

d) density changes in one-component systems (such as dense metal vapors [7] and expanded liquid metals [8]).

Most of these studies [1] monitored the electrical transport properties and the magnetic properties of the system undergoing the MNM transition. We shall discuss a spectroscopic criterion for the observation of the MNM transition. Wannier-Mott exciton states in a two-component system, consisting of open shell metallic atoms and of closed shell saturated atoms, are utilized as a spectroscopic probe to monitor the MNM transitions. These large-radius excited states are expected to persist only in the insulating state, and become unbound in the metallic state due to short range dielectric screening effects.

Our experimental approach is based on Mott's classical argument concerning the effects of long range forces on the MNM transition [9]. The long range electron-hole potential in the nonmetallic state

$$V(r) = -e^2/\varkappa r \tag{1}$$

(where $\varkappa$ is the static dielectric constant), which sustains an infinite number of excited states, is replaced in the metallic state by a short range potential, which according to the Thomas-Fermi prescription is

$$V(r) = -(e^2/\varkappa r)\exp(-qr) \tag{2}$$

where the screening length is

$$q^2 = 4m^* e^2 (3n/\pi)^{1/3}/\hbar^2 \varkappa \tag{3}$$

with n corresponding to the free electron density while m^* represents the electron effective mass. As is well known [9], the potential well (2) does not have bound states for

$$q\, a_H \gtrsim 1.0 \tag{4}$$

where the modified exciton Bohr radius [9] is

$$a_H = \frac{\hbar^2 \varkappa}{m^* e^2}. \tag{5}$$

It will be useful to recast the screening condition (4) in a more transparent form. The screening parameter q^2 is determined by the density of the free carriers and depends on the exciton radius, as utilizing Eqs. (3) and (5) we have

$$q^2 = \frac{4\left(\dfrac{3}{\pi}\right)^{1/3} n^{1/3}}{a_H}. \tag{6}$$

The screening condition (4) can be now expressed in the form

$$q^2 a_H^2 = 4\left(\frac{3}{\pi}\right)^{1/3} n^{1/3} a_H \gtrsim 1 \tag{7}$$

which leads to the familiar relation [9]

$$n^{1/3} a_H \gtrsim 0.25. \tag{8}$$

Consider the quantitative implications of these arguments for the description of Wannier-Mott type shallow and deep exciton and impurity states [10, 11]. In a nonmetallic solid [10] the Coulomb electron-hole attraction is dielectrically screened, whereupon for large radius states the microscopic variation of the crystal and of the positive hole potentials is replaced by Eq. (1). Furthermore, when the conduction band is wide and parabolic, the effects of the crystal potential can be subsummed into an effective mass. This approach has been widely utilized for shallow excited states [10] and has been also justified for deep states invoking pseudopotential theory [11]. The envelope function for large radius exciton and impurity states obeys the equation

$$\left(-\frac{\hbar^2}{2m^*} \nabla^2 + V(r) - E\right) \Psi = 0 \tag{9}$$

where the potential is given by Eq. (1). A Rydberg series converging to the bottom of the conduction band will result, characterized by the energy levels, E_n, which are given by

$$E_n = E_G - \frac{G}{n^2} \tag{10}$$

where E_G is the band gap, and G is the exciton binding energy

$$G = \frac{e^2}{2a_H \varkappa} \tag{11}$$

with a_H given by Eq. (5). Extensive experimental evidence was reported for shallow exciton and impurity states in semiconductors [12] and for deep lying states in rare gas solids [13]. It is now established that the observation of exciton states in dense rare gases is independent of symmetry arguments, and that these excited states are amenable to experimental observation in positionally disordered systems (i.e. liquid rare gases) [14] and in substitutionally disordered systems (i.e. heavily substituted rare gas alloys) [15]. Now, when an insulator (such as a rare gas solid) is gradually substituted by unsaturated metal atoms this two-component system may eventually undergo a MNM transition, whereupon $V(r)$ in Eq. (9) will take the approximate form (2). Furthermore, when the screening length obeys Eq. (7), or Eq. (8), no bound excited Wannier states will exist any more [9].

To conclude this discussion it will be useful to recast the exciton screening condition [Eqs. (7) and (8)] in terms of the exciton binding energy, G, which can be obtained directly from experimental data [see Eq. (10)] for the exciton Wannier series in the insulating state. Making use of Eqs. (8) and (11) the screening condition takes the form

$$\frac{n^{1/3}}{\varkappa G} \leqq \frac{1}{2e^2}. \tag{12}$$

The critical carrier density, n_c, required for screening of the large radius Wannier states satisfies the relation

$$\frac{n_c^{1/3}}{\varkappa G} = \frac{1}{2e^2} = 5 \times 10^7 \, \text{eV}^{-1} \, \text{cm}^{-1}. \tag{12a}$$

From these results we conclude that the critical density for exciton screening is proportional to the third power of the exciton binding energy. Because exciton binding energies vary from $\sim 10^{-3}$ eV for shallow exciton states in germanium [12] up to ~ 1 eV in rare gas solids [13], we expect that the critical carrier density required for exciton screening will vary greatly for different systems.

Two experiments concerning metallic screening of exciton and impurity states were reported previously. Asnin and Rogachev [17] observed the disappearance of Wannier impurity states in arsenic-doped germanium with an increase in the impurity doping level. These data provide direct evidence for the exciton screening effect, as the optically active impurity states in the insulating region are the source of the free carriers in the metallic state. Wilson and Yoffe observed the disappearance of Wannier states in WSe_2 doped with niobium. At 1% Nb the high $(n \leqq 2)$ Wannier states disappear while at a Nb level doping of 6% the whole excitonic structure is washed out [18]. In this experiment Nb doping fills the narrow d/p band of the group VI compound. The effect of exciton screening originating from the metal–nonmetal transition was experimentally observed by us in mercury/xenon mixed solids at low temperatures. This system is of interest because:

a) This two-component system is characterized by a wide (empty) conduction band in the insulating state, while in a metallic state a broad (filled) conduction band will result due to the overlap of the mercury s and p bands.

b) These experiments provide the first evidence from screening of deep impurity states.

We have studied the vacuum ultraviolet absorption spectra (spectral region $1200-1600\,\text{Å}$) of Xe/Hg solid films at $10-40\,°\text{K}$. Gaseous mixtures of Xe and

Hg containing $0-0.8$ mole fraction [i.e., $(0-80)$% mole percent] of *Hg* were condensed on a LiF window, mounted on a variable-temperature helium-flow cryostat. Relevant experimental details are described in Ref. 16.

In Fig. 1 we display the absorption spectra of *Xe/Hg* films. The spectrum of pure *Xe* exhibits the well-known[13] Wannier exciton series [$n=1$ ($^2P_{3/2}$) state at 8.40 eV, $n=2$ ($^2P_{3/2}$) state at 9.08 eV, and $n=3$ ($^2P_{3/2}$) state at 9.22 eV], in accord with Baldini's original work.[13] When the *Hg* concentration is increased in the range $(15-40)$% the *Xe* $n=1$ and $n=2$ states are well defined, apart from line broadening, which is attributed to the effects of substitutional disorder. The simple coherent-potential approximation [19] predicts that when cellular disorder is exhibited just by the diagonal matrix elements of the crystal

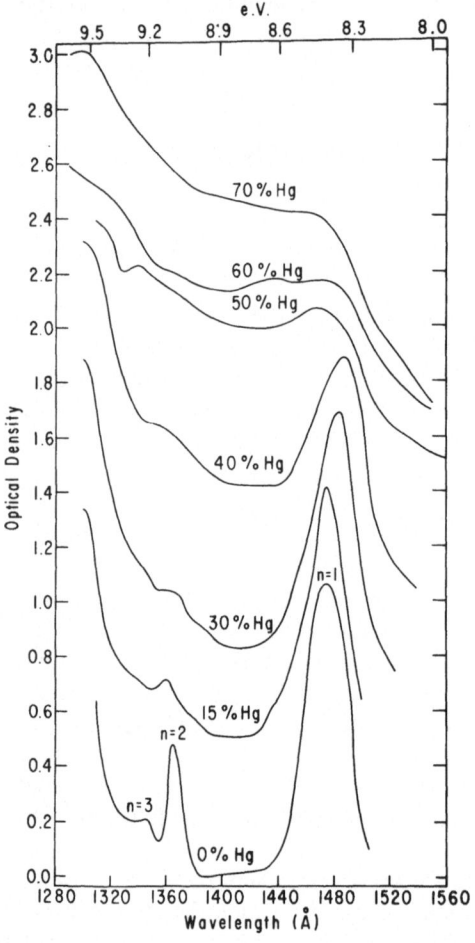

Fig. 1. Vacuum u.v. absorption spectra of mercury-xenon solid films deposited at 40 °K and measured at 10 °K. The spectra were practically temperature independent in the range 10—40 °K. The percentages represent the mole fractions of mercury. The spectra were vertically displaced on the optical density scale, the residual absorption at 1550 Å in each case was 0.1—0.2 optical density units

Hamiltonian, the optical line broadening in a binary alloy is proportional to $[x(1-x)]^{1/2}$, where x is the mole fraction of one of the components. The gradual increase in the width of the $n=1$ $(^2P_{3/2})$ and of the $n=2$ $(^2P_{3/2})$ exciton states of Xe in the range $x_{Hg}=15$—40% is qualitatively consistent with these expectations. Furthermore, provided that the system is nonmetallic we expect a minor change in the Xe line widths in the concentration range $x_{Hg}=40$—60%. However, in this concentration region a dramatic modification of the optical spectra takes place (Fig. 1), whereupon at 50% mercury appreciable line broadening occurs while at 60% mercury the excitonic spectrum is completely washed out. From these results we conclude that:

a) The disappearance of Wannier exciton states in Xe/Hg solid films occurring at $x_{Hg}=55\%\pm5\%$ is attributed to the screening of these large radius states in the metallic state.

b) A rough estimate of the critical transition density to the metallic state is in order. Assuming a close packing of Xe atoms and of Hg^{+2} ions in the metallic phase ($x_{Hg}=50\%$), the "critical" density of mercury atoms for the MNM transition is ~9 g cm^{-3} (i.e. number density 4.5×10^{22} cm^{-3}). This density is very close to the critical density for MNM transition in expanded subcritical and supercritical mercury at high temperature and pressures where Hall effect [21] data, the temperature coefficient of the electrical conductivity and the thermoelectric power [22] indicate that the MNM transition in the one-component system occurs at the density of 9 g cm^{-3}.

c) Our spectroscopic criterion for the MNM transition in Xe/Hg films at $x_{Hg}\approx55\%\pm5\%$ implies that in this concentration region an abrupt increase in the electrical conductivity should take place. This expectation is consistent with recent studies [6, 18] of electrical conductivity of some divalent metals/rare gas solids (i.e. Pb/Ar and Cu/Ar films) [20] where an abrupt decrease of the electrical conductivity was observed at $\sim60\%$ metal concentration.

d) The general condition for the disappearance of bound electron hole pairs due to short range screening [Eq. (12)] is well satisfied for the Xe/Hg system (see Table I). In this table we have assembled the available experimental data concerning exciton screening effects due to the metal insulator transition. The critical densities vary over five orders of magnitude while the exciton binding G factors vary over the region of 200, in agreement with the predictions of Eq. (12).

Table I. Exciton screening near the MNM transition

System	n_c cm^{-3}	G eV	$\varkappa$	$n_c^{1/3}/G\varkappa$ cm^{-1} eV^{-1}	Ref.
Xe/Hg	4×10^{22}	0.8	2.3	2.2×10^7	Present work
WSe$_2$/Ta	2×10^{20}	0.05		1.2×10^7	[1]
Ge/As	2×10^{17}	0.005	16	0.1×10^7	[2, 3]
Theory	—	—	—	5×10^7	—

1. Wilson, J. A., Yoffe, A. D.: Advan. Phys. **18**, 194 (1969).
2. Asnin, V. M., Rogachev, A. A.: Phys. Stat. Sol. **20**, 755 (1967).
3. The discrepancy between theory and experiment in this case supports the arguments presented in the text.

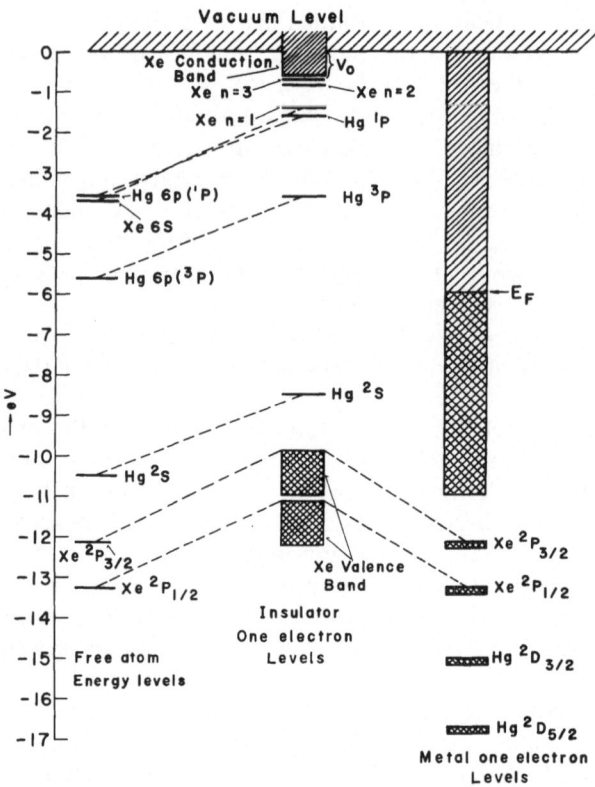

Fig. 2. Schematic one-electron energy-level diagrams for mercury/xenon solid mixtures in the insulating phase (low x_{Hg}) and in the metallic phase ($x_{Hg} \leqq 55\%$)

Finally, we turn our attention to the optical spectra of the metallic state ($x_{Hg} > 60\%$). In Fig. 2 we present a schematic (one electron) energy-level diagram for the Xe/Hg system. The atomic energy levels [23] are modified at low Hg concentrations as follows: Xe valence and conduction bands are formed, their positions being estimated from experimental spectroscopy [13] and photo-emission [24] and from the theoretical estimates of the hole polarization energy [25]. The experimental spectroscopic data for the location of Xe Wannier states have also been utilized [13]. The mercury impurity states were estimated from the atomic data, from a rough estimate of the positive hole (Hg^+) polarization energy, and from the experimental location of the mercury absorption bands in Xe/Hg at low x_{Hg}. When the system becomes metallic ($x_{Hg} > 50\%$) the mercury s and p bands overlap, the Fermi energy is estimated at 5 eV (compared to 6.9 eV in liquid Hg). The Xe^+ hole states are now lowered (relative to the insulating state) due to the loss of the long range polarization energy. The broad structureless optical transition observed for $x_{Hg} > 50\%$ (see Fig. 1) may originate from core excitations to the conduction band states above the Fermi energy. Such excitons may occur from the Xe 5p core states originating at 7 eV (see Fig. 2) and being split by the $Xe^+({}^3P_{3/2}) - Xe^+({}^2P_{1/2})$ spin orbit coupling [23] (1.2 eV), or/and from the Hg 5d levels (as is the case in liquid Hg [26]) which are expected to occur at 9 eV and to be split by 1.9 eV due to the $Hg^+(^2D_{3/2}) - Hg^+(^2D_{5/2})$ spin orbit coupling [23].

We hope that the present results provide new physical insight into the interesting problem of the MNM transition. Several interesting questions have to be elucidated concerning metal/rare gas solids: a) Is the structure of the binary system crystalline or amorphous? b) Is the dispersion of the metal in the rare gas statistical or do microscopic metallic particles form near the MNM transition? c) Provided that the system is amorphous and homogeneous, can experimental spectroscopic evidence be obtained concerning the formation of a pseudogap [27] near the MNM transition? d) What is the relation between the optical properties studied herein and electrical transport and magnetic properties of these systems?

Acknowledgement

This work was supported by a grant from the Israeli National Council for Research and Development.

References

1. For a recent review see Mott, N. F., Zinamon, Z.: Rep. Prog. Phys. **33**, 881 (1970).
2. Morin, J. F.: Phys. Rev. Letters **3**, 34 (1959).
3. Feinlieb, J., Paul, W.: Phys. Rev. **155**, 841 (1967).
4. Fritsche, H.: J. Phys. Chem. Solids **6**, 69 (1959).
5. Thompson, J. C.: Rev. Mod. Phys. **40**, 704 (1968).
6. Cate, R. C., Wright, J. G., Cusack, N. E.: Phys. Letters **32** A, 467 (1970).
7. Henzel, F., Franck, E. U.: Rev. Mod. Phys. **40**, 697 (1968).
8. Even, U., Jortner, J.: Phys. Rev. Letters **28**, 31 (1972).
9. Mott, N. F.: Phil. Mag. **6**, 287 (1967).
10. Kohn, W.: Solid state physics, Vol. 5, p. 258. Turnbull, D., Seitz, F. (Eds.). New York: Academic Press Inc. 1957.
11. Hermanson, J., Phillips, J. C.: Phys. Rev. **150**, 652 (1966).
12. Burstein, E., Bell, E. E., Davisson, J. W., Lax, M.: J. Phys. Chem. **57**, 849 (1953).
13. Baldini, G.: Phys. Rev. **128**, 1562 (1962).
14. Raz, B., Jortner, J.: Proc. Roy. Soc. (London) A **317**, 113 (1970).
15. Nagasawa, N., Karasawa, T., Miura, N., Nanba, T.: J. Phys. Soc. Jap. **32**, 1155 (1972).
16. Raz, B., Magen, J., Jortner, J.: Vacuum **19**, 571 (1969).
17. Asnin, V. M., Rogachev, A. A.: Phys. Stat. Sol. **20**, 755 (1967).
18. Wilson, J. A., Yoffe, A. D.: Advan. Phys. **18**, 193 (1969).
19a. Onodera, D., Toyozawa, Y.: J. Phys. Soc. Japan **24**, 341 (1967).
19b. Soven, P.: Phys. Rev. **156**, 809 (1967).
19c. Velicky, B., Kirkpatrick, S., Ehrenreich, H.: Phys. Rev. **175**, 747 (1968).
19d. Hoshen, J., Jortner, J.: J. Chem. Phys. **56**, 933 (1972).
20. Cusack, N. E.: Lecture presented at the Chelsea Meeting on Amorphous Semiconductors. London, December 1971.
21. Even, U., Jortner, J.: Experimental relations for the hall effect near the metal–nonmetal transition. Phil. Mag. **25**, 715 (1972).
22. Schmutzler, R. W., Hensel, F.: J. Non-crystalline Solids 1972 (in press).
23. Moore, C. E.: Atomic energy levels. Washington: Natl. Bureau of Standards 1955.
24. O'Brien, J. F., Teegarden, K. J.: Phys. Rev. Letters **17**, 919 (1966).
25. Fowler, W. B.: Phys. Rev. **151**, 657 (1966).
26. Wilson, E. G., Rice, S. A.: Optical properties and electronic structure of metals and alloys, p. 271. Abeles, F. (Ed.). Amsterdam: North Holland 1966.
27. Mott, N. F.: Phil. Mag. **13**, 989 (1966).

Discussion

J. KOMMANDEUR   Do you have any information concerning the structure and aggregation effects in the metal-rare gas films which you studied?

J. JORTNER   The choice of Xe/Hg films for the present spectroscopic study was motivated by the relatively high pressure of liquid Hg at room temperature, so that He–Hg gaseous mixtures were prepared at room temperature. The mixtures were then condensed on a cooled LiF window at 10—30 °K. In view of the high sticking coefficients of both components at 30 °K we can safely assert that no enrichment of the solid sample relative to the gas phase composition is expected.

At present, we have no direct structural information as to whether our samples are crystalline or amorphous or if metallic Hg particles are formed. These questions will be resolved by electron diffraction studies of these metal-rare gas films.

M. H. COHEN   The persistence of broad features in the optical absorption centered at the pure exciton energies or, equivalently, the $^2 5P_{3/2}$ and $^2 5P_{1/2} \rightarrow 6S_{1/2}$ in atomic Xe suggests that these are resonances. That is, an internal transition occurs on the Xe atom from either the $5P_{3/2}$ and the $5P_{1/2}$ to the $6S_{1/2}$. The latter level is degenerate with the conduction band continuum of the surrounding Hg and so is a resonance. Very crudely, the width Γ is given by

$$\Gamma = 2\pi V^2 n_0(E_{6S}) g(E_{6S}) \tag{1}$$

where V is the matrix element transferring the 6S electron from the Xe atom to the sourrounding Hg, $n_0(E_{6S})$ is the free electron density of states at the 6S level, and g is the ratio of the actual density-of-states there to the latter. $g(E_{6S})$ will increase with increasing density but not as fast as $g(E_F)$; thus the line widths should increase in the manner observed once the upper bound of the pseudogap, E_c, falls below E_{6S}. This should occur at a density lower than that corresponding to $g = 0.3$.

J. JORTNER   I expected the suggestion proposed by Cohen, that the two broad transitions in Hg/Xe mixtures in the metallic region may originate from Xe resonances. Support for our assignment of these two transitions as originating from Xe 5P core states excited to conduction band states above the Fermi energy, is obtained from high energy reflectivity data of normal pure liquid mercury [Wilson, E. G., Rice, S. A.: In: Abeles, F. (Ed.): Optical properties and electronic structure of metals and alloys, p. 271. North-Holland 1966.] Wilson and Rice have reported core excitations from the Hg 5D levels to conduction band states above the Fermi level of liquid mercury. Our interpretation for the Xe/Hg transitions in the metallic region is similar. It should however be pointed out that in addition to the two (spin orbit split) core excitations in liquid Hg, Wilson and Rice have observed an additional peak on the low energy side which may be due to a resonance in the liquid metal.

Mobility Studies of Excess Electrons in Nonpolar Hydrocarbons

H. Ted Davis, L. D. Schmidt, and Roger G. Brown

Abstract

A summary of recent studies of electron drift velocities in hydrocarbon liquids is presented. In most hydrocarbon liquids, the mobilities obey an Arrhenius plot implying an activation process of some sort. A possible explanation of this process is the conjecture that the electron can be trapped in a solvent cavity with a bound state near the quasi-free conduction state. In terms of this model, the electron will be frequently trapped for short times until thermally promoted to the quasi-free state where it moves between traps. It is suggested that the freely moving CH_2 group may play a dominant role in either the solvent cavity or the phonon-assisted hopping model.

I. Introduction

Drift velocity studies have proved very useful in understanding the behavior of excess electrons in nonpolar fluids. Probably the most important feature of the drift velocity technique is that it can easily be used to study electron concentrations as low as 10^5 electrons per cubic centimeter. By studying the field and temperature dependence of electron drift velocities (or of the mobility, which is defined as the ratio of the drift velocity to the imposed electric field), one is able to understand a great deal about an excess electron and the nature of its transport through a given liquid.

We wish in this lecture to summarize recent experimental findings that we and other investigators have made on the behavior of thermalized excess electrons in hydrocarbon liquids. Then we shall discuss the kinetic theory of quasi-free electrons in polyatomic fluids. We shall also discuss in some detail a model proposed earlier in which the electron is periodically captured for short times by collective localized traps and moves between traps as a quasi-free electron. An electron behaving as prescribed by this model would constitute a new electronic species in condensed media whose properties bear some resemblance both to solvated electrons in polar fluids and to quasi-free electrons in liquids argon, krypton and xenon. The model is shown to be consistent with the temperature dependence of the electron mobility in such liquids as propane, n-pentane, n-hexane, etc. However, the temperature dependence of the mobility data is also consistent with the phonon-assisted hopping mechanism known to apply to electronic motion in amorphous semiconductors. A Hall mobility experiment the authors are currently preparing will perhaps be definitive in determining the mechanism of charge transport in nonpolar hydrocarbon liquids.

Several techniques have been used to determine electron drift velocities in nonpolar systems. The technique we [1] use is to photoinject electrons into the fluid and measure their time of flight under an electric field between two electrically operated shutters. Schmidt and Allen [2] introduced electrons by X-ray pulse radiation and obtained drift times from the rise time of the conductivity in a parallel-plate cell. Freeman and co-workers [3, 4] deduce the mobility from transient electric current and free-ion yield measurements in pulse X-irradiated liquids. Recently, Beck and Thomas [5] have deduced relative electron mobilities from studies of the rate constants for the reaction of electrons with biphenyl (and other reactants) in dilute solutions of biphenyl in organic liquids. They produce the electrons in the liquids by two-photon laser photolysis of anthracene and pyrene which are also present in small amounts in the organic liquids studied. Of the techniques mentioned here, we believe the photoinjection-double shutter technique which we use is the most accurate for determining drift velocities. However, it has the disadvantage of requiring a greater degree of purification than the other techniques.

II. Observed Drift Velocities, Mobilities and Work Functions

In 1968 Tewari and Freeman [6] reported a rapidly decaying current "spike" which appeared in a neopentane sample subjected to a pulse of X-rays. This was perhaps the first indication that electrons with mobilities which were high compared to ionic mobilities could be observed in nonpolar hydrocarbon liquids.

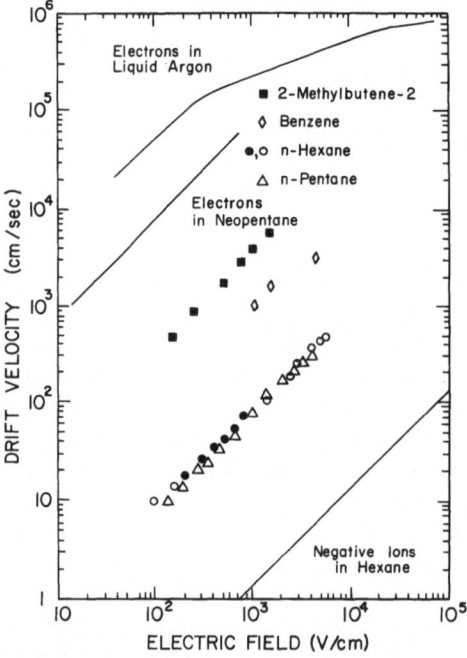

Fig. 1. Electron drift velocity versus field in several hydrocarbon liquids [1, 7] at saturation pressure at 300 °K. Also shown for comparison are curves for electron in liquid argon (87 °K) and ions in n-hexane (300 °K). Drift velocity in hydrocarbons was measured in single shutter (open data points) and double shutter (closed data points) apparatuses. The neopentane [7] and argon [9] curves were drawn from double shutter data

About a year later Minday, Schmidt, and Davis [1] and Schmidt and Allen [2] independently reported time of flight measurements of fast negative species in several nonpolar hydrocarbon liquids. The former investigators used a double shutter apparatus to measure the time of flight of photoinjected electrons across a given drift space subjected to an electric field, while the latter investigators used a parallel-plate diode cell to measure the time of flight of electrons created by pulse radiation. Recently, Fuochi and Freeman [3] and Dodelet and Freeman [4] determined electron mobilities in several nonpolar hydrocarbons by X-radiolysis free-ion yield studies, and Beck and Thomas [5] by the reaction rate studies mentioned above.

In Fig. 1, drift velocities versus electric field are displayed for negative ions in hexane and for electrons in several nonpolar hydrocarbons [1, 7]. Also plotted in Fig. 1 is the drift velocity versus electric field of electrons in liquid argon [8, 9] where it is known that the electrons are quasi-free (i.e., conduction) electrons [8—11]. Two important features of the behavior of electrons in hydrocarbons are apparent in Fig. 1. The first is that the electron mobility (drift velocity/electric field) depends very sensitively on the structure of the hydrocarbon molecule, e.g., at $300\,^{\circ}\text{K}$ the electron mobility in saturated n-pentane liquid is about $0.07\,\text{cm}^2/\text{V}$ sec while in saturated neopentane it is about $70\,\text{cm}^2/\text{V}$ sec. As the molecule becomes geometrically more compact and symmetrical, the mobility

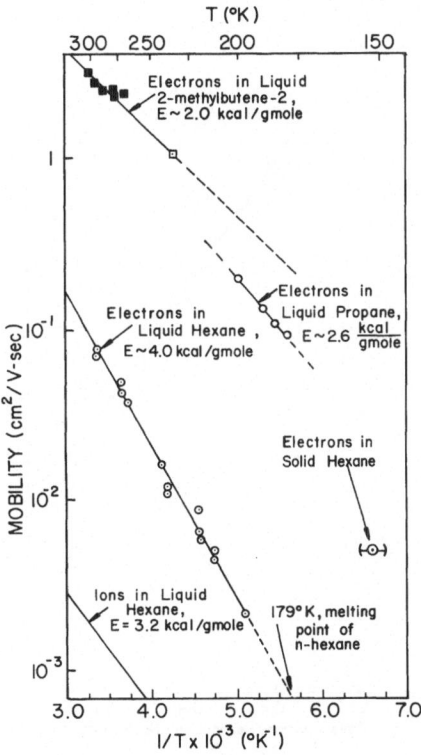

Fig. 2. Plots of the logarithm of mobility versus $1/T$ for several systems [1, 12]

increases rapidly. The other feature is that the drift velocity remains a linear function of the field (i.e., the mobility remains field independent) up to fields much higher than those at which the electron drift velocities in the inert gas liquids become nonlinear. Schmidt and Allen [2] observed that electron drift velocities remain linear functions of the electric field up to at least 10^4 V/cm in tetramethyl silane (mobility about $90\,cm^2/V\,sec$) and to at least 10^5 V/cm in n-pentane. These are the highest field results available.

Figure 2 demonstrates [12] another important feature of excess electrons in many nonpolar hydrocarbons: The mobilities as a function of temperature obey an Arrhenius, i.e.,

$$\mu = \mu_0\,e^{-E/RT}\,. \tag{1}$$

Such a temperature dependence implies some sort of localization of the excess electrons. One possibility is that the electron is periodically localized for a short time in a shallow trap and moves from trap to trap as a conduction electron. Another possibility is that the electron moves via phonon-assisted hopping from

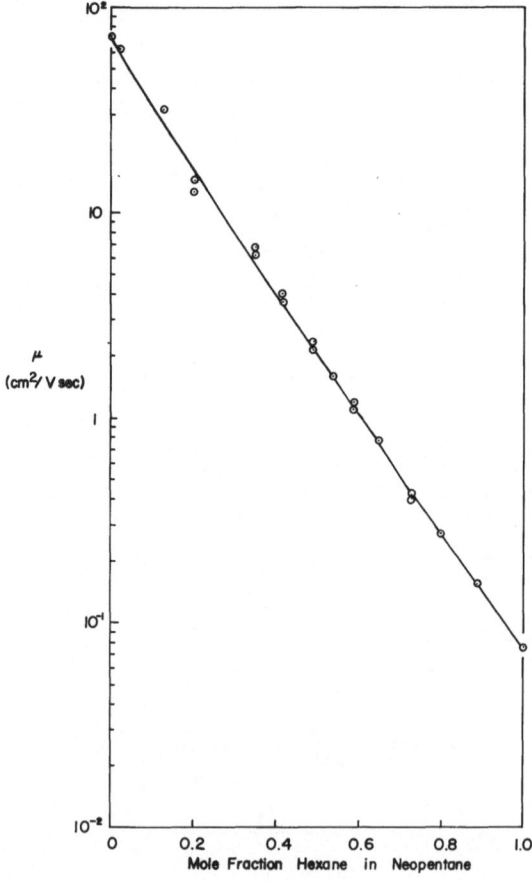

Fig. 3. Plot of the logarithm of electron mobility versus mole fraction of n-hexane in neopentane and n-hexane mixture [7] at 300 °K

one localized site to another — this sort of hopping model has been proposed to explain electron mobilities in amorphous semiconductors [13—16]. We shall discuss these possibilities further in a later section.

To investigate the possibility that the electron trap or localization site was an ion formed by a solvent molecule, Minday *et al.* (MSD) [17, 7] studied electron mobilities in mixtures of n-hexane (in which the mobility activation energy is 4.06 kcal) and neopentane (in which the mobility activation energy is almost, if not in fact, zero). If the mechanism of localization in n-hexane is ion formation, then the mobility activation energy in the mixture should be the same as that in pure n-hexane. In Fig. 3 the logarithm of the electronic mobility in the mixture at 300 °K is plotted versus the mole fraction of neopentane. The data lie on a

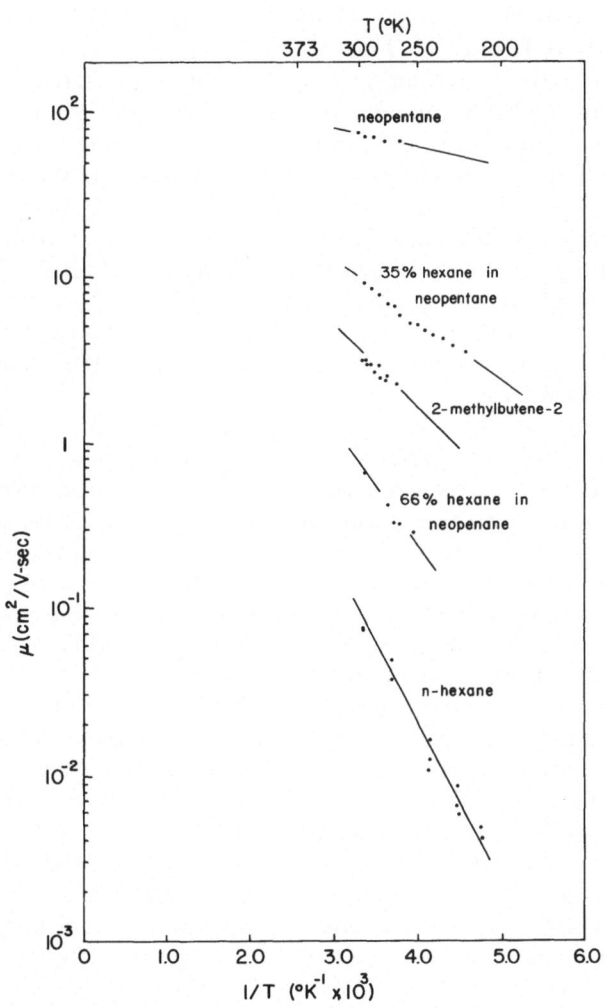

Fig. 4. Logarithm of electron mobility versus $1/T$ for several compositions of neopentane and n-hexane mixture [7]

straight line and can be expressed very well by the equation

$$\mu_{mix} = 70 \exp(-6.8\, x_h) \text{cm}^2/\text{V sec}, \tag{2}$$

where x_h is the mole fraction of n-hexane. In Fig. 4 the logarithm of the mobility is plotted versus $1/T$ for several compositions of the mixture. The data correlate well with the expression

$$\mu_{mix} = 70 \exp[-x_h E_h/RT], \tag{3}$$

where $E_h (= 4.06 \text{ kcal})$ is the mobility activation energy in pure n-hexane. From the composition dependence of the activation energy, MSD [7] concluded that the trap is a collective trap made up of an ideal (random) mixture of n-hexane and neopentane. (They assigned an activation energy of ~ 0.5 kcal to the observed mobility in neopentane. However, statistical analysis of the neopentane data suggests that this number is not too reliable and could as well be zero and that the best value of E_h is 4.06 kcal and not the value 4.3 kcal that MSD [7] used in their analysis.) Recently, Beck and Thomas [5] obtained relative electron mobilities in iso-octane and n-hexane mixtures as a function of composition. Their data, too, obey an expression of the form of Eq. (2) indicating that the mobility activation energy in the iso-octane–n-hexane mixture is also of the ideal solution form, $E_{mix} = x_h E_h + x_{i-0} E_{i-0}$. An ideal solution form for E_{mix} argues that the localized electron does not introduce substantial intermolecular structural rearrangement of the solvent molecules (the two mixtures studied are known to form essentially ideal thermodynamic mixtures). This important implication would mean that the localization states or traps must involve fluctuations of local molecular arrangements already present in the fluid rather than caused by the presence of the electron.

Another experiment done by MSD [1] was to draw an electron current from liquid n-hexane into n-hexane vapor. If the electron is in a negative energy state with respect to the vacuum, then one should observe very little current in such an experiment. To perform the experiment a triode cell was constructed whose grid arrangement is shown in Fig. 5. Charge carriers were injected from the cathode and drawn upward to the grid, both of which were submerged in the liquid. A field was also applied between the grid and the collector such that some of the negative carriers could pass through the grid and be drawn towards the collector, which was above the liquid level.

Figure 5 shows the current to the collector in hexane as a function of the nominal field between the grid and collector for a constant emitter-grid field of 500 V/cm. The open circles indicate the currents with the liquid level between the grid and the collector, while the solid circles indicate the currents with the collector submerged in the liquid. The latter should be a measure of the fraction of charge carriers transmitted through the grid, while the former depends on both the transmission through the grid and the liquid-vapor interface. It is seen that at fields greater than 100 V/cm the currents are very similar, indicating that grid transmission is controlling, while for lower fields the current through the vapor is definitely lower.

Corresponding measurements with the liquid level between the grid and collector and no field in the emitter region (no photocurrent from the cathode) gave

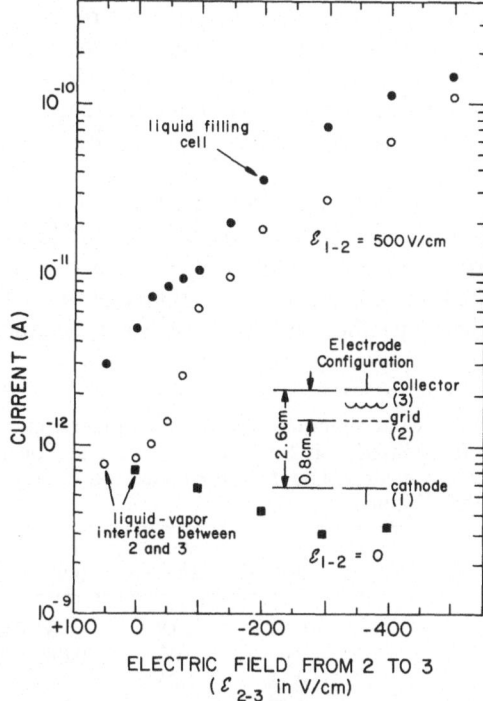

Fig. 5. Transmission of photoinjected electrons from liquid to vapor across interface [1]. A cell with the electrode configuration indicated was used to determine the current which could be drawn out of the liquid and into the vapor. The low currents observed with $E_{1-2} = 0$ indicate that most of the charge carriers originated at the cathode

a very small photocurrent, as indicated by the squares in Fig. 5. This shows that the current to the collector was indeed originating at the cathode.

In a sample containing electrons and ions, it was observed that both reached the collector when the collector was submerged, but only the fast carriers reached the collector through the vapor as expected.

From the field at which the onset of current flow occurs (of the order of 100 V/cm in Fig. 5) across the liquid-vapor interface, we can estimate, roughly, the energy barrier to electron emission from the liquid. Suppose the energy barrier is $-V_0$, where V_0 is the work function of the liquid. Then the field necessary to draw electrons out of the liquid may be estimated as $e \varepsilon \Lambda \simeq -V_0$, where e is the electronic charge and Λ is the thermalization mean free path (or jump distance) of the electron. In view of the field independence of the electron mobility in n-hexane for fields greater than 10^4 V/cm, the quantity Λ will be appreciably less than 100 Å. Thus, taking $\varepsilon \sim 100$ V/cm and $\Lambda = 100$ Å, we obtain the bound $-V_0 < 10^{-4}$ eV. On the other hand, there is a positive barrier to emission from the liquid since a finite field was required to draw a current, so that we have the bounds $0 \leqq -V_0 < 10^{-4}$ eV. The implication of this conclusion is that the energy state from which the electron is emitted is essentially the vacuum energy $V_0 = 0$. If we assume that the electron in n-hexane liquid moves as a conduction electron between short-lived trapping states, then V_0 is the energy of the bottom of the conduction band. If on the other hand we assume that the electron moves by hopping from one localized state to another across a barrier equal to the mobility activation energy, then the bottom of the conduction band may lie above the vacuum and the electron may be emitted directly from the localized state into the vacuum.

By studying the difference between the work functions of a metallic surface in a vacuum and in contact with a liquid, Holroyd and Allen [18] measured the energy V_0 of injection of electrons from the vacuum into several nonpolar hydrocarbon liquids. Their data (which has an average uncertainty of about ± 0.04 eV) for V_0 are given in Table I. The value they obtain for V_0 for n-hexane is ~ 0.04 eV (or

Table I. Measured work function shifts [18] and electron-molecule hard core cut-offs deduced from Springett, Jortner, and Cohen's [23] quasi-free electron theory. The measured electron mobilities are also given for these liquids

Liquid	V_0 (eV)	$\tilde{a}$ (Å)	μ (cm^2 V^{-1} sec^{-1})
n-hexane	$+0.04$	2.18	0.08[a], 0.09[b]
n-pentane	-0.01	2.03	0.08[a], 0.16[b]
Cyclohexane	-0.28	1.85	1.1[b]
2,2,4-trimethylpentane	-0.18	2.36	7[b]
Neopentane	-0.43	1.93	70[a], 55[b]
Tetramethylsilane	-0.62		90[b]

[a] Mobilities taken from Ref. [1, 7]
[b] Mobilities taken from Ref. [2].

approximately zero to the order of the uncertainty of the data) which is consistent with the value $V_0 \sim 0$ found by MSD in their liquid-liquid emission experiment. Thus, the state the electron was pulled from in MSD's experiment is the same state that Holroyd and Allen inject the electron into in their photoinjection experiment. An increasingly negative value of V_0 correlates with an increasing mobility, as shown in Table I.

Let us close this section of the talk by summarizing in Table II the mobility data available for various nonpolar hydrocarbons. Except for the entries with asterisks, the mobility is given for each fluid at a temperature equal to 0.9 times the normal boiling point of that fluid. For those substances in which temperature studies of the mobilities are available and in which the mobilities obey an Arrhenius expression of the form of Eq. (1), the pre-exponential factor μ_0 and activation energy E are also given in Table II. The high sensitivity of the electron mobility to the geometric symmetry of the molecule is apparent in this table. As remarked by Dodelet and Freeman [4], the mobilities appear to decrease with increasing anisotropy α^a of the polarizability of the molecules of the fluid. The correlation, however, is weak in the sense that in going from 2-2-4-4-tetramethylpentance for which $\alpha^a \equiv 0$ to 2-2-5-5-tetramethylhexane for which $\alpha^a = 0.7$ Å^3, the mobility

Table II. Typical values of electron mobilities of several hydrocarbon liquids. Where the mobility is known as a function of temperature the values of μ_0 and E are determined by a least-squares fit of the data to the expression $\mu = \mu_0\, e^{-E/RT}$. Using this expression the mobility in a given liquid is predicted at a temperature corresponding to 0.9 times the boiling point of the liquid. The mobility data entries with asterisks are not known as a function of temperature and are given at the temperature of the experiment. The isotropic $\bar{\alpha}$ and anisotropic α^a polarizabilities entered in the table were calculated in Ref. [4] except for the n-hexane values which the authors estimated

Liquid	$\mu(T)$ cm^2/V sec	T (°K)	μ_0 cm^2/V sec	E kcal/mole	$\bar{\alpha}$ Å^3	α^a Å^3
Methane	300*[a]	120				
Tetramethyl silane	90*[b]	296				
Neopentane	55*[b]	296			9.8	0.0
	67*[c]	296		$\leqq 0.5$		
	50*[d]	296		$\leqq 0.5$		
2-2-4-4-tetramethyl pentane	37.5[d]	347	308	1.48	17.0	0.0
2-2-5-5-tetramethyl hexane	17.4[d]	371	78	1.1	18.8	0.7
2-2-dimethyl butane	12.6[b]	290	84	1.2	11.6	0.7
2-2-3-3-tetramethyl pentane	8.5[d]	373	62	1.46		
2-methyl butene-2	2.4[e]	281	90	2.0		
2-methyl propane	1.42[a]	236	$\leqq 860$	$\leqq 2.0$		
Cyclopentane	1.1*[b]	296				
Cyclohexane	0.67[d]	329	78	3.0		
	0.38*[b] (0.48[d])	296				
Benzene	0.6*[e]	298				
Toluene	0.54*[c]	298				
Propane	0.4[f]	208	210	2.6	6.2	0.9
n-butane	0.4*[b]	296				
n-pentane	0.075*[e]	298				
	0.16*[b]	296				
n-hexane	0.09[e]	307	72	4.06	11.8	~ 1
	0.118[b]	307	319	4.82		

Key to sources of mobility data: a, b, c, d, e and f = Ref. [3, 2, 7, 4, 1 and 12].

Table III. Comparison of electron mobilities in several fluids. T_b denotes the normal boiling point of fluid

Liquid	T °K	μ cm^2/V sec	$\dfrac{T}{T_b}$
Argon	90	~475	1.034
Methane	120	300	1.075
Neopentane	305	72	1.075
Isobutane	283	4.2	1.075
Propane	248	1.1	1.075
Butane	296	0.4	1.084
n-hexane	367	0.27	1.075

is reduced by about a factor of two, while the mobilities of 2-2-5-5-TMH and n-hexane with $\alpha^a \sim 1 \, \text{Å}^3$ differ by a factor of over 100. We believe a more important correlation may be a decreasing mobility with an increasing availability of $-CH_2-$ groups whose local dipole helps, through orientation fluctuations, to create the electronic localization states. Data given in Table III illustrate the trend of mobility with $-CH_2-$ groups. We shall discuss further the possible role these groups play in trapping the electron in a later section of the talk. In this section, we have tried to review the data with minimum speculation about the mechanism of electron transport in nonpolar liquid hydrocarbons. The following section will be devoted to such speculation.

III. Quasi-Free Electron Conduction

Quasi-free electron conduction in liquids argon, krypton, and xenon has been adequately understood in terms of the Cohen-Lekner theory [10, 11] in which the electron is assumed to obey a Boltzmann equation in which the many-body collision cross section is derived from coherent single scattering events. The Cohen-Lekner theory as originally developed neglected incoherent scattering contributions arising from the internal degrees of freedom of polyatomic molecules. Davis, Schmidt, and Minday [19] extended the Cohen-Lekner theory to account for these incoherent contributions. Their extension will be used in the present discussion of quasi-free electron conduction. The version given here is slightly more general than the original MSD work.

Under the single-scatterer approximation the scattering amplitude $F_{l|i}$ for the scattering by a many-molecule system, of a plane wave electron from the momentum state $\hbar k_0$ through a solid angle Ω fixed in the element of solid angle $d\Omega = \sin \Theta \, d\Theta \, d\Phi$ and into the momentum state $\hbar k$ is of the form

$$F_{l|i} = \sum_{v=1}^{N} F_{l_v|i_v} e^{i\mathbf{K} \cdot \mathbf{r}_v}, \tag{4}$$

where $F_{l_v|i_v}$ is the scattering amplitude of the vth molecule which goes from state i_v to l_v in scattering the plane wave electron from state $\mathbf{k}_0$ into state $\mathbf{k}$. The momentum vector $\mathbf{K}$ is by definition $\mathbf{K} = \mathbf{k}_0 - \mathbf{k}$. The cross section per scatterer per unit solid angle for the $i \rightarrow l$ transition is

$$\left(\frac{d\sigma}{d\Omega} \right)_{l|i} = \frac{k}{k_0 N} |F_{l|i}|^2, \tag{5}$$

so that the total cross section per scatterer per unit solid angle arising from energy conserving transitions of a system initially in state i is

$$\left(\frac{d\sigma}{d\Omega} \right)_i = \sum_l \int d(\hbar \omega) \, \delta(E_i - E_l - \hbar \omega) \left(\frac{d\sigma}{d\Omega} \right)_{l|i}. \tag{6}$$

Here $\hbar \omega (= \hbar^2 k_0^2/2m - \hbar^2 k^2/2m)$ is the energy lost by the electron to the system. From Eq. (6) it follows that the differential cross section corresponding to the state i of the system is

$$\left(\frac{d^2\sigma}{d\Omega\,d\hbar\omega}\right)_i = \sum_l \delta(E_i - E_l - \hbar\omega)\left(\frac{d\sigma}{d\Omega}\right)_{l|i}$$

$$\equiv (2\pi\hbar)^{-1} \sum_l \int_{-\infty}^{\infty} dt\, e^{-i(\hbar\omega - E_i + E_l)t/\hbar}\left(\frac{d\sigma}{d\Omega}\right)_{l|i},$$

(7)

where the second equality arising from a representation of the Dirac delta function. If H_0 denotes the Hamiltonian of the system (excluding the electron) and $|j\rangle$ the eigenfunction of H_0 corresponding to the eigenvalue E_j, then Eq. (7) can be rewritten in the form

$$\left(\frac{d^2\sigma}{d\Omega\,d\hbar\omega}\right)_i = \frac{k}{2\pi\hbar k_0 N} \sum_{\nu=1}^{N} \sum_{\nu'=1}^{N} \int_{-\infty}^{\infty} dt\, e^{-i\omega t}$$

$$\cdot \langle i|\hat{F}_\nu(t)\, e^{\,i\boldsymbol{K}\cdot\boldsymbol{r}_\nu(t)}\, \hat{F}_{\nu'}^*(0)\, e^{-i\boldsymbol{K}\cdot\boldsymbol{r}_{\nu'}(0)}|i\rangle,$$

(8)

where $\hat{F}_\nu$ is a scattering amplitude operator whose matrix elements are defined by the relation

$$\langle l|\hat{F}_\nu|i\rangle = F_{l_\nu|i_\nu} \prod_{\alpha \neq \nu}^{N} \delta_{l_\alpha, i_\alpha}.$$

(9)

$\hat{F}^*$ is the Hermitian conjugate of $\hat{F}$. In the spirit of the single scatter approximation, Eq. (9) supposes one can describe the state of the system in terms of individual molecule quantum numbers. The Heisenberg operator notation has been used in Eq. (8), i.e., time dependent operators are defined as follows:

$$\hat{F}_\nu(t)\, e^{\,i\boldsymbol{K}\cdot\boldsymbol{r}_\nu(t)} \equiv e^{\,iH_0 t/\hbar}\, \hat{F}_\nu\, e^{\,i\boldsymbol{K}\cdot\boldsymbol{r}_\nu}\, e^{-iH_0 t/\hbar}.$$

(10)

Finally, in a system at thermal equilibrium the appropriate differential cross section is the following canonical ensemble average of Eq. (8):

$$\frac{d^2\sigma}{d\Omega\,d\hbar\omega} = \frac{k}{2\pi\hbar k_0 N} \sum_{\nu=1}^{N} \sum_{\nu'=1}^{N} \int_{-\infty}^{\infty} dt\, e^{-i\omega t}$$

$$\cdot \langle \hat{F}_\nu(t)\, e^{\,i\boldsymbol{K}\cdot\boldsymbol{r}_\nu(t)}\, \hat{F}_{\nu'}^*(0)\, e^{-i\boldsymbol{K}\cdot\boldsymbol{r}_{\nu'}(0)}\rangle_T,$$

(11)

where

$$\langle\alpha\rangle_T \equiv \sum_i e^{-E_i/kT}\frac{\langle i|\alpha|i\rangle}{\sum_i e^{-E_i/kT}} \equiv \frac{\mathrm{Tr}\{e^{-H_0/kT}\alpha\}}{\mathrm{Tr}\{e^{-H_0/kT}\}}.$$

(12)

If the scattering amplitudes $F_{l_\nu|i_\nu}$ are independent of the internal state of the molecules, then the operators $\hat{F}_\nu$ may be replaced in Eq. (11) by numbers which may be taken out of the ensemble average. In this case — which is the case treated by Cohen and Lekner — the differential cross section is proportional to the van Hove spectral function [20] $S(\boldsymbol{K}, \omega)$, which represents the 4-dimensional Fourier transform of the space-time correlation function $G(\boldsymbol{r}, t)$. In a molecular system $G(\boldsymbol{r}, t)$ represents (clasically) the probability density that the center of mass of a molecule will be at position $\boldsymbol{r}$ at time t if the center of mass of a molecule was at $\boldsymbol{r} = 0$ at time $t = 0$. In the more general case treated here, $\hat{F}_\nu$ depends on the internal coordinates (vibrational and rotational coordinates, e.g.) of the

v^{th} molecule. Thus, the cross section expressed by Eq. (11) depends on the time correlations of internal as well as center of mass molecular coordinates. This can give rise to incoherent scattering. An easy physical picture of this phenomenon is that of a system of rigid rotators which rotate with a period that is long com-compared to the time it takes an electron to enter and leave the influence of the electron rotators potential of interaction. In this case the electron "sees" what appears to be a system of different molecules (since a different orientation will give a different electron-rotator potential) which will interfere with the coherence of the electron wave scattered from the intermolecular correlation of the centers of mass of the molecules. Explicit examples of this phenomenon are discussed by DSM [19]. Impurity scattering and scattering amplitude fluctuations as treated by Lekner [21] give rise to similar incoherence effects.

The steady state Boltzmann equation for the distribution function f of an electron obeying Eq. (11) and in the presence of an electric field E is [10, 19]

$$c\boldsymbol{E} \cdot \frac{\partial f}{\partial \boldsymbol{p}} = n \int d\Omega \, d(\hbar\omega) \frac{p'}{m} \mathcal{H}(\boldsymbol{K}, \omega) \left[e^{-\beta\hbar\psi} f(\boldsymbol{p}') - f(\boldsymbol{p}) \right], \tag{13}$$

where $\boldsymbol{p}' \equiv \hbar \mathbf{k}$, $\mathrm{p} \equiv \hbar \boldsymbol{k}_0$, and

$$\mathcal{H}(\boldsymbol{K}, \omega) \equiv \frac{p}{p'} \frac{d^2\sigma}{d\Omega \, d(\hbar\omega)}. \tag{14}$$

The quantity $f(\boldsymbol{p}) \, d^3\boldsymbol{p}$ is the probability that an electron will be located in a momentum state between $\boldsymbol{p}$ and $\boldsymbol{p} + d\boldsymbol{p}$. n is the number density of the molecules of the system. In terms of f the drift velocity is defined by the expression

$$\boldsymbol{v}_{\mathrm{d}} = \int f(\boldsymbol{p}) \frac{\boldsymbol{p}}{m} \, d^3\boldsymbol{p}. \tag{15}$$

To a good approximation the low field solution of Eq. (13) is

$$f = f_0 + \frac{\boldsymbol{E} \cdot \boldsymbol{p} f_1}{E p}, \tag{16}$$

where

$$f_0 = \left(\frac{1}{2\pi k T} \right)^{3/2} \exp(-\beta p^2 / 2m)$$

and

$$f_1 = \frac{e E \Lambda_1}{k T} f_0 \tag{17}$$

with

$$\Lambda_1^{-1} = \frac{2\pi n}{N} \sum_{v=1}^{N} \sum_{v'=1}^{N} \int d\Theta \, \sin\Theta \, (1 - \cos\Theta)$$
$$\cdot \left\langle \hat{F}_v(0) \, e^{i\boldsymbol{K}_0 \cdot \boldsymbol{r}_v(0)} \, \hat{F}_{v'}^*(0) \, e^{-i\boldsymbol{K}_0 \cdot \boldsymbol{r}_{v'}(0)} \right\rangle_T, \tag{18}$$

where $\boldsymbol{K}_0 = \boldsymbol{K}|_{k=k_0}$. The drift velocity computed from Eq. (16) is

$$\boldsymbol{v}_{\mathrm{d}} = e\boldsymbol{E}/3mkT \int \Lambda_1 \, p f_0 \, d^3\boldsymbol{p}$$

giving the following low field quasi-free electron mobility:

$$\mu^f = \frac{e}{3mkT} \int A_1 \, p f_0 \, \mathrm{d}^3 \boldsymbol{p} \, . \tag{19}$$

By further analysis of Eq. (13) we can also show that the reduced parameter

$$b = \frac{1}{3} \frac{(e \overline{A}_0 E)(e \overline{A}_1 E)}{\overline{\lambda}(kT)^2} \tag{20}$$

approximately characterizes the onset of field dependent mobilities. The quantity $\overline{\lambda}$ represents the mean fractional energy loss per electron molecule collision (for a monatomic system with no internal degrees of freedom $\overline{\lambda} \approx 2m/M$, where m and M are the electron and molecular masses, respectively). The definitions of A_0 and λ are

$$A_0^{-1} = 2\pi n \int \mathrm{d}\Theta \, \sin\Theta \, (1 - \cos\Theta) \, \langle |\hat{F}_1(0)|^2 \rangle_T \tag{21}$$

and

$$\lambda = \frac{n A_0}{N(p^2/2m)} \sum_{v=1}^{N} \sum_{v'=1}^{N} 2\pi \int \mathrm{d}\Theta \, \sin\Theta \tag{22}$$

$$\cdot \langle [H_0, \hat{F}_v(0) \, e^{i\boldsymbol{K}_0 \cdot \boldsymbol{r}_v(0)}] \, \hat{F}_v^*(0) \, e^{-i\boldsymbol{K}_0 \cdot \boldsymbol{r}_{v'}(0)} \rangle_T$$

where $[A, B]$ denotes the commutator and A and B. The meaning of the bar on λ, A_0 and A_1 is that the momentum of the electron appearing in these quantities is to be approximated by the thermal momentum $p_{th} = \sqrt{8mkT/\pi}$.

In the estimates we shall make of μ^f in the next section we shall considerably simplify the expressions for A_1, A_0 and λ. First we shall asume that the Hamiltonian describing the center of mass motion of the molecules commutes with the Hamiltonian describing the internal motions of the molecules and that the internal motions (rotations and vibrations) of the molecules are independent. The scattering amplitude operator can be written in the form

$$\hat{F}_v = a + \Delta a(I_v) + \hat{F}_v^{\mathrm{In}} \, , \tag{23}$$

where a is the mean elastic scattering length defined by $a = \langle \hat{F}_v \rangle_T$; $\Delta a(I_v)$ is the angular dependent elastic scattering length which is a function of those internal coordinates I_v (such as Euler angles of orientation of the molecule) which do not change appreciably during an electron-molecule collision; and $\hat{F}_v^{\mathrm{In}}$ represents inelastic electron-molecule collisions. An example of $\Delta a(I)$ could be the scattering amplitude of an electron beam scattered by a rigid ellipsoid as a function of the orientation of the ellipsoid relative to the initial direction of the electron beam. An example of $\hat{F}_v^{\mathrm{In}}$ is the scattering amplitude of an electron inducing a rotational or vibrational transition of the target molecule. Under the assumption of separability of the internal Hamiltonian of the molecules and with the division of $\hat{F}_v$ given in Eq. (23), we obtain

$$A_1^{-1} \simeq 4\pi a^2 n \, S(0) + 2\pi n \int \mathrm{d}\Theta \, \sin\Theta \, (1 - \cos\Theta) \, \langle (\Delta a(I_1) + \hat{F}_1)(\Delta a(I_1) + \hat{F}_1^*) \rangle_T \tag{24}$$

where $S(0)$ ($= nkT\varkappa_T$, with $\varkappa_T$ the isothermal compressibility) is the low wave-vector (small $\boldsymbol{K}_0$) limit of the X-ray structure factor. The quantity $4\pi a^2 S(0)$ is the coherent scattering cross section obtained by Cohen and Lekner [11], while the second term on the *rhs* of Eq. (24) is an incoherent contribution characteristic of polyatomic systems with internal molecular motions whose periods are long compared to electron-molecule collision times. Λ_0^{-1} can be obtained by setting $S(0) \equiv 1$ in Eq. (24) and λ can be shown to be roughly approximated as $\lambda \approx 2m/M + \lambda^{\text{In}}$, where λ^{In} is the mean fractional electron energy loss per inelastic collision.

In obtaining the above results we have neglected multiple scattering effects. As has been noted by Lekner [11], partial correction for multiple scattering can be provided by viewing the electron as an excitation with an effective mass in a conduction band and by computing the electron-molecule scattering length from a potential determined self-consistently for the state of the system. Unfortunately, we do not know the effective mass of an electron in liquid hydrocarbons so we shall ignore this correction in our discussion.

IV. Discussion

In concluding this lecture let us pursue, mostly qualitatively, the idea that the mechanism of elctron transport in hydrocarbons is quasi-free conduction between short-lived traps. The major problem in applying the quasi-free mobility theory developed in the preceding section is the computation of electron-molecule scattering amplitudes. Although such amplitudes are not presently available, we can estimate them indirectly for very symmetrical molecules — such as methane, neopentane and tetramethyl silane — in which Δa and $\hat{F}_v^{\text{In}}$ are expected to give very small contributions to Λ_1^{-1} in the thermal electron energy range. One should keep in mind, however, that even when $\hat{F}_v^{\text{In}}$ gives a small contribution to Λ_1^{-1} one cannot generally neglect the contribution of $\hat{F}_v^{\text{In}}$ to the fractional energy loss λ, because inelastic scattering is much more efficient than elastic scattering in absorbing the electron energy. For example [22], although for thermal electron scattering by nitrogen, the inelastic electron-quadrupole scattering amplitude is only about 0.09 Å compared to an elastic scattering amplitude of about 0.6 Å, the contribution of the quadrupole scattering to λ is about 50 times the elastic contribution $2m/M$.

The indirect method of estimating low energy elastic scattering amplitudes makes use of work function data and Springett *et al.* [23] theory of the ground state of the conduction band of quasi-free electrons. These investigators combine the Wigner-Seitz boundary condition on an average lattice representation of the liquid with a pseudopotential model in which the electron potential is a sum of the mean-field polarization potential and a hard-core potential representing the repulsive Hartree-Fock pseudopotential. Their final result for the ground state energy, V_0, of a quasi-free electron is

$$V_0 = T_0 + U_{\text{p}}, \tag{25}$$

with

$$U_{\text{p}} = -\frac{3\bar{\alpha}e^2}{2r_{\text{s}}^4} \left[\tfrac{8}{7} + (1 + \tfrac{8}{3}\pi\bar{\alpha}n)^{-1} \right] \tag{26}$$

$$T_0 = \hbar^2 \varkappa_0^2 / 2m, \tag{27}$$

where r_s and $\varkappa_0$ are determined by the relations

$$\tfrac{4}{3}\pi r_s^3 n = 1 \tag{28}$$

and

$$\tan\varkappa_0(r_s - \tilde{a}) = \varkappa_0 r_s. \tag{29}$$

$\bar{\alpha}$ denotes the mean (isotropic) polarizability and $\tilde{\alpha}$ the mean hard sphere cut-off length. If there are no localization states (which we are assuming to be the case for methane, neopentane and tetramethyl silane) then the work functions measured by Holroyd and Allen will equal V_0. Using Eq. (25) and the measured values of V_0 we calculate the values for $\tilde{a}$ shown in Table I. On the basis of Lekner's [11] self-consistent electron-atom cross section calculations for liquid argon, we expect the polarization potential contributions to cancel out so that the scattering amplitude used in computing Λ_1^{-1} for electrons in neopentane may be approximated as [24] $\hat{F}_v \simeq \tilde{a}$; the quantities Δa and $\hat{F}_v^{\text{ln}}$ are assumed to give negligible contribution to Λ_1^{-1}. Using this approximation for $\hat{F}$ in Eq. (24) for Λ_1^{-1} and computing μ^f from Eq. (19) for saturated liquid at 300 °K, we predict for neo pentane the value $\mu^f \simeq 93$ cm^2/volt sec in good agreement with the experimental value $\mu_{\text{exp}} \simeq 72$ cm^2/volt sec. Similar agreement may be obtained between μ^f and μ_{exp} for tetramethyl silane by the same procedure.

Determining from liquid argon data [9] the value of b, Eq. (20), for which the drift velocity begins to become nonlinear in the field (this occurs at $E \sim 500$ V/cm in Ar) and estimating λ by assuming that the electrons collides inelastically with the local quadrupole moment associated with the freely rotating CH_3 groups on the neopentane molecule, we estimate that the electron drift velocity in neopentane should be linear in the field up to $E \sim 30000$ V/cm, consistent with experiment which has been done up to $E = 15000$ V/cm.

Because we presently have no good estimates of Δa and $\hat{F}_v^{\text{ln}}$ for the unsymmetrical hydrocarbon molecules and have no work function data for methane, we shall not attempt to calculate μ^f for any other liquids [25]. However, we can examine further the qualitative aspects of the traps that may interrupt the quasi-free motion of the electron. If the quasi-free electron is assumed to be trapped with a frequency v and to remain trapped in an immobile trap for a time τ, then the overall mobility of the electron will be [1]

$$\mu = \frac{\mu_f}{1 + v\tau}. \tag{30}$$

The trapping frequency could, for example, be the frequency of a resonance collision which localizes the electron long enough for it to relax into its trap. If the trap is not too deep, then the electron may be thermally promoted back into the conduction band so that the lifetime is given by $\tau = \tau_0 e^{E/RT}$, where $\tau_0 \simeq 10^{-13}$ sec is the characteristic phonon time and the activation energy E is the difference between the energies of the localization state and the quasi-free state. For sufficiently large E, we expect $v\tau_0 e^{E/RT} \gg 1$ and $\mu \simeq (\mu \simeq (\mu_f/v\tau_0) e^{-E/RT}$ in qualitative agreement with experimentation on unsymmetrical nonpolar hydrocarbons. From the observed activation energies of propane and n-hexane, we estimate the lifetimes $\tau_p(T = 200 \,°K)$ and $\tau_h(T = 300 \,°K)$ to be 6.7×10^{-11} and 7.8×10^{-11} sec, respectively.

A possible model for the electron trap is the cavity model introduced by Cope-land, Kestner, and Jortner [26] in describing solvated electrons in polar solvents — the only difference being that the long-ranged Landau potential is absent in nonpolar fluids. We assume that the electron is trapped in the cavity shown in Fig. 6 whose potential energy is of the form

$$V(r) = -\frac{N\bar{\alpha}e^2}{2r_d^4} - \frac{ND_{local}e}{r_d^2}, \quad r < R \tag{31}$$

$$= V_0 \quad , \quad r > R.$$

This is a square well potential which may lead to a bound state as shown in Fig. 6. N represents the coordination number of the trapped electron, $\bar{\alpha}$ the isotropic

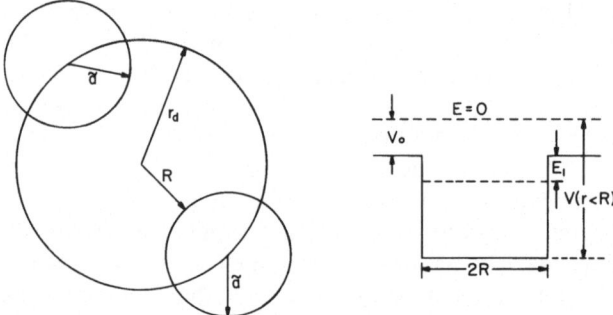

Fig. 6. Schematics illustrating the solvated cavity in which the electron is trapped and the square well potential and bound state E_1 associated with the cavity

molecular polarizability (we neglect the anisotropic contribution), r_d the distance from the center of the cavity to the center of mass of the nearest neighbor molecules, and $R = r_d - \tilde{a}$. The quantity D_{local} denotes a local dipole moment arising from CH_2 groups which may be intramolecular rotations which add up to a non-zero dipole moment of the solvent molecule in the coordination sphere. Estimating $r_d \approx 4\,\text{Å}$ for *n*-hexane and equating the energy E_1 shown in Fig. 6 to the activation energy $E = 4.06\,\text{kcal/mole}$ for *n*-hexane, we find from the solution of the Schrodinger equation for a particle in a square well that the potential energy $V(r)$, $r < R$, must be equal to $-2.4\,\text{eV}$. Assuming next that $N = 6$ and taking the *n*-hexane value of $\bar{\alpha} (= 11.8\,\text{Å}^3)$, we see that D_{local} must be 0.78 Debye to provide the trap. A single C–H bond has a dipole moment of the order of 0.4 Debye so that our required D_{local} does not seem to be an impossibility.

The preceding considerations are in the nature of plausibility arguments and do not, of course, prove that the motion of the electron involves quasi-free conduction interrupted by short-lived traps. If the motion involves phonon-assisted hopping from one localized state to another, we might expect the mobility to be of the form [13, 14]

$$\mu = \frac{eL^2}{\tau_0 kT}\, e^{-E/RT}, \tag{32}$$

where L is the mean hopping distance and $\tau_0 \sim 10^{-13}$ sec. Equating $eL^2/\tau_0 kT$ to $70\,\text{cm}^2/\text{V sec}$, the observed pre-exponential mobility of n-hexane, we find $L \simeq 43\,\text{Å}$, a jump distance which seems rather large if the phonon-assisted hopping mechanism is to be operative.

To try to establish definitively the electron transport mechanism in nonpolar hydrocarbons, we are currently pursuing three courses of action. First, we are preparing to measure the Hall mobilities of injected electrons. If the electron moves by quasi-free conduction between short-lived immobile traps, then the Hall mobility will be equal to the quasi-free mobility and will consequently be several orders of magnitude larger than the drift mobility. On the other hand, if the electron moves by phonon-assisted hopping from one localized state to another the mobility will still be activated [13, 15], although the Hall mobility activation energy may only be about 1/3 of the drift mobility activation energy. Thus, the Hall mobility should certainly distinguish between the two mechanisms considered here. Our second approach to the problem will be to do drift velocity experiments in the high-temperature dense gas region where it is expected that trapping will become weak and quasi-free conduction should be the dominant mechanisms of charge transport. This type of experiment should give us a better idea of the nature of the scattering amplitude operator for dense nonpolar hydrocarbon systems. Finally, we are trying to incorporate multiple scattering effects in the quasi-free electron kinetic theory and to develop a uniform description (á la Kubo-Greenwood [27]) of conduction in which quasi-free and localized contributions to the mobility enter as thermal averages of the density-of-states weighted mobilities [28]. The theoretical studies necessitate a much more refined statement of the electron-molecule interactions than we have considered in this talk.

Acknowledgements

The authors are grateful to the National Science Foundation and the Petroleum Research Fund for financial support for this work.

References

1. Minday, R. M., Schmidt, L. D., Davis, H. T.: J. Chem. Phys. **50**, 1473 (1969); **54**, 3112 (1971).
2. Schmidt, W. F., Allen, A. O.: J. Chem. Phys. **50**, 5037 (1969); **52**, 4788 (1970).
3. Fuochi, P. G., Freeman, G. R.: J. Chem. Phys. **56**, 2333 (1972).
4. Dodelet, J.-P., Freeman, G. R.: Can. J. Chem. (in press).
5. Beck, G., Thomas, J. K.: Dept. Chem., University of Notre Dame (private communication to be published).
6. Tewari, P. H., Freeman, G. R.: J. Chem. Phys. **49**, 4394 (1968).
7. Minday, R. M., Schmidt, L. D., Davis, H. T.: J. Phys. Chem. **76**, 442 (1972).
8. Schnyders, H., Rice, S. A., Meyer, L.: Phys. Rev. **150**, 127 (1966).
9. Miller, L. S., Howe, S., Spear, W. E.: Phys. Rev. **166**, 871 (1968).
10. Cohen, M., Lekner, J.: Phys. Rev. **158**, 305 (1967).
11. Lekner, J.: Phys. Rev. **158**, 130 (1967).
12. Data in Fig. 2 are from Ref. [1], except the propane data used to construct the propane curve were recently obtained by one of us (RGB).
13. Cutler, M., Mott, N. F.: Phys. Rev. **181**, 1336 (1969).

14. Holstein, T.: Ann. Phys. (N.Y.) **8**, 325, 343 (1959).
15. Holstein, T., Friedman, L.: Phys. Rev. **165**, 1019 (1968).
16. A good review of the theory of amorphous semiconductors has been written by Cohen, M. H.: In: J. Non-Crystalline Solids **4**, 391 (1970).
17. Minday, R. M., Schmidt, L. D., Davis, H. T.: Phys. Rev. Lett. **26**, 360 (1971).
18. Holroyd, R. A., Allen, M.: J. Chem. Phys. **54**, 5014 (1971).
19. Davis, H. T., Schmidt, L. D., Minday, R. M.: Phys. Rev. **3**, 1027 (1971).
20. Van Hove, L.: Phys. Rev. **95**, 249 (1954).
21. Lekner, J.: Phys. Letters **27** A, 341 (1968); Phil. Mag. **18**, 1281 (1968).
22. Gerjuoy, E., Stein, S.: Phys. Rev. **97**, 1671 (1955).
23. Springett, B. E., Jortner, J., Cohen, M. H.: J. Chem. Phys. **48**, 2720 (1968).
24. This estimate of the scattering length was used by Fueki, K., Feng, D.-F., Kevan, L.: Chem. Phys. Letters (to appear) in computing quasi-free electron mobilities from the work function data of Holroyd and Allen with the aid of the Cohen-Lekner mobility formula. Davis, H. T., Schmidt, L. D., Minday, R. M.: Chem. Phys. Letters **13**, 413 (1972) also suggested that this estimate was a reasonable approximation.
25. Estimates of μ^f for the hydrocarbons in Table 1 are given by Fueki et al. in their paper mentioned in Ref. [24]. They estimate Λ_1 by assuming $\hat{F}_v \approx \tilde{a}$.
26. Copeland, D. A., Kestner, N. R., Jortner, J.: J. Chem. Phys. **53**, 1189 (1970).
27. Kubo, R.: Can. J. Phys. **34**, 1274 (1956).
28. The theory of amorphous systems as outlined by Cohen, M. H., in his lecture to this colloquium and in Ref. [16] should prove useful in ultimately describing the range of behavior observed in hydrocarbon systems.

Discussion

M. J. SIENKO   I should like to offer an experimental suggestion. You may be able to test out your ideas of electron trapping due to orientation fluctuations by studying the perfluorocarbons. They are also nonpolar but have a bigger bond dipole moment and are much stiffer to conformational deformation.

P. DELAHAY   There is a similarity between your experiment in which you have emission in the gas phase and thermionic emission by bona fide solvated electrons in hexamethyl phosphoric triamide which I mentioned in my talk.
I have a question: How does photoelectron injection, which you used, compare with the use of a tunnel diode as reported by Marvin Silver? Surely, this is an experimental point; we can perhaps discuss the matter privately.

H. T. DAVIS   I think the tunnel diode method is troubled more with impurity problems and capable of lower operating currents then the photoelectron injection technique.

P. DELAHAY   With reference to Dr. Raz' suggestion of determining EDC's, the relatively high vapor pressure of the compounds you used may prevent such determinations.

J. JORTNER   I would like to attempt to present a new picture for electron localization in liquid hydrocarbons and in hydrocarbon glasses. The interpretation of the mobility data implies electron binding energies ~ 0.15 eV while the optical absorption spectra imply that the binding energy is much higher (~ 0.6 eV). It is reasonable to suggest that these systems contain a distribution of electron traps of varying energies. The deep traps will substantially contribute to the optical spectra (in the energy region studied so far) while the shallow traps will contribute to the conductivity. Now one can attempt to apply the

Anderson model for localization of an excess electron in hydrocarbons. When the ratio between the potential fluctuation ΔV and the mean value $\langle V \rangle$ of the potential exceeds a magic number $\Delta V/\langle V \rangle > 5.5$ localization will occur. Such localization mechanism may be operating in liquid hydrocarbons where rotational and radial fluctuations may result in a wide distribution of traps.

H. T. DAVIS   There certainly must be significant fluctuations in the potential due to internal molecular motions, and these could give rise to Anderson localization. Experimentally, the Hall mobility determined as a function of temperature should establish whether or not the Anderson mechanism maintains. We are attempting the Hall mobility experiment presently.

Excitons in Liquids and the Energy of the Quasi-Free Electron State

BARUCH RAZ and JOSHUA JORTNER

Abstract

We review the available spectroscopic information concerning Wannier impurity states originating from an atomic and a molecular positive ion core in solid and in liquid rare gases. The impurity band gap in these dense media yields a spectroscopic value for the energy of the quasi-free electron state, which in turn yields direct information concerning electron localization in liquid rare gases.

The energetically stable ground state of an excess electron in an insulating nonpolar liquid can correspond to one of the following two extreme physical situations [1]: a) A quasi-free electron state in the conduction band of the liquid, where the excess electron does not perturb the liquid structure; b) a localized excess electron in a bubble, whereupon a large modification of the liquid structure results in a cavity formation. The nature of the excess electron state in any particular nonpolar liquid is determined by the energy, V_0, of the quasi-free electron state (or rather the energy of the botton of the conduction band) in that system [2, 3]. From the study of Springett et al. [3] the localization criteria for an excess electron in a nonpolar liquid can be summarized as follows:

a) The quasi-free electron state is favored provided that

$$V_0 < 0 \tag{1a}$$

or alternatively,

$$V_0 > 0 \quad \text{and} \quad \frac{2\pi\hbar^2\gamma}{mV_0^2} \geq 0.047 . \tag{1b}$$

b) The localized excess electron state is energetically stable when

$$V_0 > 0 \quad \text{and} \quad \frac{2\pi\hbar^2\gamma}{mV_0^2} \leq 0.047 \tag{2}$$

where γ represents the surface tension of the liquid. Information concerning V_0 values in liquid rare gases was obtained in some cases from adiabatic electron injection experiments [4] and theoretical estimates [3]. In this paper we shall demonstrate how spectroscopic data for large radius Wannier impurity states in solid and liquid insulators can be utilized to obtain V_0 in simple liquids, and how these data pertain to the problem of electron localization in dense fluids.

We shall consider the one-electron energy level diagram for the excited states of an atomic or molecular impurity in a solid (Fig. 1). When the conduction band is parabolic, the extravalence highly excited impurity levels bear no relation to the extravalence excitations of the atom or the isolated impurity in the low pressure

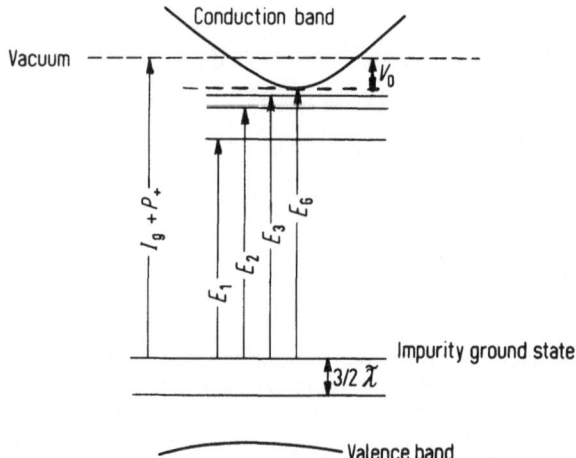

Fig. 1. Energy level diagram for an impurity in an insulating solid where the conduction band is parabolic

gas phase [5—8]. Stated in different terms these excited states cannot be described in terms of the tight binding approximation. The Wannier model [9] is applicable to handle these states. When the host conduction band is parabolic, the highly excited impurity levels, E_n, are given by a hydrogenic series [9]

$$E_n = E_G - G/n^2 \tag{3}$$

where the excitation binding energy, G, is given by the modified Rydberg formula.

$$G = 13.6 \, m^*/\varkappa^2 \text{ eV} \tag{4}$$

where n is an integer, m^* is the electron effective mass in the conduction band and $\varkappa$ is the dielectric constant of the medium. The excitonic series converges to the bottom of the conduction band (see Fig. 1). E_G represents the impurity energy gap, or stated in simpler terms it corresponds just to the ionization potential of the impurity in the solid. This cardinal term for an impurity state is given by [6]

$$E_G = I_g + P_+ + V_0 \tag{5}$$

where I_g is the impurity gas phase ionization potential while P_+ represents the electrostatic polarization energy of the solid by the positive ion core. Reliable theoretical upper limits for P_+ in solid rare gases are available [10].

Up to this point we have been concerned with impurity states. In pure insulators such as solid rare gases, an exciton series is observed resulting from excitations of an electron from the valence band to bound states located below the conduction band [9]. The Wannier exciton series are given by Eq. (3) except that the effective mass in Eq. (4) corresponds now to the reduced mass of the electron in the conduction band and of the hole in the valence band, and the exciton band gap E_G is modified for the pure solid being given by [11]

$$E_G = I_g + P_+ + V_0 + E_V \tag{6}$$

where E_V is half the width of the valence band. For solid Xe $E_V = -1.0$ eV while for solid Ar $E_V \approx 0.1$ eV and for solid Ne $E_V \approx 0$.

Turning now to a real life situation we present in Figs. 2 and 3 the exciton series in pure solid rare gases obtained from the work of Baldini [12, 13] and of

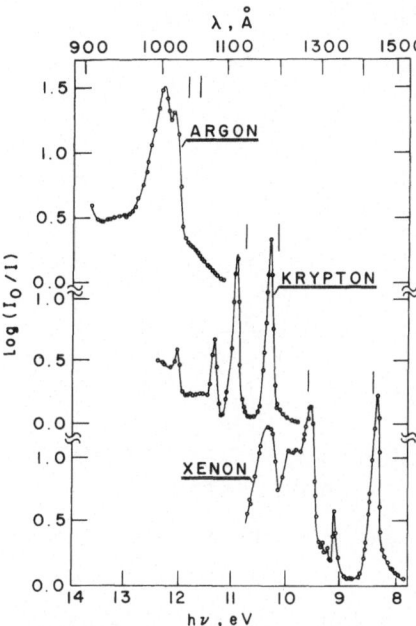

Fig. 2. Wannier exciton states in pure rare gases [Baldini, G.: Phys. Rev. **128**, 1562 (1962)]. The atomic transitions corresponding to the $^1S_0 \rightarrow {}^3P_1$ and $^1S_0 \rightarrow {}^1P_1$ are marked by vertical lines

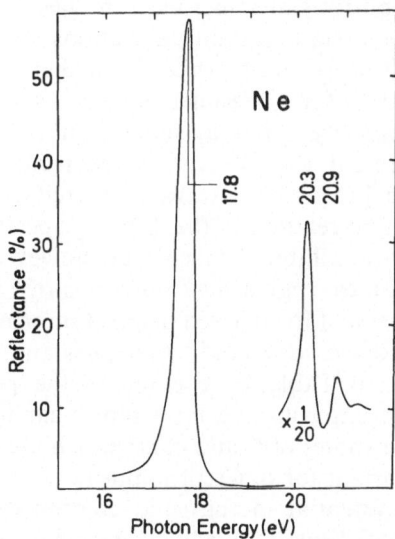

Fig. 3. Wannier exciton states in solid Ne (Ref. [14a])

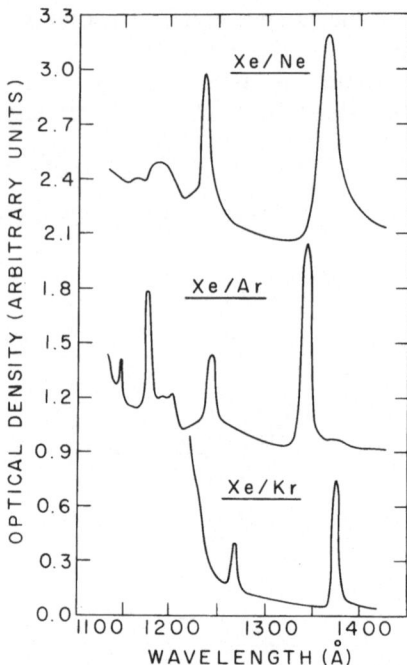

Fig. 4. Wannier impurity states of Xe in solid rare gases [Gedanken, A., Raz, B., Jortner, J.: J. Chem. Phys. (1972)]

Hensel *et al.* [14], while in Fig. 4 we display the absorption spectra of Xe impurity states in solid rare gases first studied by Baldini [13]. Both in the case of exciton and atomic impurity states in solid rare gases the Wannier series is clearly exhibited. The $n = 1$ state exhibits deviations from E_1 calculated by Eq. (3) due to central cell corrections, while the higher $n \geqq 2$ member follows faithfully the energetic pattern for a Wannier series.

The study of Wannier impurity states in solids is not limited to atomic impurities. In Fig. 5 we present the absorption spectrum of the methyliodide molecule in solid krypton [15]. The extravalence excitations of this molecule in the solid rare gas exhibit no relation to the Rydberg spectrum in the gas phase. The Wannier series are clearly exhibited. One should however note the vibrational structure of the Wannier states originating from a positive ion core. From these and other data Gedanken *et al.* [15] derived detailed information concerning the impurity band gaps, the excited state binding energies and the electron effective mass in solid rare gases. In Table I we compare the pertinent information obtained for an atomic Xe impurity and for a molecular impurity (CH_3I). It is comforting to note that the values of G and of m^* which are an intrinsic property of the host matrix are invariant for different impurities.

The impurity ionization potential in the solid combined with the gas phase ionization potential [Eq. (5)] leads to a reliable estimate for $P_+ + V_0$. From the available estimates of P_+ for rare gases reasonable estimates [10] for V_0 in

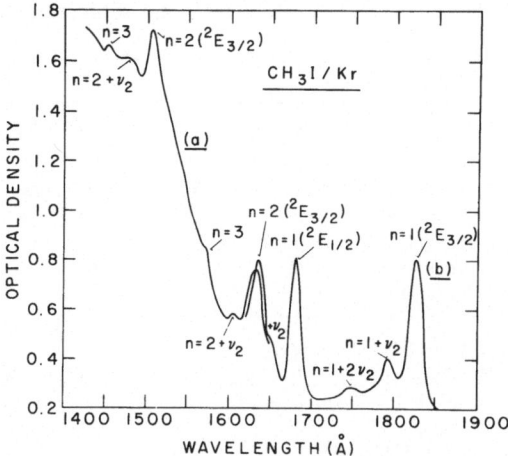

Fig. 5. Absorption spectrum of CH_3I in solid krypton [Gedanken, A., Raz, B., Jortner, J.: J. Chem. Phys. (1972)]. The Wannier states exhibit two spin orbit split series originating from the $^2E_{3/2}$, $^2E_{1/2}$ positive molecular ion cores

Table I. Energetic data for large impurity states of CH_3I in rare gas solids[a]

	Ne matrix	Ar matrix	Kr matrix	Guest molecule
E_G	12.15	10.54	10.40	Xe
G	4.4	2.4	1.72	
m^*/m	0.5	0.5	0.6	$Ig = 12.13\ eV$
$P_+ + V_0$	0.02	-1.59	-1.73	
P_+[b]	-0.58	-1.29	-1.11	
V_0[c]	0.6	-0.3	-0.62	
E_G	9.51	8.58	8.13	CH_3I
G	4.4	2.2	-2.10	
m^*/m	0.5	0.45	0.50	$Ig = 9.49\ eV$
$P_+ + V_0$	0.02	-0.91	-1.36	
V_0[d]	0.60	-0.3	-0.62	
P_+[e]	-0.62	-0.6	-0.74	

[a] All energies in eV.
[b] Fowler, W.: Phys. Rev. **151**, 657 (1966).
[c] Calculated from the P_+ data.
[d] Adopted from Xe impurity data.
[e] Estimated using[d].

solid Ne, Ar, Kr and Xe were obtained (see Table I). Estimates of V_0 from the exciton spectra in pure solid rare gases are summarized in Table II.

Up to this point we have been concerned with the spectroscopic determination of V_0 in solid rare gases. The theoretical treatment of Wannier exciton and impurity states rests on the translational symmetry of the host lattice. Recent theoretical work has established that symmetry restrictions can be relaxed in this context and that Wannier impurity states should be detected in liquid rare

Table II. Energy of the quasi-free electron state in solid rare gases derived from spectroscopic data. All energy values in eV

Source of data	Medium	E_g	I_g	P_+[a]	E_v	V_0
Solid neon[b,c]	Ne (solid)	21.42	21.56	−0.68	0±0.1	+0.54±0.1
Xe impurity in neon[d]	Ne (solid)	12.15[f]	12.13	−0.68	−	+0.66±0.1
Solid argon[e]	Ar (solid)	14.17	15.76	−1.10	0±0.1	−0.49±0.1
Xe impurity in Ar[d]	Ar (solid)	10.54	12.13	−1.10	−	−0.49±0.1
Solid krypton[e]	Kr (solid)	10.67	14.00	−1.10	−0.57	−0.61±0.2[g]
Xe impurity in Kr[d]	Kr (solid)	10.40	12.13	−1.10	−	−0.63±0.2
Solid Xe[e]	Xe (solid)	9.28	12.13	−1.3	−0.95	−0.6 ±0.2[g]

[a] Beall Fowler,W.: Phys. Rev. **151**, 657 (1966).

[b] Boursey,E., Roncin,J.Y., Damany,H.: Phys. Rev. Letters **25**, 1279 (1970).

[c] Hensel,R., Keitel,G., Koch,E.E., Kosuch,N., Skibowsky,M.: Phys. Rev. Letters **25**, 1281 (1970).

[d] Baldini,G.: Phys. Rev. **137**, A508 (1965).

[e] Baldini,G.: Phys. Rev. **128**, 1562 (1962).

[f] Computed according to the assignment of the 10.95 eV transition to $n=2$ state.

[g] Derived [Raz,B., Jortner,J.: Chem. Phys. Letters **4**, 155 (1969)] from X-ray exciton spectrum in solid krypton and xenon [Hensel,R., Keitel,G., Koch,E.E., Skibowsky,M., Schreiber,P.: Phys. Rev. Letters **23**, 1160 (1969)].

gases [16]. We have observed [5, 6] Xe impurity Wannier states in liquid Ar and in liquid Kr. From the comparison of the liquid and solid spectra (see Figs. 6 and 7) we conclude that the absorption spectra of Xe impurity states in liquid rare gases exhibit the $n=1$ and the $n=2$ Wannier states. Conclusive evidence

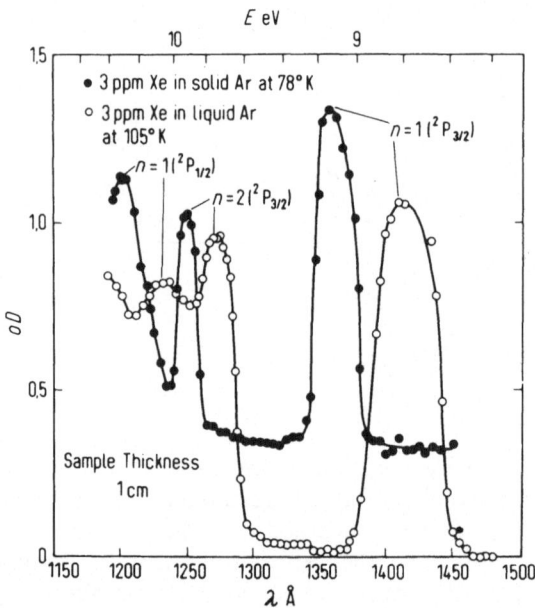

Fig. 6. Absorption spectrum of Xe impurity in solid and in liquid argon [Raz,B., Jortner,J.: Proc. Roy. Soc. A **317**, 113 (1970)]

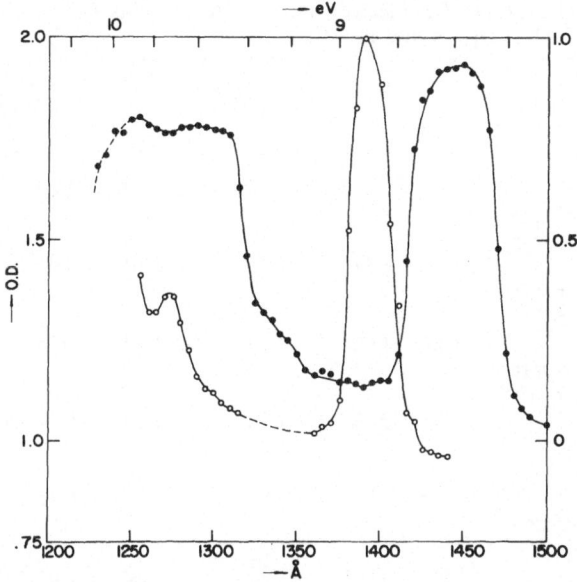

Fig. 7. Absorption spectrum of Xe impurity in solid and in liquid krypton [Raz,B., Jortner,J.: Proc. Roy. Soc. A **317**, 113 (1970)]

for the existence of a long ($n > 2$) Wannier series in liquid rare gases was obtained from the study [15] of the optical spectrum of CH_3I in liquid krypton, where a long Wannier series, up to $n = 4$, was observed. From these results it is apparent that large radius Wannier states are amenable to experimental observation in liquid rare gases (Fig. 8).

Utilizing Eq. (3) we have derived [6, 8] a spectroscopic estimate of V_0 for liquid Ar and for liquid Kr. These results (Table III) indicate that the energy of the

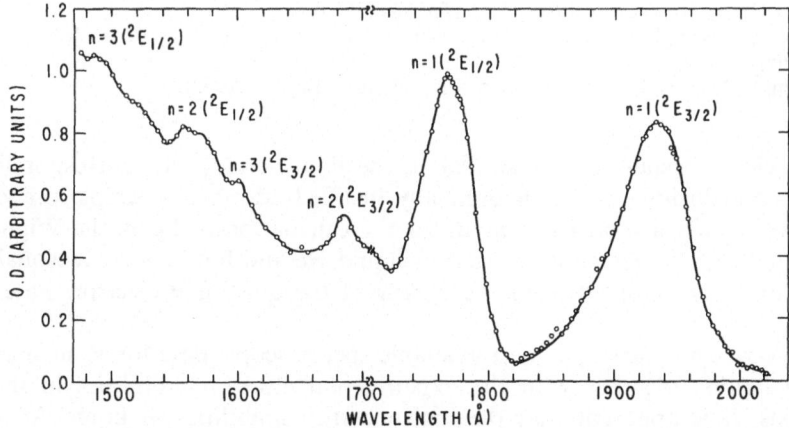

Fig. 8. Absorption spectrum of CH_3I impurity in liquid krypton [Karsch,Z., Raz,B., Jortner,J.: Chem. Phys. Letters (1972)]

Table III. Energy of the quasi-free electron state in simple solids and liquids as derived from spectroscopic data. All energy values in eV

Medium	Source of data for calculating E_G	E_G	I_g	$P_+ + V_0$ Eq. (3)	P_+[f]	V_0 (Spectroscopic)	V_0 (other sources)
Ar (s) 80 °K	$n = 2$ Wannier state in Xe/Ar(s)[a, b]	10.65[a]	12.13	− 1.45	− 1.10	−0.35 ± 0.1	
Ar(l) 105 °K	$n = 2$ Wannier state in Xe/Ar(l)[b]	10.51[a]	12.13	− 1.55	− 1.10	−0.45 ± 0.2	−0.33 + 0.1[g] −0.45[h]
Kr(s) 70 °K	$n = 2$ Wannier state in Xe/Kr(s)[a,b]	10.40[d]	12.13	− 1.73	− 1.10	−0.63 ± 0.1	
Kr(s) 20 °K	X-ray Excitons in Kr(s)[c]	92.2[e]	93.9	− 1.70	− 1.10	−0.60 ± 0.1	
Kr(l) 135 °K	$n = 2$ Wannier state in Xe/Kr(l)[b]	10.25[d]	12.13	− 1.88	− 1.10	−0.78 ± 0.2	
Xe(s) 20 °K	X-ray Excitons in Xe(s)[c]	65.6[e]	67.55	− 1.95	− 1.32	−0.63 ± 0.2	−0.39[i]
$H_2(s)$ 6 °K	Wannier states in Xe/H_2(s)	10.11[j]	12.13	− 2	− 1	− 1	
$D_2(s)$ 6 °K	Wannier states in Xe/D_2(s)	10.20[j]	12.13	− 2	− 1	− 1	

[a] Baldini, G.: Phys. Rev. A **508**, 137 (1965).
[b] Raz, B., Jortner, J.: Proc. Roy. Soc. A **137**, 113 (1970).
[c] Hensel, R., Keitel, G., Koch, E. E., Skibowsky, M., Schreiber, P.: Optics Comm. **2**, 59 (1970).
[d] Estimated from Eq. (1) (see text).
[e] d shell ionization potentials, Bearden, J. A., Burr, A. F.: Rev. Mod. Phys. **39**, 125 (1967).
[f] Fowler, W. B.: Phys. Rev. **151**, 657 (1966) and Ref. j.
[g] Experimental value from electron injection experiments, Halpern, B., Lekner, J., Rice, S. A., Gomer, R.: Phys. Rev. **156**, 351 (1967).
[h] Theoretical value, Lekner, J.: Phys. Rev. **158**, 130 (1967).
[i] Experimental value photoemission data, O'Brien, J. F., Teegarden, K. J.: Phys. Rev. Letters **17**, 919 (1966).
[j] Gedanken, A., Raz, B., Jortner, J.: Chem. Phys. Letters **14**, 326 (1972).

quasi-free electron state is very similar in the liquid and in the corresponding solid. This conclusion can be theoretically justified adopting a simple pseudo-potential (or rather a model potential) for the calculation of V_0 by the Wigner-Seitz method [3, 17]. Thus for the case of liquid He and liquid Ne a reasonable estimate for V_0 is obtained from the energy of the quasi-free electron state in the solids.

In Table IV we summarize the best available spectroscopic data for V_0 in simple insulating liquids together with the experimental electron mobilities in these dense fluids. It is apparent that the high electron mobilities in liquid Ar, Kr and Xe are consistent with the energetic stability condition [Eq. (1a)] of the quasi-free electron state. The case of liquid Ne where $V_0 \approx +0.5$ eV is of particular

Table IV. Electron mobility data in nonpolar liquids

Liquid	Temperature °K	$cm^2 Volt^{-1} sec^{-1}$	$V_0 eV$	Ref.
He³	2.25	4.06×10^{-2}	$+0.9$	a
He⁴	4.2	2.16×10^{-2}	$+1.05$	a
Ne	22	1.6×10^{-3}	$+0.5$	f
Ar	82	475, 440	$-0.33, -0.45$	b, c
Kr	117	1800	-0.78	c
Xe	163	2200	-0.63	c
H_2	20	5×10^{-3}	-1.0	d, e

[a] Meyer, L., Davis, H. T., Rice, S. A., Donnelly, R. J.: Phys. Rev. **126**, 1927 (1962).
[b] Schnyders, H., Rice, S. A., Meyer, L.: Phys. Rev. Letters **15**, 187 (1965).
[c] Miller, L. S., Howe, S., Spear, W. E.: Phys. Rev. **166**, 861 (1968).
[d] Halpern, B., Gomer, R.: J. Chem. Phys. **43**, 1069 (1965).
[e] Gedanken, A., Raz, B., Jortner, J.: Chem. Phys. Letters (1972).
[f] Loveland, R. J., Le Combre, P. G., Spear, W. E.: to be published.

interest, because in this case the localized excess electron state is on the verge of energetic stability [3, 11]. We have predicted [11] that the excess electron in liquid Ne should be localized in a bubble. Recent experimental studies [18] have recorded a low electron mobility in this system, providing conclusive proof for electron localization in liquid Ne.

We thus conclude that the nature of excess electron states in liquid rare gases is well understood. Molecular nonpolar liquids pose some serious problems. We would like to point out the discrepancy between the negative spectroscopic V_0 value for H_2 and D_2 and the low electron mobility in liquid H_2 and D_2. Spectroscopic studies [19] of Xe impurity states in solid H_2 and D_2 (Fig. 9) lead to $(P_+ + V_0) \approx -2 eV$. A lower limit of $P_+ \gtrsim -1 eV$ was estimated so that $V_0 \approx -1 eV$. This spectroscopic value is in considerable variance with the

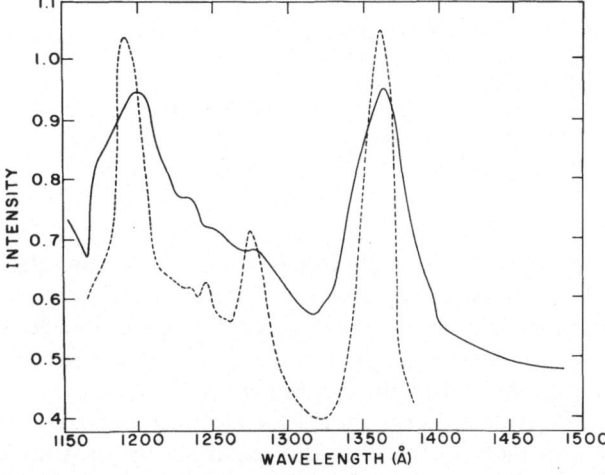

Fig. 9. Absorption spectra of Xe impurity in solid H_2 and D_2 [Gedanken, A., Raz, B., Jortner, J.: Chem. Phys. Letters (1972)]

theoretical estimate [3] $V_0 \simeq +2.2$ eV obtained utilizing a simple model potential specified by a hard core radius, obtained from electron scattering calculations and a long range screened polarization field. It should however be pointed out that the theoretical value [3] for V_0 is very sensitive to the hard core radius, and a reliable estimate of this parameter is not yet available. The spectroscopic value of $V_0 = -1$ eV does not contradict the experimental upper limit $V_0 \lesssim 0.5$ eV obtained by Halpern and Gomer [20] from electron(injection) experiments. The negative spectroscopic value of V_0 implies that the localized bubble state in liquid H_2 and D_2 will be energetically unstable. The low electron mobility in this system may originate from polaron coupling effects.

Acknowledgement

We are grateful to A. Gedanken for fruitful collaboration. This report was supported in part by the Israeli National Council for Research and Development.

References

1. See for example: Jortner, J., Kestner, N. R., Rice, S. A., Cohen, M. H.: J. Chem. Phys. **43**, 2614 (1965).
2. Springett, B. E., Cohen, M. H., Jortner, J.: Phys. Rev. **159**, 183 (1957).
3. Springett, B. E., Cohen, M. H., Jortner, J.: J. Chem. Phys. **48**, 2720 (1968).
4. Halpern, B., Lekner, J., Rice, S. A., Gomer, R.: Phys. Rev. **156**, 351 (1967).
5. Raz, B., Jortner, J.: Chem. Phys. Letters **4**, 511 (1970).
6. Raz, B., Jortner, J.: Proc. Roy. Soc. A **317**, 113 (1970).
7. Katz, B., Brith, M., Sharf, B., Jortner, J.: J. Chem. Phys. **50**, 5195 (1969).
8. Raz, B., Jortner, J.: Chem. Phys. Letters **4**, 155 (1969).
9. Knox, R. S.: Theory of excitons, New York: Academic Press 1962.
10. Fowler, W. B.: Phys. Rev. **151**, 657 (1966).
11. Raz, B., Jortner, J.: Chem. Phys. Letters **9**, 224 (1971).
12. Baldini, G.: Phys. Rev. **128**, 1562 (1962).
13. Baldini, G.: Phys. Rev. **137**, A 508 (1965).
14a. Hensel, R., Keitel, G., Koch, E. E., Kosuch, N., Skibowsky, M.: Phys. Rev. Letters **25**, 1281 (1970).
14b. Hensel, R., Keitel, G., Koch, E. E., Skibowsky, M., Schreiber, P.: Optics Commun. **2**, 59 (1970).
15. Gedanken, A., Raz, B., Jortner, J.: J. Chem. Phys. (in press).
16. Rice, S. A., Jortner, J.: J. Chem. Phys. **44**, 4470 (1966).
17. Lekner, J.: Phys. Rev. **158**, 130 (1967).
18. Loveland, R. J., Le Combre, P. G., Spear, W. E.: Phys. Letters **39** A, 225 (1972).
19. Gedanken, A., Raz, B., Jortner, J.: Chem. Phys. Letters (1972).
20. Halpern, B., Gomer, R.: J. Chem. Phys. **43**, 1069 (1968).

Discussion

M. H. COHEN Considering the high value of V_0, 1.2 eV, and the low binding energy of liquid H_2, a possible explanation of the observed low mobility is that of "iceberg" formation, that is, the electron is self-trapped in a local increase of density which may be great enough for freezing. This explanation requires a sufficiently rapid increase of V_0 with increasing its density.

J. JORTNER The "Hydrogen Berg" model for electron localization corresponds to the small polaron picture. In other molecular solids such as nitrogen, low electron mobilities (10^{-3} cm^2 Volt^{-1} sec^{-1}) were observed by Spear *et al.*, indicating the same kind of electron mobility mechanism.

Electron Localization in Dense He Gas

ITZCHAK WEBMAN and JOSHUA JORTNER

Abstract

We provide a localization criterion for the formation of a bound bubble state of an excess electron in gaseous He. The optical properties of a localized electron state in this system are explored, utilizing a simple configuration diagram model for bound-bound transitions and the nuclear Breit-Condon model for the bound-continuum transitions. In some cases the appearance of resonances results in nuclear motion narrowing of the bound-continuum intensity distribution.

In this work we present the results of some calculations on the properties of excess electrons in dense He gas. The calculations concern the critical densities and pressures for the stability of the electron-excess electron in a bubble state [1, 3], and some of the optical properties of such bubbles. He gas under high pressures might prove to be a good system for the study of the optical properties of electron bubbles in a nonpolar fluid, because by varying pressure and density we can obtain bubbles with a wide range of physical properties. Optical absorption should occur in the range 0.1—0.6 eV, and it might be more easily amenable to direct experimental observation than for the bubbles in the liquid phase. As is now well established [2], He gas may be considered as a fluid of hard core scatterers, and under appropriate conditions, it exhibits electron localization originating from short range repulsions.

Figure 1 exhibits the range of densities of He gas at 133 °K under pressures up to 1000 atm. The density at high pressures approaches (and even slightly exceeds) the density of liquid He at 4.2 °K. The equilibrium radius of bubbles in the gas is usually smaller than in the liquid. Figure 2 displays a plot of the bubble radius

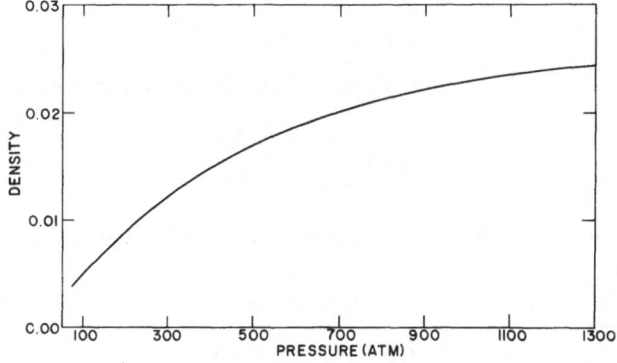

Fig. 1. Density of He gas under high pressures at $T = 133\,°K$ (density units $Å^{-3}$)

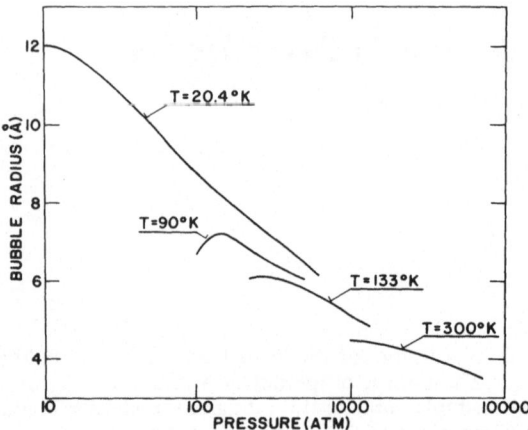

Fig. 2. Equilibrium radius of the bubble at several temperatures

vs. pressure for several temperatures. The radius becomes as small as $3\,\text{Å}$ in the high pressures. The bubble state is stable if the total energy, i.e. electronic energy, E_{el}, plus medium reorganization energy, E_M, is lower than the energy of the quasi-free state, V_0, so that

$$E_{el}(R, V_0) + E_M(R) < V_0 .$$

By defining the parameters X and K (for $V_0 > 0$) as:

$$E_{el}(R, V_0) = X^2 V_0$$
$$V_0 = \beta K^2 \tag{1}$$
$$\beta = \frac{\hbar^2}{2m}$$

the condition for the occurrence of a minimum in the $E_{tot}(R)$ diagram is

$$\gamma \equiv \beta^{3/2} p \, V_0^{-5/6} = \bar{f}(X) \tag{2}$$

where

$$\bar{f}(X) = \frac{X^2 \sqrt{1 - X^2}}{2\pi (KR)^2 \, (KR \sqrt{1 - X^2} + 1)} \tag{3}$$

and in order to achieve energetic stability γ must also fulfill the condition

$$\gamma < \bar{g}(X) \, \frac{3}{4\pi} \, \frac{(1 - X^2)}{(KR)^3} . \tag{4}$$

From the relations (2)—(4) one can deduce the criterion for energetic stability in the gas phase

$$\gamma < 4.3 \cdot 10^{-3} . \tag{5}$$

For γ in the range

$$4.3 \cdot 10^{-3} < \gamma < 7.2 \cdot 10^{-3} \tag{6}$$

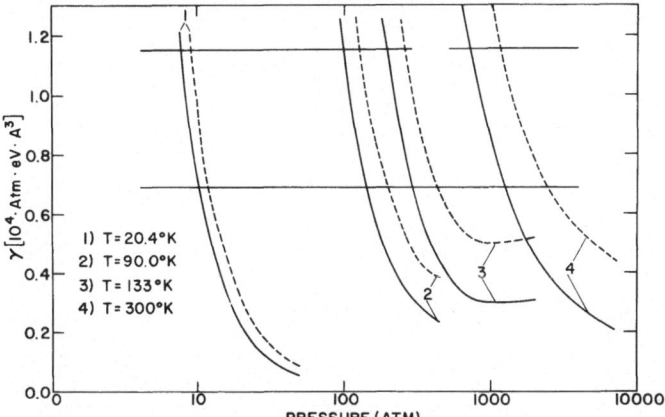

Fig. 3. The stability parameter $\gamma = \beta^{3/2} p V^{-5/2}$ vs pressure. The lower horizontal line is the upper limit of γ for an energetic stable bubble. The higher line is the upper limit for γ of a metastable bubble

only condition (2) is fulfilled and the bubble it metastable. Critical pressures for stability at several temperatures may be deduced from Fig. 3 which displays γ vs. the pressure at various temperatures.

The study of optical properties is based on a model of a spherical potential well of depth V_0 proposed by Springett et al. [2]. Under the conditions of pressure and density considered above, the well contains either a single 1s bound state, or two bound states: 1s and 1p. In the former case optical transitions occur from the bound state to continuum states only. The theory of Breit and Condon [5] for the photodisintegration of the deuteron has been utilized to study the bound-continuum transitions. Lineshapes were calculated by applying the semiclassical configuration coordinate model [6] taking into account the spherical symmetric vibration and the quadripole distortion. Most of the lineshapes contain the contributions of both the 1s→1p and 1s→continuum components as can be

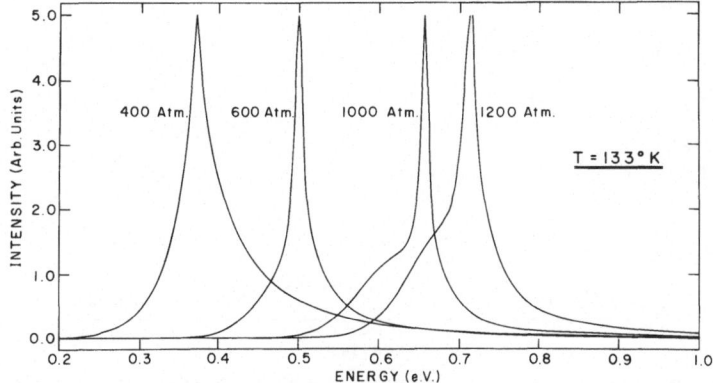

Fig. 4. Absorption lineshapes at 133 °K

I. Webman and J. Jortner:

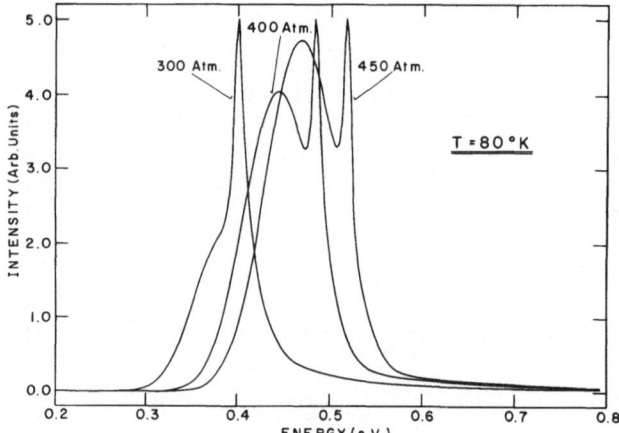

Fig. 5. Absorption lineshapes at 80 °K

seen in Figs. 4 and 5. The final formula for the general lineshape is

$$I(\hbar\omega) = \int\limits_{-\infty}^{\infty} e^{-K_Q Q^2/KT} F(\hbar\omega, Q)\, dQ \qquad (7)$$

where Q is the nuclear coordinate for quadrupole distortion and

$$K_Q = \left(\frac{\partial^2 E_{\text{tot}}(R, Q)}{\partial Q^2}\right)_{R=R_{\text{equilibrium}}}$$

is the force constant for this distortion. Finally $F(\hbar\omega, Q)$ is given by

$$F(\hbar\omega, Q) = \int\limits_{-\infty}^{\infty} S(R, Q, \hbar\omega)\, e^{-(\varepsilon_{1s}(R) - \varepsilon_{1s}(R_0))/KT}\, dR$$

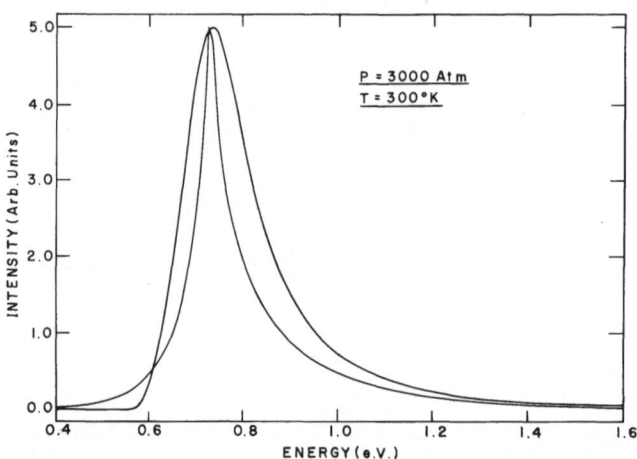

Fig. 6. Absorption lineshapes of 1s→continuum at 300 °K and 3000 atm for a static bubble and for a vibrating one. Only the breathing mode has been taken into account

where R is the cavity radius. The function S in Eq. (8) is either equal to $\delta(E_{1p}(R) - E_{1s}(R) - \hbar\omega)$ for the bound-bound (1s→1p) transition, or to $\sigma(R, \hbar\omega)$, the Breit-Condon lineshape [5] for a static bubble in the case of bound-continuum transitions. Typical optical lineshapes containing both bound-bound and bound-continuum contributions are displayed in Figs. 5 and 6. It is interesting to note that some of the 1s→continuum lineshapes in Figs. 4 and 5 and also the 1s→continuum components of the combined lineshapes are rather narrow. Figure 6 exhibits the 1s→continuum lineshape of a static bubble composed with the lineshape of a vibrating bubble (at $P = 3000$ atm, $T = 300\,°$K). In this case nuclear motion leads to a decrease in the line width. The reason for the line narrowing originates from the observation that for a bubble radius somewhat larger than the radius for which the 1p state becomes bound, there is a resonance in the p wave-scattering as E approaches zero [7]. Under these conditions one may consider a 1p state which is quasi-bound by the centrifugal barrier, and transitions to this state have a very narrow line width. This line width is compatible with the lifetime of such a quasi-bound state. In the process of calculating the lineshape for a vibrating bubble, contributions from bubble radii that give rise to resonance will occur and reduce the total line width.

References

1. Kuper, G.: Phys. Rev. **122**, 1007 (1961).
2. Springett, B. E., Cohen, M. H., Jortner, J.: Phys. Rev. **159**, 183 (1967).
3. For a detailed review and references see: Jortner, J.: Action chimique et biologique de radiations, 14th Series, Haissinsky, M. (Ed.). Paris: Mason et Cie 1970.
4. Fowler, W. B., Dexter, D. L.: Phys. Rev. **176**, 337 (1968).
5. Breit, G., Condon, E. V.: Phys. Rev. **49**, 904 (1936).
6. Markham, J. J.: F Centres in Alkali Halides, Solid State Physics Supplement 8. New York: Academic Press (1966).
7. Downs, B. W.: Am. J. Phys. **30**, 248 (1962).

Discussion

J. JORTNER – The calculations of the bound-continuum transitions on the basis of the nuclear Breit-Condon model and incorporation of phonon broadening provide an assessment of the relative contributions of the 1s→2p (when the 2p state is bound) and the 1s→continuum transition and their lineshapes. This treatment is general and should be applicable for any electron trapping center in nonpolar liquids and glasses where the potential can be described in terms of a square well. Delahay's results for the optical spectra of the localized electron in 3MP where the bound-bound and bound-continuum transitions were separated is amenable to theoretical study of our model.

Excess Electron States in Dense Polar Vapors

ARIEL GAATHON and JOSHUA JORTNER

Abstract

We report the results of a theoretical study of the energy levels and the optical properties of excess electrons in water and ammonia polar vapors. A molecular model for short range interactions and continuum model for long range interactions were adopted. Variational self-consistent solutions were obtained for the ground and for the first-excited states. We have calculated the density dependence of the configurational diagrams, the equilibrium ground state radius, the heat of solution, the vertical excitation energy, the temperature coefficient of the absorption maximum and the optical line width. The localized excess electron is stable in ammonia and in water vapors down to the lowest calculated density of 0.1 g cm^{-3}, while the transition energy exhibits a weak density dependence in this region. To assess the stability of the localized excess electron at low densities we have handled electron binding by a "supermolecule" consisting of a cluster of $N = 4$ solvent molecules, where in the case of NH_3 the localized electron state is unstable while for H_2O it is on the verge of stability. New experimental data are reported in the absorption spectrum of the localized excess electron in subcritical and supercritical D_2O in the density range $0.5 - 0.15 \text{ g cm}^{-3}$. The results of the theoretical calculations compare favorably with the available experimental data.

I. Introduction

Localization of excess electrons in polar and in nonpolar liquids [1, 2] is a general phenomenon which was observed in a large number of systems that can be subdivided into two general categories:

a) nonpolar liquids consisting of light and saturated molecules and atoms (i.e. liquid He and liquid Ne) where electron localization is due to short range repulsive interactions [2];

b) a large number of polar liquids where electron localization is attributed to short range and long range attractive interactions [1].

Consider now a gas consisting of molecules which are characterized by: a) a negative electron affinity, so that the isolated molecular negative ion is energetically unstable; and b) the localized excess electron state is stable in the corresponding liquid. At sufficiently low density an excess electron will be quasi-free so that its mobility will be high and properly accounted for in terms of elastic scattering of a plane wave [3]. When the fluid density is increased the localized excess electron state is expected to become energetically stable. Electron localization at a certain "critical" density can be monitored by the observation of an abrupt drop in the electron mobility and the appearance of a typical optical absorption spectrum characteristic of the localized state. Electron localization originating from short range repulsions was experimentally observed in the He gas where an abrupt drop in the electron mobility over a narrow density range was recorded [4]. Electron localization in polar vapors (at sufficiently high densities) is expected to be a general phenomenon. The following problems are relevant for the understanding of electron localization and the nature of excess electron states in polar vapors:

a) Can one define a "critical" density for the transition between the localized and in the quasi-free excess electron state in a polar vapor in analogy to the case of a repulsive nonpolar gas? From the theoretical point of view two extreme possibilities should be considered: a.1) an excess electron is not bound to a cluster of solvent molecules, and in this case a finite fluid density is required to form an attractive long potential which will lead to electron localization, a.2) a macromolecule consisting of an electron bound to a solvent cluster is energetically stable, so that electron localization in the polar gas will be governed by the probability of cluster formation.

b) What are the physical properties of localized excess electron states in polar vapors in the density range where the localized state is energetically stable?

c) What is the proper theoretical description of a localized excess electron in polar vapor? The current theoretical schemes [2, 5, 6] for the solvated electron (involving many variations on the same theme) involve a combination of a molecular model to specify the short range interactions and a continuum model for the long range interactions. Density changes are expected to affect the long range potential. Thus the combination of theoretical and experimental studies ef excess electrons in polar vapors will provide a crucial test for the relative roles of short range and long range interactions in the stabilization of the solvated electron.

Before alluding to theoretical arguments it might be safer to consider the available experimental data on excess electrons in polar vapors:

1. Concerning point (a), Michael et al. [7] observed the absorption spectrum of the localized electron in water down to the thermodynamic critical point ($\varrho = 0.32$ g cm^{-3}) while Olinger and Schindewolf [8] recorded the optical spectrum of the localized excess in supercritical ammonia down to a density of $\varrho = 0.1$ g cm^{-3}.

2. Concerning point (b), Michael et al. [7] demonstrated that the absorption peak of localized electrons in H_2O shifts from 1.72 eV ($\varrho = 1$ g cm^{-3} $T = 300\,°K$) to 0.9 eV ($\varrho = 0.32$ g cm^{-3} $T = 661\,°K$). In this experiment the temperature effect and the density effect were not separated. Olinger and Schindewolf [8] observed that the peak of the absorption spectrum of localized electrons in supercritical ammonia is practically density independent in the range 0.5—0.15 g cm^{-3}, and that the absorption band may be narrowed relative to the spectrum of the electron in liquid NH_3.

In this paper we report the results of a theoretical study of excess electron states in polar vapors focussing attention on the popular H_2O and NH_3 systems. The results of the theoretical study are compared with the available experimental optical data [7, 8] and with the spectra of excess electrons in supercritical water vapor recently measured by us [9].

II. A Molecular Model for the Electron Localization Center

To treat the electronic state of an excess electron in polar vapors we shall consider a somewhat modified version of the molecular model presented by Copeland et al. [5]. The present model includes the following ingredients:

a) The trapping center for the localized excess electron is decsribed in terms of an oriented first coordination layer consisting of N solvent molecules, and a polarizable polar medium beyond it. The structure of the first solvation layer is density independent, while the external medium is characterized by the normal density, ϱ, of the polar fluid.

b) An electrostatic microscopic potential [5] will be adopted to account for the attractive interactions with the first coordination layer.

c) The Landau polaron potential [10] is retained for long range attractive interactions exerted by the polar molecules beyond the first coordination layer.

d) The short range repulsive electron-medium interactions and the long range electronic polarization interactions outside the first coordination layer will be subsummed into a constant V_0 term, specifying the energy of the conduction band [11].

e) The short range electron solvent repulsions and the long range electronic polarization contribution in the first solvation layer will be specified in terms of a constant energy parameter V_{os}. Note that the "background" potential is different to the first coordination layer and for the continuum. It should be also pointed out that both V_0 and V_{os} are density dependent, and that V_{os} also depends on the radius of the first coordination layer.

f) As the contribution of the long range electronic polarization was incorporated into V_0 and into V_{os} we shall avoid double bookkeeping and refrain from adding an additional electronic polarization term to the electronic energies calculated using the molecular potential.

The specific electron-medium potentials (Fig. 1) take the form

$$V(r) = -\frac{N e \mu_0 \langle \cos \Theta \rangle}{r_{d^2}} - \frac{\beta e^2}{r_c}; \qquad r < r_d - \tilde{a}$$

$$V(r) = -\frac{N e \mu_0 \langle \cos \Theta \rangle}{r_{d^2}} - \frac{\beta e^2}{r_c} + V_{os}; \qquad r_d - \tilde{a} < r < r_d$$

$$V(r) = -\frac{\beta e^2}{r_c} + V_{os}; \qquad r_d < r < r_c$$

$$V(r) = -\frac{\beta e^2}{r} + V_0; \qquad r > r_c \tag{1}$$

N is the coordination number, μ_0 is the (gas phase) dipole moment and the thermally averaged value of $\langle \cos \Theta \rangle = \langle \mu_0 \cdot r_d / |\mu_0| |r_d| \rangle$ is obtained from the Langevin function. r_d represents the distance to the center of the nearest dipole, and r_c corresponds to the radius of the first coordination layer. Finally $\tilde{a}$ is the effective hard core repulsive radius of a single solvent molecule; this effective length determines both V_{os} and V_0 and constitutes a basic parameter in the present theory.

The medium background energy V_0 was estimated utilizing the conventional Wigner-Seitz model and recast in the usual form [2, 11]

$$V_0 = T + U_p \tag{2}$$

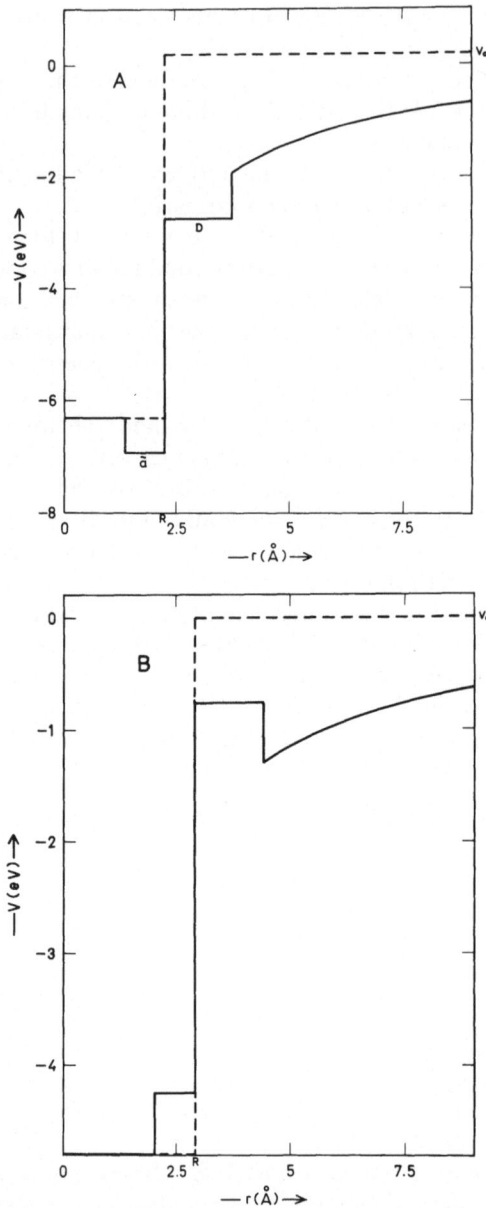

Fig. 1. Typical potentials for the localized electron in water at various densities. The broken lines represent the appropriate finite square well potential well. a Water $\varrho = 1.0 \, \text{g cm}^{-3}$. b Water $\varrho = 0.1 \, \text{g cm}^{-3}$

where T is the repulsive kinetic energy term while U_p corresponds to the attractive electronic polarization contribution. Both terms are determined by the Wigner-Seitz radius $r_s = (3/4\pi\varrho)^{1/3}$. The strong dependence of V_0 on the parameter $\tilde{a}$ is displayed in Fig. 2. The first coordination layer is characterized by the local

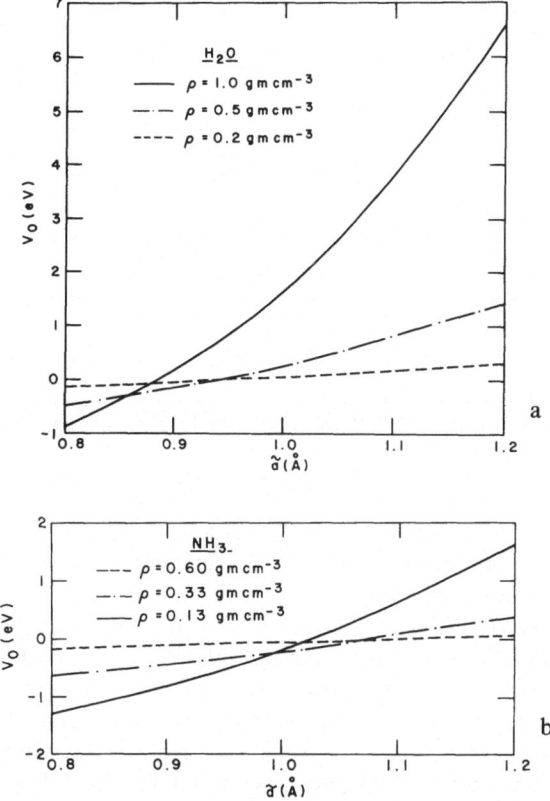

Fig. 2. The dependence of V_0 on the hard core radius $\tilde{a}$. Data were obtained for different densities. a Water. b Ammonia

density ϱ_1, so that the "background" energy in the range can be approximated by

$$V_{os} = T^{(1)} + U_p^{(1)} + U_p^{(2)} \tag{3}$$

where $U_p^{(1)}$ is the polarization interaction with the molecule on the boundary of the Wigner-Seitz cell (determined by ϱ) while $U_p^{(2)}$ is the electronic polarization of the external medium (determined by ϱ_1) [12]. The short range repulsive term $T^{(1)}$ is calculated using the local density. In Fig. 3 we present the density dependence of the energy parameters V_0 and V_{os} in ammonia and water for "reasonable" values of $\tilde{a} = +0.8 - 1.0 \text{ Å}$. The V_{os} values depend on the coordination number and were computed for the equilibrium radius. It should be noticed that in view of the large density of normal water V_0 takes larger positive values in this system. It should be also noted that in general $V_0 < V_{os}$ over the low density range, while for normal water $\varrho_1 < \varrho$ so that $V_0 > V_{os}$.

The third density-dependent "background" parameter involves the long range medium polarization term $\beta = (D_{op}^{-1} - D_s^{-1})$. The optical dielectric constant, D_{op}, was evaluated from the Clausius Mossoti equation. For the static dielectric constant D_s the available experimental data for water were utilized [13], while

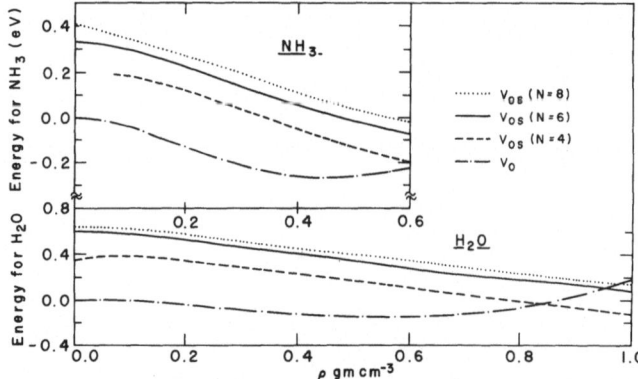

Fig. 3. The density dependence of the "background" potentials V_0 and V_{os}. These data were calculated taking $\tilde{a} = 0.9\,\text{Å}$ for H_2O and $\tilde{a} = 1.0\,\text{Å}$ for NH_3 (see text). The values of V_{os} were calculated at $R = \bar{r}_d$ for $N = 4, 6$ and 8

for ammonia the Onsager equation was used. In Fig. 4 we display the density dependence of β for water and for ammonia. Surprisingly enough this parameter varies slowly down to $\varrho = 0.2\,\text{g cm}^{-3}$ for both systems and exhibits a very weak temperature dependence. Only for $\varrho < 0.1\,\text{g cm}^{-3}$ β exhibits a sharp drop. Without referring to any specific theoretical calculations we can assert that in the density range $\varrho = 1.0 - 0.15\,\text{g cm}^{-3}$ for H_2O and $\varrho = 0.6 - 0.15\,\text{g cm}^{-3}$ for NH_3, the long range electron-medium attractive interaction will not be greatly modified. For reasonable values of V_0 ($\tilde{a} \approx 1.0 - 0.8\,\text{Å}$) the resulting V_0 values can be negative (of the order of a few tenths of eV) however V_0 (Figs. 2 and 3) is still higher than the total ground state energy whereupon the excess electron is localized over the above-mentioned density region.

The electronic energies for the 1s state (E_e^g) and for the first optically allowed 2p state (E_e^u) were determined by the self-consistent variational procedure. The only modification introduced by us involved the use of two variational wave-

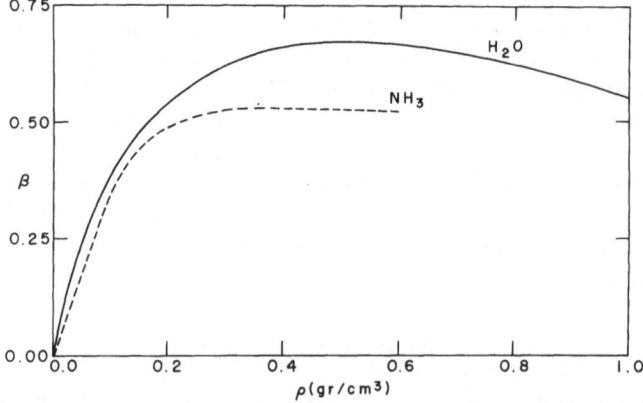

Fig. 4. The density dependence of the long range medium polarization term

functions: a) the conventional hydrogenic functions; b) the spherical Bessel and Hankel wavefunctions which are appropriate for the spherical well,

$$v(r) = -\frac{Ne\mu_0 \langle \cos\Theta \rangle}{r_{d^2}} - \frac{\beta e^2}{r_c} \qquad r < r_d \tag{5}$$

$$v(r) = V_0 ; \qquad r > r_d .$$

These trial wavefunctions do not involve a variable parameter. Calculations were performed for different values of $N = 4, 6, 8, 12$ and for variable r_d. For the ground 1s state the spherical well wavefunctions led to lower electronic energies in all cases. For the excited 2p state the lowest energy was obtained with hydrogenic functions for high densities and low r_d values and with the spherical well wavefunction for high r_d and low densities.

The total energies E_T of the ground state and of the excited state

$$E_T^i(r_d) = E_e^i(r_d) + E_M(r_d); \qquad i = 1s, 2p \tag{4}$$

consist of the electronic energy and of the medium rearrangement energy, $E_m(r_d)$. The following contributions were incorporated into the medium rearrangement energy:

1. the dipole-dipole repulsion, Edd., between the permanent oriented dipoles in the first coordination layer;
2. the medium polarization energy

$$\Pi = \frac{\beta e^2 C_g(r_c)}{2r_c} ; \tag{6}$$

3. the short range solvent-solvent repulsions. In the case of ammonia we utilized the exponential repulsive potential proposed by Copeland et al. [5]

$$E_{HH} = A_H(N) \exp[-B_H(N) r_d] . \tag{7}$$

For water we applied Kamb's empirical formula [14]

$$E_{HH} = I(N) A(\sigma/C(N) r_d) \tag{8}$$

where $\sigma = 2.8$ Å, $I(N)$ is the number of interacting pairs, $C(N)$ is a distance scaling factor, and $A = 4$ kcal/mole^{-1} is an empirical energy parameter.

The total medium rearrangement energy is:

$$E_M(r_d) = E_{dd} + \Pi + E_{HH} \tag{9}$$

where we have disregarded the surface tension term (which is meaningless in the gas phase) and the pressure volume work (which is negligible for the small void up to 500 atm).

III. Predictions of the Molecular Model

The following physical properties of the solvated electron in normal liquids and of the excess electron in polar vapors were determined:

a) the configurational diagrams $E_T^{1s}(r_d)$ and $E_T^{2p}(r_d)$ for the ground and for the first excited state (see Fig. 5);

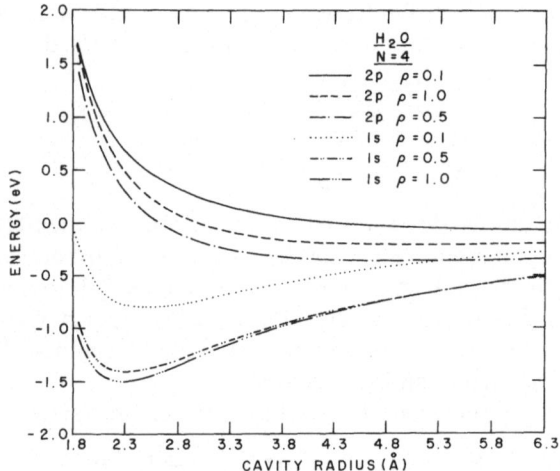

Fig. 5. Configurational diagrams of the ground state and of the first excited state in H_2O fluid calculated for $\tilde{a} = 0.9$ Å, $N = 4$ at various densities. The densities ϱ are in units of g cm^{-3}

b) the equilibrium ground state radius $\bar{r}_d$;

c) the "heat of solution" (i.e. stabilization energy of the localized state relative to the vacuum state) given by $-E_T^{1s}(\bar{r}_d)$;

d) ther vertical excitation energy

$$h\nu = E_T^{2p}(\bar{r}_d) - E_T^{1s}(\bar{r}_d);\tag{10}$$

e) the photoconductivity threshold $-E_e^{is}(\bar{r}_d) + V_0$;

f) the temperature dependence of the absorption maximum at constant volume $(\partial h\nu/\partial T)\varrho$;

g) the absorption lineshape (i.e. the intensity distribution in absorption) and the optical line width at half maximum, Δ.

In order to facilitate comparison between theory and experiment we have to provide an "intelligent guess" for the hard core radius $\tilde{a}$ which enters as a parameter into our theory. The results of our calculations for liquid H_2O and for liquid NH_3 at normal densities (Table I) led to the reasonable agreement between theory and experiment for $\tilde{a} = 0.9$ Å $- 0.8$ Å for H_2O and $\tilde{a} = 1.0$ Å $- 0.9$ Å for NH_3. The calculations reported herein are based on the values $\tilde{a} = 0.9$ Å for H_2O and $\tilde{a} = 1.0$ Å for NH_3 together with all the assumptions inherent in our physical model.

From the configurational diagrams (Fig. 5) we conclude that:

a) The ground state potential curves for the localized electron both in water and in ammonia exhibit a minimum down to the lowest calculated density of 0.1 g cm^{-3}.

b) The density dependence of the heats of solution of an excess electron in these two polar fluids (see Figs. 6 and 7) exhibit a gradual decrease with decreasing density. In the density range 1.0—0.2 g cm^{-3} for H_2O and 0.6—0.2 g cm^{-3} for NH_3 where the long range medium polarization interactions (expressed in

Table I. Experimentally measured and theoretically predicted energetic data for solvated electrons at the normal liquid density

	Ammonia		Water	
	Experiment	Theory ($\tilde{a} = 1.0$ Å; $N = 4$) $T = 300\,°$K	Experiment	Theory ($\tilde{a} = 0.9$ Å; $N = 4$) $T = 300\,°$K
Transition energy (eV)	0.8[a]	0.89	1.72[b]	2.0
Line width (eV)	0.46[a]	0.18	0.92[b]	0.42
Temperature coefficient of $h\nu_{max}$	$-0.2 \cdot 10^{-3}$[a,c]	$-0.3 \cdot 10^{-3}$	$-1.0 \cdot 10^{-3}$[b,c]	$-0.3 \cdot 10^{-3}$
Equilibrium radius (Å)	—	2.87	—	2.29
Heat of solution ΔH (eV)	1.7 0.7[d]	1.02	1.7[e]	1.5
Photoelectric threshold (eV)	1.6[f]	1.94	—	3.0
Photoconductive threshold (eV)	—	1.72	—	3.2
Background potential V_0 (eV)	—	-0.22	—	$+0.19$

[a] Quinn,R. K., Lagowski,J.J.: J. Phys. Chem. **73**, 2326 (1969).
[b] Hart,E.J., Gottschall,W. C.: J. Am. Chem. Soc. **71**, 2102 (1967).
[c] The measured $(\partial h\nu/\partial T)_p$ were corrected using Eqs. (11) and (12) to obtain $(\partial h\nu/\partial T)_e$.
[d] Ref. [10].
[e] Ref. [1].
[f] Teal,G. V.: Phys. Rev. **71**, 138 (1948).

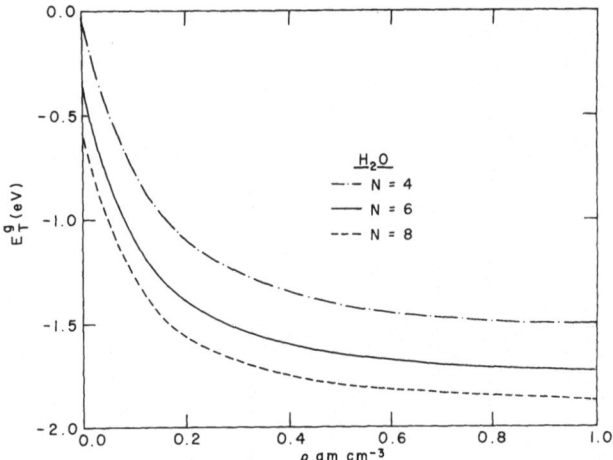

Fig. 6. The density dependence of the "heat of solution" of an electron in dense water vapor. Calculations were performed for $\tilde{a} = 0.9\,\text{Å}$ and for several N values

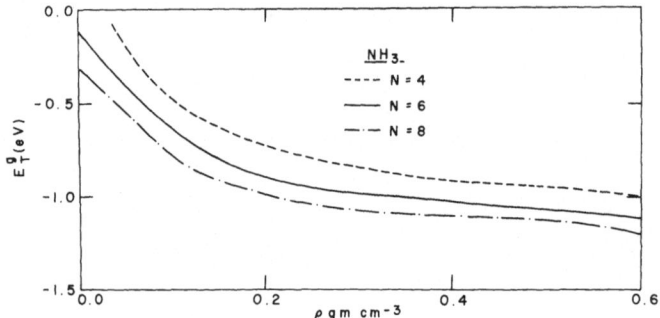

Fig. 7. The density dependence of the "heat of solution" of an electron in ammonia vapor. Calculations were performed for $\tilde{a} = 0.9\,\text{Å}$ and for several N values

terms of the parameter β) are slowly varying, the decrease in E_T^g originates from the density dependence of the short range "background" terms V_0 and V_{os}.

c) The ground state energetic stability condition $E_T^g(\bar{r}_d) < V_0$ is satisfied both for NH_3 and for H_2O down to the lowest calculated density of $\varrho = 0.1\,\text{g cm}^{-3}$.

A partial exposition of the pertinent energetic data for the ground and for the excited state of the localized electron in polar vapors is given in Tables II and III. In Figs. 8 and 9 we display the density dependence of the absorption maximum at $T = 406\,°K$ for NH_3 and at $T = 660\,°K$ for H_2O. Typical optical absorption lineshapes are given in Fig. 10. The following comments are in order:

a) The vertical excitation energies calculated at a constant temperature exhibit a relatively weak density dependence down to $\varrho = 0.1\,\text{g cm}^{-3}$ both for H_2O and for NH_3.

b) The calculated temperature dependence of the band maximum at constant density is $\left(\dfrac{dh\nu}{dT}\right)_\varrho = 4 \times 10^{-4}\,\text{eV deg}^{-1}$ for water and $\left(\dfrac{dh\nu}{dT}\right)_\varrho = 3 \times 10^{-4}\,\text{eV deg}^{-1}$

Table II. Energetic data for the ground state and first excited state of the localized excess electron in fluid. Water = 0.9 Å. $T = 661\,°K$[a], $N = 4$

$\varrho\left(\dfrac{g}{cm^3}\right)$	V_0 (eV)	R_0 (Å)	V_{os} (eV)	E_e^g (eV)	E_e^u (eV)	E_M (eV)	E_T^g (eV)	E_T^u (eV)	$h\nu$ (eV)	Δ (eV)
1.0	0.19	2.29	−0.12	−2.98	−0.97	1.48	−1.50	0.52	2.01	0.42
0.5	−0.15	2.36	0.16	−2.95	−1.31	1.54	−1.41	0.23	1.64	0.61
0.4	−0.13	2.39	0.23	−2.85	−1.28	1.51	−1.34	0.23	1.57	0.61
0.3	−0.1	2.41	0.30	−2.69	−1.15	1.44	−1.25	0.29	1.53	0.58
0.2	−0.05	2.46	0.36	−2.37	−0.90	1.27	−1.10	0.38	1.47	0.60
0.1	−0.004	2.58	0.39	−1.72	−0.47	0.92	−0.80	0.45	1.25	0.54
0.0	0.00	3.01	0.35	−0.17	0.0	0.13	−0.04	0.13	0.17	0.29

[a] Except for $\varrho = 1$ where the temperature is 300 °K.

Table III. Energetic data for the ground and first excited states of the localized electron in ammonia, $\tilde{a} = 1.0\,Å$; $N = 4$; $T = 406\,°K$[a]

$\varrho\left(\dfrac{g}{cm^3}\right)$	V_0 (eV)	R_0 (Å)	V_{os} (eV)	E_e^g (eV)	E_e^u (eV)	E_M (eV)	E_T^g (eV)	E_T^u (eV)	$h\nu$ (eV)	Δ (eV)
0.60	−0.22	2.87	−0.20	−1.94	−1.06	0.92	−1.02	−0.13	0.89	0.18
0.41	−0.26	2.92	−0.06	−1.82	−1.05	0.90	−0.93	−0.15	0.771	0.21
0.33	−0.23	2.95	0.01	−1.76	−1.00	0.89	−0.89	−0.12	0.76	0.21
0.28	−0.20	2.97	0.05	−1.71	−0.96	0.88	−0.84	−0.08	0.75	0.22
0.20	−0.13	3.02	0.12	−1.53	−0.81	0.81	−0.73	0.00	0.73	0.22
0.13	−0.07	3.11	0.18	−1.23	−0.56	0.65	−0.58	0.09	0.66	0.22
0.07	−0.02	3.38	—	−0.68	0.22	0.35	−0.33	0.14	0.47	0.19
0.00	0.0	—	—	—	—	>0	>0	—	—	—

[a] Except for $\varrho = 0.6\,g/cm^3$ where $T = 300\,°K$.

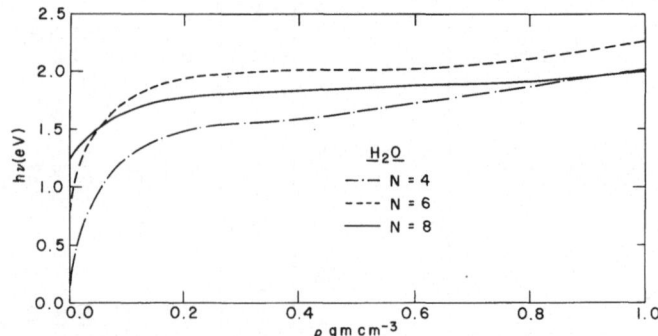

Fig. 8. Density dependence of the vertical transition energy for $N = 4, 6, 8$ coordination numbers $\tilde{a} = 0.9$ Å in H_2O

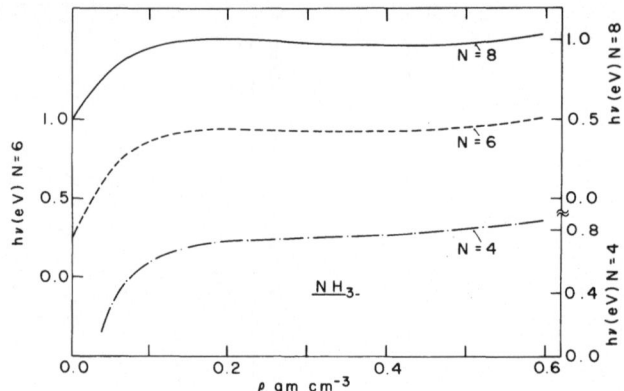

Fig. 9. Density of the vertical transition energy for $N = 4, 6, 8$ coordination numbers $\tilde{a} = 1.0$ Å in NH_3

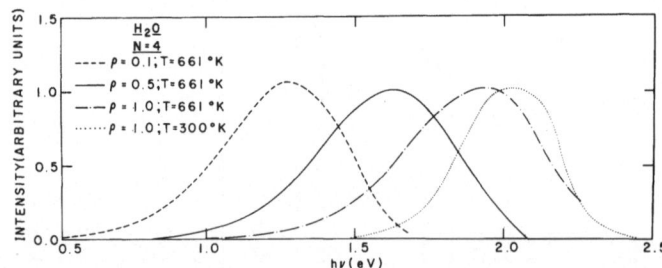

Fig. 10. Calculated absorption line shapes for an excess electron in H_2O for different densities. The high density data were calculated for two temperatures

for NH_3 (calculated for $N = 4$) being almost density independent down to $= 0.1 \text{ g cm}^{-3}$.

c) The optical line widths calculated for a single configuration ($N = 4$) are $\Delta \approx 0.2$—0.3 eV for NH_3 at 406 °K and $\Delta \approx 0.5$—0.6 eV for H_2O at $T = 661$ °K and vary slowly over the whole density range.

d) The calculated half line width exhibits the temperature dependence $\Delta \propto \sqrt{T}$ as expected from the configurational diagram model at the high temperature limit.

e) The calculated line widths calculated at room temperature (for a single $N = 4$) $\Delta = 0.18$ eV for liquid NH_3 and $\Delta = 0.42$ eV for liquid H_2O are lower by a numerical factor of ~ 2 compared to the experimental values [1]. This discrepancy is attributed to the fact that several configurations of the localized electron (specified within the framework of our model by different N values) are characterized by similar E_T^g values and contribute to the absorption spectrum in the normal liquid. At lower densities and higher temperatures a single configuration may become dominating whereupon in such a case the experimental line widths will be narrowed in the supercritical polar fluid.

Up to this point we have been concerned with the properties of the localized electron at moderate densities $\varrho \gtreqqless 0.1$ g cm^{-3}. It is interesting to inquire whether the localized excess electron state will be stable at low densities (say $\varrho \sim 0.01$ g cm^{-3}). In an attempt to resolve this question we have considered a "macromolecule" consisting of a cluster of N solvent molecules constituting the localization center while at these low densities the long range Landau potential is switched off setting $\beta = 0$. Obviously in this case $V_0 = 0$ (while $V_{os} \neq 0$). The potential exerted on the electron (Fig. 11) is calculated from Eq. (1) with $\beta = 0$ and $V_0 = 0$. Energetic data for the macromolecule are presented in Table III and in Fig. 12. The preliminary conclusions originating from the present simple model are:

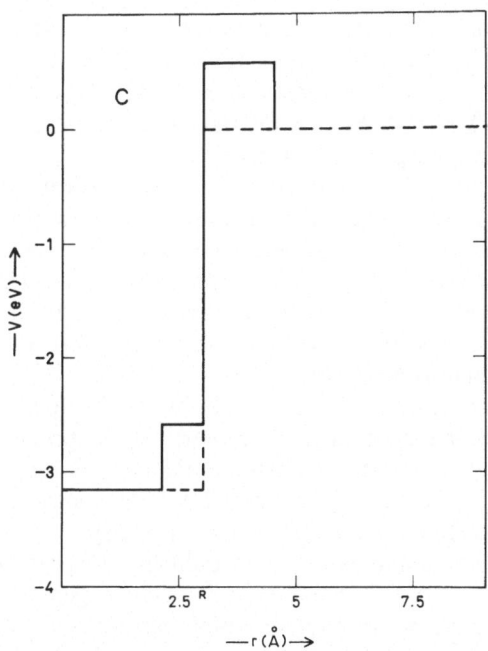

Fig. 11. The potential well for an electron in field of $N = 4$ water molecules

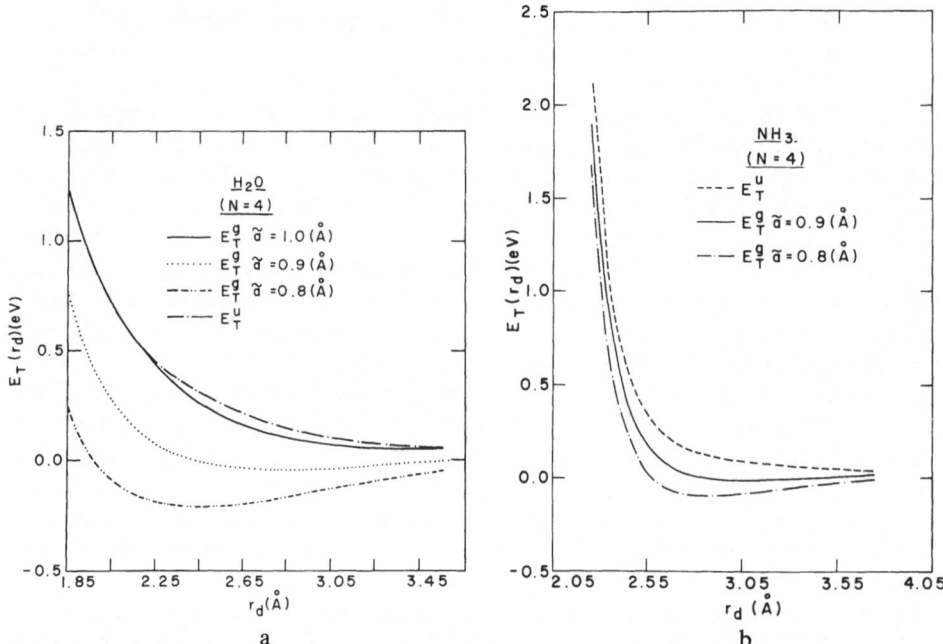

Fig. 12. Configurational diagrams for an electron trapping center consisting of $N = 4$ molecules. Calculations were performed for different $\tilde{a}$ values. a The macromolecule in H_2O. b The macromolecule in NH_3

a) In the case of H_2O (for $\tilde{a} = 0.9$ Å and $N = 4$) the ground state potential curve exhibits a minimum, and $E_T^g \approx k_B T$ (at 660 °K). Thus the localized excess electron state is marginally stable at low supercritical H_2O densities.

b) In NH_3 (for $\tilde{a} = 1.0$ Å $N = 4$) the ground state potential curve is repulsive and the localized state is not energetically stable.

c) For higher N values the macromolecule is stable; however, we assume that such clusters ($N > 4$) are not expected to persist in the low pressure gas phase. These tentative conclusions should be taken with a grain of salt as they are very sensitive to the parameters $\tilde{a}$ employed in the calculation. Thus for $\tilde{a} = 1.0$ Å the macromolecule does not localize the excess electron both in H_2O and in NH_3. More elaborate calculations are required to establish the stability of the localized electron in a macromolecule.

A crucial point concerning the energetic stability of the localized electron (e_l) at low densities involves the fact that the free energy of the process $e_{gas} \rightarrow e_l$ (rather than the internal energy calculated by us) should be negative. The free energy change rather than the internal energy change should be considered. This problem was preliminary handled by us but deserves further study.

To conclude this theoretical discussion we would like to point out that in conventional pulse radiolysis and flash photolysis electron generating experiments in low density polar vapors where relatively high densities of excess electrons are monitored by various (i.e. optical and conductivity) methods, the stability of the localized electron state may be determined by kinetic factors rather than by energetic considerations.

IV. Analysis of Experimental Data

The experimental data of Michael *et al.* [7] and of Olinger and Schindewolf [8] were already discussed. We recently started an experimental study [9] of the absorption spectra of excess electrons in subcritical and supercritical D_2O. Our experiments were performed in a high pressure (up to 400 atm) cell (optical path length 8 cm) equipped with parallel saphire windows. The cell was placed in an oven and heated in the range 200—390° C. The temperature was kept constant within $\pm 1°$ C and determined within $\pm 3°$ C. Pulse radiolysis of the fluid was performed by a linear accelerator delivering 1.3 µsec 6 MeV pulses (current 160 mA). A conventional optical detection system perpendicular to the accelerator beam was employed utilizing a np Ge photodiode detector which is adequate up to 17000 Å. The D_2O solvent was used to eliminate interference by infrared overtones up to 15500 Å. The solvent was introduced into the cell with the addition of a trace (10^{-2}—10^{-4} M) of ammonia to remove H^+ ions and saturated with H_2. The initial decay half-lifeperiod of the infrared absorption is density dependent and for $\varrho = 0.2\,\mathrm{g\,cm}^{-3}$ was found to be of the order of 0.5 µsec at 390° C. At high densities the half-lifeperiods were longer. The short decay times reported by Michael *et al.* [7] presumably originate from the reaction of the dense water vapor with quartz. Our preliminary absorption spectra are presented in Fig. 13.

Our experimental spectroscopic data demonstrate that in dense H_2O vapor:
a) The electron is localized down to $\varrho = 0.2\,\mathrm{g\,cm}^{-3}$ (the lowest density studied by us to date). b) The absorption maximum is weakly density-dependent in the range 0.5 g cm^{-3}— 0.2 g cm^{-3}.

In order to analyze the spectroscopic data it is important to separate the temperature dependence and the energy dependence. In previous analyses of experimental data for normal liquids [1, 7, 5], the quantity $dh\nu/dT$ was considered without

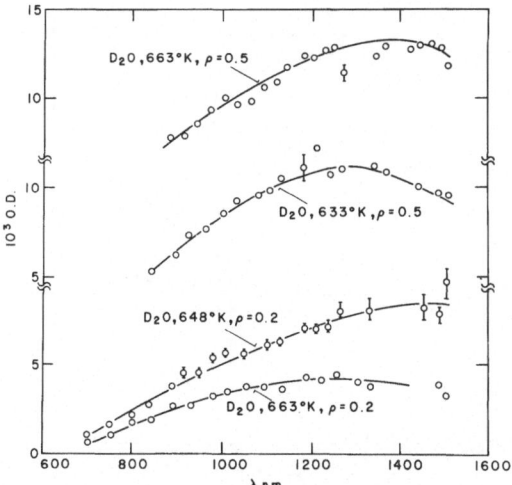

Fig. 13. Experimental absorption spectra of an excess electron in subcritical and supercritical D_2O [9]. These spectra were obtained by pulse radiolysis (see text). The spectra for $\varrho = 0.5\,\mathrm{g\,cm}^{-3}$ were obtained 3.5 µsec after the pulse, while the spectra for $\varrho = 0.2\,\mathrm{g\,cm}^{-3}$ were measured 2 µsec after the pulse

referring to the conditions under which this temperature coefficient was determined, and the large discrepancy between theory [5] and experiment delete [1, 7] was discouraging. Obviously the experimental data for the normal liquid were obtained at constant pressure, p, while the theoretical calculations were performed at constant density. From the experimental point of view, in order to compare the supercritical water data (at the temperature T) with the subcritical liquid data (say at T_0) with theory, we have to correct the subcritical data for the temperature dependence. Taking just the linear term we have for a given subcritical density

$$(hv)_T = (hv)_{T_0} + \left(\frac{\partial hv}{\partial T}\right)_\varrho (T - T_0) \tag{11}$$

where on the basis of theoretical calculations we assert that $(\partial hv/\partial T)_\varrho$ is practically temperature independent. The acquired temperature coefficient is obtained from the relation

$$\left(\frac{\partial hv}{\partial T}\right)_\varrho = \left(\frac{\partial hv}{\partial T}\right)_p - \left(\frac{\partial hv}{\partial \varrho}\right)_T \left(\frac{\partial \varrho}{\partial T}\right)_p. \tag{12}$$

Utilizing the experimental value for liquid water [7]

$$(\partial hv/\partial T)_p = -2.8 \times 10^{-3} \text{ eV deg}^{-1},$$

$$(\partial hv/\partial \varrho)_T = 2.5 \text{ eV g}^{-1} \text{ cm}^3$$

(obtained from the pressure dependence of hv) [15] and the average value [16] $(\partial \varrho/\partial T)_p = 5.5 \times 10^{-4} \text{ cm}^{-3}/\text{deg}$ (obtained from thermodynamic data) we have $(\partial hv/\partial T)_\varrho = -1.4 \times 10^{-3} \text{ eV deg}^{-1}$ which is in somewhat better agreement with theory (see Table I). The experimental data [7] for subcritical water at $T = 300\,^\circ\text{K}$

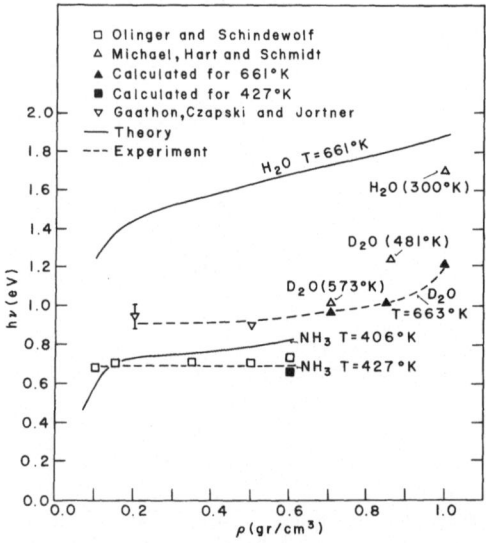

Fig. 14. Comparison of experiment and theory for the peak of the spectra of an excess electron in simple polar fluids. The experimental data were corrected for the temperature dependence at constant density (see text). The theoretical data were calculated for $N = 4$

($\varrho = 1$ g cm^{-3}) and at 508 °K ($\varrho = 0.8$ g cm^{-3}) are now corrected utilizing Eq. (11) and compared with the theoretical predictions for $N = 4$ (Fig. 14). A similar analysis for liquid NH_3 using [16, 17] $(\partial h\nu/\partial T)_p = -1.5 \times 10^{-3}$ eV deg^{-1}, $(\partial h\nu/\partial \varrho)_T = 1$ eV g^{-1} cm^3 and $(\partial \varrho/\partial T)_p = 1.3 \times 10^{-3}$ g cm^{-3} deg^{-1}, yields $(\partial h\nu/\partial T)_\varrho = -0.3 \times 10^{-3}$ eV deg^{-1} and $h\nu = 0.70$ eV at 150° C.

The comparison between the theoretical predictions for the density dependence of $h\nu$ in NH_3 and in H_2O (for $N = 4$) and the experimental facts of life is presented in Fig. 14. The following points are now in order:

a) The weak density dependence [8] of $h\nu$ for NH_3 in the density range 0.6—0.15 g cm^{-3} is adequately interpreted by theory. At somewhat lower densities ($\varrho < 0.1$ g cm^{-3}) a decrease of $h\nu$ is expected.

b) The decrease (20%) of $h\nu$ in H_2O with decreasing density (for constant T) in the range 1.0—0.5 g cm^{-3} and the slow variation of $h\nu$ in the range 0.5—0.2 g cm^{-3} [8, 9] is adequately reproduced by the theoretical calculations. Again as in the case of NH_3 a marked decrease of $h\nu$ is expected for $\varrho < 0.1$ g cm^{-3}.

c) The possible sharpening of the absorption spectra in NH_3 with decreasing density [8] (relative to the liquid spectrum) may be due to the dominating role of a single cluster configuration at low densities.

In view of the ubiquity of solvated electrons in a large number of polar liquids it is expected that localization of excess electron states in supercritical polar vapors is a general phenomenon and that localized excess electron states will be soon observed and studied in a variety of dense polar vapors. We hope that the present study will provide some useful and coherent guidelines for the understanding of this new field.

Acknowledgements

We are grateful to Professor G. Czapski for rewarding joint work on the pulse radiolysis of subcritical and supercritical D_2O. We are indebted to Professor U. Schindewolf for fruitful collaboration on the general problem of electrons in polar gases. We wish to thank Professor N. R. Kestner for his comments on the manuscript.

References

1a. Lepoutre, G., Sienko, M. J.: Metal ammonia solutions. New York: Benjamin 1964.
1b. Metal-Ammonia Solution, Proc. Inter. Conf. Colloque Weyl II, Ithaca, New York, 1969. London: Butterworth 1970.
2a. Jortner, J., Kestner, N. R.: Ref. 2b, p. 49.
2b. Jortner, J.: Ber. Bunsenges. Physik. Chem. **75**, 696 (1971).
3. Cohen, M. H., Lekner, J.: Phys. Rev. **158**, 305 (1967).
4. Levine, J. L., Sanders, T. M.: Phys. Rev. Letters **8**, 159 (1962).
5. Copeland, D. A., Kestner, N. R., Jortner, J.: J. Chem. Phys. **53**, 1189 (1970).
6. Kestner, N. R.: This issue.
7. Michael, B. D., Hart, E. J., Schmidt, K. H.: J. Phys. Chem. **75**, 2798 (1971).
8. Olinger, R., Schindewolf, U., Gaathon, A., Jortner, J.: Ber. Bunsenges. Phys. Chem. **75**, 690 (1971).
9. Gaathon, A., Czapski, G., Jortner, J.: J. Chem. Phys. **58**, 2648 (1973).
10. Jortner, J.: J. Chem. Phys. **30**, 839 (1959).
11. Springett, B. E., Cohen, M. H., Jortner, J.: Phys. Rev. **159**, 183 (1967).

12. Springett, B. E., Cohen, M. H., Jortner, J.: J. Chem. Phys. **48**, 2720 (1968).
13. Fogo, J. K., Benson, S. W., Copeland, C. S.: J. Chem. Phys. **22**, 209 (1957).
14. Kamb, B.: J. Chem. Phys. 3917 (1965).
15. Hentz, R. R., Farhataziz, Hansen, E. M.: J. Phys. Chem. **75**, 4974 (1971).
16. Int. Critical Tables, Vol. III, pp. 23—26. New York-London: McGraw-Hill Book Company Inc. 1933.
17. Schindewolf, U.: Ref. [1 b], p. 199.

Discussion

P. DELAHAY   I am puzzled as to how you meet your first requirement for water since there is electron attachment to water in the vapor phases (at low pressures at least).

A. GAATHON   Old experimental data by Bradbury and Tatel (1934) seemed to indicate electron attachment to water at low pressures. However, subsequent work indicated that this observation was due to impurity effects (probably oxygen). It is now well established (Christophonov, L. G.: Atomic and molecular radiation physics, p. 469. New York: Wiley-Interscience (1971)) that electron attachment to a single water molecule does not take place in low density water vapor.

K. BAR-ELI   How is the geometrical shape of the dipoles taken into account? Or are you using a point dipole?

J. JORTNER   Our model for the solvent molecules in the first coordination layer involves point dipoles located at the centre of a hard-core sphere. This model is grossly oversimplified being based essentially on the approach of pseudopotential theory. Elaborate calculations can be performed on the basis of the SCF scheme taking into account the real charge distribution on the solvent molecules. If the role of long range interactions can be disregarded such a SCF study will provide evidence concerning the stability of the "supermolecule" which we handled by our approximate model. It is gratifying that Newton's theoretical SCF results for the instability of the supermolecule concur with our conclusions. In order to stabilize, the long range Landau field is essential. This is a real victory for Landau's theory.

M. H. COHEN   It should be noted that in the one nonpolar vapor understood in detail, He, there is a transition in the excess electron mobility from the free-electron value to a value approaching that of the bubble in the liquid, a change of 6 orders of magnitude, within less than a decade in density. Moreover, the transition is complete before the bubble becomes metastable in the vapor, let alone stable. As described by Eggarter and myself, the transition is associated with the formation of pseudobubbles within density fluctuations of varying sizes and depths. Eggarter's calculations fit the data remarkably well, the essential point in the present context being that a mobility transition can occur in polar vapors as well, at densities lower than the densities at which clearly marked optical absorption occurs. The presence of pseudobubbles may also lead to optical absorption, but this would be quite broad because of the range of amplitudes and sizes of the fluctuations. In polar vapors, fluctuations in orientation as well as density should be included. It would be a simple matter to generalize Eggarter's theory to include them.

Optical Absorption Spectrum of the Solvated Electron in Ethers and in Binary Liquid Systems*

Leon M. Dorfman and F. Y. Jou

Abstract

Optical properties of the solvated electron produced by pulse radiolysis in ethers and binary liquids provide new evidence concerning the structure and properties of the solvated electron. A range of behavior is observed in mixed solvent systems which requires that more attention be paid to the short range solvent structure.

Introduction

Optical properties of excess electrons in liquids are most extensively known for compounds for which the absorption spectrum of the solvated electron is found in the visible to near infrared region of the spectrum. These compounds are, for the most part, fairly strongly polar molecules. Depending upon the type of compound and its dielectric properties, the absorption band of the solvated electron may, however, lie well into the infrared. The development of a fast infrared detector [1, 2] with a rise-time of less than 100 nsec, with which we have been able to record data to 2300 nm, has made it possible for us to determine the absorption spectrum of the solvated electron formed in the pulse radiolysis of a series of ethers, all of which are weakly polar liquids. A preliminary report of these data for tetrahydrofuran has been published [1], and the comprehensive results are presented in this report. In addition, optical absorption spectra have now been determined in a variety of binary liquid systems comprising solutions of two different strongly polar compounds, a strongly polar and a weakly polar compound, and a strongly polar and a nonpolar [3] compound.

These optical absorption band data have been obtained with several objectives in mind: a) The spectra for the ethers provide data for another type of compound (in addition to amines and hydroxy-compounds) which may serve to draw attention to the dependence of the absorption maximum upon the structural class of the compound. b) The values for λ_{max} for these additional compounds should provide a more comprehensive test of the validity of theoretical models than tests which rest on the predictive value of such models for water and ammonia alone. c) The absorption bands of the electron in the *pure* liquid, readily obtained by pulse radiolysis, since no alkali metal need be present, provide necessary data to assess the extent of perturbation (if any) of the spectrum by metal cation-electron pairing. d) The concentration dependence of λ_{max} in binary liquids may provide some clue as to the role of solvent structure in determining the optical properties of excess electrons in·liquids.

* This work was supported by the United States Atomic Energy Commission under Contract No. AT (11 − 1) − 1763.

In addition to the foregoing observations, a brief summary of results will be presented, if time permits, of some of our recent observations on some reactions of solvated electrons.

Results and Discussion

All the spectra obtained in pure liquids and in the various binary liquid systems were determined from pulse radiolysis of these liquids using a 3—4 MeV electron pulse from our Varian V–7715 electron linear accelerator. The experimental details of our experimental arrangement have been reported [4] as have the details of our infrared detector [1, 2], and the various methods of purification of the liquids used [1].

Solvated Electron Spectrum in Ethers

The optical absorption spectrum of the solvated electron in the pure liquid ethers was determined over a wavelength range extending to our detection limit of 2300 nm for the following ethers: tetrahydrofuran (THF), methyltetrahydro-furan (MTHF), dimethoxyethane (DME), diglyme and diethyl ether (DEE). In addition, the spectrum was also determined for diethylamine. These ethers all have macroscopic dielectric constants in the range 4.3—7.9 at room temperature. The solvated electron spectrum for all of the foregoing compounds is found to exhibit a maximum well into the infrared in the region 1900—2300 nm. The spectra for the ethers, extending from 700—2300 nm, are shown in Fig. 1.

Since the absorption maximum is fairly broad and is, in at least two cases, not far removed from our long-wavelength detection limit, its location, within an uncertainty of only ±50 nm, was determined not only by inspection of the

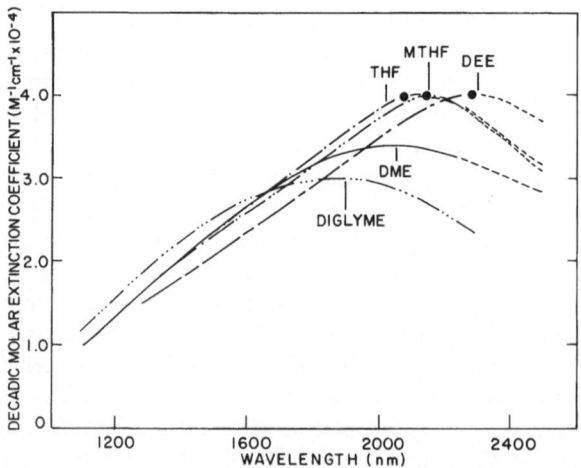

Fig. 1. Absorption spectra of the solvated electron in ethers. The solid circle represents the absorption maximum in the ether obtained by extrapolation of the data for ether–EDA solutions to pure ether. The extinction coefficients for THF and MTHF were obtained from pyrene scavenger experiments, giving an oscillator strength of 0.8. The others were calculated using the value $f = 0.8$. The uncertainty in ε may be ±20%

optical absorption curve obtained by point-to-point mapping, but also from an extrapolation procedure (as an independent check) involving the spectrum in binary liquid systems. The binary liquid system ether–ethylenediamine exhibits a solvated electron spectrum with a maximum intermediate [5] to that in the ether and in ethylenediamine, for which λ_{max} is at 1350 nm. The dependence of λ_{max} (in cm^{-1}) as a function of concentration for such systems is not far from linear, permitting an independent estimate of λ_{max} for the pure ethers by a short extrapolation from wavelengths for which λ_{max} in the binary system is not quite so close to our detection limit.

The optical absorption data for the electron in the foregoing ethers is shown in Table I, which contains the absorption maximum, at 25° C, the width at half-height of the absorption band, and an estimate of the molar extinction coefficient [6] at the maximum.

Table I. Optical absorption data for the solvated electron in ethers

Ethers	Absorption band maximum		Width at half-height	Extinction coef., at max
	(eV)	(cm^{-1})	(cm^{-1})	($M^{-1}cm^{-1}$)
Tetrahydrofuran	0.58	4720	3450	4.0×10^4
Methyltetrahydrofuran	0.57	4650	3600	4.0×10^4
Dimethoxyethane	0.60	4880	4050	3.4×10^4
Diethyl ether	0.54	4350	3500	4.0×10^4
Diglyme	0.65	5220	4650	3.0×10^4

A brief word of explanation concerning the determination of the width at half-height and of the molar extinction coefficient is in order. The value for the width at half-height is, of necessity, an approximation since with a long-wavelength detection limit at about 2300 nm it was not possible to measure the entire absorption band. The extrapolation of the absorption band in the low energy portion was obtained by using a shape function to fit the data. A combined Gaussian-Lorentzian function was used, with the Lorentzian function fitted to the high energy half of the curve and the Gaussian function fitted to the low energy half. Such shape functions had, in previous cases [5], been found to fit the data very well. The experimental data and the fitted curve for diglyme are shown in Fig. 2. It is from such curves that the width at half-height was determined.

The molar extinction coefficient was estimated [6] by using pyrene as a scavenger for the solvated electron and observing the pyrenide radical anion at 493 nm where the molar extinction coefficient is $4.9 \times 10^4\,M^{-1}\,cm^{-1}$. The uncertainty in the estimated molar extinction coefficient may be as large as $\pm 20\%$. The oscillator strength was determined for tetrahydrofuran and was found to have a value of $f = 0.8$, which may have an uncertainty as high as $\pm 25\%$.

From the data in Table I it may be seen that all of the ethers investigated exhibit a solvated electron absorption maximum at about the same frequency, corresponding to about 0.6 eV, well into the infrared and at substantially longer wavelengths than had been indicated in earlier work [7], or more recently where

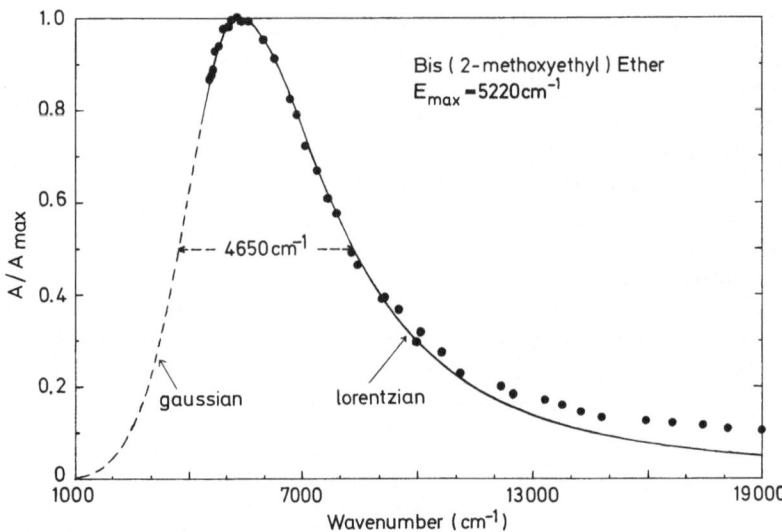

Fig. 2. Gaussian-Lorentzian fit to experimental data for the absorption band of the solvated electron in diglyme

the maximum in MTHF is given [8] as 6500 cm^{-1}. The absorption maximum for the ethers, while similar for all the compounds in this table, is drastically different from the absorption maximum of the hydroxy compounds (including water) and lies at a significantly longer wavelength than the maxima for ammonia and ethylenediamine. This marked separation in terms of types of compounds suggests a dependence of λ_{max} upon the structural class of the compound, which will be discussed in the context of correlation of these data with theory.

A comparison of the spectra for the solvated electron in the pure liquid with the spectra obtained with alkali metal solutions containing cyclic polyethers as solublizing agents [9, 10] reveals a significant difference in both the shape and absorption maxima in the latter case. The absorption maxima in the pure liquid are at a significantly lower frequency, shifted by about 100 cm^{-1} for THF, about 300 cm^{-1} for DME and about 1100 cm^{-1} for DEE. The absorption band with the solublizing agent and alkali metal present appears to be substantially broader judging from somewhat incomplete spectra for these solutions [10], the width at half-height of the absorption band being substantially greater than for the pure liquid. In diethyl ether, for example, the width at half-height for the pure liquid is 3500 cm^{-1} compared with over 5500 cm^{-1} for the solution containing crown or cryptate as the complexing agent. While the effect of the solublizing agent on the spectrum may be of interest, these differences indicate that the optical data for the electron in the pure liquid and in the metal-complex solutions are not to be regarded as a unified set of data.

In several ethers containing alkali metals it has been shown [8, 11] that, except at very low temperature, the solvated electron absorption band is not seen, but that a band appears at 12000 cm^{-1} for sodium solutions and 9000 cm^{-1} for potassium solutions. This absorption is identified [11] as an electron-metal cation pair, the equilibrium:

$$e^-_{sol} + Na^+ \rightleftharpoons (e^-_{sol} \cdot Na^+) \tag{1}$$

lying far to the right at room temperature. Comparison of the maxima at $12000\,cm^{-1}$ and $9000\,cm^{-1}$ with our spectra for the electron in pure THF and MTHF indicates a very drastic shift indeed. For sodium, this shift amounts to an energy difference of fully 0.9 eV. The magnitude of this perturbation of the solvated electron spectrum leads us to question whether the species at $12000\,cm^{-1}$ can be simply a loosely coupled electron-cation pair.

Solvated Electron Spectrum in Binary Liquids

The nature of the spectrum in binary liquids and the dependence of λ_{max} upon concentration in such two-component systems are now known for a variety of compounds because the availability of fast infrared detection has extended the observations to systems of weakly polar and nonpolar liquids. A summary of new results from our laboratory along with earlier published data will be presented. Binary liquid systems in which the spectra have been observed include: a) solutions of two strongly polar or moderately strongly polar liquids such as a number of alcohol–water solutions [12] (in which the absorption maxima of the individual components are not very dissimilar), ammonia–water [5], ethylene-diamine–water [5] and ethylenediamine–ethanol [3] (in all of which the individual adsorption maxima are widely separated); b) a solution of a strongly polar and a weakly polar liquid such as ethylenediamine–THF [1], or more particularly water–THF [1] (in which the individual maxima are at 1.7 eV and 0.58 eV, a very wide separation); c) solutions of a strongly polar with a nonpolar liquid such as ethanol with various hydrocarbons.

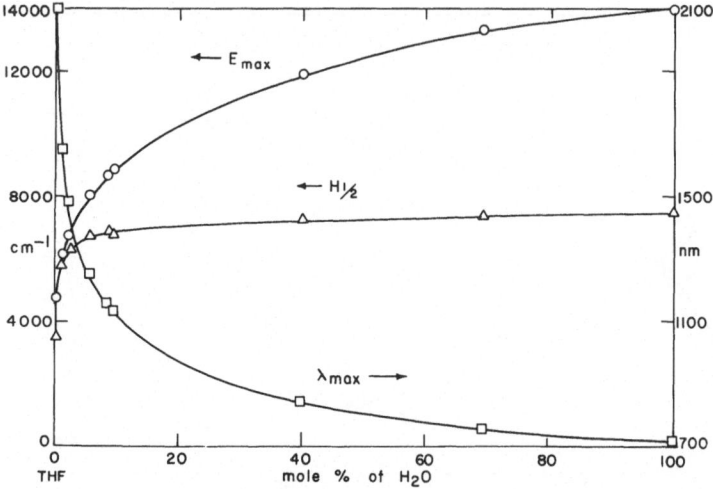

Fig. 3. Absorption maximum (in units of cm^{-1} and nm) and width at half-height of the band for the solvated electron in THF-water as a function of composition, at room temperature

The spectra of all of these binary systems exhibit one common characteristic, namely, that a single absorption band is always seen with a maximum intermediate to the maxima of the pure components, and with that λ_{max} continuously variable with composition between the two limits. This observation, on the face of it, would seem to suggest that the electron samples the aggregate properties (to a varying extent) of the solution and that selective solvation by the individual components does not occur. The dependence of λ_{max} and of the width at half-height of the band exhibits a very broad range of behavior which is, in some systems, suggestive of selective solvation.

A few examples of these results are shown here. Figure 3 shows the dependence upon composition in the water–THF solution of both λ_{max} (in units of nm as well as cm^{-1}) and the width at half-height. The values of the maximum and of the width at half-height, for this binary solution, are clearly dominated by the water.

This composition dependence does not follow the change in dielectric constant with composition which, in contrast, is dominated by THF, as may be seen in Fig. 4. Figure 5 shows the dependence of E_{max} upon $(1/D_{op} - 1/D_s)$ and upon $(1/D_{op} - 1/D_s)^2$ for the THF–water system. The dependence is anything but linear.

Two interesting examples of ethanol–hydrocarbon solutions from recent, as yet unpublished work of Brandon and Firestone [3], are shown in Fig. 6. The absorption band in 3-methylpentane–ethanol shows a slight red shift with increasing hydrocarbon content as the maximum shifts from 700 nm to 780 nm for 6 mole-% ethanol and 2 mole-% ethanol respectively. The maximum is, however, strongly dominated by the ethanol since the absorption maximum in pure ethanol is at 700 nm. A similar effect, with virtually no shift, is seen in 2,2-dimethylbutane–ethanol. The data seem to suggest selective solvation by the

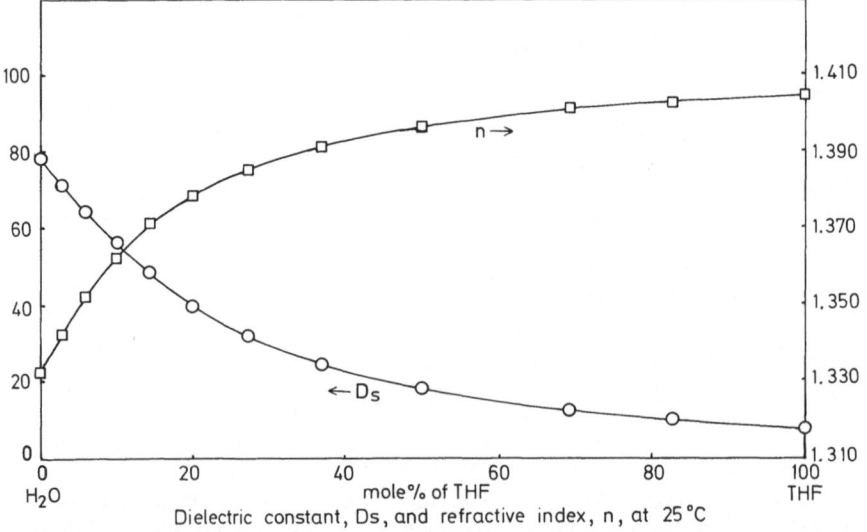

Fig. 4. Macroscopic dielectric constant of THF–water solutions at 25° C. From Ref. [13]

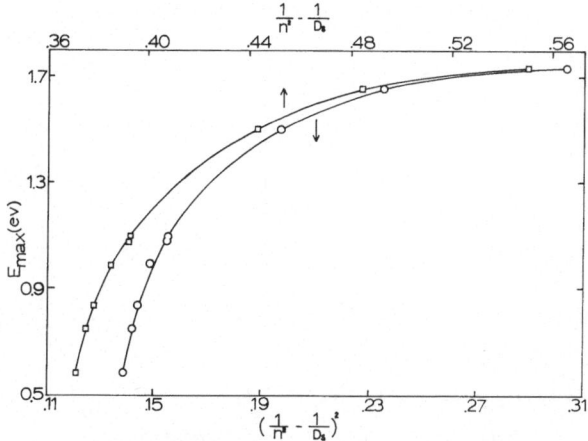

Fig. 5. Absorption maximum (in units of eV) for THF–water solutions versus $(1/D_{op} - 1/D_s)$ and $(1/D_{op} - 1/D_s)^2$

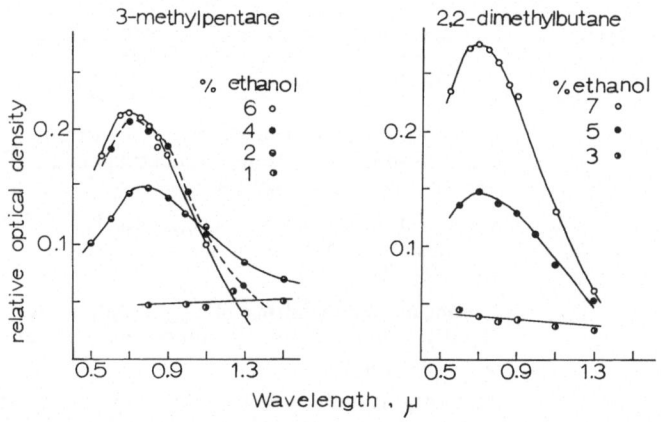

Fig. 6. Solvated electron spectra in ethanol–hydrocarbon solutions. From Brandon and Firestone [3]

ethanol. In an ethanol–ethylenediamine solution, on the other hand, the absorption maximum is strongly dominated by the ethylenediamine. In Fig. 7, the absorption band in a solution containing 56 mole-% ethanol is very little different from that observed in pure ethylenediamine.

In the ether–ethylenediamine solutions [1, 6], on the other hand, the dependence of the maximum upon composition does not deviate sharply from linearity. In the amine–water systems reported [5], the position of λ_{max} is similarly not dominated by one component. A summary, in qualitative terms, of the results for these binary liquids is given in Table II.

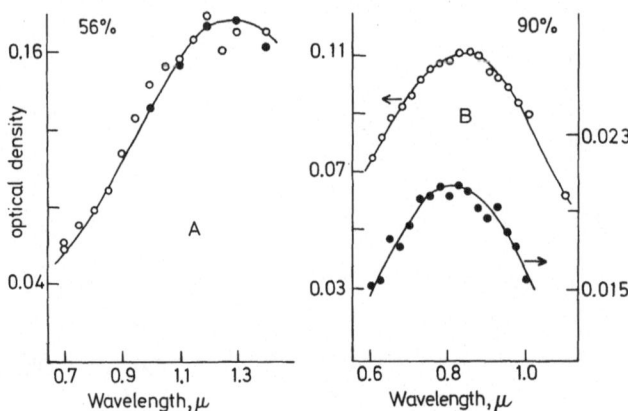

Fig. 7. The absorption spectrum of the solvated electron in ethanol–ethylenediamine solution. a The solid line is for pure ethylenediamine, from Ref. [5]. The open circles are for 56 mole-% ethanol in EDA. The closed circles are for pure EDA. b Spectra in 90 mole-% ethanol taken at the end of a 100 nsec pulse (open circles) and two decay half-times later. From Brandon and Firestone [3]

Table II. Solvent dependence of absorption maximum of e_{sol}^- in binary liquids

Binary liquid	Characteristics of e_{sol}^- absorption band	Ref.
Water–alcohol	Composition dependence of E_{max} not far from linear in some cases; some domination of λ_{max} and $W_{1/2}$ by water in others.	[12]
Water–ammonia	Composition dependence of E_{max} and $W_{1/2}$ deviates only slightly from linearity (slight sigmoid curve).	[5]
Water–EDA	Composition dependence of E_{max} deviates only slightly from linearity; $W_{1/2}$ almost independent of composition.	[5]
Ethanol–EDA	E_{max} very strongly dominated by EDA.	[3]
Water–ether	E_{max} and $W_{1/2}$ strongly dominated by water;	[1, 6]
EDA–ether	Composition dependence of E_{max} very nearly linear.	[1, 6]
Ethanol–hydrocarbon	E_{max} very strongly dominated by ethanol.	[3]

Correlation with Theory

The absorption spectrum of the solvated electron in a variety of pure liquids is shown in Fig. 8. These liquids include the alcohols and water, ammonia and some amines and five ethers. The available data indicate a marked spectral grouping of the absorption maxima on the basis of the type of compound. The hydroxy compounds shown have maxima in the range of about 580—820 nm; the ethers shown all have maxima in the range of 1900—2300 nm. Ammonia and

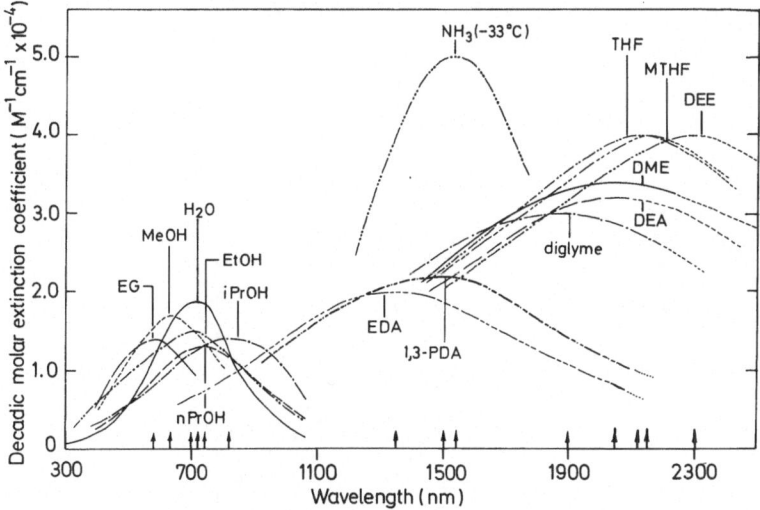

Fig. 8. Absorption spectra of the solvated electron in various pure solvents at room temperature. The arrows on the abscissa (nm) indicate the position of the maxima

the two amines lie in a somewhat broader intermediate range. Our earlier suggestion, that such a dependence of λ_{max} upon the structural class of the compound may find a useful interpretation for cavity-continuum models, in terms of the cavity radius, R_0, as the variable parameter, has been tested for the earlier, somewhat simpler model proposed in 1959 by Jortner [14]. The cavity radius required by this model to explain the values of the absorption maxima for all of the compounds in Fig. 8 has been calculated and the values shown in Fig. 9, which shows a plot of E_{max} versus R_0. The cavity radius for the hydroxy compounds lies in the range 1.0—1.6 Å. The cavity radius for the five ethers lies in the range 3.2—3.8 Å.

This simpler, earlier model [14], however, has limited validity, and more extensive calculations, using the model of Copeland et al. [15] are, at the time of writing, being, attempted, with greater importance attached to short-range interactions.

The range of behavior of λ_{max} for the variety of binary systems studied would seem to present a considerable challenge for theoretical interpretation. At first glance it would seem that the amine–water, amine–ether and alcohol–water systems may be amenable to explanation by a cavity-continuum model. The ether–water systems apparently call for a substantial variation of R_0 with composition, as the data in Fig. 5 would suggest. The alcohol–hydrocarbon data shown suggest selective solvation with a marked dependence upon short-range interaction. The most dramatic illustration of this seems to lie in the ethanol–ethylenediamine data. While the absorption maximum in the ethanol–hydrocarbon system is strongly dominated by the ethanol, the absorption band in an ethanol–ethylenediamine solution containing 56 mole-% ethanol appears to be only slightly different from the band *in pure ethylenediamine*. Perhaps more atten-

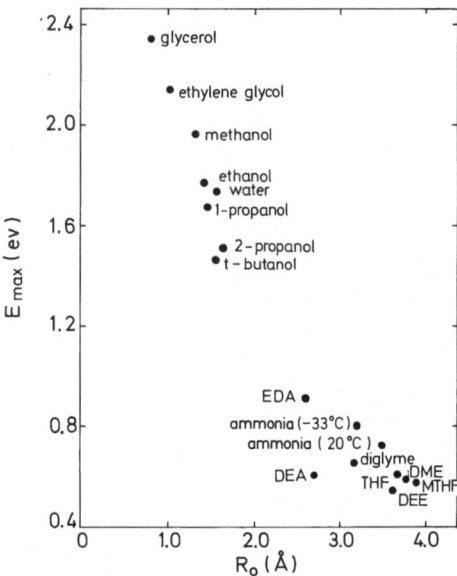

Fig. 9. Values of the cavity radius R_0, calculated from Jortner (1959) [14], to fit the values for E_{max}

tion needs to be paid to the role of hydrogen-bonded structure in the phenomenon of electron solvation.

Some Reactions of e_{sol}^-

A brief mention of the kinetics of two particular reactions of the solvated electron may be of interest in connection with considerations of the nature of this species. The first concerns the combination of e_{sol}^- with sodium ion in THF, reaction (1), giving the cation-paired form of the solvated electron. Bockrath [16], in our laboratory, has recently determined by pulse radiolysis observations, the value of the forward rate constant at room temperature, $k_{1f} = 7.9 \times 10^{11} M^{-1} sec^{-1}$. The reaction has a very low activation energy, amounting to only 1.9 kcal/mole. Thus, the half life for cation-electron pairing, at a sodium ion concentration as low as $10^{-4} M$, will be less than 10 nsecs. Taking the estimated dissociation constant for (e_{sol}^-, Na^+) to be $K_d = 3 \times 10^{-8}$ mol, from Fisher et al. [17], we obtain $k_{1r} = 2.4 \times 10^4 sec^{-1}$ for the reverse rate constant.

The second reaction concerns the bimolecular reaction of e_{sol}^- with another single-electron species, notably with another solvated electron, which had been shown [18, 19] some years ago to proceed rapidly in water:

$$e_{aq}^- + e_{aq}^- \xrightarrow{H_2O} H_2 + 2OH^- . \tag{2}$$

Recently published work [20] has indicated that the reaction of e_{sol}^- with Na^+ in ethylenediamine to form Na^-, proceeds through the analogous reaction:

$$e_{sol}^-, Na^+ + e_{sol}^-, Na^+ \rightarrow (e_{sol}^-, Na^+)_2 , \tag{3}$$

$$(e_{sol}^-, Na^+)_2 + Na^+ \rightarrow Na^- + 2Na^+ , \tag{4}$$

with the intermediate formation of a two-electron species being consistent with the observed kinetics, and $k_3 = 1.7 \times 10^9 \, \text{mol}^{-1} \, \text{sec}^{-1}$ at room temperature.

Acknowledgement

This manuscript was written while one of us (LMD) was on sabbatical leave from the Ohio State University as a Fellow of the John Simon Guggenheim Foundation at the Laboratorio di Fotochimica e Radiazioni d'Alta Energia in Bologna, Italy. I am indebted to the Ohio State University and the Guggenheim Foundation for their support, and to Professor A. Breccia for the kind hospitality extended during my brief, but pleasant stay in Bologna.

References

1. Dorfman, L.M., Jou, F.Y., Wageman, R.: Ber. Bunsenges. Phys. Chem. **75**, 681 (1971).
2. Dorfman, L.M., Pulse Radiolysis, Chap. XI. In: Hammes, G., Weissberger, A. (Eds.) Investigation of Rates and Mechanisms of Reaction. Part II. Elementary Reactions in Solution and Very Fast Reactions, New York, John Wiley and Sons (1973).
3. Brandon, J.R., Firestone, R.F.: Unpublished results.
4. Felix, W.D., Gall, B.L., Dorfman, L.M.: J. Phys. Chem. **71**, 384 (1967).
5. Dye, J.L., DeBacker, M.G., Dorfman, L.M.: J. Chem. Phys. **52**, 6251 (1970).
6. Jou, F.Y., Dorfman, L.M.: J. Chem. Phys. **58**, 4715 (1973).
7. Eloranta, J., Linschitz, H.: J. Chem. Phys. **38**, 2214 (1963).
8. Giling, L., Kloosterboer, J.G., Rettschnick, R.P.H., VanVoorst, J.D.W.: Chem. Phys. Letters **8**, 457 (1971).
9. Dye, J.L., Lok, M.T., Tehan, F.J., Coolen, R.B., Papadakis, N., Ceraso, J.M., DeBacker, M.G.: Ber. Bunsenges. Phys. Chem. **75**, 629 (1971).
10. Dye, J.L.: Private communication.
11. Kloosterboer, J.G., Giling, L.J., Rettschnick, R.P.H., Van Voorst, J.D.W.: Chem. Phys. Letters **8**, 462 (1971).
12. Arai, S., Sauer, M.C., Jr.: J. Chem. Phys. **44**, 2297 (1966).
13. Critchfield, F.E., Gibson, J.A., Jr., Hall, J.L.: J. Am. Chem. Soc. **75**, 6044 (1953).
14. Jortner, J.: J. Chem. Phys. **30**, 839 (1959).
15. Copeland, D.A., Kestner, N.R., Jortner, J.: J. Chem. Phys. **53**, 1189 (1970).
16. Bockrath, B., Dorfman, L.M.: L.M.: J. Phys. Chem. **77**, 1002 (1973).
17. Fisher, M., Rämme, G., Claesson, S., Szwarc, M.: Proc. Roy. Soc. (London) A**327**, 481 (1972).
18. Dorfman, L.M., Taub, I.A.: J. Am. Chem. Soc. **85**, 2370 (1963).
19. Matheson, M.S., Rabani, J.: J. Phys. Chem. **69**, 1324 (1965).
20. Dye, J.L., De Backer, M.G., Eyre, J.A., Dorfman, L.M.: J. Phys. Chem. **76**, 839 (1972).

Discussion

M. H. COHEN The results for optical absorption by the solvated electron in binary solutions were very interesting in that both extreme kinds of concentration dependence were observed, a quasi-linear concentration dependence and an absorption characteristic of only one constituent. The latter can be understood in two quite different ways: First, if there is a pair correlation which *favors variety* and if the range of the correlation function is larger than the distance needed to build up the polarization energy, then the electron will find the constituent giving the greater energy of solvation. Second, if the difference in energies of solvation in the two pure constituents is large enough, solvation of the electron can give rise to a demixing in the immediate environment of the cavity irrespective of the structure of the solution. Sufficiently short time resolution may separate these two cases if solvation time is shorter than demixing time. Calculation can be made if the phase diagram and some additional information is known so that the free energy of demixing can be added to the solvation energy calculation. The quantum mechanics aspects can be handled by a suitable variant of the coherent potential approximation. Finally, one should be able to observe a transition from one kind of behavior to another in experiments near a consolute point. Well above it can have nonspecific solvation, and as one approaches it, specific.

N. R. KESTNER 1a. Following comments of Morrell Cohen, it might be possible to consider the Copeland, Kestner and Jortner model in a binary liquid. One simply does calculations first of all on the pure liquids and then on cavities with various mixed first coordination layers. The actual experimental results then should be the statistical average of these separate calculations. This would correspond to case 2 of Cohen's.

1b. In accord with these comments and those of Morrell Cohen, temperature dependent measurements are extremely important. Temperature could greatly affect the radial distribution functions and correlation lengths of solvent-solvent interactions in the pure and mixed solvents.

2. The ether results seem to be in accord with simple ideas of steric effects, especially when compared with the amines and alcohols. Steric effects and the dipole moment (and polarizability) should dominate considerations; within one type of solvent, steric effects seem to explain (at a quick glance) most general trends.

J. JORTNER I would like to respond to the comments of Cohen and Kestner. You cannot use the old continuum dielectric model, as there are too many parameters in this crude treatment, and you will have to use at least three R parameters in such a treatment. In the mixed solvent most of the action originates from the role of the first coordination layer which has to be treated on the basis of a molecular model.

In this context one has to invoke the old chemical term of specific solvation. There are now different configurations of the electron localization center involving both different compositions and different coordination numbers in the first solvation layer, which in the final relaxed state may differ from the composition of the unperturbed solvent. There can be specific solvent exchange in the first coordination layer after the electron was initially localized. Some evidence for specific solvation may be obtained from the energy shifts and half line widths in mixed solvents. When the transition energy changes linearly with composition, $W_{1/2}$ exhibits a similar trend and that indicates a nonspecific statistical solvation. On the other hand, in the specific solvation case, strong variations are exhibited both in E_{max} and in $W_{1/2}$.

M. S. MATHESON Morrell Cohen remarked that in some cases of mixed solvents where the components are well mixed on the molecular scale, an electron could preferentially solvate itself with one type of solvent molecule by reorganizing the solvent with a reorganization time of about 10^{-10} sec. In pure water the solvation process is faster than this, and is apparently complete in less than 10^{-11} sec. This suggests that it may be possible to distinguish the case a) where electrons are solvated by pre-existing clusters of one type of solvent molecule from the case b) where the electron solvates by reorganizing the solvent mixture. The latter case b) will be slower and may be distinguished by the rise of picosecond pulse radiolysis of laser flash photolysis.

U. SCHINDEWOLF   From the absorption spectra it is concluded that the cavity size of the solvated electrons in ethers would be even larger than in ammonia. So far we have no further evidence for this. In our investigations on the equilibrium of the reaction: solvated electron + benzene ⇌ benzene radical anion in ammonia we found a pressure effect in accordance with an electron volume of 60—70 ml/mole, whereas in tetrahydrofuran we got no pressure effect, indicating very small electron volume in this solvent. Perhaps density measurements of the ether solutions with crown or cryptate, in which the metal solubility can be increased up to almost 1 mole/1, should give a direct answer about the size of the electrons.

J. L. DYE   The appearance of new optical bands of $Na^+ \cdot e^-$ in ethers suggests a comparison with ESR spectra in which the hyperfine contact of e^-_{solv} with the metal nucleus increases with a decrease in polarity. Perhaps we have a continuum from weakly coupled ion-pairs in the more polar amine solvents to "monomers" in the ethers.

M. MATHESON   According to Prof. Dorfman, the association of alkali metal cation and electron in ethers is closer than that of an ion pair. In semiconductors, S type orbitals have been calculated for example for the extra electron of a phosphorus atom in germanium using a central field reduced by the dielectric constant of germanium.

J. BELLONI   Concerning high rate constants in low dielectric liquids, which we recently observed at Orsay in pulse radiolysis experiments carried out in collaboration with P. Cordier and J. Delaire, $k_{e^-_{am} + Cu^{2+}}$ was $3 \cdot 10^{11}\,\mathrm{mol^{-1}sec^{-1}}$ at low temperature ($-48°$ C). It would be only a lower limit if e^-_{am} would react with free ions excluding pairs.

Radiolytic and Photolytic Formation of Stable e_{am}^- in Amide Solutions

J. Belloni and E. Saito

Abstract

The yields of formation of solvated electrons in the radiolysis of liquid ammonia obtained by varying the dose, the concentration of KNH_2 and the hydrogen pressure are interpreted on the basis of a competition between the ionization of the radical NH_2, giving NH^-, and its reaction with e_{am}^-. This mechanism implies that the reactivity of NH_2 towards H_2 is much weaker than that of its ionized form, NH^-. One can then determine the ratio of the rate constants $k_{(e^- + NH_2)}/k'_{(NH^- + NH_2)} = 1.5 \times 10^5$ and, from the limiting value of $G(e_{am}^-)$ at high amide concentrations, the primary radiolytic yield $G_{e_{am}^-} = 1.9$.

Introduction

It was shown previously [1] that in the presence of amide ions and hydrogen, the ammoniated electrons formed by γ radiolysis of liquid ammonia were protected from reactions with NH_4^+ and with oxidizing radicals NH_2 to such an extent that, at room temperature, they were as stable as ammoniated electrons obtained by the dissolution of alkali metals. H and NH_2 radicals formed directly by radiolysis of such solutions also disappeared with the formation of e_{am}^-.

We also showed [2] that the measured electron yield resulting from these various reactions $G(e_{am}^-)$ is independent of the hydrogen pressure above 0.6 atm for the amide concentration employed (5×10^{-2} mole l^{-1}). This fact suggested that the reactions of H_2 with the radicals were not competing with any other reaction. Nevertheless, as the presence of NH_2^- seemed to be necessary for the capture of NH_2 by H_2, it would appear that the ionization of these radicals had an influence on their reactivities. A systematic study of the influence of the amide concentration was thus of particular interest.

The total capture by NH_2^- or H_2, of all of the primary species resulting from the radiolysis of NH_3 and their quantitative replacement by e_{am}^- (easily measured because it is stable) offered the possibility, through the use of other independent relations between the yields, of determining the yields of the primary species. In particular, it seemed possible by this method to evaluate the primary yield of the solvated electron $G_{e_{am}^-}$, on which there is little agreement [3—9].

Knowledge of this G value, which is a measure of those electrons ejected in the primary ionization process which have escaped recombination with the parent positive ion, can also in a more general framework, provide information concerning the influence of physical parameters of the solvents on the probability of electron escape.

Experimental

In order to study the radiolytic formation of the ammoniated electron in the presence of potassium amide and hydrogen, we used cells of the type shown in Fig. 1. The whole cell is made of silica with tubes of inner diameter not exceeding

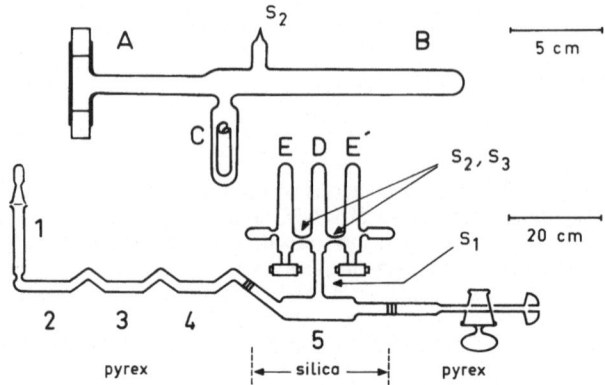

Fig. 1. Silica cell. *A* Optical cell, *B* Irradiation tube, *C* Platinum foil

10 mm. Such cells appear to withstand pressures up to 20 atmospheres. The optical section of these cells (*A*) has windows which are cylinders of high grade suprasil, diameter 10.0 mm and length 10.0 or 15.0 mm. These windows are fused into silica tubes for part of their length, leaving at least 3 mm free so that the cells can be placed in holders with 10.1 mm diameter supports and positioned in the beam of the spectrophotometer. The other end (*B*) of the cell is a round-bottomed tube in which the ammonia solution is irradiated. A side arm (*C*) contains a piece of platinum foil which serves as a catalyst for the reaction of NH_3 with the alkali metal. We used bright platinum foil instead of *Pt* black to eliminate particles from the solution.

The stability of the ammoniated electrons requires the utmost cleanliness of the tubes into which the potassium metal is distilled as well as of the cell-irradiation tube described above. The inner surface is cleaned with sulfochromic acid, rinsed with distilled water, and ions are desorbed in triply distilled water over several days. The water on the inner surfaces is removed by heating the tubes under vacuum, the silica parts being heated to about 800° C.

The lower drawing of Fig. 1 shows the standard set-up for the distillation of the potassium metal.

After being opened at both ends, the Pyrex tube into which potassium was previously distilled and stored is inserted in part 1 while being flushed with dry argon. The argon is pumped off and when the vacuum is established, the potassium is distilled successively from compartments 2, 3, 4 to 5, the preceding tube being sealed off after each transfer. Finally, the tube is rotated around the horizontal axis and a fraction of the metal flows into tube D. Purified ammonia is then condensed into tube D, and the set of silica tubes is removed after being sealed off at S_1. The ammonia is first allowed to warm up and dissolve the metal, and then aliquots are poured into tubes E and E'. E, D and E' are immersed in liquid nitrogen, and the three tubes are separated by sealing off at S_2 and S_3. In E and E' the solution is transferred to the side arm containing the platinum foil to catalyze the decomposition of the metal-ammonia solution. The solution in D is used to determine the concentration of KNH_2 in the final solution. The amount of ammonia is measured by weighing the solution and the quantity of metal by conductometric titration with HCl after transformation to KOH.

The pressure of H_2 in the cell is calculated from the stoichiometry of the reaction and the volume of space in the cells. The solubility of H_2 in ammonia is about 3×10^{-3} mole l^{-1} atm^{-1} at $+25°$ C.

The ammonia is purified by condensing it in a flask containing potassium metal and letting the solution stand for a few hours at $-40°$ C. The purified ammonia is stored as a gas in 10 l flasks from which requisite amounts are condensed in the various cells.

When the cell is properly cleaned and the potassium and ammonia are highly purified, the dark blue solution takes months to decompose even in contact with the platinum foil. We rejected solutions which decomposed in less than a month because we observed that in these tubes the decay rate of electrons induced by γ or UV irradiation was not negligible compared to their rate of formation.

Optical Measurements

The spectra of the solvated electron were recorded with a Beckman DK—1A double beam spectrophotometer. In order to determine relatively low concentrations of e_{am}^-, cells with optical lengths of 2.5 cm were used. These had the disadvantage of presenting an anomaly at wavelengths at which the solvent itself strongly absorbs. This subject is discussed in another paper [10].

When the optical density of the 1750 peak of the e_{am}^- at 20° C increased above the value of 2.0, measurements were made at $\lambda = 1400$ nm. It was found that the ratio OD (1400 nm)/OD (1750 nm) has an average value of 0.69 in concentrated amide solutions [10].

With the pair of cells described above, the amounts of solution in the cells, leading to different quantities hence pressures of H_2 evolved. As a result, the equilibrium concentrations of e_{am}^- were not identical.

Irradiation

The 1500 Ci ^{60}Co source used for γ irradiation gave dose rates of 10^{18} to 10^{19} eV g^{-1} h^{-1}. They were determined using the Fricke dosimeter and taking into account the relative electronic density of the media at each temperature. During all operations except the spectrophotometric measurements the solutions were kept in the dark.

The photochemical experiments were carried out using a mercury high-pressure lamp (Hanau Q 600) and an interference filter, Intervex A ($\lambda_{max} = 254$ nm, $\Delta\lambda = 19$ nm, transmission $T = 0.21$). Actinometry was effected using the Parker and Hatchard method   with potassium ferrioxalate in 0.1 N sulfuric acid.

Results

With high amide concentrations (0.5—2.5 mole l^{-1}) and hydrogen pressures of a few atmospheres, the thermal equilibrium[12] concentration of e_{am}^- is not negligible and must be measured before irradiation:

$$NH_2^- + 1/2 H_2 \rightleftharpoons e_{am}^- + NH_3 . \tag{1}$$

The optical density for zero dose at the absorption maximum was therefore measured using the pure solvent as reference. However, the spectra after irradiation were recorded relative to a reference cell containing the same amide solution in order to decrease the perturbations of the absorption curve due to this ion.

Curves of the optical density measured at room temperature as a function of absorbed dose are given in Fig. 2 for concentrations of KNH$_2$ (0.58, 0.87, 2.35 mole l^{-1}), and pressures of H$_2$ (0.83, 3.3, 8 atm). We obtained identical results when after a first irradiation run, the solution was left in contact with the *Pt* foil until thermal equilibrium was attained and then the irradiation experiment was repeated.

The concentrations of e_{am}^- were calculated along the whole temperature range ($-40°$ C to $+30°$ C) using a single value of the extinction coefficient $\varepsilon = 4.8 \times 10^4$ l mole^{-1} cm^{-1} [13], at the respective wavelengths of the maxima: 1,800 nm (30° C), 1,750 nm (20° C), 1550 nm ($-15°$ C) and 1400 nm ($-40°$ C). The radiolytic yield of e_{am}^- decreased with dose, but the solvated electron concentration was far from reaching a limiting value even at the highest OD which could be measured with our instrument. At a given temperature, the rate

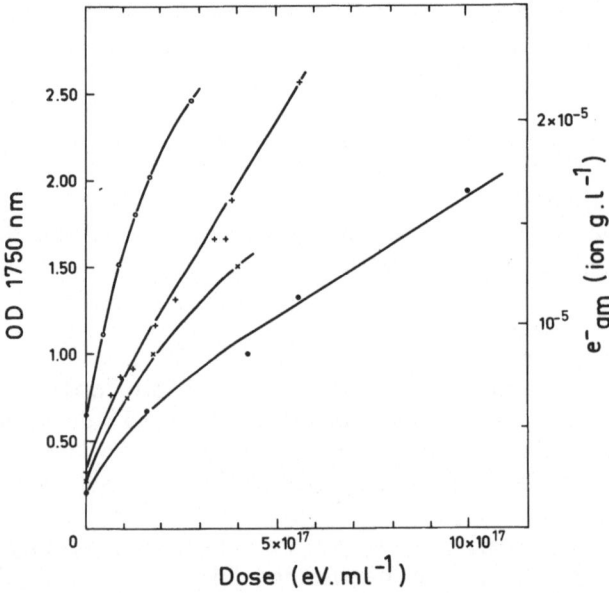

Fig. 2. Radiolytic formation of e_{am}^- at 20° C for different amide concentrations. ○ 0.58 mole l^{-1}, 0.83 atm H$_2$. + 0.87 mole l^{-1}, 3.3 atm H$_2$. ● 2.35 mole l^{-1}, 8 atm H$_2$

of formation of e_{am}^- was greater the higher the amide concentration (Fig. 2). For all amide concentrations, this rate decreased at low temperatures (Fig. 3) and one could observe, as shown previously[1] for a concentration of 0.05 mole l^{-1} at 20° C, a plateau in the range of measurable OD's.

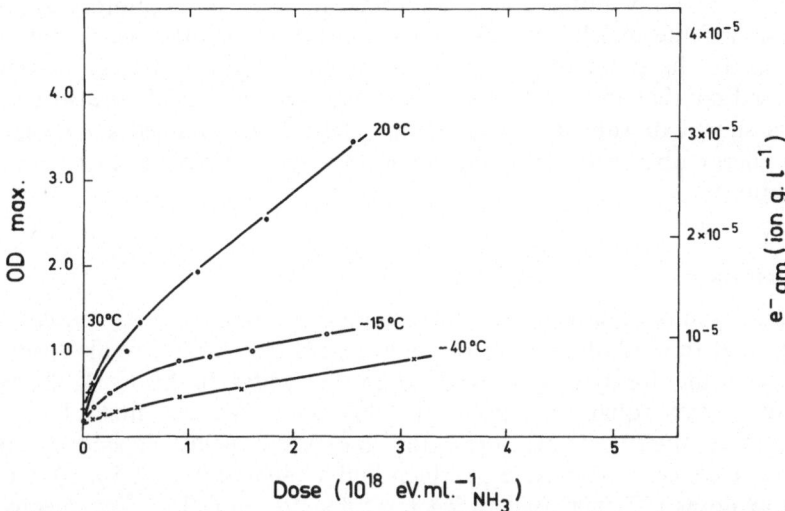

Fig. 3. Radiolytic formation of e_{am}^- at different temperatures. The cell contains 1.16 g NH_3, 1.1×10^{-3} mole KNH_2 (0.58 mole l^{-1} at 20° C), 0.83 atm H_2 at 20° C. Optical path length $= 2.43$ cm; concentrations calculated with $\varepsilon_{max} = 4.8 \times 10^4$ l mole^{-1} cm^{-1}. × $T = -40°$ C ($\lambda_{max} = 1400$ nm). ○ $T = -15°$ C ($\lambda_{max} = 1550$ nm); ● $T = 20°$ C ($\lambda_{max} = 1750$ nm), + $T = 30°$ C ($\lambda_{max} = 1800$ nm)

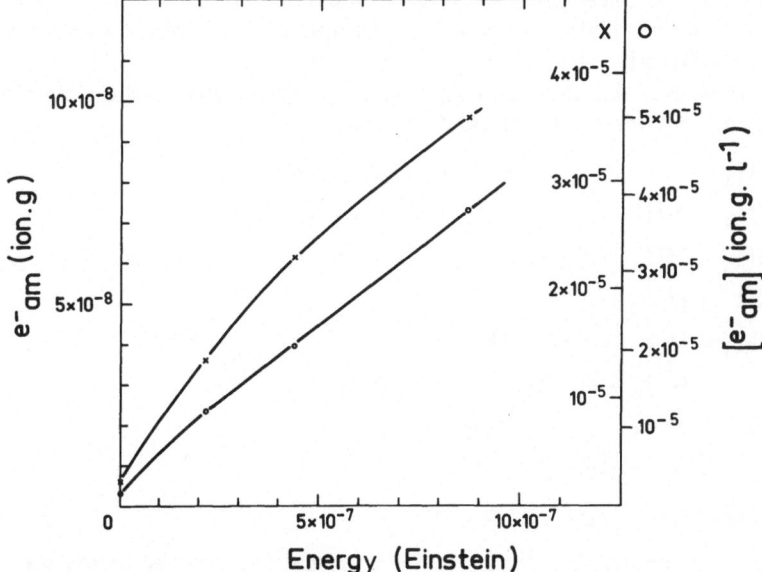

Fig. 4. Photolytic formation of e_{am}^- at 25° C; ○ 0.58 mole l^{-1}, 0.83 atm H_2, vol. 1.91 ml; × 0.87 mole l^{-1}, 3.3 atm H_2, vol. 2.67 ml

The results for the photolytic formation of a stable ammoniated electron are not essentially different from those presented above (Fig. 4). Here also, hydrogen plays the role of a scavenger of radicals which would otherwise recombine with e_{am}^- resulting from the photoionization of NH_2^-. Indeed, with solutions of amide alone and in opposition with bleached metal-ethylamine solutions the solvated electron was only observable as a transient after flash photolysis [14]. The longest half life which concerns a small fraction of the species was 1 min at $-65°$ C. In the presence of hydrogen, on the contrary, the e_{am}^- concentration increased notably and the OD constant with time was easily measured at $20°$ C after a short exposure to a weak lamp. The results obtained show that for the same energy absorbed, the quantum yield is smaller for the lower amide concentration.

Discussion

Previous results obtained with dilute amide solutions (5×10^{-2} mole l^{-1}) have shown that the radiolytic yield of ammoniated electrons at weak doses did not increase further for hydrogen pressures above 0.6 atm. In this work, the hydrogen pressures were equal or superior to this limit. We see, first of all, that at comparable hydrogen pressures, the measured yields depend strongly on the amide concentration: $G(e_{am}^-)$ which had a value of 0.63 at 5×10^{-2} mole l^{-1} and low doses [2], increased to 2.4 for 5.8×10^{-1} mole l^{-1}. This means that in contrast to the reactions of capture of radicals by H_2 which are complete above 0.6 atm, the reactions of NH_2^- are not at 6×10^{-1} mole l^{-1}.

The decrease of $G(e_{am}^-)$ as a function of dose can be attributed to the gradual accumulation of e_{am}^- thus favoring its disappearance by reactions with radicals. At a given hydrogen pressure and amide concentration, the radiolytic yield $G(e_{am}^-)$ depends chiefly on the concentration of e_{am}^- already present produced either radiolytically or by thermal equilibrium [1].

In the presence of amide and hydrogen the reactions of the different species formed in the primary radiolytic process are:

(0) $NH_3 \xrightarrow{\quad\quad} H, e_{am}^-, NH_4^+, H_2, N_2H_4, NH_2, NH$;

(2) $NH_4^+ + NH_2^- \rightarrow 2NH_3$;

(3) $NH + NH_3 \rightarrow N_2H_4$;

(4) $NH + H_2 \rightarrow NH_3$;

(5) $NH_2 + H_2 \rightarrow NH_3 + H$;

(6) $H + NH_2^- \rightarrow e_{am}^-$;

(7) $NH_2 + NH_2^- \rightarrow NH^- + NH_3$;

(8) $NH^- + H_2 \rightarrow e_{am}^-$;

(9) $NH_2 + e_{am}^- \rightarrow NH_2^-$.

In basic as in neutral solutions, the NH radical leads to the formation of N_2H_4 [reaction (3)], but in the presence of high pressures of hydrogen it is not excluded that it recombines with H_2 [reaction (4)]. Nevertheless, these two reactions do not influence the measured value of $G(e_{am}^-)$, since in both cases the products formed do not react with e_{am}^-.

Concerning the radical NH_2, the three reactions (5), (7) and (9) are possible. If reactions (5) and (8) were each much more efficient than (9), the formation of e_{am}^- by (5) followed by (6) and by (8) would be independent of the neutral or ionized form of the radical NH_2, and would not vary with the basicity of the solution. The measured yield $G(e_{am}^-)$ which includes the primary solvated electrons (reaction 0) as well as those coming from reaction (6), increases with the amide concentration, so one must admit that in its neutral form, NH_2 is much less reactive with H_2 than it is in its ionized form NH^-. In this case, the most probable fate of the neutral radicals is that they recombine with e_{am}^- [reaction (9)]. This competition between (7) and (9) may be the origin of both the concentration effects of amide and the variations of $G(e_{am}^-)$ with dose.

To verify that the results correspond quantitatively to the proposed mechanism, the measured yields at different doses (curves of Fig. 2) have been plotted as a function of the ratio of amide concentration to the ammoniated electron concentration already existing at each value considered. The curve obtained, corresponding to various concentrations of KNH_2 between 0.58 and 2.35 mole l^{-1} is given in Fig. 5a. There is no distinction made for the concentrations of e_{am}^- indicated on the figure between those electrons arising from thermal equilibrium and those from radiolysis.

We have also shown on this figure, the result previously established for 0.05 mole l^{-1} on KNH_2 and 0.8 atm of H_2: for a concentration* of 3×10^{-6} in e_{am}^-, $G(e_{am}^-) = 0.63$. The curve shows that the yields seem to depend only on the relative concentrations of KNH_2 and e_{am}^- and not on the concentration of KNH_2 or on the pressure of H_2. The measured $G(e_{am}^-)$ increases markedly until $C_{KNH_2}/C_{\theta_{am}} \sim 2 \times 10^5$, then approaches a limit in agreement with the mechanism

---

* Concentrations of e_{am}^- and G values of Ref. [2] were corrected for the change of extinction coefficient $\varepsilon_{max} = 4.8 \times 10^4 \, l \, mole^{-1} \, cm^{-1}$ instead of $4.1 \times 10^4 \, l \, mole^{-1} \, cm^{-1}$ [2].

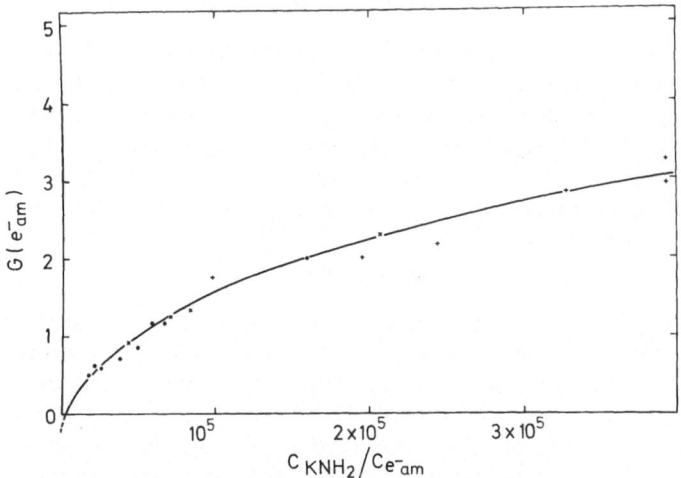

Fig. 5a. Variation of $G(e_{am}^-)$ as a function of the ratio $C_{KNH_2}/C_{e_{am}^-}$

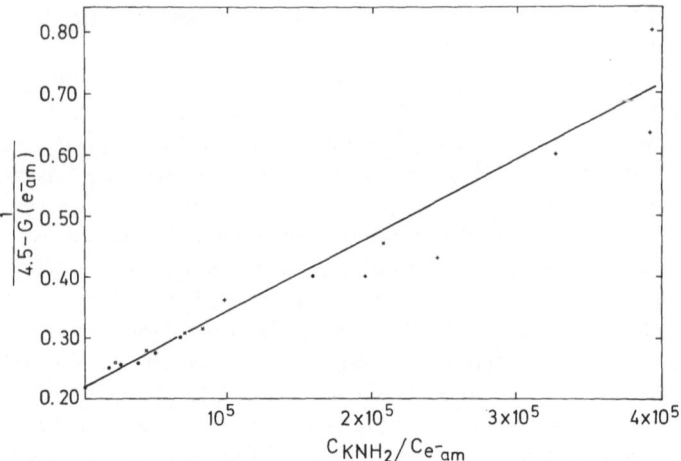

Fig. 5b. Plot of $1/G_{\text{lim}} - G(e_{\text{am}}^-)$ vs. $C_{\text{KNH}_2}/C_{e_{\text{am}}^-}$

given by the reactions (0) to (9). According to this mechanism, the yield should vary from the value of $G_{e_{\text{am}}^-} + G_H - G_{NH_2}$ in the absence of amide that is, starting from a negative value $G(e_{\text{am}}^-) = -0.12$, [3], [15], [16] up to the value of $Ge_{\text{am}}^- + G_H + G_{NH_2}$. In order to determine this last value more precisely, the results have been plotted as an inverse function (Fig. 5b):

$$\frac{1}{G_{e_{\text{am}}^-} + G_H + G_{NH_2} - G(e_{\text{am}}^-)} = \frac{1}{2G_{NH_2}} + \frac{1}{2G_{NH_2}} \times \frac{k_7 C_{\text{KNH}_2}}{k_9 C_{e_{\text{am}}^-}}$$

This function is linear if the value of $G_{e_{\text{am}}^-} + G_H + G_{NH_2}$ is adjusted to 4.5 ± 0.3 which at the same time determines the value of G_{NH_2} equal to 2.3. Hence the yield of the reducing species:

$$G_{e_{\text{am}}^-} + G_H = G_{NH_2} - 0.12 = 2.2 \,.$$

If this value of the yield is compared to the yield $G(H_2) - G_H = 0.30$ measured in the presence of solutes known to form H_2 with atomic H (alcohols, hydrazine) [16, 17], it becomes evident that even in acid media, the totality of the reducing species does not participate in this mechanism of H_2 formation and that most probably the yield corresponds to the capture of H alone: $G_H = 0.30$. Pugsley [8] arrived at the same conclusion measuring $G(HD)$ in solutions of deuterated formate. The final result is

$$G_{e_{\text{am}}^-} = 1.9 \pm 0.2 \,.$$

The accuracy of this value depends in the first place on the method of calculating the asymptote. Secondly, since the amide concentrations are relatively high and correspond to a ratio of solute/solvent of $\sim 3\%$ for 1 mole l^{-1}, the energy absorbed directly by the solute may be at the origin of a supplementary formation of e_{am}^- as observed in photolysis. Furthermore, amide or H_2 present at high concentrations in the solution may react in the dense ionization tracks, for example with the NH_2 radicals, with precursors of hydrazine, or with H atoms and hence increase the $G(e_{am}^-)$ above the value calculated on the basis of the mechanisms (0) to (9). Nevertheless, since the results obtained do not vary noticeably for a given value of the ratio $C_{KNH_2}/C_{e_{am}^-}$ and for amide concentrations up to 2.35 mole l^{-1}, it follows that the direct effect of irradiation and the capture of precursors can be neglected in this range.

We also examined the possibility that the radicals arising from the over-all reaction (0) react much faster with the isolated NH_2^- ions or ammoniated electrons, than with the complexes of these species with the cation K^+. Using the association constants given in the literature, we plotted our $G(e_{am}^-)$ values as a function of the ratio of the concentration of free ions, NH_2^- and e_{am}^-. However, since the equilibrium between e_{am}^- and K^+ depends on the K^+ concentration, and so on the association $K^+ NH_2^-$, the form of the curve is not very different from that of Fig. 1. For concentrations of 0.05—0.87 mole l^{-1} we obtained a lower limiting value than that previously found, $G(e_{am}^-) \sim 4.0$. However, the results concerning the solution 2.35 mole l^{-1} now lie on another slightly displaced curve. It is possible that this effect results simply from uncertainty in the values of the associated constants. The value $G_{e_{am}^-} = 1.7$ is near the lower limit of error of the value obtained without taking association of ions into account.

At 30° C, the results are similar to those at room temperatures; at low temperature the determination of $G(e_{am}^-)$ as a function of the ratio of concentrations $C_{KNH_2}/C_{\theta_{am}}$ does not lead to a limiting value as for 20° C. For example, at $-15°$ C and high concentrations of KNH_2 (for which at zero dose $C_{KNH_2}/C_{e_{am}^-} = 2 \times 10^6$), the highest value of $G(e_{am}^-)$ is 2.85, but the asymptotic value may be much higher. This difference in behavior with temperature may arise from a displacement in the competition between reactions (7) and (9), such that reaction (9) becomes relatively important at low doses, without it being necessary to suppose a great variation of $G(e_{am}^-)_{max}$ with temperature.

We collected in Table I the published values for $G_{e_{am}^-}$ and the corresponding experimental conditions. One must remark that direct determinations by spectro-photometry using pulsed radiolysis techniques depend on the value chosen for the extinction coefficient of e_{am}^-. The accuracy of such values is also dependent

Table I. Radiolytic yield of the solvated electron in liquid ammonia

Authors	$G_{e_{am}^-}$	T		Method
Cleaver et al. (1958) [3]	$1.75 - G_H$	20° C		γ, $G(-$ Meth. blue$)_{MB}$, $G(-e_{am}^-)_{Na}$
Dainton et al. (1964) [4]	$0.34 - G_H$	20° C		γ, $G(N_2)_{N_2H_4}$
Compton et al. (1965) [5]	0.45	$-45°$ C	p.r.[a]	
Dobo (1967) [6]	$1.74 - G_H$	20° C		γ, $G(2$–amino–4–nitrophenol$)$ paranitrophenol
Ward (1968) [7]	3		p.r.	
Pugsley (1968) [8]	1.95	20° C		γ, $G(N_2)_{N_2O}$, $G_H = 0.30$
Khaïkin et al. (1971) [9]	1.2 ± 0.3	$-60°$ C	p.r.	
This work (1972)	1.9 ± 0.2	20° C		γ, $G(e_{am}^-)_{NH_2^-, H_2}$, $G_H = 0.30$

[a] Pulse radiolysis.

on that of the dosimetry. In steady state radiolysis, values obtained, even when they are based on results confirmed by several authors, greatly depend on the supposed mechanisms and the relations linking the primary yields. All of the papers referred to in Table I admit that $0 \leqq G_{NH_2} - G_{\theta_{am}} - G_{\theta_{am}} - G_H \leqq 0.12$ but that the value of G_H is not known.

We can compare these data with the value recently obtained at Orsay [18] by pulsed radiolysis of pure ammonia at $-48°$ C using 3 ns electron pulses from a Febetron 706 accelerator of 600keV, where a ratio $G_{\theta_{am}}(-48° C)/G_{\theta_{aq}}(20° C) = 0.92$ was obtained taking $e_{am}^- = 13000 \, 1 \, mole^{-1} cm^{-1}$ at 1000 nm and at $-48°$ C.

The dose absorbed was of the order of 10^{18} eV ml^{-1} per pulse. Taking the value of $G_{\theta_{aq}} = 3.3$ we obtain $G_{\theta_{am}} = 3.0$ at $-48°$ C. Considering the temperature difference, this is not incompatible with the value deduced from the study of amide solutions.

These values, as well as the value of Pugsley [8] and even more so that of Ward [7], are high compared to the free electron yields in other polar solvents [19—21].

This peculiarity which distinguishes NH_3 from other solvents of comparable dielectric constants, may on one hand arise from the very short relaxation time of the solvent ($\tau = 1.3 \times 10^{-12}$s) allowing an efficient escape by solvation of the ejected primary electron, and on the other hand, from the very low chemical reactivity of e_{am}^-, particularly with respect to the acidic ion NH_4^+ ($k_{e_{am}^- + NH_4^+} \sim 10^6 1$ mole^{-1}s^{-1} [22]) which may thus reduce the yield of recombination of the solvated ions.

Another point which the present results seem to establish, to the extent that the mechanisms (0) to (9) are verified, is the ratio of the rate constants $k_9/k_7 = 1.7 \times 10^5$. This is deduced from the slope of the curve of Fig. 6 corresponding to the case for which the ions are considered to react whether they are associated or not. If we suppose that only the free ions react in (7) and (9), the ratio k_9/k_7 is 8×10^4. As Ward [7] has shown, it is very likely that the constant k_9 has a high value: $k_9 = 2.5 \times 10^{10} 1$ mole^{-1}s^{-1}.

It would have been interesting to determine k_9/k_7 from the photolysis results for which e_{am}^- and NH_2^- should also play the role of scavengers of the radical NH_2:

(00) $\quad NH_2^- \quad \xrightarrow[NH_3]{hv} NH_2 + e_{am}^- \; ;$

(7) $\quad NH_2 + NH_2^- \rightarrow \quad NH^- + NH_3 \; ;$

(8) $\quad NH^- + H_2 \quad \rightarrow \quad e_{am}^- \; ;$

(9) $\quad NH_2 + e_{am}^- \quad \rightarrow \quad NH_2^- \; .$

Indeed, we see from Fig. 4 that the quantum yields decrease with increasing e_{am}^- concentrations, in agreement with the competition between reactions (7) and (9). Nevertheless, the highest values of $\Phi_{(e_{am}^-)}$ at the origin (0.09—0.11) are a long way from the theoretical limiting value of $\Phi_{e_{am}^-} = 2$ when reaction (9) is negligible. They are even lower than the quantum yields of photodetachment found for other systems [23]. In fact, for the present system in which NH_2^- is both the source of e_{am}^- and the scavenger of NH_2, we are using NH_2 concentrations such that the absorption is complete in a very small thickness of the solution. Aside from cage effects which decrease the yield of the reaction (00), the homogeneous recombination (9) must be a factor which reduces $\Phi_{e_{am}^-}$ in the absorption zone, particularly because the local concentration of the species is quite high. Although it is impossible, under such circumstances, to apply a treatment for homogeneous kinetics, nevertheless, these experiments confirm that amide solutions containing hydrogen can stabilize the ammoniated electron formed by photodetachment, whereas in the presence of amide alone, the electron reacts rapidly with the oxidizing radicals in the form of NH_2 and perhaps also NH^-.

References

1. Belloni,J., Fradin de la Renaudiere,J.: Nature Phys. Sc. **232**, 173 (1971).
2. Belloni,J., Fradin de la Renaudiere,J.: Int. J. Radiat. Phys. Chem. **5**, 23 and 31 (1973).
3. Cleaver,D., Collinson,E., Dainton,F.S.: Trans. Faraday Soc. **56**, 1640 (1960).
4. Dainton,F.S., Skwarski,T., Smithies,D., Wezranowski,E.: Trans. Faraday Soc. **60**, 1068 (1964).
5. Compton,D.M.J., Bryant,J.F., Cesena,R.A., Gehman,B.L.: Pulse Radiolysis, p. 43. Ebert, Keene, Swallow, Baxendale (Eds.). New York: Academic Press (1965).
6. Dobo,M., Schulte-Frohlinde,D.: Report Kernforschungszentrum, Karlsruhe (1967).
7. Ward,B.: Advan. Chem. Ser. **81**, 601 (1968).
8. Pugsley,R.G.: Thesis, Leeds (1968).
9. Khaïkin,G.I., Zhigunov,V.A., Dolin,P.I.: Khim. Vys. Energ. **5**, 54 (1971).
10. Saïto,E.: Colloquium Weyl III.
11. Parker,C.A., Hatchard,C.G.: Proc. Roy. Soc. (London) A**235**, 518 (1956).
12. Kirschke,J.J., Jolly,W.L.: Inorg. Chem. **6**, 855 (1967).
13. Douthit,R.C., Dye,J.L.: J. Am. Chem. Soc. **82**, 4472 (1960). Jolly,W.L., Hallada,C.J., Gold,H.: Metal-ammoniac solutions. In: Lepoutre,G., Sienko,M.J. (Eds.): New York: Benjamin 1964. – Quinn,R.K., Lagowski,J.J.: J. Phys. Chem. **73**, 2326 (1969).
14. Ottolenghi,M., Linschitz,H.: Advan. Chem. Ser. **56**, 149 (1965).
15. Belloni,J.: Radiochem. Radioanal. Letters **3**, 305 (1970).
16. Belloni,J.: Actions Chim. Biol. Rad., Haissinsky,M., Ed., **15**, 47 (1971).
17. Schischkoff,D., Schulte-Frohlinde,D.: Z. Phys. Chem. **44**, 112 (1965).
18. Belloni,J., Cordier,P., Delaire,J.: Preliminary results.
19. Freeman,G.R.: Actions Chim. Biol. Rad., Haissinsky,M., Ed., **14**, 73 (1970).
20. Hayon,E.: J. Chem. Phys. **53**, 2353 (1970).
21. Fletcher,J.W., Richards,P.J., Seddon,W.A.: Can. J. Chem. **48**, 1645 (1970).
22. Brooks,J.M., Dewald,R.R.: J. Phys. Chem. **75**, 986 (1971).
23. Blandamer,M.J., Fox,H.F.: Chem. Rev. 59 (1970).

Kinetic Studies of Reactions Involving the Ammoniated Electron

R. R. DEWALD, R. L. JONES, and H. BOLL

Abstract

Second-order rate constants are reported for the reactions of the ammoniated electron with triphenylmethane, dimethyl sulfide, diethyl sulfide, imidazole, pyrrole, dimethyl sulfoxide, DL-norleucine, thiophene and pyridine. The second-order rate constants were found to vary from 5×10^5 to $6.5 \times 10^{-2} \, M^{-1} \sec^{-1}$ while activation energies for the reactions in which present data allowed determination, were invariant at 4—6 kcal/mole. The kinetic behavior of the ammoniated electron is compared with results obtained for the hydrated electron.

Introduction

Hart and Anbar [1] concluded to the first approximation that the rates of solvated electron reactions are independent of the dielectric constant and the viscosity of the solvating medium. Moreover, they suggested that hydrated electron, e_{aq}^-, reactions are in accord with an electron tunneling mechanism. Their hypothesis is based on the surprisingly comparable rates of the reactions of aromatic compounds with solvated electrons in water and methanol [2], the relative rates of mobile electrons in ice with different substrates [3], the invariance of the activation energy [1], and demonstrated free energy correlations [1]. However, these authors carefully point out that with the available experimental data, it is still not possible to unequivocally state that e_{aq}^- reactions involve electron tunneling.

Using arguments based on the transition state theory, Schindewolf [4, 5] has suggested that a marked difference should exist between the kinetic behavior of e_{aq}^- and ammoniated electrons, e_{am}^-. His argument is that the activation volume is much more negative in e_{am}^- reactions than in e_{aq}^- reactions because of the cavity size difference [6] between e_{aq}^- and e_{am}^-. Consequently if the reaction involves a transition state, $k_{(e_{aq}^- + R)}$ should be considerably greater than $k_{(e_{am}^- + R)}$, where R is any reactant.

Kinetic data for the ammoniated electron are limited compared with data reported for the aqueous electron [1, 7, 8]. The reactions of sodium with water [9], urea [10], hydrazine [11], ethanol [12], tert-butyl alcohol [9], and ammonium ion [13] have been studied kinetically. In addition, the kinetic properties of solvated electrons in water–ammonia mixtures have been reported [5]. Some competitive kinetic studies have been reported [14, 15] but the interpretation of results of these studies is probably open to question. The purpose of this paper is to report some second-order rate constants for reactions involving the ammoniated electrons and to compare these results with similar systems involving the aqueous electron. We are hesitant to speculate on possible mechanisms for these reactions since in most cases the reaction products are not yet known with certainty and the kinetic data should be considered preliminary in nature.

Experimental Section

Ammonia (Matheson) and sodium (United Mineral and Chemical Co.) were purified and stored using procedures described elsewhere [9, 16]. Dimethyl sulfide, sodium oxalate, sodium acetate, dimethyl sulfoxide, and pyridine were obtained from Fisher Scientific. Triphenylmethane, diethyl sulfide, imidazole, DL-norleucine and thiophene were obtained from Eastman. Sodium cyanide, sodium thiocyanide, methyl disulfide and n-butyl disulfide were obtained from K&K Laboratories. Fragile glass ampoules containing the compounds having an appreciable vapor pressure were prepared using a repeated freezing and evacuation technique described elsewhere [9, 12]. Compounds having negligible volatility at room temperature were placed into fragile glass ampoules or break-seal tubes and sealed off under high vacuum.

The progress of slow reactions was followed conductometrically by using equipment similar to that described elsewhere [9, 16]. A number of reaction apparatuses, similar to those described elsewhere [9, 16], were used in the present study. First a fragile ampoule or break-seal tube containing a weighed amount of reactant was sealed to the reaction apparatus. The apparatus was then scrupulously cleaned using an established procedure [17]. After evacuation, sodium was distilled into the reaction vessel *via* a side arm and ammonia condensed into the calibrated bulb of the vessel. The apparatus was next placed into (Harris Mfg. Co.) a convection fluid test chamber capable of maintaining temperatures (within the range $-80°$ to $25°$ C) to better than $0.1°$ C at the selected temperature. The conductance of the sodium-ammonia solution was determined and the initial sodium concentration was calculated from reported conductance data [16]. The resistance of the solution was subsequently measured as a function of time for approximately an hour to insure the stability of the metal solution. Finally, the fragile ampoule or break-seal tube containing the reactant was broken, the solution mixed, and resistance of the reacting system measured as a function of time. The resistance-time data were evaluated following methods outlined elsewhere [18].

For reactions too fast to permit study by conventional methods, a thermostated stopped-flow apparatus described elsewhere [13] was used. All operations including solution make-up, solution transfer, rapid mixing and observation were performed *in vacuo*. The progress of the reaction was followed by monitoring the absorbance decay of e_{am}^- at 1000 nm where the extinction coefficient is about 10^4 [19].

In order to verify the stoichiometry of the reaction in all cases, a conductometric titrator [20] capable of performing titrations *in vacuo* was developed. Work concerned with verification of stoichiometry and identification of products for the reactions reported in the present study is currently in progress.

Results and Discussion

All of the kinetic data in this study were analyzed by assuming second-order kinetics, i.e., the reaction being first-order with respect to each of the reactants. Second-order plots of $[1/(a-b)] \ln [b(a-x)/a(b-x)]$ vs. t, where a and b are the initial concentrations of reactant and e_{am}^-, respectively, and x is the e_{am}^- concentration at time t gave a reasonable fit of the kinetic data in most cases. Table I contains a summary of the kinetic results of this study. It should be noted that the second-order rate constants varied by about seven orders of magnitude for the different reactants while the activation energies were essentially invariant at about $4-6$ kcal/mole.

Table II is a comparison of some rate constants for reactions involving e_{am}^- and e_{aq}^-. The rate constants for the reaction of e_{am}^- in Table II are those calculated for $25°$ C using the experimentally determined activation energies or an estimated 4.5 kcal/mole for the reactions in which data did not permit evaluation of the activation energy. It should be noted from Table II that the oxalate, acetate, cyanide, and thiocyanide ions do not react with e_{am}^-. These observations are in accord with results reported for e_{aq}^-. Moreover, the reactions of e_{am}^- and e_{aq}^- with imidazole and ammonium ions have comparable rates. However, for the other reactants given in Table II, the rate constants of e_{aq}^- reactions are considerably greater than the corresponding e_{am}^- reactions.

Table I. Summary of the kinetic data for reactions of the ammoniated electron

Reactant	Method	k mol^{-1}sec^{-1}	Temperature °C	Ea[a] kcal/mole	Ref.[b] concerning reaction products
Dimethyl sulfide	Cond.[c]	1.7	$-65°$	3.8	[21−23]
	Flow[d]	5.5	$-34°$		
Diethyl sulfide	Cond.	1.2×10^{-2}	$-65°$	6.3	[21−23]
	Cond.	4.1×10^{-2}	$-45°$		
	Cond.	8.5×10^{-2}	$-34°$		
Dimethyl sulfoxide	Cond.	8.8×10^{-1}	$-65°$	5.5	[24]
	Flow	5.0	$-34°$		
Pyrrole	Cond.	1.7×10^{-1}	$-65°$	2.4	[25, 26]
	Cond.	2.8×10^{-1}	$-45°$		
Thiophene[e]	Cond.	3×10^{-1}	$-65°$	5.6	[27, 28]
	Cond.	1.0	$-45°$		
DL-norleucine	Flow	1.7×10^4	$-32°$	−	[29]
Imidazole	Flow	5.0×10^5	$-34°$	−	[29]
Pyridine	Flow	$\sim 4 \times 10^4$	$-34°$	−	[30]
Triphenylmethane	Cond.	$\sim 1 \times 10^{-5}$	$-34°$	−	[31, 32]

[a] Estimated using the Arrhenius rate equation.
[b] References for some reactants refer to similar compounds other than the specific reactant.
[c] Conventional conductometric method.
[d] Stopped-flow method.
[e] Poor fit to second-order analysis.

Table II. Comparison of some rate constants for e_{am}^- and e_{aq}^-

Reagent	Rate constants	
	$e_{aq}^{-\ a}$ $k\,mol^{-1}sec^{-1}$	$e_{am}^{-\ b}$ $k\,mol^{-1}sec^{-1}$
Oxalate ion	$< 10^7$	No reaction
Acetate ion	$< 10^6$	No reaction
CNS$^-$	$< 10^6$	No reaction
CN$^-$	$< 10^6$	No reaction
NH$_4^+$	1.3×10^6	$7.8 \times 10^{6\ c}$
Dimethyl sulfide	2.0×10^7	2.7×10^1
Imidazole	2.4×10^7	3.3×10^6
Dimethyl sulfoxide	1.7×10^6	5.0×10^1
Pyridine	1×10^9	2.6×10^5
Pyrrole	6.0×10^5	1.0
Thiophene	6.5×10^7	18
DL-norleucine	3.3×10^6	1.1×10^5

[a] Rate constants for e_{aq}^- were taken from Ref. [1].
[b] Rate constants calculated for 25° using appropriate activation energies or an estimated 4.5 kcal/mole, where data were not available to calculate E_a.
[c] Ref. [13] for $k_{(NH_4^+ + e_{aq}^-)}$, all other e_{am}^- reactions, this work.

Table III contains a comparison of some relative rate constants for e_{am}^- reactions at $-34°$ C with corresponding e_{aq}^- reactions at 25° C. Again, there is a marked difference in the relative rates of reactions in water and ammonia. This is in contrast to the observation that the relative reaction rates of mobile electrons in ice with different solutes at 77° K are identical, within experimental error, to the relative rates of the same solutes with e_{aq}^- in water at 300° K[3].
The reactions of methyl disulfide and n-butyl disulfide with e_{am}^- were found to be too fast at $-34°$ to permit study by the flow method, i.e., $k < 10^6\,M^{-1}sec^{-1}$. These observations are consistent with the reactivity reported [33] for e_{aq}^- towards disulfides, where rates tend to approach the diffusion limit. The reaction

Table III. Some relative rates of e_{aq}^- compared with corresponding relative rates of e_{am}^-

Reagent (R)	Relative rates[a]	
	$k_{(e_{aq}^- + R)}$	$k_{(e_{am}^- + R)}$[b]
Ammonium ion	1.0	1.0
Imidazole	18.0	4.2×10^{-1}
Dimethyl sulfide	15.0	4.5×10^{-6}
Pyrrole	0.46	9.4×10^{-7}
Dimethyl sulfoxide	1.3	4.2×10^{-6}
Pyridine	770.0	3.3×10^{-2}
DL-norleucine	2.5	1.4×10^{-2}
Thiophene	50.0	1.5×10^{-6}

[a] Normalized to $k_{(e_{sol}^- + NH_4^+)} = 1$.
[b] Estimated rate constants at $-34°$ using data summarized in Table I.

of e_{am}^- with acetone [31] was also found to be fast ($k < 5 \times 10^4 \, M^{-1} sec^{-1}$) and is probably too rapid for study by the flow technique.

For encounter controlled reactions, the simple interpretation of the Arrhenius rate equation is known to break down [34]. In this study we have reported some second-order rate constants of $5 - 10^{-2} \, M^{-1} sec^{-1}$ for reactions with corresponding activation energies of $4 - 6$ kcal/mole. These observations are not in accord with Arrhenius type reactions in solution since the activation energies are exceptionally small [34]. Also, as a consequence of the magnitude of the rate constants and low activation energies, the frequency factors for these reactions are abnormally small [35].

One possible explanation of these observations has been suggested by Schindewolf et al. [4, 5] in that a large negative entropy of activation is associated with the formation of the transition state. Moreover, according to these authors, if a transition state is formed, then $k_{(e_{aq}^- + R)}$ would be much greater than $k_{(e_{am}^- + R)}$. This does not appear to be the case for reactions involving ammonium ions and imidazole.

A correlation between relative rates of e_{aq}^- and e_{am}^- reactions is not apparent from Table III. This fact would not be expected if an electron tunneling mechanism was operative for both types of solvated electrons. According to all theoretical treatments, tunneling becomes relatively more important as the temperature is lowered [35]. Unfortunately, our present rate-temperature data are too limited to draw any conclusions about the above point.

Another possible explanation for the small activation energies reported in Table I is that the reaction mechanisms could involve one or more equilibria and intermediates prior to the rate determining step [36]. Indeed, kinetic studies of reactions of sodium with weak acids [9 – 12] in liquid ammonia have been interpreted in terms of a two-step mechanism rather than direct reaction between e_{am}^- and the weak acid.

Acknowledgement

This research was supported by the National Science Foundation.

References

 1. Hart,E.J., Anbar,M.: The hydrated electron, Chapter 8. New York: Wiley-Interscience 1970.
 2. Sherm,W.V.: J. Am. Chem. Soc. **88**, 1567 (1966).
 3. Kevan,L.: J. Am. Chem. Soc. **89**, 4238 (1967).
 4. Schindewolf,U.: Angew. Chem. internat. Edit. **7**, 190 (1968).
 5. Olinger,R., Schindewolf,U.: Ber. Bunsenges. Phys. Chem. **75**, 693 (1971).
 6. Hart,E.J., Anbar,M.: Ref. [1], p. 58.
 7. Anbar,M., Neta,P.: Intern. J. Appl. Radiat. Isotopes **18**, 493 (1967).
 8. Anbar,M.: Quart. Rev. **22**, 578 (1968).
 9. Dewald,R.R., Tsina,R.V.: J. Phys. Chem. **72**, 4520 (1968).
10. Jolly,W.L., Prizant,L.: Chem. Commun. **1968**, 1345.
11. Belloni,J.: Int. J. Radiat. Phys. Chem. **1**, 411 (1969).
12. Dewald,R.R.: Metal-ammonia solutions, International Conference, p. 193. London: Butterworths 1970.
13. Brooks,J.M., Dewald,R.R.: J. Phys. Chem. **75**, 986 (1971).
14. Jacobus,J., Eastham,J.F.: Chem. Commun. **1969**, 138.

15. Dewald, R. R., Brooks, J. M., Trickey, M. A.: Chem. Commun. **1970**, 963.
16. Dewald, R. R., Roberts, J. H.: J. Phys. Chem. **72**, 4224 (1968).
17. Feldman, L. H., Dewald, R. R., Dye, J. L.: Advan. in Chem. Ser. No. 50, p. 163. Washington: American Chemical Society, 1965.
18. Frost, A. A., Pearson, R. G.: Kinetics and mechanism, 2nd Ed., p. 29, New York: Wiley Interscience 1961.
19. Gold, M., Jolly, W. L.: Inorg. Chem. **1**, 818 (1962).
20. Jones, R. L., Dewald, R. R.: Details to be published elsewhere.
21. Krug, R. C., Tocker, S.: J. Org. Chem. **20**, 218 (1955).
22. Williams, F. E., Gebauer-Fuelnegg, E.: J. Am. Chem. Soc. **53**, 352 (1931).
23. Truce, W. E., Breiter, J. J.: J. Am. Chem. Soc. **84**, 1621 (1962).
24. O'Connor, D. E., Lyness, W. I.: J. Org. Chem. **30**, 1620 (1965).
25. Granklin, E. C.: J. Phys. Chem. **24**, 81 (1920).
26. O'Brien, S., Smith, D. C. C.: J. Chem. Soc. **1960**, 4609.
27. Birch, S. F., McAllen, D. T.: Nature **165**, 899 (1950).
28. Krug, R. C., Tocker, S.: J. Org. Chem. **20**, 1 (1955).
29. Watt, G. W.: Chem. Reviews **46**, 317 (1950).
30. Lebeau, P., Picon, M.: Compt. Rend. **173**, 1178 (1921).
31. Kraus, C. A., White, G. F.: J. Am. Chem. Soc. **45**, 769 (1923).
32. Kraus, C. A., Rosen, R.: J. Am. Chem. Soc. **47**, 2739 (1925).
33. Braams, R.: Radiation Res. **27**, 319 (1966).
34. Caldin, E. F.: Fast reactions in solution, p. 7. New York: Wiley Interscience, 1964.
35. Laidler, K. J.: Theories of chemical reaction rates, Chapter 4. New York: McGraw-Hill, 1969.
36. Freeman, G. R.: Chem. Phys. Letters **8**, 241 (1971).

Comments on Electron Transfer Reactions

N. R. KESTNER and J. LOGAN

Abstract

The success of the models for excess electrons in ammonia suggests that the continuum theories of Levich and Dogonadze can and should be extended to include discrete solvation layers, as well as the continuum. These discrete layers will have an additional set of vibrations which should be very important in determining relaxation rates since these will involve fundamentals with higher energies than those which characterize the polaron fluid. Various approaches to this problem are summarized using the ideas which have been applied to solvated electrons.

The most successful and complete theory of electron transfer reactions, especially those not involving bond breaking, is that of Levich, Dogonadze, and the Russian group [1]. It has been applied to many situations including the familiar reactions of ions.

$$A^{+N} + B^{+M} \rightleftharpoons A^{+N+1} + B^{+M-1} \tag{1}$$

where A and B are cations or even the solvated electron. In the Russian work the solvent is treated as a continuum with a solvent vibrational spectrum given by that of the pure solvent. The ions are assumed to displace the normal coordinates but not to change their frequencies or to introduce any new vibrational frequencies. Using these ideas of polaron theory, very detailed theories of electron transfer reactions can be derived. It is useful in these studies to separate the quantum and classical coordinates. Those dealing with the electron are treated quantum mechanically, while those which concern solvent motion or ion motion can be treated classically, i.e. the energy can be calculated for fixed values of these coordinates.

The kinetics depend on two factors, one essentially temperature independent and the other essentially an exponential in $1/T$. The temperature dependence is the easiest term to calculate since in the low temperature limit it is dominated by the separation in energies of the initial and final states, while in the high temperature limit the energy to reorganize the solvent enters into the effective activation energy expression. The temperature independent part is harder to calculate as it contains the coupling matrix element. The coupling operator consists of two terms, one due to the breakdown of the Born-Oppenheimer approximation (i.e. the assumption that the solvent modes are classical), and the second due to an electron exchange integral between the two centers. It is assumed that the second dominates, thus making a calculation of the matrix element possible, at least in principle.

However, we are discovering that the short range molecular order is extremely important in calculating the energy and various properties of trapped electrons and of solvated ions. When one has strong short range ordering there are new vibrational frequencies to consider. These correspond to collective vibrations of the solvent molecules which occupy the first coordination layer, assuming that these have a reasonably long lifetime and that the exchange of molecules with the continuum is not too fast. Thus, in the Hamiltonians for the initial and final state of the electron, one must add a separate term for the vibrations in the first solvation layer. These vibrations should be strongly influenced by the electron charge distribution of ion or solvated electron, i.e. the force constants for the various modes will depend on the charge enclosed. The energy of the initial state will depend not only on the solvent coordinate, η, but also on the various first solvation layer normal mode coordinates, $\{Q_i\}$. Likewise the final state will depend on the solvent coordinate, η, and on the first coordination layer coordinates $\{Q_f\}$. These solvation coordinates depend on the vibrational coordinates around each ion. Contrary to what happens as a function of the solvent coordinate, it is no longer obvious that the initial and final states ever cross for various values of the Q's. Thus the mechanism may be strongly modified. In other radiationless processes the higher energy modes usually dominate [2].

A simpler approach but involving some questionable assumptions is to assume that all vibrations are harmonic and some function of the charge density. It is not clear that the harmonic approximation is good. An even simpler approximation is that the electron is localized and all force constants have one value in the initial state and another in the final state. Then using quantized vibrations, we can simply find the manifold of energy levels in the two cases, and once the coupling terms are determined the entire reaction rate is easily calculated. Estimating the force constants is not easy as these are not usually experimentally available.

There is much concern as to the types of vibrational modes of the solvation layer we need to include. It is our opinion that one must introduce the nontotally symmetric modes as well as the totally symmetric ones. This opinion is based on a belief that the solvation layer can essentially open to provide low energy paths for the electron to transfer. However, this point can only be checked with further calculations. It is likely that the same factors which influence lineshape so strongly also influence electron transfer reactions.

Using the above modifications of the theory of Levich and Dogonadze we believe it should be easy to obtain a good theory of electron transfer reactions which contains all of the relevant factors. There are important points which can be answered by these model calculations. We should be able to compare solvated electron reactions with those of other ions and compare these with observed trends.

Nevertheless our present models are limited and can probably never yield precise quantitative agreement with experiment. Model calculations should not be expected to do this. It is known [3] that multipole calculations yield poor solvation energies for ions and thus our present model for the solvated electron which includes only dipole interactions cannot be sufficient. It is hoped that the theory can, however, be put into a semi-empirical form in which the gross features can be introduced using experimental data and that the theory will only be needed for the relaxation phenomena.

References

1. See, for example, Levich, V. G.: Advan. Electrochem. and Electrochem. Eng. **4**, 249 (1966). or: Dogonadze, R. R.: Ber. Bunsenges. Physik. Chem. **75**, 628 (1971).
2. See, for example, Englman, R., Jortner, J.: Mol. Phys. **18**, 145 (1970).
3. Desnoyers, J. E., Jolicoeur, C., In: Modern Aspects of Electrochemistry, Vol. 5, B. E. Conway, J. O'M. Bockris, Eds., New York: Plenum Press, 1969.

Discussion

J. JORTNER - I would like to make several comments concerning the theory of thermal electron transfer process. This problem is analogous to a wide class of nonradiative processes in solid state and molecular physics, such as thermal electron capture by holes in semiconductors, energy transfer between a donor and an acceptor in solids and in solutions and the nonradiative decay process of excited states of large molecules. In all of these cases an initial electronic state is coupled to a final dissipative manifold. While in the case of energy transfer the coupling involves dipole-dipole interaction, and in the case of molecular electronic nonradiative relaxation the coupling terms involve the adiabatic nuclear kinetic energy and/or spin orbit coupling, the relevant perturbation in the case of thermal electron transfer reactions involves the electronic exchange term, as pointed out clearly by Levich. Following Levich, one calculates the transition probability between a pair of ions at fixed separation. For this calculation, one needs the general multidimensional nuclear (adiabatic) potential surfaces of the system. In the early work of Levich and Dogonadze, a continuum model specified by a single frequency was utilized. Subsequent work by Levich *et al.* extended the continuum model using a general frequency and wave number-dependent dielectric function. The form of this function in the vicinity of a solvated ion is unknown. Following current theoretical ideas for the structure and spectrum of the solvated electron discussed at this meeting, it might be interesting and useful to apply a molecular model for electron transfer. The first solvation layer surrounding each ion will be described in terms of a molecular model, while the continuum model will be applied for the solvent outside this layer. The relevant nuclear modes are those whose origins (and frequencies) change between the initial and the final stages. These modes involve the radial displacements of the first coordination layer and the continuum polarization modes. I do not think that the angular distortion modes of the first coordination layer are important in determining the electron transfer probability. Within the framework of the harmonic approximation one can utilize the generating (i.e. Green's) function method first proposed by Kubo and Toyozawa to evaluate the nonradiative electron transfer probability. Such a theory will by no means be quantitative, but it will be useful in providing general relations and correlations, and in rationalizing some recent interesting observations such as the anomalously large cross sections for some reactions of the solvated electron in water. We should bear in mind that the purpose of theory is not to reproduce experiment but rather to provide a deeper understanding of the real world.

"Spin off" of Solvated Electron Research

U. SCHINDEWOLF

As scientists we repeatedly have to evaluate the relevance of the work we are doing. How do our studies influence other fields of research, what is the impact on society, is there any practical application of our results?

The study of the metal–ammonia solutions or solvated electrons undoubtedly has improved our knowledge: we have gained a better understanding of the interaction of charged particles with polar and nonpolar solvents, we have contributed to the phenomenon of the nonmetal/metal transition. Our research may provide deeper insight into reaction kinetics, especially of charge transfer reactions. The development of transistors also might benefit from the theory of electron states and electron transport properties in noncrystal materials as deduced from the properties of the metal–ammonia solutions. One day we perhaps will find that our work has obtained attention in other important fields of science.

But this is, as yet, of no practical application, just as we do not find any practical use of the knowledge about the moon's history gained from space exploration. Sarcastically, however, some people claim that the teflon-covered frying pan is the most important "spin off" of the space adventure which led us to the moon.

Over the past years, the solvated electron research has also had its "spin off" in two applied fields, not the least related to solvated electrons and not envisaged at the time the work was done. One "spin off" concerns isotope separation, the other, the destruction of a dangerous poison.

In 1952 Wilmarth and Dayton were studying the conversion of para-hydrogen in a dilute metal–ammonia solution:

$$p - H_2 \overset{e^-_{sol}}{\rightleftharpoons} o - H_2 .$$

The conversion is catalyzed by the paramagnetic solvated electrons in these solutions, the rate of catalysis being in fair agreement with the theory of Wigner. In some experiments, however, the conversion rate was higher than expected by several powers of ten. Wilmarth and Dayton then recognized that in these experiments part of the metal had decomposed to metal amide which turned out to be a powerful catalyst not only for the para-hydrogen conversion but also for the isotope exchange between molecular hydrogen and ammonia, by which the heavy hydrogen isotope is enriched in the liquid ammonia:

$$HD + NH_3 \overset{KNH_2}{\rightleftharpoons} H_2 + NH_2D .$$

This exchange reaction can be used for the large-scale production of heavy water by multiplication of the separation effect of the single equilibrium in a countercurrent column. The highly enriched deutero–ammonia obtained is then converted to heavy water, e.g. by combustion with oxygen.

This new heavy water process was developed by the Centre d'Etudes Nucléaires de Saclay, with Prof. Lepoutre contributing to it, and independently – following a different scheme – by our laboratory in the Kernforschungszentrum Karlsruhe in cooperation with German industry.

Calculations based on the experiences with the pilot plant in Mazingarbe, France and in Hoechst, Germany show that this process is more economical then the so-called hydrogen sulfide process (countercurrent exchange $HDS + H_2O \rightleftarrows H_2S + HDO$) by which heavy water has thus far been produced in the United States, supplying the heavy water needs of the western countries. The advantages of the new process lie in the better thermodynamics of the hydrogen–ammonia system in connection with the good kinetics of the afore-mentioned new and powerful potassium amide catalyst. In addition, the ammonia process is free of the corrosion problems with which the hydrogen sulfide process is heavily burdened.

One heavy water plant, which will utilize the French process, is now under construction in India; the construction of a second plant, also in India, which will operate according to the German process, will begin very soon. There are also plans in Israel to set up a small production unit adopting the ammonia process.

The cheaper production of heavy water, a multimillion dollar business, is one of the by-products of the research on metal–ammonia solutions. The other by-product does not concern production, but rather destruction of an industrial product, i.e. the neutralization of poisonous cyanide salts.

Startling reports in the German press during the last year have told of cyanide wastes in the kiloton range, enough to wipe out the entire European population, which have been found almost everywhere in the country, and which threaten to poison the water reservoirs of the large cities. These wastes, originating e.g. from metal-hardening works, were deposited secretly and with no controls because their safe destruction (burning in an oil flame, oxidation with chloride, complexing with iron) seemed to be to expensive.

The impulses for the solution of this urgent problem came from Colloque Weyl II. At that conference, we presented a new convenient laboratory procedure for the preparation of deutero-ammonia which, instead of normal ammonia, must be used for some NMR- and IR-spectroscopy of metal–ammonia solutions. The procedure is based on the hydrolysis of sodium cyanide with heavy water at increased temperature and pressure:

$$NaCN + 2\,D_2O \rightarrow ND_3 + DCOONa\,.$$

This procedure provides a method for the inexpensive and complete destruction of the cyanide wastes: With water as the only reagent and only a little heat, more than 99.9998% of the poisonous cyanide can be transformed to harmless ammonia and formiate in a continuous flow type apparatus. German industry is now adapting this process.

These two examples indicate that our research on metal–ammonia solutions also has resulted in relevant "spin off" in technical fields which indeed have no relationship to our original research interest but which perhaps might lead to cheaper nuclear energy or might prevent harm to people.

Subject Index

Color Plates

Figure 19 of paper by P. M. RENTZEPIS:

Ultrafast Optical Processes

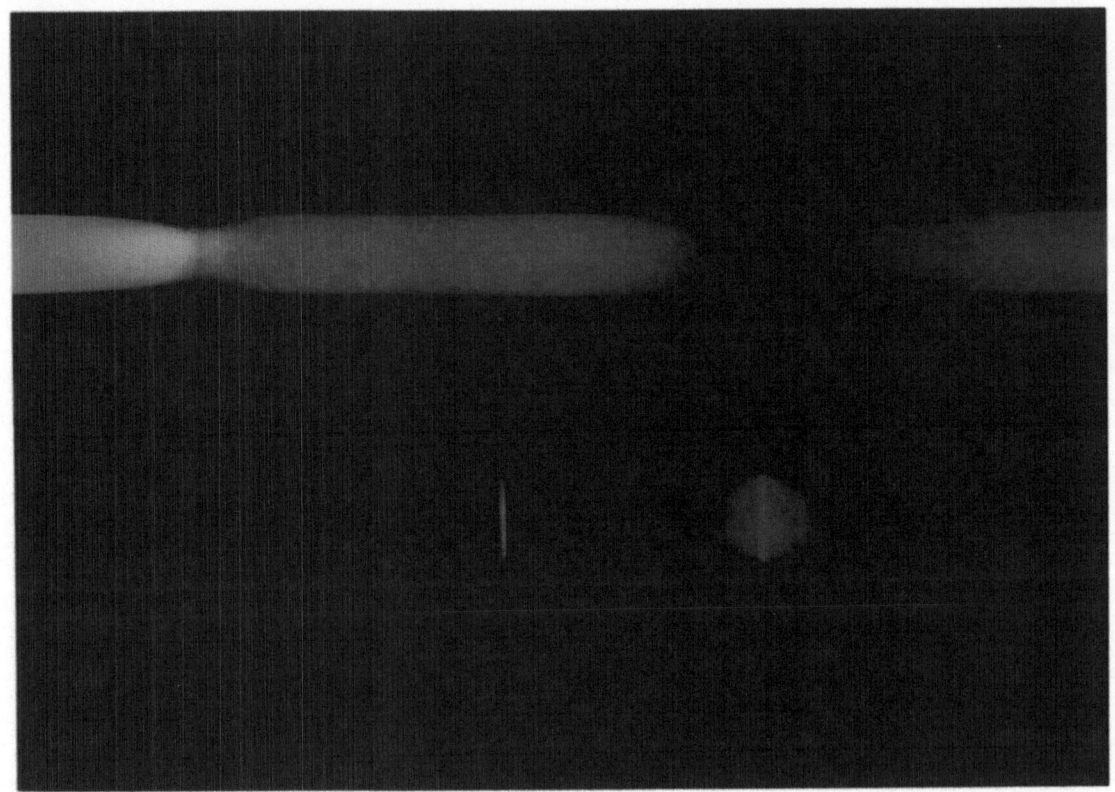

Fig. 19. Spectral resolution of a continuum picosecond pulse

Figure 2a and b of paper by F. HENSEL:

Metallic Vapors

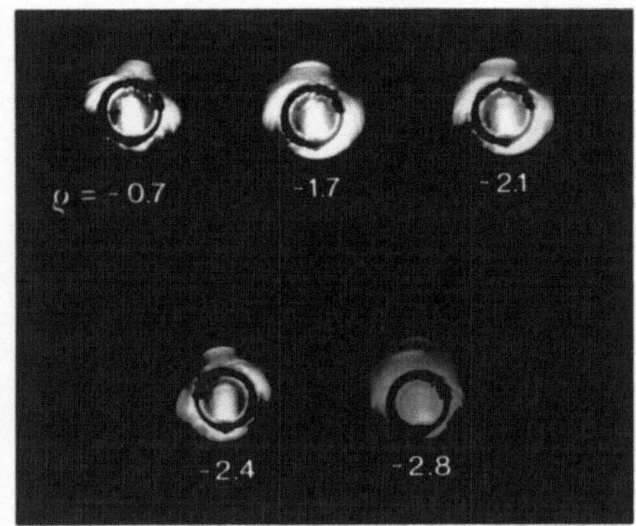

Fig. 2a. Transparency of a 10 μ-mercury film at different densities (the small numbers give the density in g/cm³). The temperature is 1800 °K

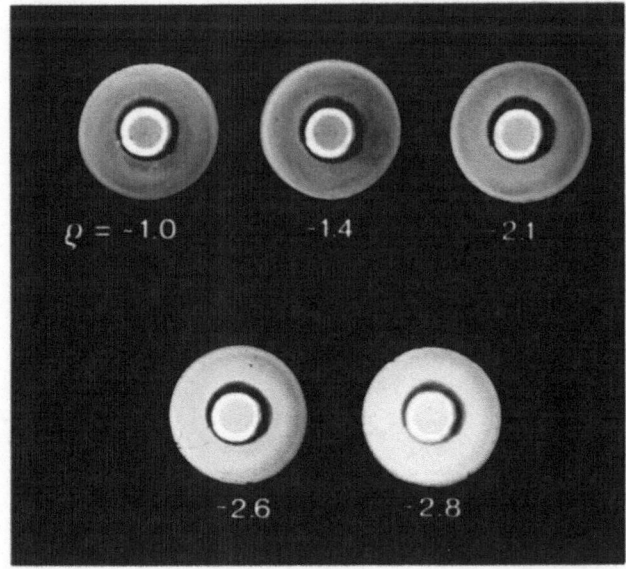

Fig. 2b. Radiation of a 100 μ-mercury film at different densities (density in g/cm³; temperature 1800 °K)